T0156063

Bryophyte Biology

Second Edition

Bryophyte Biology provides a comprehensive yet succinct overview of the hornworts, liverworts, and mosses: diverse groups of land plants that occupy a great variety of habitats throughout the world. This new edition covers essential aspects of bryophyte biology, from morphology, physiological ecology and conservation, to speciation and genomics. Revised classifications incorporate contributions from recent phylogenetic studies. Six new chapters complement fully updated chapters from the original book to provide a completely up-to-date resource. New chapters focus on the contributions of *Physcomitrella* to plant genomic research, population ecology of bryophytes, mechanisms of drought tolerance, a phylogenomic perspective on land plant evolution, and problems and progress of bryophyte speciation and conservation. Written by leaders in the field, this book offers an authoritative treatment of bryophyte biology, with rich citation of the current literature, suitable for advanced students and researchers.

BERNARD GOFFINET is an Associate Professor in Ecology and Evolutionary Biology at the University of Connecticut and has contributed to nearly 80 publications. His current research spans from chloroplast genome evolution in liverworts and the phylogeny of mosses, to the systematics of lichen-forming fungi.

A. JONATHAN SHAW is a Professor at the Biology Department at Duke University, an Associate Editor for several scientific journals, and Chairman for the Board of Directors, Highlands Biological Station. He has published over 130 scientific papers and book chapters. His research interests include the systematics and phylogenetics of mosses and liverworts and population genetics of peat mosses.

Bryophyte Biology

Second Edition

BERNARD GOFFINET
University of Connecticut, USA

AND

A. JONATHAN SHAW
Duke University, USA

CAMBRIDGE
UNIVERSITY PRESS

University Printing House, Cambridge CB2 8BS, United Kingdom

Cambridge University Press is part of the University of Cambridge.

It furthers the University's mission by disseminating knowledge in the pursuit of education, learning and research at the highest international levels of excellence.

www.cambridge.org
Information on this title: www.cambridge.org/9780521693226

First published 2000
Second edition 2009

A catalogue record for this publication is available from the British Library

Library of Congress Cataloguing in Publication data
Bryophyte biology / [edited by] Bernard Goffinet & A. Jonathan Shaw. – 2nd ed.
p. cm.
ISBN 978-0-521-87225-6
1. Bryophytes. I. Goffinet, Bernard. II. Shaw, A. Jonathan (Arthur Jonathan) III. Title.
QK533.B715 2008
588–dc22

2008021975

ISBN 978-0-521-87225-6 Hardback
ISBN 978-0-521-69322-6 Paperback

To Lewis Anderson

Contents

Contributors

J. W. Bates
Department of Biology, Imperial College at Silwood Park, Ascot, Berkshire SL5 7PY, UK.

W. R. Buck
New York Botanical Garden, Bronx, NY 10458-5126, USA.

B. Crandall-Stotler
Department of Plant Biology, Southern Illinois University, Carbondale, IL 62901-6509, USA.

A. C. Cuming
Centre for Plant Sciences, Faculty of Biological Sciences, Leeds University, Leeds LS2 9JT, UK.

R. J. Duff
Department of Biology, ASEC 185, University of Akron, Akron, OH 44325-3908, USA.

B. Goffinet
Department of Ecology and Evolutionary Biology, 75 North Eagleville Road, University of Connecticut, Storrs, CT 06269-3043, USA.

T. Hallingbäck
Swedish Species Information Centre, Swedish University of Agricultural Sciences, PO Box 7007, SE-750 07 Uppsala, Sweden.

D. G. Kelch
California Department of Food and Agriculture, Plant Pest Diagnostics Laboratory, CDA Herbarium, 3294 Meadowview Road, Sacramento, CA 95832-1448, USA.

D. G. Long
Bryology Section, Royal Botanic Garden, Edinburgh EH3 5LR, UK.

B. D. Mishler
University Herbarium, Jepson Herbarium, and Department of Integrative Biology, University of California, Berkeley, 1001 Valley Life Sciences Building #2465, Berkeley, CA 94720-2465, USA.

M. J. Oliver
USDA-ARS-MWA-PGRU, 205 Curtis Hall, University of Missouri, Columbia, MO 65211, USA.

M. C. F. Proctor
School of Biosciences, University of Exeter, The Geoffrey Pope Building, Stocker Road, Exeter EX4 4QD, UK.

K. S. Renzaglia
Department of Plant Biology, Southern Illinois University, Carbondale, IL 62901-6509, USA.

H. Rydin
Department of Plant Ecology, Evolutionary Biology Centre, Uppsala University, Villavagen 14, SE-752 36 Uppsala, Sweden.

A. J. Shaw
Department of Biology, Duke University, Durham, NC 27708, USA.

R. E. Stotler
Department of Plant Biology, Southern Illinois University, Carbondale, IL 62901-6509, USA.

A. Vanderpoorten
Département des Sciences de la Vie Université de Liége, Sart Tilman B22, B-4000 Liége, Belgium.

J. C. Villarreal
Department of Ecology and Evolutionary Biology, 75 North Eagleville Road, University of Connecticut, Storrs, CT 06269-3043, USA.

D. H. Vitt
Department of Plant Biology, Southern Illinois University, Carbondale, IL 62901-6509, USA.

R. K. Wieder
Room 105, St. Augustine Center, Villanova University, 800 Lancaster Avenue, Villanova, PA 19085, USA.

Preface

Bryophytes have gained a lot of publicity in the past 10–15 years, at least among scientists. While there have always been those who for inexplicable reasons have had a particular fondness for bryophytes, in academic circles these organisms were generally viewed as just "poor relatives" of the more flashy and exciting angiosperms. The bryophytes include fewer species, of smaller stature, with more subdued colors, of less obvious ecological significance, and with apparently simpler and less exciting evolutionary stories to tell. That view has changed.

The three major groups of bryophytes – mosses, liverworts, and hornworts – comprise the earliest lineages of land plants derived from green algal ancestors. Although we still do not know with certainty which of the three lineages is the sister group to all other land plants, we do know that the earliest history of plants in terrestrial environments is inextricably bound to the history of bryophytes. If we wish to understand fundamental aspects of land plant structure and function, we should turn to the bryophytes for insights. These aspects include the origin and nature of three-dimensional plant growth from apical cells and meristems, the evolution of cellular mitotic mechanisms and machinery, the development of thick, water- and decomposition-resistant spore (and later pollen) walls, the molecular and biochemical mechanisms underlying desiccation tolerance, and plant genome structure, function, and evolution. Even if our ultimate goal is to understand the structure and function of angiosperms because it is indeed those plants that feed the human world as agricultural crops, we are nevertheless wise to look more deeply into plant history for a thorough understanding of plant unity and diversity. We cannot fully understand how evolution has tinkered with structure and function in angiosperms without a sense of history. Although the angiosperms are impressively diverse in numbers and structure, they are, we now know from phylogenetic insights into plant evolution, just glorified bryophytes!

Although it is well established that the bryophytes do not constitute a single monophyletic lineage, these organisms share a fundamentally similar life cycle with a perennial and free-living, photosynthetic gametophyte alternating with a short-lived sporophyte that completes its entire development attached to the maternal gametophyte. There are a number of bryophytes that have variously reduced gametophytes and/or sporophytes, and at least one liverwort that is parasitic and non-photosynthetic, but however much the morphological details vary from species to species, the basic bryophyte life cycle is shared among mosses, liverworts, and hornworts. The gametophytes of many species have the ability to replicate clonally either through specialized asexual propagules or by fragmentation, and at sexual maturity they form multicellular female and male gametangia, archegonia and antheridia, respectively. Water is required for fertilization, as bryophyte sperm are flagellated and must swim to reach an egg. Because of their life cycles, bryophytes are ideally and uniquely suited to address some questions of fundamental significance in biology.

Sporophytes and gametophytes differ greatly in morphology, yet under some circumstances (e.g. bryophytes with bisexual gametophytes that self-fertilize) they differ only in ploidy: the sporophyte has the exact but duplicated genome of the gametophyte. This alternation of haploid gametophytes and diploid sporophytes that differ in morphology and function is one of the most basic aspects of plant (and indeed organismal) life cycles, and control of morphological and functional differences between gametophyte and sporophyte generations has intrigued scientists since these alternating life cycles were discovered in the nineteenth century. Given the identity in genome sequence between isogenic sporophytes and gametophytes, differences between the generations obviously derive from differences in gene expression rather than genetic composition. Technological advances during the past 20 years have for the first time allowed us to begin to understand molecular processes that underlie the alternation of generations in plants, and bryophytes have proven to be invaluable organisms for this sort of research. Yet we are only now scratching the surface in this area of inquiry: bryophytes will continue to play a central role in new developments.

For many years, bryophytes had a reputation of being "unmoving, unchanging sphinxes of the past" with little going on in terms of current evolutionary activity. In other words, evolutionarily boring! This view has proven inaccurate. Bryophyte species show local adaptation to heterogeneous environments, demonstrating their responsiveness to natural selection, and have engaged in complex speciation processes that include hybridization, polyploidization, and morphologically cryptic genetic differentiation. Indeed, the homosporous life cycle of bryophytes provides an opportunity for these organisms to exhibit more – not fewer – variations in reproductive biology than is possible in heterosporous seed plants, including angiosperms. Bryophyte species with bisexual gametophytes, those that produce both archegonia and antheridia, can undergo true or intragametophytic self-fertilization, which results in a completely homozygous sporophyte in a single generation. This is not possible in heterosporous plants because, unlike bryophytes, they form male and female gametes meiotically rather than mitotically. "Self-fertilization" in a seed plant describes the situation in which two genetically different (albeit related) gametophytes produced from the same sporophyte mate to form the next sporophyte generation. Bryophytes can engage in such sexual behavior as well, in addition to true self-fertilization. This reproductive mode, mating between different but related gametophytes, is commonly referred to as "selfing" in the seed plant literature because of a bias in the way we view plant life cycles. Coming from an angiosperm point of view, gametophytes (e.g. pollen, embryo sacs) are seen as part of the reproductive apparatus of the "individual" or "self", which is the sporophyte. There is nothing objectively accurate about viewing sexual crosses between genetically different gametophytes as "selfing", even if those gametophytes came from the same sporophyte. The common perception of sporophytes as individuals or "selfs" and gametophytes as simply parts of those "selfs" is an example of *ploidy-ism*, which can cloud our ability for insight akin to the way racism clouds our perceptions in humanistic issues. It is just as correct to think of a chicken as an egg's way of reproducing itself, as the reverse! Bryophytes offer a fresh perspective in plant reproductive biology that can loosen the intellectual shackles of an angiosperm-centered worldview.

The second edition of *Bryophyte Biology* is thoroughly revised and should be viewed as complementary to, rather than as a substitute for, the first edition. Our goal when the first edition of *Bryophyte Biology* was being developed was to produce a volume that could serve simultaneously as an intermediate to advanced text for a bryology course, and as a reference for scientists dealing with bryophytes in physiological, biochemical, molecular, or ecological research. In retrospect we felt that we only partly fulfilled our goal in making a hybrid book that serves both of these sometimes conflicting purposes. The second edition of *Bryophyte Biology* is also designed to serve both functions, and we feel that we have come closer to our goal by including new and revised chapters that cover the breadth of subjects that should be included in a bryology course, and that are also relevant to researchers working in other fields. As in the first edition, every chapter provides extensive bibliographic citations to primary literature. We consider this resource important, both for the developing student of bryology and for established scientists in some more specialized field who want to learn more about bryophytes. The first three chapters dealing with the morphology and classification of liverworts (Chapter 1), mosses (Chapter 2), and hornworts (Chapter 3) have expanded coverage of morphology as appropriate for a textbook, and also have revised classifications that reflect developments since the first edition was published. We include a new chapter (Chapter 4) on phylogenomics, reviewing relatively recent developments from using whole-genome characters to resolve phylogenetic relationships among early land plants. With the growing importance of *Physcomitrella patens* for molecular genetic research, Chapter 5 provides a timely overview of mosses as model organisms. Chapters 6–12 deal with the physiology, biochemistry, ecology, evolution, and conservation of bryophytes. A new chapter (Chapter 7) focused on desiccation tolerance in bryophytes reflects the importance of these organisms for modern molecular and biochemical research in this area. Desiccation tolerance is arguably the most thoroughly studied physiological adaptation in plants, and mosses have proven to be an invaluable group of organisms for such research. This value derives both from the relative structural simplicity of mosses and their phylogenetic position in the land plant tree of life. All chapters in the second edition of *Bryophyte Biology* are either completely new or completely revised relative to those included in the first edition.

We hope that *Bryophyte Biology*, edition 2, will provide an entry for established scientists into the literature dealing with bryophytes, and will stimulate enthusiasm among young bryology students for careers focusing on these humble but fantastic organisms.

1

Morphology and classification of the Marchantiophyta

BARBARA CRANDALL-STOTLER, RAYMOND E. STOTLER AND DAVID G. LONG

1.1 Introduction

Liverworts are a diverse phylum of small, herbaceous, terrestrial plants, estimated to comprise about 5000 species in 391 genera. They occupy an assortment of habitats, including disturbed soil along stream banks, road cuts and trails, as well as rocks, logs and trees in natural landscapes. They occur on all continents, including Antarctica, but are most diversified in the montane rain forests of the southern hemisphere. Many species are quite tolerant of repeated cycles of drying and wetting (Clausen 1964, Wood 2007), a feature that has allowed them also to exploit epiphytic substrates, including leaves and branches of the forest canopy. Like mosses and hornworts, they have a heteromorphic life cycle with a sporophyte that is comparatively short-lived and nutritionally dependent on the free-living, usually perennial gametophyte. However, they differ from both of these groups in numerous cytological, biochemical, and anatomical features as detailed by Crandall-Stotler (1984). Significant diagnostic characters of the phylum include the following: they tend to have a flattened appearance, even when leafy, because their leaves are always arranged in rows, never in spiral phyllotaxis; rhizoids are unicellular, thin-walled, and usually hyaline; both leafy and thalloid forms frequently develop endosymbiotic associations with fungi; sporophytes mature completely enclosed by gametophytic tissue and are incapable of self-sustaining photosynthesis; sporophyte setae are parenchymatous and elongate by cell expansion, rather than cell division; and capsules lack the stomates, cuticle, and columella that are common in mosses and hornworts.

Liverworts occupy a critical position in land plant evolution, forming the sister group to all other extant land plants (e.g. Groth-Malonek *et al.* 2005, Qiu

Bryophyte Biology: Second Edition, ed. B. Goffinet & A. J. Shaw. Published by Cambridge University Press.

et al. 2006). Fossil spores that are comparable to liverwort spores date back to 475 million years before present (Wellman *et al.* 2003), and estimates of divergence times based on molecular evidence suggest a Late Ordovician origin for the phylum (Heinrichs *et al.* 2007). Despite rather sparse representation in the fossil record of the Paleozoic, all major (backbone) lineages of hepatics appear to have been established by the Permian (Oostendorp 1987, Heinrichs *et al.* 2007).

Traditionally, liverworts have been subdivided into the marchantioid group, or complex thalloids, and the jungermannioid group, which comprises two morphological subgroups, the anacrogynous, simple thalloids and the acrogynous, leafy hepatics. These groups have been defined in the hierarchy of most classification schemes and have long been viewed as natural phylogenetic units. For example, in Crandall-Stotler & Stotler (2000) they are recognized as classes, Marchantiopsida and Jungermanniopsida, with the latter comprising two subclasses, Metzgeriidae (simple thalloids) and Jungermanniidae (leafy hepatics). A large suite of anatomical and ontogenetic characters differentiates the two classes, including different patterns of gametangial development, spermatid architecture, capsule wall anatomy (Crandall-Stotler & Stotler 2000), and mechanisms involved in defining cytokinetic planes during meiosis (Shimamura *et al.* 2004, Brown & Lemmon 2006). Recent molecular phylogenetic studies (e.g. Heinrichs *et al.* 2005, 2007, Forrest *et al.* 2006, He-Nygrén *et al.* 2006, Qiu *et al.* 2006) have greatly modified this morphology-based concept, especially as regards the simple thalloid group. Whereas the monophyly of the complex thalloids and the leafy hepatics is broadly supported in all of these analyses, the simple thalloids are paraphyletic with representatives in four of the six backbone clades. One of these, comprising the Haplomitriaceae and Treubiaceae, has been identified as the earliest diverging lineage of the hepatics and relegated to a third class, Haplomitriopsida (Forrest *et al.* 2006). Liverworts are unambiguously resolved in these more recent, comprehensive multilocus analyses as monophyletic, in contrast to earlier postulates that they are polyphyletic (Capesius & Bopp 1997, Bopp & Capesius 1998).

This chapter provides a conspectus of liverwort morphology, with emphasis on the defining characters of the major lineages (clades) currently recognized. Although our knowledge of morphological character diversity has changed little since the first edition of this book, many interpretations of character evolution within the group have been modified (e.g. He-Nygrén *et al.* 2004, 2006, Crandall-Stotler *et al.* 2005). A classification scheme that links morphological data with the well-supported relationships generated in recent molecular phylogenetic analyses is provided, with brief morphological diagnoses for the taxon ranks above the level of family. Unless otherwise indicated, class, subclass, ordinal and family names used in the text refer to these ranks as they are defined and circumscribed in this classification.

1.2 Conspectus of liverwort morphology

The foundations for morphological studies in hepatics were laid in the nineteenth century with the seminal publications of Hofmeister (1851) and Leitgeb (1874–1881), whose comparative studies clarified the homologies among embryophytes, and documented the structural diversity and complexity of hepatics, respectively. Later workers, including Goebel (1893, 1895, 1912), Douin (1912), Evans (1912), Knapp (1930), Crandall (1969) and Renzaglia (1982) among others, have contributed additional anatomical descriptions of selected structures across broad groups of hepatics. Nevertheless, many gaps persist in our knowledge, with the vast majority of liverwort taxa known only at the level of a taxonomic description. This conspectus serves to provide a general overview of what is currently known about the structural organization and diversity of liverworts. To date, few reconstructions of morphological character state evolution have been published, so definitive statements about evolutionary trends in many characters cannot yet be made. Comprehensive reviews of the comparative anatomy and morphology of hepatics can be found in Schuster (1966, 1984a) and Crandall-Stotler (1981).

1.2.1 *Apical cells and gametophyte growth*

Whether leafy or thalloid, liverwort gametophytes display modular organization, with each module composed of a series of merophytes that trace their origin back to a single apical cell, the dynamic generative center of the gametophyte (Hallet 1978, Crandall-Stotler 1981). All metamers derived from a single apical cell compose a module that is a single branch or shoot (Mishler & DeLuna 1991). Since branching is common, most plants are composed of more than one module.

Four geometrically different types of apical cell occur within hepatics, namely, tetrahedral (or pyramidal), cuneate (or wedge-shaped), lenticular (or lens-shaped) and hemidiscoid types (Crandall-Stotler 1981: Fig. 1.1). As the name suggests, a tetrahedral apical cell has four somewhat curved, triangular surfaces, one of which forms the external or free surface of the cell. The other three surfaces, referred to as the cutting faces, are surrounded by the ranks of daughter cells generated from division of the apical cell. This type of apical cell has a triangular outline in all planes of section (Fig. 1.1A, B), and produces merophytes in three ranks. A lenticular apical cell has a lens-shaped free surface and two triangular cutting faces. It produces merophytes in two lateral ranks, has a triangular outline in both vertical and horizontal longitudinal sections, and is shaped like a convex lens in transverse section (Fig. 1.1C, D). A cuneate apical cell is wedge-shaped with five surfaces, a narrow, rectangular free surface, two

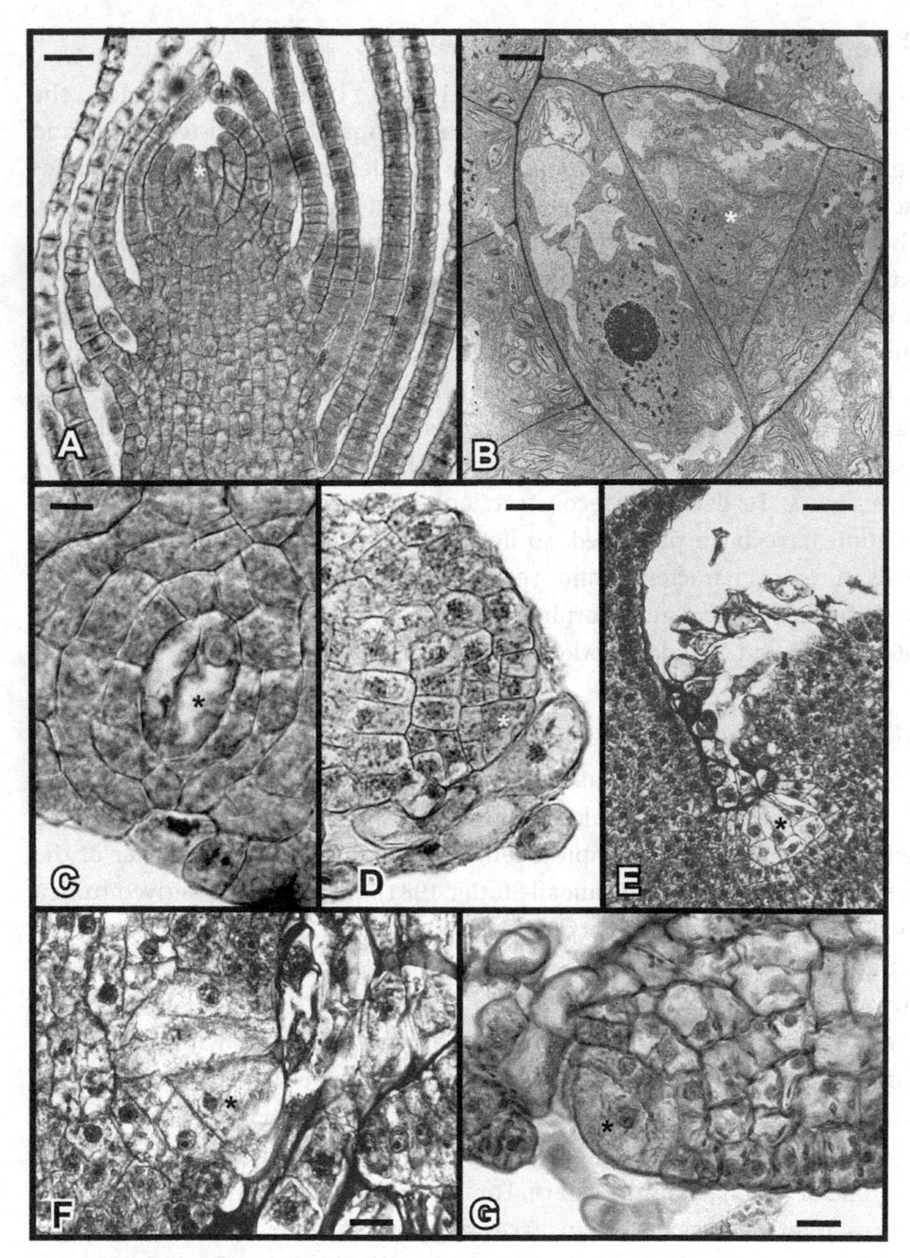

Fig. 1.1. Apical cell diversity in liverworts; apical cells are marked with asterisks. (A, B) Apices with tetrahedral apical cells; (A) *Porella platyphylla*, horizontal longitudinal section, bar = 25 μm; (B) *Haplomitrium hookeri*, transverse section, bar = 5.4 μm. (C, D) Apices with lenticular apical cells; (C) *Pallavicinia ambigua*, transverse section, bar = 10 μm; (D) *Aneura pinguis*, vertical longitudinal section, bar = 25 μm. (E, F) Apices with cuneate apical cells, *Phyllothallia nivicola*; (E) horizontal longitudinal section, bar = 50 μm; (F) vertical longitudinal section, bar = 18 μm. (G) Apex with a hemidiscoid apical cell, *Pellia epiphylla*, vertical longitudinal section, bar = 18 μm. Note the slime cells overarching the apical cell in (D–G); in *Aneura* (D) they form only on the ventral surface, but in *Phyllothallia* and *Pellia* they arise from both dorsal and ventral surfaces.

vertically aligned, triangular surfaces and two horizontally aligned, broad rectangular surfaces. Apical cells of this type have rectangular outlines in transverse and horizontal longitudinal sectioning planes, but triangular outlines in vertical longitudinal section (Fig. 1.1E, F). They produce merophytes in four ranks: dorsal, ventral, and two lateral. Finally, the rather specialized hemidiscoid apical cell appears rectangular in both transverse and horizontal longitudinal sections, but has a prismatic to semicircular outline in vertical longitudinal section (Fig. 1.1G). This type of apical cell has two lateral cutting faces and a single basal cutting face, rather than a dorsal and a ventral face as in the cuneate form. According to Hutchinson (1915) and Campbell (1913), respectively, in *Pellia epiphylla* and *Sandeothallus radiculosus* (= *Calycularia radiculosa*), the hemidiscoid geometry is developmentally derived from a cuneate form by a rounding out of the dorsal and ventral faces into a single, curved basal face.

Although there is substantial variation in apical cell dimensions as well as pattern and rate of merophyte formation, typically apical cell geometry is conserved within taxa. A tetrahedral apical cell, which has been reconstructed as the plesiomorphic state in hepatics (Crandall-Stotler *et al.* 2005), is characteristic of the Haplomitriopsida and all of the Jungermanniidae, as well as select genera of the Pelliidae. The assumption by He-Nygrén *et al.* (2004, 2006) that a cuneate geometry is the plesiomorphic state and that tetrahedral geometries have been derived independently in several lineages is not supported by analyses of character evolution. All Marchantiopsida possess cuneate apical cells, often with lenticular types in early stages of ontogeny (Leitgeb 1881). Lenticular apical cells are characteristic of all genera of the Metzgeriidae, with the lenticular apical cell of *Pleurozia*, in fact, providing the sole morphological signal of its relationship with the Metzgeriales. Only the Pelliidae exhibit multiple apical cell types: tetrahedral in *Noteroclada*, *Petalophyllum*, and *Sewardiella*; cuneate in *Makinoa*, *Allisonia*, the *Pellia endiviifolia* species complex, *Phyllothallia*, *Moerckia*, and *Symphyogyna*; and hemidiscoid in *Calycularia*, *Sandeothallus*, and the *Pellia epiphylla* species complex.

There is no absolute correlation between plant form and apical cell type but taxa with tetrahedral apical cells do tend to have "leafy" morphologies, and taxa with hemidiscoid apical cells are always thalloid. Lenticular and cuneate apical cells typically occur in thalloid taxa, but some leafy plants, e.g. *Fossombronia* and *Pleurozia*, possess lenticular apical cells and others, like *Phyllothallia*, have cuneate apical cells (Renzaglia 1982).

Distinctive patterns of early merophyte division lead to the formation of leaves in the Jungermanniidae and the foliar appendages and thallus wings in the rest of the hepatics (Crandall-Stotler 1981). As verified by many workers (e.g. Leitgeb 1875, Evans 1912, Crandall 1969), in the Jungermanniidae the first division of a lateral merophyte is perpendicular to its free surface (i.e. anticlinal), partitioning the merophyte into two halves. Subsequent divisions that are parallel to the free surface (i.e. periclinal) generate a five-celled merophyte comprising two primary leaf and three primary stem initials; in many groups one or both leaf initials divide again to form three or four secondary leaf initials. Leaf growth occurs first from apical cells delimited from each of the leaf initials, which establish the segments or lobes of the leaf, and then from a basal meristematic zone that forms the undivided lamina (Bopp & Feger 1960). The number and size of lobes that occur in a leaf are dependent on the number of apical cells differentiated and the relative proportion of apical to basal growth that occurs. Leaf apical growth is terminated with the conversion of the apical cells to club-shaped papillae. If apical growth is pronounced, these papillae occur at the tips of the leaf lobes, as in *Lepidozia* or *Lophocolea*; however, if growth is mostly from the basal meristem, they are found near the base of the leaf, as in *Jungermannia* or the dorsal lobe of *Porella* (Bopp & Feger 1960, Fig. 23).

In groups other than the Jungermanniidae, two successive anticlinal divisions partition the lateral merophyte into three cells, the middle one of which forms the single wedge-shaped initial from which the thallus wing or foliar appendages are derived (Renzaglia 1982, Bartholomew-Began 1991). The cells to either side generate the tissues of the stem, midrib, or central portion of the thallus. Since there is only a single foliar (or wing) apical cell per merophyte, leaves in taxa such as *Haplomitrium*, *Noteroclada*, and *Fossombronia* are never deeply lobed although they can be marginally incised owing to the activity of secondarily produced centers of marginal growth (Bartholomew-Began 1991). The basal meristem is established early in leaf or wing ontogeny. Leaves of this type are polystratose at the base, often have scattered marginal papillae, and are homologous to the wings of both simple and complex thalloid taxa. A modification of this pattern occurs in *Treubia* and perhaps *Pleurozia*. In the former, the large lobes or "leaves" of the plant develop from the wedge-shaped central cell of the three-celled merophyte and a small lobule develops from the cell dorsal to it (Renzaglia 1982). In *Pleurozia* early divisions appear to produce a wedge-shaped central cell, but subsequent leaf development is like that of a true leafy liverwort, involving multiple initials and apical cells (Crandall-Stotler 1976).

1.2.2 *Oil bodies*

Liverworts are distinguished from all other embryophytes by their almost universal production of oil bodies, unique membrane-bound organelles that synthesize and sequester a vast array of terpenoids and other aromatic compounds (Flegel & Becker 2000, Suire *et al.* 2000). Oil bodies are formed during early stages of cell maturation (Crandall-Stotler 1981) as dilatations of the endoplasmic reticulum (Duckett & Ligrone 1995, Suire 2000) or from dictyosome vesicle fusion (Galatis *et al.* 1978, Apostolakos & Galatis 1998). The enclosing membrane of the oil body resembles the tonoplast in having an asymmetric, tripartite appearance but differs from it in enzyme composition and transport capabilities (Suire 2000). The oil body interior consists of small osmiophilic droplets suspended in a granular stroma that is rich in proteins and carbohydrates (Pihakaski 1972, Suire 2000). Frequently, in addition to oil bodies, cells contain dispersed lipid droplets (oleosomes) in their cytoplasm. These are droplets of triacylglycerides and neither these, nor the plastoglobules common in plastids, are involved in or part of oil body development (Suire 2000).

In Jungermanniopsida and *Haplomitrium*, oil bodies are usually produced in all cells of both the sporophyte and gametophyte generations. In these taxa variations in oil body size, shape, color, number and distribution are taxonomically informative (Pfeffer 1874, Müller 1939, Schuster 1966, 1992a, Gradstein *et al.* 1977), with five broadly defined categories recognized (Fig. 1.2). *Massula*- and *Bazzania*-type oil bodies are shiny, homogeneous, and either very small and abundant (*Massula*-type) (Fig. 1.2B) or larger and fewer per cell (*Bazzania*-type) (Fig. 1.2C). Oil bodies of the *Calypogeia* type are botryoidal, consisting of grape-like clusters of discrete, shiny globules; they can be translucent or pigmented, of small to medium size and usually many per cell (Fig. 1.2D). Oil bodies that are opaque, gray to gray-brown and granulose to papillose in texture (*Jungermannia*-type) are the most common type in the Jungermanniopsida; these can be small and numerous per cell, or very large and solitary as in *Radula* (Fig. 1.2E). In *Treubia* and most genera of the Marchantiopsida, oil bodies occur only in scattered idioblastic cells of the gametophyte; they are large, solitary, granular and opaque, gray to gray-brown (Fig. 1.2F). These idioblastic "oil cells" differ from the surrounding vegetative cells only by the presence of the large oil bodies in them (Suire 2000), in contrast to earlier views that they lack chloroplasts (Schuster 1984b).

Unfortunately, because of the volatility of the oils contained in them, oil bodies rapidly "disappear" in dried specimens. In fact, their morphology is often modified even during short-term storage in the dark, so observations of oil body morphology must be conducted only on freshly collected samples. Ultrastructural evidence confirms that the oil body membrane and internal matrix remain intact for up to six weeks in dark-stored specimens, but the oil droplets within the matrix disappear within a few days (B. Crandall-Stotler, unpublished data).

In phylogenetic reconstructions, the presence of oil bodies is a synapomorphy of the Marchantiophyta, and oil bodies of the *Massula*-type are reconstructed as the plesiomorphic state (Crandall-Stotler *et al.* 2005). Oil bodies have been

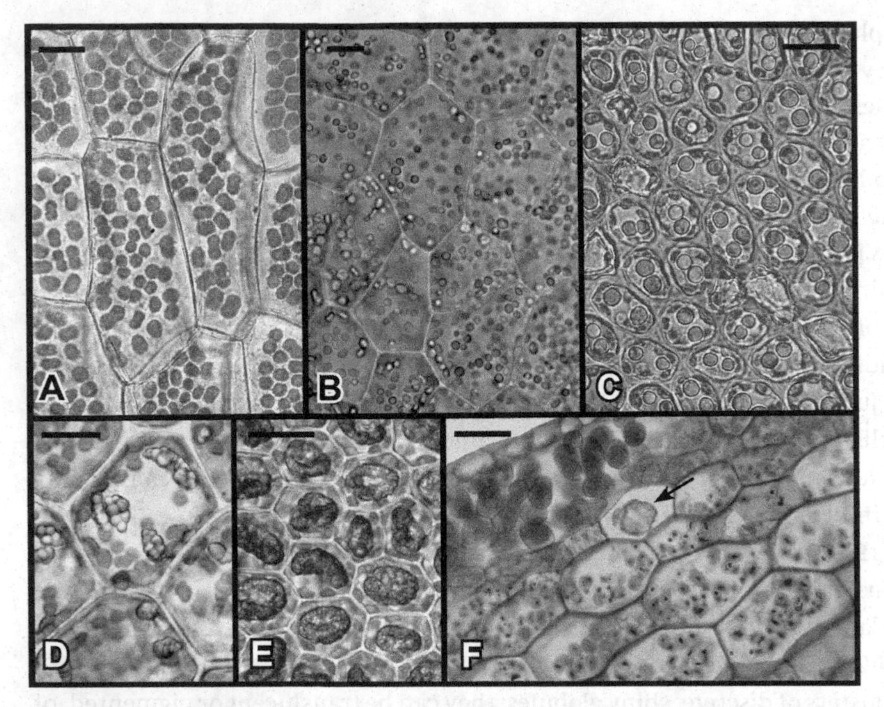

Fig. 1.2. Cells and oil bodies of liverworts. (A) Thallus wing cells of *Blasia pusilla*, thin-walled and lacking oil bodies, bar = 15 μm. (B) Leaf cells of *Austrofossombronia peruviana*, with inconspicuous trigones and numerous small, homogeneous oil bodies of the *Massula*-type, bar = 25 μm. (C) Leaf cells of *Marsupella emarginata*, with large, triangular trigones and large homogeneous oil bodies of the *Bazzania*-type, bar = 20 μm. (D) Leaf cells of *Calypogeia muelleriana*, with medium triangular trigones, and botryoidal oil bodies of the *Calypogeia*-type, bar = 20 μm. (E) Leaf cells of *Radula obconica*, with inconspicuous trigones and large, solitary, papillose oil bodies of the *Jungermannia*-type, bar = 15 μm. (F) Longitudinal section of the thallus of *Marchantia polymorpha*, showing an idioblastic oil cell at the arrow, bar = 15 μm.

independently lost in several families, including the Antheliaceae, Cephaloziaceae, Lepidoziaceae, and Metzgeriaceae of the Jungermanniopsida, and the Blasiaceae (Fig. 1.2A), Sphaerocarpaceae, and Ricciaceae of the Marchantiopsida (Schuster 1966, Crandall-Stotler *et al.* 2005).

Various hypotheses have been formulated regarding the adaptive value of oil bodies, including suggestions that oil bodies deter herbivores and provide cold and/or UV protection (Schuster 1966). Immunolabeling techniques have shown that oil bodies contain enzymes involved in isoprenoid biosynthesis (Suire *et al.* 2000), confirming that they are active metabolic compartments of the liverwort cell. In addition, they sequester terpenoids and other secondary aromatics, much like the secretory glands of vascular plants (Flegel & Becker 2000).

1.2.3 *Gametophyte organizations*

Three very different types of gametophyte organization occur within the phylum. The most widespread morphology is the leafy shoot, or nodal type organization, in which the gametophyte is composed of a stem and two or three rows of leaves. This type of organization is distributed across the phylogeny, occurring in all of the subclasses delineated in this work (Figs. 1.3–1.7), and characterizes almost all of the genera of the Jungermanniidae (Fig. 1.7). Simple thalloid morphology is common in the Pelliidae (Fig. 1.5) and Metzgeriidae (Fig. 1.6), but is also found in a few Marchantiopsida (e.g. Blasiales and *Monoclea* and *Monosolenium* in the Marchantiales), and Jungermanniidae (e.g. *Pteropsiella* and *Schiffneria*). In this morphology, plants consist of an unspecialized, planate thallus that is usually composed of a somewhat thickened central midrib and two lateral wings. In contrast, a dorsiventrally differentiated thallus, bearing a system of dorsal air pores and air chambers and a ventral storage zone, characterizes complex thalloid organization. This is the most restricted type of morphology, occurring only in the Marchantiidae (Fig. 1.4). Since the variation that occurs in each of these morphological categories employs a different suite of descriptors, they will be discussed separately. It should be noted, however, that leafy, simple thalloid and complex thalloid categories do not necessarily imply natural groupings, but simply refer to a type of morphological organization.

Variation in leafy morphologies

Multiplicity in the distribution, form, size and insertion of leaves provides many of the characters that define genera and species of foliose liverworts. In *Haplomitrium* (Fig. 1.3) and a few genera of the Jungermanniidae, e.g., *Herbertus* and *Lepicolea*, plants are erect and radially symmetric with three ranks of identical leaves (isophylly). The vast majority of leafy forms, however, display bilateral symmetry in which plants bear two rows of lateral leaves with or without a single row of smaller ventral leaves or underleaves (= amphigastria). In anisophyllous taxa the underleaves can be morphologically like the leaves, but smaller, or differ both in size and morphology. In traditional classification schemes, isophyllous taxa were considered primitive and evolution was presumed to progress toward planation and anisophylly (e.g. Evans 1939, Stotler & Crandall-Stotler 1977, Schuster 1984b). Recent phylogenetic hypotheses derived from sequence data suggest, however, that isophylly is a derived state (Davis 2004, Crandall-Stotler *et al.* 2005, He-Nygrén *et al.* 2006).

With a few exceptions, such as *Pachyglossa* and *Herzogiaria*, leaves in the Jungermanniidae are completely unistratose, whereas those of *Treubia*, *Haplomitrium*, and leafy taxa of the Pelliidae and Marchantiopsida are polystratose for

Fig. 1.3. Characters of the Haplomitriopsida. (A) *Treubia lacunosoides*, dorsal view, showing the lobate thallus and small dorsal lobules (at arrow) associated with each thallus lobe, bar = 4 mm. (B–D) Representatives of three lineages of *Haplomitrium*. (B) Male shoots of *Haplomitrium gibbsiae* (subg. *Haplomitrium* sect. *Archibryum*), arising from a slime-covered stolon system; antheridia are clustered at the apices of the leafy shoots. This species is sister to all other species in the genus, bar = 1.5 mm. (C) Male shoot of *Haplomitrium hookeri* (subg. *Haplomitrium* sect. *Haplomitrium*), showing antheridia (arrow) in the axils of unmodified leaves just below the shoot apex, bar = 300 μm. (D) *Haplomitrium mnioides* (subg. *Calobryum*), dorsal view; note the branched, leafless stolon system and anisophyllous shoots, with the smaller third row of leaves on the dorsal side, bar = 2 mm.

Fig. 1.4. Characters of the Marchantiopsida. (A) *Asterella tenella*, lateral view of a carpocephalum, showing the pseudoperianth (at arrow) emerging from a tubular involucre, bar = 2.6 mm. (B) *Conocephalum conicum*, dorsal view, showing hexagonal outlines of the air chambers and a conical carpocephalum; a tubular involucre (at arrow) encloses the nearly mature sporophyte, bar = 2 mm. (C) *Monoclea gottschei*, dorsal view; oil cells appear as scattered white dots throughout the thallus, bar = 5 mm. (D, E) *Marchantia polymorpha*, longitudinal sections of gametangiophores, bars = 1 mm: (D) antheridiophore with antheridium indicated by arrow; (E) archegoniophore with archegonium indicated by arrow.

several cell rows at the base, gradually becoming unistratose distally. They are generally composed of a uniform network of isodiametric to slightly elongate chlorophyllous cells with thin or unevenly thickened walls (Fig. 1.2). Trigones, the corner wall thickenings between leaf cells (Fig. 1.2C), consist mostly of hemicelluloses (Zwickel 1932) and are important in the apoplastic conduction of water (Proctor 1979). In some taxa the surface walls are roughened with papillae, granulae, or striae. Although these are treated as cuticle markings in taxonomic descriptions, they are actually projections of the wall proper, not waxy deposits (Duckett & Soni 1972). To date, there is no unequivocal evidence that a true cuticle exists in jungermannioid liverworts (Cook & Graham 1998). Occasionally, idioblastic oil cells, or ocelli, are interspersed among the normal leaf cells, e.g., some species of *Frullania*, or a line of highly elongate, thick-walled cells forms a vitta or nerve in the leaf as in *Herbertus*.

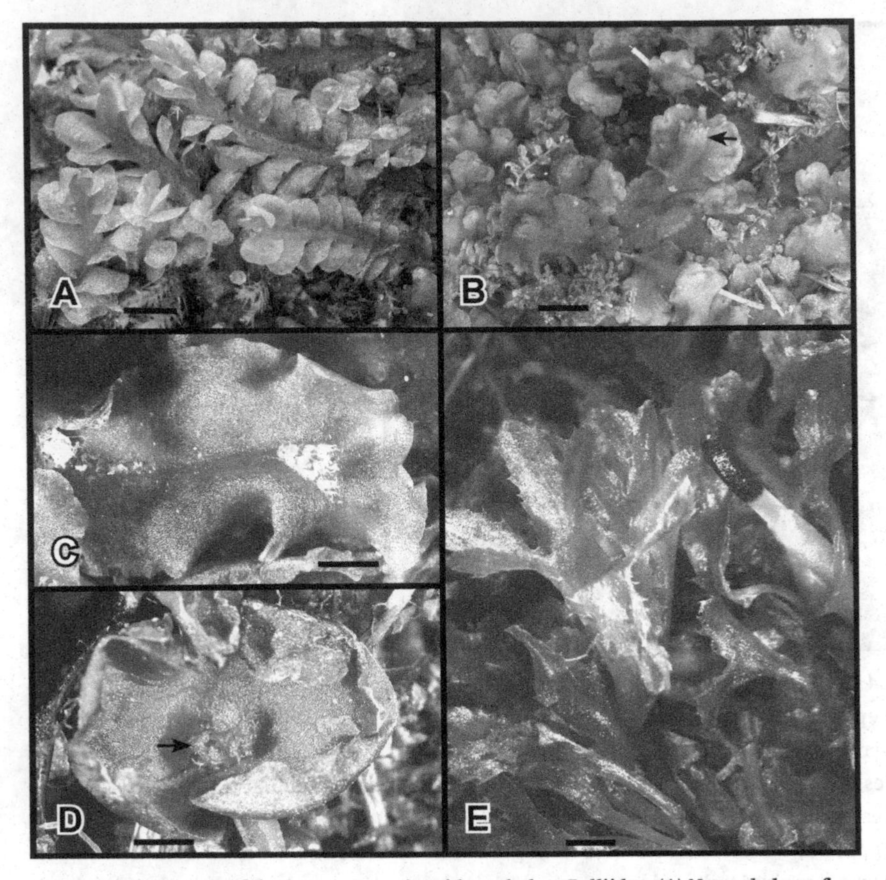

Fig. 1.5. Characters of the Jungermanniopsida, subclass Pelliidae. (A) *Noteroclada confluens*, dorsal view, plants with undivided succubous leaves and two rows of naked archegonia on the midrib, bar = 2.3 mm. (B) *Pellia epiphylla*, dorsal view, showing acrogynous perichaetium (at arrow) positioned between furcate thallus branches, bar = 3 mm. (C) *Allisonia cockaynei*, dorsal view, male plant with a cluster of antheridia and perigonial scales near thallus apex, bar = 2 mm. (D) *Phyllothallia nivicola*, dorsal view, with developing perichaetia and archegonia (at arrow) at node, between a pair of opposite leaves, bar = 1.5 mm. (E) *Jensenia connivens*, illustrating a dendroid thallus habit; note the sporophyte emerging from a perichaetial pseudoperianth to the right, bar = 1.1 mm.

Leaves are commonly lobed or divided in the Jungermanniidae, but undivided leaves characterize some families, such as the Jungermanniaceae and Plagiochilaceae, as well as *Haplomitrium* (Fig. 1.3) and the foliose taxa of the Pelliidae (Fig. 1.5) and Marchantiopsida. Divided leaves can be bifid, trifid, quadrifid, multifid, or bisbifid, i.e., having the two lobes of a bifid leaf themselves less deeply divided into two lobes (see e.g. Schuster 1984a, Fig. 15). In addition, lobe margins may be ciliated or toothed, as in *Trichocolea*, which

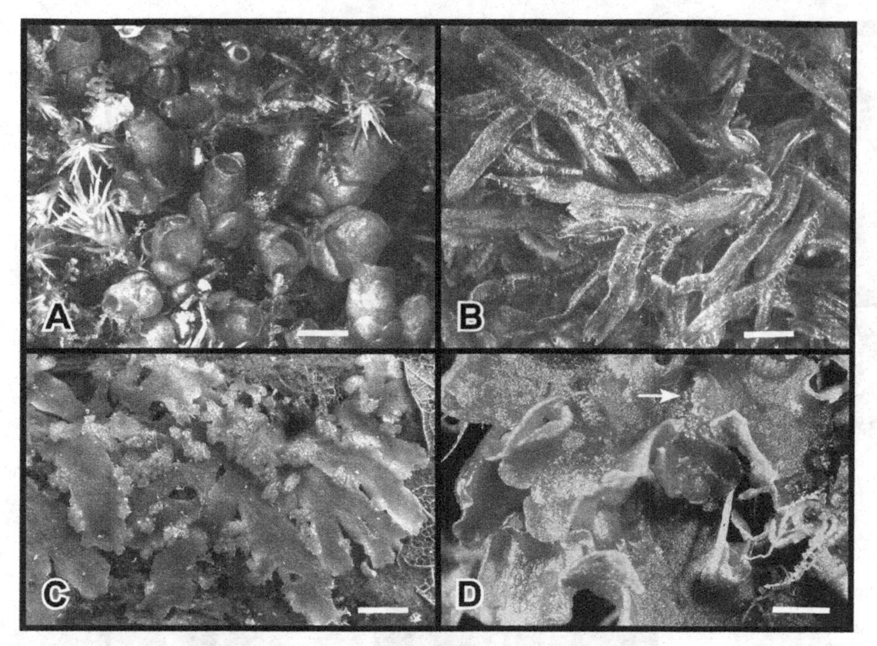

Fig. 1.6. Characters of the Jungermanniopsida, subclass Metzgeriidae. (A) *Pleurozia acinosa*, showing leafy shoots with abundant tubes, often referred to as sterile gynoecia, on abbreviated lateral branches, bar = 1.8 mm. (B) *Metzgeria leptoneura*, ventral view, illustrating hyaline hairs on the involute wing margins and midrib, bar = 2.1 mm. (C) *Aneura pinguis*, male plants bearing numerous androecia, each on an abbreviated lateral branch, bar = 6 mm. (D) *Verdoornia succulenta*, female plants with a gynoecium on the dorsal surface of the thallus (at arrow), bar = 4 mm.

enhances the uptake and ectohydric transport of water by creating capillary spaces between leaves. Lobes can be equal in size and symmetric, as in many genera of the Lepidoziaceae, or different in size, shape and even form, as in most genera of the Porellales. Lobe number and size are established early in leaf ontogeny and may prove to be phylogenetically informative (Bopp & Feger 1960, Schuster 1984a). In the Porellales and some Jungermanniales, e.g., Schistochilaceae and Scapaniaceae, leaves are complicate-bilobed, meaning that the asymmetrically bifid leaves are longitudinally folded so that the smaller lobe, or lobule, is appressed to either the dorsal or ventral surface of the larger lobe. Although usually described as complicate-bilobed, leaves in many, but not all, of the Porellales are actually trifid, consisting of a dorsal lobe, ventral lobule and small ventral stylus. According to Heinrichs *et al.* (2005) and He-Nygrén *et al.* (2006), this trifid type of organization is fundamental to the Porellales. However, leaves in both *Porella* and *Radula* are truly bifid, as demonstrated by Bopp & Feger (1960) and Leitgeb (1871a), respectively.

Fig. 1.7. Characters of the Jungermanniopsida, subclass Jungermanniidae. (A) *Bazzania novae-zelandiae*, illustrating an incubous leaf insertion, bar = 2.5 mm. (B) *Proskauera pleurata*, illustrating a succubous leaf insertion and terminal, pluriplicate perianths (at arrow), bar = 1.4 mm. (C) *Balantiopsis rosea*, lateral view, immature, hollow marsupium of the *Calypogeia*-type, bar = 800 μm. (D) *Megalembidium insulanum*, dendroid plant with extensively branched rhizome system, bar = 3 mm. (E) Male plants of *Tylimanthus saccatus*, illustrating a terminal androecium and succubous leaf arrangement, bar = 1.25 mm. (F) *Isotachis lyellii*, with an erect stem perigynium bearing a highly reduced perianth at its apex (at arrow), bar = 750 μm.

Frequently, in complicate leaves the lobules form inflated water sacs, which are of two developmentally different types. In the *Lejeunea*-type water sac, cell divisions restricted to just below the free margin of the lobule enroll it against the lobe. The mouth or opening of the water sac is directed towards the leaf apex. The water sac is confluent with the lobe for most of its length and has a long, vertical line of insertion on the stem (Crandall 1969, Fig. 91). This type of water sac is characteristic of the Lejeuneaceae, the largest family of liverworts, but is also found in *Trichocoleopsis* (Neotrichocoleaceae), *Nowellia* (Cephaloziaceae), *Tetracymbaliella* (Lophocoleaceae) and *Delavayella* (Delavayellaceae). The galeate or *Frullania*-type water sac, in contrast, is inflated medially, like a balloon, from cell divisions restricted to the middle of the lobule. The free margins are not enrolled, but instead are constricted around the mouth, which is usually directed downwards. The water sac is scarcely confluent with the lobe but is joined to the stylus that in turn is attached to the stem (Crandall 1969, p. 97). This second type of water sac occurs in several families of the Porellales, including the Lepidolaenaceae and Frullaniaceae, and in *Neotrichocolea* (Neotrichocoleaceae, Ptilidiales). The water sac of *Pleurozia* is unique in developing through a combination of marginal enrolling and ballooning processes and differentiating a flap-like valve that opens or closes the mouth in response to hydration levels (Crandall-Stotler 1976). Water sacs enhance the uptake of water during periods of hydration, but appear to quickly lose water with drying (Blomquist 1929, Proctor 1979), so they probably do not function in water storage, except perhaps in the valvate types found in *Pleurozia* and *Colura* (Schuster 1966). It has also been suggested that small invertebrates that inhabit the water sacs may provide nitrogenous compounds to the mostly epiphytic taxa that bear them (Verdoorn 1930, Hess *et al.* 2005).

Underleaves are always transversely inserted and have their laminae appressed to the ventral surface of the stem. Lateral leaf insertions are variable, with the fundamental line of insertion dependent upon the orientation of the shoot apical cell (Buch 1930, Crandall 1969). If the medial axis of the apical cell is vertically aligned with the center of the shoot, leaves will be transversely inserted as occurs in erect-growing, isophyllous taxa, e.g. *Haplomitrium* (Fig. 1.3B, C) and *Herbertus*, as well as a few prostrate taxa, e.g. *Cephaloziella*. In most prostrate taxa the medial axis of the apical cell is tilted and leaves are obliquely inserted; a dorsal tilt results in a succubous insertion and a ventral tilt in an incubous insertion. In both of these insertions, the leaf lamina extends horizontally out from the stem. In plants with a succubous insertion, the shoot apex bends up, away from the substrate, the lower or basiscopic margin of the leaf is inserted on the dorsal side of the stem, and the adaxial surface of the leaf is dorsal (Fig. 1.7B, E). In contrast, in plants with an incubous insertion, the shoot apex bends down toward the substrate, the upper or acroscopic margin of the leaf is inserted on the dorsal side of the stem, and the abaxial surface of the leaf is dorsal (Fig. 1.7A). In complicate leaves, lobe and lobule insert differently, resulting in an oblique J- or U-shaped line of insertion. Succubous insertions tend to occur in taxa that grow on moist, soil substrates, whereas incubous insertions seem to be more common in epiphytes. Clee (1937) postulated that succubous insertions favor water movement from below, whereas plants with incubous insertions are better adapted to capturing water flowing from above. Schuster (1966) suggests that incubous and especially incubous-complicate leaves can be more tightly overlapped or shingled than succubous leaves and are hence better adapted to retaining water in their ventral lobules. This inherent ability of incubous taxa to limit water loss may contribute to the widespread success of the Porellales as epiphytes.

The stems of leafy liverworts are relatively unspecialized, consisting mostly of parenchymatous cells. However, in some genera, like *Plagiochila* and *Herbertus*, the outer three or four layers of cells, including the epidermis, are thick-walled and prosenchymatous as compared to the cells of the interior, and in others, like *Lejeunea* and *Cephalozia*, the epidermis is replaced by a hyalodermis of highly inflated, thin-walled cells. In the Treubiaceae and Scapaniaceae, cells in the ventral part of the stem form a distinct mycorrhizal zone, and in *Goebeliella* all cells of the stem are prosenchymatous. Among leafy forms, only *Haplomitrium* possesses a central strand of elongate, hydrolyzed "conducting" cells; these are smaller in diameter than the surrounding cells of the cortex, thin-walled, and minutely perforate. Note that in the absence of substantial anatomical differentiation the outer zone of shorter stem cells is often referred to as a cortex and the inner zone of more elongated cells as a medulla. For consistency with other plant groups, however, we suggest that the terms epidermis, cortex, and central strand be applied to hepatics, as defined in Magill (1990).

Rhizoids are typically found on the ventral surface of the plant, developing from either specialized cells of the underleaves or the stem epidermis. They are generally hyaline, although deeply pigmented rhizoids are diagnostic of some genera, e.g. *Fossombronia, Herzogianthus*, and *Schistochila*. In *Radula* rhizoids are formed from the center of the ventral lobules of the leaf, but in most taxa that lack underleaves, rhizoids are widely scattered on the ventral surface of the stem, as in *Jungermannia*. In some taxa, like *Megalembidium* and *Pleurozia*, they are restricted to stoloniferous or rhizomatous branches (Fig. 1.7D). In many epiphyllous Lejeuneaceae, the rhizoid initials form a sucker-like disc on the underleaf that firmly attaches the liverwort to its substrate. In fact, the primary function of rhizoids seems to be substrate attachment, but an important secondary function may be to host symbiotic fungi, as demonstrated by Duckett *et al.* (1991). Such mycorrhizal rhizoids have swollen, branched tips. Among liverworts, only *Haplomitrium* completely lacks rhizoids.

Branching systems can be furcate or superficially dichotomous (e.g. *Bazzania* and *Fossombronia*) or sympodial with formation of subfloral innovations (e.g. Scapaniaceae) or more commonly monopodial (Buchloh 1951). Often branch modules are heteroblastic so that the first-formed leaves or appendages at the base of the branch are morphologically distinct from those differentiated from subsequent merophytes. The form of these modified first-branch leaves and underleaves can be systematically informative, as demonstrated repeatedly in studies of the Frullaniaceae (Verdoorn 1930, Stotler 1969, von Konrat & Braggins 2001). Branches may resemble the main stem, or be differentiated as microphyllous shoots, flagellae, or stolons. In a few taxa, e.g. *Bryopteris* and *Megalembidium* (Fig. 1.7D), a dendroid, monopodial leafy shoot system arises sympodially from a creeping rhizome or caudex that is attached to the substrate.

Twelve patterns of branch ontogeny, based on differences in the delineation of the branch initial and/or early stages of bud differentiation, have been described (Crandall-Stotler 1972, Thiers 1982). As first recognized by Leitgeb (1871b, 1872), these patterns can be classified into three groups based on the spatial relationship between the branch and the shoot apex, namely (1) terminal with stem leaf modified, (2) terminal with stem leaf unmodified, and (3) intercalary. In the first group, which includes branches of the *Frullania*-, *Kurzia*- (= *Microlepidozia*) and *Acromastigum*-types, the branch apical cell is formed very near the shoot apex from an outer cell, or "segment half", of the three-celled merophyte, thereby restricting leaf development to half of its usual initials. Consequently, a half-leaf (or -underleaf) develops on the stem at the position of branch emergence. In *Frullania*-type branches, which are the most common type in this group, the branch replaces the ventral part of the leaf. The degree of half-leaf modification varies greatly among taxa; e.g. in *Frullania* the half-leaf lacks a lobule and stylus, whereas in *Chiloscyphus*, it is only slightly reduced in size but lacks insertion on the ventral side of the stem (Crandall 1969, Fig. 248).

In the second group of terminal branches, the branch begins development very close to the apex but from an epidermal cell that is basiscopic to a leaf primordium. Leitgeb (1871b) described this as the *Radula* pattern, but branches of the *Bryopteris*-, *Lejeunea*-, *Aphanolejeunea*- and *Fontinalis*-types [*Fontinalis*-type = *Haplomitrium*-type of Schuster (1966)] also belong to this group. Since the branch initial is differentiated later in merophyte development, leaf morphology is unaffected by branch formation. In *Fontinalis*-type branching, one or more stem cells occur between the branch primordium and the basiscopic insertion of the stem leaf, but in the other types in this group, the stem leaf partially inserts on the upper side of the branch. In *Radula*-type branching, branch growth by apical cell segmentation begins near the shoot apex so there is synchronous maturation of stem and branch tissues along a 45–60° angle of branch divergence (Crandall-Stotler 1972, Fig. 23). In the other types, the branch usually remains as a bud or primordium to some distance below the shoot apex. Since branch maturation occurs after the stem cells have elongated, the tissues of the branch appear to abut those of the stem at a 90° angle (Crandall-Stotler 1972, Fig. 52). In branches of the *Lejeunea*- and *Bryopteris*-types a layer of tissue derived from the basiscopic part of the leaf in the former, or by longitudinal division of the branch initial in the latter, internalizes the branch primordium prior to branch apical cell formation. When growth resumes, the branch pushes through this tissue, leaving it as a collar at its base. These branches mimic intercalary branches in their production of basal collars even though they originate from epidermal cells.

In contrast to *Lejeunea-* and *Bryopteris*-type branches, intercalary branches develop from undifferentiated cells of the stem cortex usually near the axils of leaves (lateral intercalary or *Plagiochila-* and *Anomoclada*-types) or underleaves (ventral intercalary or *Bazzania*-type). Grolle (1964) first applied type names to these patterns; they are equivalent to the *Lophozia-*, *Andrewsianthus-* and postical intercalary types of Schuster (1966, p. 445). Although the enlarged, usually rounded initials of intercalary branches are actually formed near the shoot apex, they remain dormant until some distance below the meristematic zone. In some cases, in fact, they do not break dormancy unless the stem apex is physically destroyed or replaced by a gynoecium (Crandall 1969). When dormancy is broken and the branch bud begins to grow, the contiguous epidermal cells divide to form a bulging cover layer. This epidermal tissue is subsequently torn when the branch elongates, forming a collar at the branch base. The tissues of the branch abut the main axis at a 90° angle, but are more deeply inserted into the cortex than occurs in the exogenous types described above.

Dormant branch initials and/or primordia provide a mechanism for replacing a damaged shoot apex, but in *Cephalozia*, *Blepharostoma*, and *Radula* this potential is supplemented by formation of adventive branches (Evans 1912, Hollensen 1973, Crandall-Stotler 1981). Adventive branches originate from mature epidermal cells that dedifferentiate. They resemble stem regenerants in being easily detached from the stem, but otherwise look like normal branches. To date, they are known from only a few taxa.

The majority of leafy liverworts are dioicous, but there are also monoicous taxa, especially those with Laurasian distributions (Longton & Schuster 1983). Both archegonia and antheridia develop from superficial cells near the apex of a stem or branch. In general, antheridia consist of a spheroidal to ovoidal or occasionally ellipsoidal body and subtending stalk. The body is usually white, but bright orange to yellow antheridia are characteristic of some genera, e.g. *Fossombronia* and *Haplomitrium* (Fig. 1.3C). Typically the jacket cells are randomly arranged and the stalk is short, straight and biseriate. Systematically important variation does, however, occur as detailed in Müller (1948). For example, the stalk is four- to seven-seriate in several genera, including *Schistochila*, *Haplomitrium*, and *Fossombronia*, and is characteristically uniseriate in most Porellales. Tiered jacket cells are diagnostic of the Cephaloziaceae and some Calypogeiaceae and Lepidoziaceae as well as select species of *Haplomitrium* (Schuster 1966, Bartholomew-Began 1991).

In the Jungermanniidae, Haplomitriopsida, and *Pleurozia*, antheridia develop in the axils of modified perigonial leaves and rarely underleaves, referred to as male bracts and bracteoles, respectively. The number of antheridia per bract varies from one or two in many groups, e.g. Lejeuneaceae, to over 100 in Schistochilaceae; sometimes the antheridia are intermixed with paraphyllia, e.g. Scapaniaceae. In *Treubia* and some species of *Haplomitrium*, androecia are rather loosely organized, with bracts dispersed along the stem and scarcely modified from the vegetative leaves (Fig. 1.3C). In most leafy taxa, however, male bracts are smaller than vegetative leaves and ventricose at the antical base (Fig. 1.7E). Androecia may terminate the main stem or leading branch, or be intercalated between vegetative segments, as in the Plagiochilaceae, or be restricted to spicate to capitate branches, e.g. *Pleurozia* and most Porellales. A disciform, splash-cup type of androecium, consisting of three enlarged bracts surrounding up to 100 antheridia and intermixed slime hairs on a terminal receptacle, occurs only in *Haplomitrium* subg. *Calobryum*. In the leafy members of the Pelliidae (*Fossombronia*, *Noteroclada*, and *Phyllothallia*) and Marchantiopsida (Sphaerocarpales), the androecium is usually diffuse on the dorsal surface of the stem and there is no association between the antheridia and leaves. For example, in *Fossombronia* the antheridia are spread out along the stem and are naked or subtended individually by a single perigonial scale. *Noteroclada* and *Sphaerocarpos* produce rows of antheridia, each enclosed in a flask-shaped involucre, and *Phyllothallia* forms clusters of antheridia intermixed with perigonial scales and slime hairs.

There are two schemes of gynoecial formation in hepatics. In acrogynous liverworts the apical cell of the reproductive module is eliminated during archegonial formation and consequently, the gynoecium terminates further growth of the module. In anacrogynous liverworts, the apical cell of the reproductive module is unaffected by archegonial production, and the module continues to grow past the gynoecium. Acrogynous taxa produce a single gynoecium per module, while anacrogynous taxa can produce a succession of gynoecia along the dorsal surface of the module. Within acrogynous taxa, gynoecia can terminate a sparingly branched main stem, or normal leafy branches, or be restricted to short branches that lack vegetative leaves, simulating acrocarpy, cladocarpy, and pleurocarpy, respectively.

All Jungermanniidae are acrogynous and most Haplomitriopsida are anacrogynous, the only exception being *Haplomitrium* subg. *Calobryum*. Of the leafy taxa in other clades, only *Pleurozia* is acrogynous. In *Fossombronia* and *Noteroclada*, archegonia are naked and dispersed dorsally on the stem (Fig. 1.5A), but in most liverworts they are clustered and protected by foliar structures (Fig. 1.5D). In Treubiaceae and *Haplomitrium* subg. *Haplomitrium*, small groups of archegonia are scattered in the axils of unmodified leaves near the shoot apex, but in other species of *Haplomitrium* 20–100 archegonia are intermixed with slime hairs and scales on a terminal disc-like receptacle, surrounded in turn by one or two cycles of small inner bracts, and enlarged outer bracts. In the Jungermanniidae several cycles of modified leaves and underleaves, the perichaetial or female bracts and bracteoles, are produced prior to archegonial development. The perichaetia bear numerous marginal slime papillae and are more highly divided than the vegetative leaves. In most Jungermanniidae, the last cycle of perichaetial leaves/underleaves fuse, forming a very short tubular perianth just to the outside of the archegonial cluster (Schuster 1966, Fig. 9–9). Prior to fertilization, the perichaetia shield the archegonia and provide capillary channels for water movement into the gynoecium. Although bracts and bracteoles usually enlarge without fertilization, perianths typically remain vestigial in the absence of embryo formation.

Most leafy liverworts are capable of regenerating from fragments of leaves or stems but many liverworts also produce special asexual diaspores for vegetative reproduction and dispersal. These include caducous leaves, small branchlets, or cladia, multicellular discoid gemmae, and one- to few-celled, catenate gemmae that arise in fascicles from embryonic leaves. The first three types of broodbodies occur primarily in epiphytic taxa and are consequently most prevalent in the Porellales; the last type is restricted to the Jungermanniales (Schuster 1966).

Variation in simple thalloid morphologies

Simple thalloid morphologies vary from broad thalli with a distinct, multistratose midrib and unistratose wings, as in most of the Pallaviciniaceae, to strap-shaped thalli that are multistratose throughout, e.g. *Riccardia*, or rarely, thalli that are completely unistratose, e.g. *Mizutania* (Figs. 1.5, 1.6). Thalli are bilaterally symmetric and usually prostrate, but can also be ascending to erect, as in *Jensenia* (Fig. 1.5E) and *Hymenophyton*. In the Pallaviciniales, the thallus often arises from a wingless, cylindrical stipe that is embedded in the substrate. Rhizoids that are structurally like those of leafy liverworts are usually dispersed along the ventral surface of the midrib, but may also be produced from cells of the wing margin, especially in the Metzgeriidae.

The ventral surface of the midrib at the thallus apex frequently elaborates two or more rows of foliose scales (e.g. Petalophyllaceae, Calyculariaceae, and Blasiaceae), uniseriate hairs (e.g. Allisoniaceae, Moerckiaceae, and Makinoaceae), or stalked slime papillae (e.g. Hymenophytaceae and Metzgeriaceae). Scales and hairs typically persist for a considerable distance below the apex, but slime papillae are usually seen only near the apex (e.g. *Monoclea* and *Pellia*, Fig. 1.1G); they may be distributed in two rows on the midrib, as in the Hymenophytaceae and Metzgeriaceae, or be widely dispersed on both dorsal and ventral surfaces of the thallus as in some Pallaviciniaceae. In a few taxa of the latter, slime papillae may also occur along the wing margins. The Blasiidae are unique in having two rows of *Nostoc*-containing auricles derived from slime hairs that lie to the inside of the two rows of persistent ventral scales.

In most simple thalloid taxa, cells of both the central polystratose midrib (= costa), and the thallus wings are uniformly chlorophyllose and thin-walled, without conspicuous trigones. However, in *Cavicularia* (Marchantiopsida) and most taxa of the Pallaviciniales strands of differentiated, elongate cells are formed in the thallus midrib. These strand cells are smaller in diameter than the surrounding thallus cells, devoid of protoplasm at maturity, and hypothesized to function as water reservoirs or conduits. In Hymenophytaceae and Pallaviciniaceae the strand cells possess thick, pitted, and finely perforate walls (Ligrone & Duckett 1996), whereas in *Moerckia* and *Cavicularia* they are thin-walled and unperforated (Hébant 1977, Kobiyama 2003). Strand cells in *Hattorianthus* have uniquely thickened walls, but lack both pits and perforations like *Moerckia* (Kobiyama 2003, Murray & Crandall-Stotler 2005).

Branching is predominantly terminal, with thallus apices appearing to bifurcate or dichotomize (Crandall-Stotler 1981, Renzaglia 1982). In only a few simple thalloid taxa (e.g. *Pellia*) are these bifurcations true dichotomies as in the Marchantiopsida. In most taxa the branch apical cell arises exogenously from the central cell of the three-celled merophyte, yielding a false dichotomy. In many Pallavicinaceae, in addition to terminal furcations, dormant branch primordia are produced just ventral to the wing from epidermal cells of the midrib. Such branch primordia can generate a sympodial branching habit, with ventral exogenous intercalary branches arising near the base of the main thallus, as in *Pallavicinia* and *Jensenia*, or they may form short ventral androecial or gynoecial branches, as in *Podomitrium* and *Hymenophyton*. Among simple thalloid taxa, only Metzgeriaceae form ventral endogenous branches as well as ventral exogenous branches and terminal furcations. Monopodial branching habits of vegetative thalli are common only in the Aneuraceae (Fig. 1.6C).

Simple thalloid taxa are often sexually dimorphic, with male thalli smaller than the female (Renzaglia 1982, Table 2). As in leafy taxa, antheridia vary in color, stalk size and jacket cell orientation. They are arranged in two or more rows (e.g. *Pallavicinia*), or aggregated in clusters (e.g. *Allisonia*, Fig. 1.5C), on the dorsal surface of the midrib, and are either associated with perigonial scales or contained in flask-shaped, ostiolate perigonial chambers. Androecia of *Podomitrium*, *Hymenophyton*, Metzgeriaceae, and Aneuraceae (except *Verdoornia*) are restricted to short exogenous branches that are lateral in Aneuraceae, but otherwise ventral. *Monoclea* (Fig. 1.4C) and *Monosolenium*, which are best interpreted as complex thalloid plants that have lost their air chambers, have androecial organizations comparable to those of other complex thalloid taxa.

Most simple thalloid liverworts are anacrogynous, having their archegonia and perichaetial scales aggregated on small receptacles posterior to the thallus apex along the dorsal surface of the midrib. In *Pellia* most species are anacrogynous, but in *Pellia epiphylla* the gynoecium and subsequently formed sporophyte terminate the growth of the original thallus module (Fig. 1.5B) and a new apical cell is formed to either side of the young gynoecium (Renzaglia 1982). Thus, this species is actually acrogynous. Other acrogynous simple thalloid taxa include those that produce their gynoecia on short, determinate branches. This feature is characteristic of Metzgeriidae (except *Verdoornia*), as well as *Podomitrium* and *Hymenophyton* of the Pelliidae and the simple thalloid taxa of the Jungermanniidae. Archegonial characters are mostly taxonomically uninformative, but distinctive archegonia having extremely short, thick necks are diagnostic of the Metzgeriales as here defined (Crandall-Stotler *et al.* 1994). In most simple thalloid taxa the archegonia and intermixed slime hairs or papillae are subtended by one or more series of laciniate to dentate, leaf-like scales or perichaetia. In some, e.g. *Pallavicinia* and *Moerckia*, an additional short, tubular perichaetium encircles the archegonial cluster, just to the inside of the much larger perichaetial scales. This tubular structure, deemed analogous to the perianth of the Jungermanniidae, has been termed an inner involucre, inner perichaetium, or more commonly a pseudoperianth (Schuster 1992b). The term pseudoperianth has, however, also been applied to two non-homologous structures, including perianth-like enclosures that originate only after fertilization as in *Calycularia*, and the envelope that develops from the archegonial stalk in Marchantiidae, e.g. *Asterella* (Fig. 1.4A). The term involucre has also been loosely applied, and in complex thalloid liverworts refers to tissue of thalline origin that encloses the cluster of archegonia (Bischler 1998) (Fig. 1.4B). Crandall-Stotler *et al.* (2002) resurrected the term caulocalyx from Chalaud (1928) to replace the term pseudoperianth for post-fertilization structures of thalline origin and reserved the term pseudoperianth for structures derived from the inner perichaetium. Until an ontology is completed for the Marchantiophyta, we suggest that in simple thalloids the term pseudoperianth be accompanied by the modifier "perichaetial" to differentiate it from the pseudoperianth of the Marchantiidae and that the term involucre should be restricted to gynoecial structures of thalline origin in the Marchantiopsida.

Special asexual diaspores are formed by only a few simple thalloid taxa. In *Riccardia* 1- or 2-celled gemmae are formed endogenously, i.e. within the walls of existing cells, at the thallus apex. This rare type of asexual diaspore has also been described in *Jungermannia caespiticia* (Buch 1911). Exogenously produced, shortly stalked, multicellular, bulbous gemmae occur in Blasiales, Treubiaceae, *Aneura*, and *Xenothallus*; fragile brood branches, or cladia (sometimes referred to as gemmae) are common in *Metzgeria* and *Greeneothallus*. Subterranean tubers provide a means of both perennation and vegetative propagation in Petalophyllaceae, as well as in the related leafy taxa, *Fossombronia* and *Noteroclada*.

Variation in complex thalloid morphologies

Liverworts with complex thalloid morphologies are normally terrestrial, with a few aquatic species (*Riccia* and *Ricciocarpos*). Many are xeromorphic, drought-resistant and able to withstand strong insolation. Many gametophyte characters seem to be environmentally modulated. The thallus is a dorsiventrally flattened, prostrate to suberect, usually bilaterally symmetric shoot (Fig. 1.4); it is mostly multistratose apart from the wings, which may be marginally unistratose. The lower surface often has a prominent midrib on which are borne rhizoids and ventral scales. The thallus shows three types of branching: terminal innovations that appear as constrictions in the thallus due to seasonal cessation of growth and subsequent new growth from the same or a new apical cell; terminal dichotomies or furcations, which may be symmetric or asymmetric; and ventral intercalary branches, which arise exogenously from the lower epidermis of the midrib as in simple thalloids and have a characteristic stipitate base.

Complex thalloids by definition show internal differentiation into a dorsal epidermis with air pores, an assimilatory (photosynthetic) layer with air chambers and a ventral non-photosynthetic layer without air chambers. The epidermis is unistratose, with or without chloroplasts, and the air pores can be simple, with surrounding cells undifferentiated, e.g. *Riccia*, highly differentiated with several rings of narrow cells, e.g. *Conocephalum* (Fig. 1.4B), or compound (barrel-shaped) as in *Neohodgsonia* and Marchantiaceae. In *Dumortiera* air pores are vestigial. The assimilatory layer contains air chambers, which may be tall and columnar, broad and spreading, or irregular and spongy, with more than one type in a single thallus in some genera. They are bounded by unistratose walls of chlorophyllose cells and sometimes contain free-standing chlorophyllose filaments (Fig. 1.2F). Evans (1918) classified them into three types, *Riccia*-type (columnar), *Reboulia*-type (spreading, more than one layer) and *Marchantia*-type (spreading with free-standing filaments). Bischler (1998) showed that different types of air chambers can occur in a single genus; for example, in *Riccia* the assimilatory layer can be absent and air chambers, when present, can be either *Riccia*-type or *Reboulia*-type. In *Monocarpus* the air chambers are completely open dorsally and in *Dumortiera* they are vestigial.

The ventral tissue is usually multistratose in the thallus midrib, but can be reduced to only ventral epidermis in the thallus wings, or in a few taxa, e.g. *Cyathodium*, reduced throughout. This layer is composed of achlorophyllose cells with thin, sometimes pitted walls, and may contain mucilage cavities and mycorrhizal fungal hyphae. The ventral epidermis is scarcely differentiated and in the median part of the midrib bears both unicellular rhizoids and ventral scales. In most genera ventral rhizoids are dimorphic, with typical smooth rhizoids and pegged rhizoids with intracellular wall projections.

Ventral scales are unistratose, foliose structures borne mostly in two rows (up to eight rows in *Bucegia* and *Marchantia* or in ill-defined rows in Cleveaceae, *Ricciocarpos*, and *Corsinia*), with or without an apical appendage and often with oil cells and marginal slime papillae (Bischler 1998). Their function may be water conduction by capillary action as well as affording protection to the apical cell and, in some xeromorphic taxa (e.g. *Riccia*, *Targionia*, *Plagiochasma*), to the rolled-up thallus.

Complex thalloid species are dioicous and monoicous in approximately equal proportions (Bischler 1998). Many genera such as *Asterella*, *Athalamia*, *Cyathodium*, and *Riccia* contain both monoicous and dioicous species. Monoicous species usually bear antheridia and archegonia on different branches, and display distinctive sexual conditions that are species-specific, as in *Asterella* (Long 2006a).

Gametangia are exogenous and develop acropetally on the dorsal surface of the thallus. They are scattered or arranged in groups or cushions on the main thallus or its branches, or on specialized receptacles on highly modified branches (antheridiophores and archegoniophores) (Fig. 1.4). The stalks of these modified branches show vestigial features of vegetative branches, such as one or more "rhizoid furrows" containing pegged rhizoids in most genera and vestigial air chambers in others (*Asterella*, *Neohodgsonia*, *Marchantia*, and *Reboulia*). However, they lack air pores and ventral scales. Similarly, the sporophyte-containing disks, or carpocephala, retain some vegetative traits, particularly air chambers and air pores; some taxa (Aytoniaceae, *Conocephalum*, and *Wiesnerella*) with "simple" air pores in the thallus display "compound" air pores in the carpocephalum.

In complex thalloids the antheridial chambers are formed initially by divisions in more than one plane, then later in a single plane to form a protruding ostiole; these can be in scattered cavities on the thallus (*Riccia*) or along the midline of the thallus (e.g. *Corsinia*, *Cronisia*, *Oxymitra*, and *Ricciocarpos*), loosely aggregated in groups (e.g. *Cyathodium*, *Mannia*, and *Targionia*), or in sessile cushions that may be bounded by scales and contain simple air-pores (e.g. *Asterella*, *Conocephalum*, *Lunularia*, *Monosolenium*, *Plagiochasma*, *Reboulia*, and *Wiesnerella*), or aggregated into stalked antheridiophores (*Dumortiera*, *Neohodgsonia*, and Marchantiaceae) that have one or two rhizoid furrows (or up to four in some *Marchantia* species) (Fig. 1.4D). In *Monocarpus* antheridia are borne on the floor of open air chambers.

Archegonia also may be borne in several ways: embedded in the thallus along its mid-line (*Riccia* and *Ricciocarpos*), in a dorsal group becoming ventrally displaced (*Cyathodium* and *Targionia*), loosely aggregated on the thallus (*Oxymitra*), in cavities on the thallus (*Corsinia* and *Cronisia*), in cushions on the thallus which later become elevated on a stalk (*Aitchisoniella, Exormotheca, Cleveaceae, Aytoniaceae, Wiesnerella*, and *Conocephalum*), or under the lobes of a stalked receptacle (*Dumortiera, Lunularia, Neohodgsonia*, and Marchantiaceae). In carpocephalate taxa the stalk may elongate before or after fertilization. The archegonia are protected by an involucre (= perichaetium of some authors, e.g. Goebel 1930) that is scale-like, cup-shaped, bivalved, tubular or pyriform (Fig. 1.4B). A pseudoperianth (Fig. 1.4A) is developed around the sporophyte only in *Asterella, Marchantia*, and *Neohodgsonia*.

As in leafy and simple thalloid forms, many complex thalloid forms can regenerate from fragments of thallus, but a few genera distributed in several families produce specialized vegetative reproductive structures such as cup-shaped or crescent-shaped gemmae receptacles on the thallus (*Lunularia, Marchantia*, and *Neohodgsonia*), or specialized fragmenting thallus apices (*Conocephalum* and *Cyathodium*). The gemmae are always pluricellular and discoid to lenticular in form. In others, perennating tubers may be produced ventrally (*Conocephalum*) or as xeromorphic thallus tips (*Asterella*).

1.2.4 *Sporophytes and associated structures*

In all liverworts the sporophyte is enclosed by and physiologically dependent on the gametophyte until just prior to spore release (Thomas *et al.* 1979). Early embryology is known for relatively few taxa, but among these, a three- or four-celled filamentous embryo is most common (Schuster 1984a). Several complex thalloid liverworts, including members of the Cleveaceae, Marchantiaceae, Corsiniaceae and Ricciaceae (Müller 1954, p. 324, Schuster 1992c, p. 15), are reported to have octant-type embryos and *Monoclea* has been shown to have a free nuclear pattern of embryogeny (Campbell 1954, Ligrone *et al.* 1993). As the embryo develops, it becomes embedded in tissues derived solely from the archegonium (= true calyptra or epigonium), or solely from the female gametophore (= coelocaule or solid perigynium), or from a combination of the two (= shoot calyptra) (Knapp 1930). In a true calyptra, only tissue just below or of the archegonial venter divides, so there is little penetration of the sporophyte foot into the gametophore and the associated perichaetial structures insert below the calyptra; in the Jubulineae the calyptrae are stalked and the sporophyte foot is in contact only with tissue of venter origin. In shoot calyptrae and coelocaules, gametophore cells below the archegonial cluster are stimulated to divide after fertilization. If this meristematic zone is active for only a short period, the foot and basal part of the seta will be embedded in gametophore tissue, while the upper part of the seta and capsule will be surrounded by cells derived from the archegonial venter. A shoot-calyptra can usually be differentiated from a true calyptra by the presence of unfertilized archegonia part way up the outer surface of the calyptra. When a coelocaule is formed, in contrast, the sporophyte is completely embedded in tissue derived from the meristematic zone below the gynoecium; unfertilized archegonia, foliar elements of the perichaetium, and even vegetative leaves are distributed over the outer surface of the thick, fleshy coelocaule up to the apex of the embedded sporophyte (see Bartholomew-Began 1991, Fig. 311). In leafy forms with coelocaules, e.g. *Trichocolea* and *Lepicolea*, the sporophyte appears to be buried in a swollen stem apex (= the coelocaule), whereas in simple thalloid forms, e.g. *Symphyogyna*, *Xenothallus* (Fig. 1.8A), and *Aneura*, the coelocaule is a fleshy, club-shaped structure, ornamented with scales, papillae and archegonia.

Since the entire gynoecial receptacle is involved in forming a coelocaule, it alone protects the sporophyte. However, additional structures of perichaetial and/or axis origin frequently develop to the outside of calyptrae and shoot calyptrae. These include perianths (Fig. 1.7B), perichaetial pseudoperianths and caulocalyces, all of which are uni- or bistratose sheath-like structures that enclose the developing sporophyte. Variations in the shape and size, the number and position of keels, and the form and ornamentation of the mouth of these structures provide a suite of important taxonomic characters (e.g. see Schuster 1966, Figs. 51, 52). In the Jungermanniidae, the perianth in turn is basally ensheathed by enlarged bracts and bracteoles. In some taxa, e.g., *Isotachis*, *Marsupella*, and *Nardia*, an additional structure, referred to as a stem-perigynium, essentially replaces the perianth (Fig. 1.7F). A stem-perigynium is derived from a peripheral ring of meristematic cells to the outside of the archegonial cluster and just below the perichaetium. It can superficially resemble a perianth, but actually is a multistratose, fleshy sheath of axis origin that bears the bracts on its surface and the reduced perianth at its apex. Coelocaules, shoot-calyptrae and stem-perigynia in prostrate taxa of Jungermanniidae can be further modified by more growth of the ventral tissues beneath the gynoecial receptacle than the dorsal (Knapp 1930, Fig. 212). This asymmetric pattern of shoot growth, or geocauly, results in the formation of a pendant marsupium and the reorientation of the sporophyte to a vertical axis (Fig. 1.7C). There are two ontogenetically distinct types of marsupia, the *Tylimanthus*-type derived from a coelocaule and the *Calypogeia*-type formed from a stem-perigynium with a shoot-calyptra.

In most Marchantiopsida, protection of the sporophyte is afforded by a true calyptra and the involucre (Fig. 1.4B). In *Blasia*, *Cavicularia*, and *Monoclea* a thick tubular involucre is confluent with and indistinguishable from the thallus and

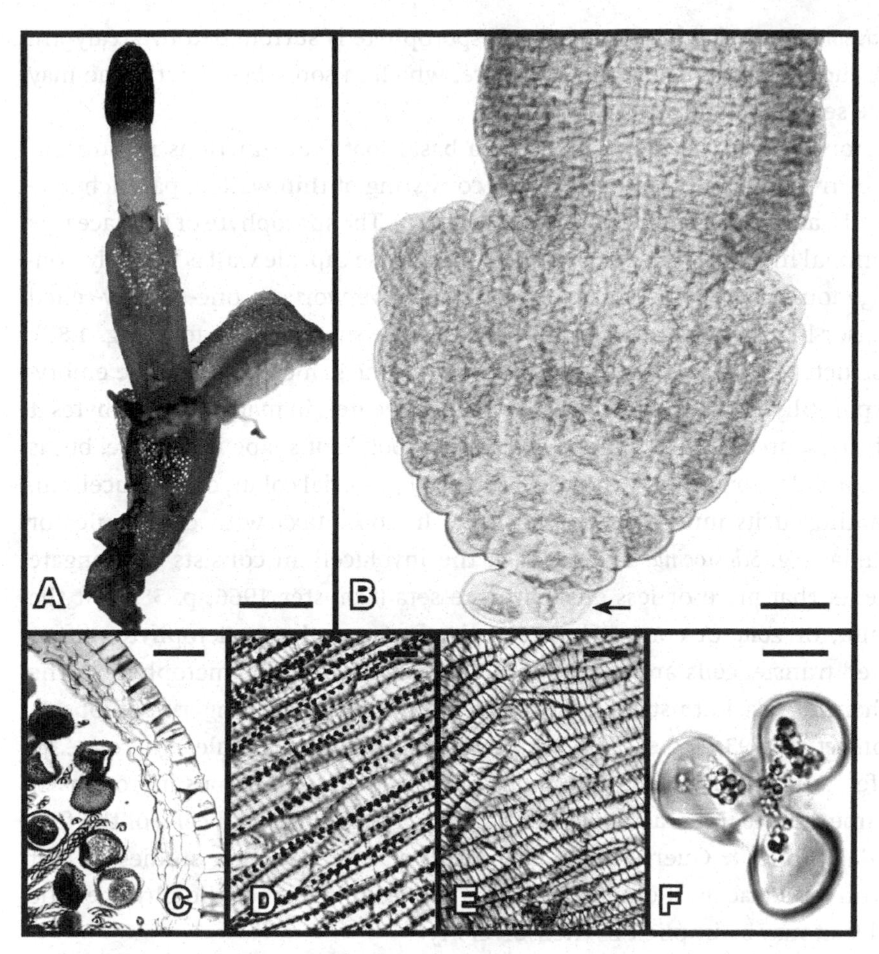

Fig. 1.8. Sporophyte generation, general structure. (A) *Xenothallus vulcanicola*, sporophyte emerging from a fleshy coelocaule, bar = 1 mm. (B) *Paracromastigum bifidum*, intact sporophyte foot, showing the elongate placental cells of the haustorial collar and the basal 2-celled haustorium (at arrow), bar = 40 µm. (C) *Porella platyphylla*, transverse section of multistratose capsule wall, showing I-band thickenings in the epidermal cells; note the elaters dispersed among the spores, bar = 75 µm. (D, E) *Aneura maxima*, capsule wall in surface view, showing wall thickening bands; (D) outer wall surface; (E) inner wall surface, bars = 50 µm. (F) *Radula obconica*, deeply furrowed sporocyte, bar = 10 µm.

consequently the sporophyte appears to be embedded in thallus tissue. In the Sphaerocarpales the sporophyte and calyptra develop within bottle-shaped pseudoperianths, and in the Ricciaceae and *Monocarpus* only calyptrae are formed (Bischler 1998). Usually the calyptrae are hyaline and 2- or 3-layered, but in *Corsinia* they can be fleshy, green and tuberculate. In a few taxa, e.g. *Asterella*,

Neohodgsonia and Marchantiaceae, each sporophyte is surrounded by a calyptra and pseudoperianth, inside the involucre, which in some Marchantiaceae may include several sporophytes.

Sporophytes are differentiated into a basal foot that functions in nutrient transfer from the gametophyte, a seta consisting of thin walled, parenchymatous cells, and a sporangium or capsule (Fig. 1.8). The sporophyte of Ricciaceae is exceptional in lacking a foot and seta; a unistratose capsule wall is the only non-sporogenous tissue it possesses. In all other liverworts, a one- to few-celled suspensor-like structure, called the haustorium, subtends the foot (Fig. 1.8B). This structure mediates nutrient transfer from the gametophyte to the embryonic sporophyte and orients the embryo in the venter; in mature sporophytes it may be obscured by the enlarged cells of the foot. Foot shape is variable, but is often conoidal, sometimes with a unistratose haustorial collar, or involucellum, originating at its juncture with the seta. In some taxa with coelocaules or marsupia, e.g. *Schistochila* and *Jackiella*, the involucellum consists of elongate filaments that more or less ensheath the seta (Schuster 1966, p. 584). At the placenta, or zone of contact between the foot and the gametophyte, differentiated transfer cells are formed in both sporophyte and gametophyte in the Marchantiopsida, but usually only in the sporophyte in the Jungermanniopsida (Ligrone *et al.* 1993, Table 3). In *Jubula* and *Radula* of the Porellales, there are no transfer cells in either generation, but instead small filamentous ingrowths of the gametophyte intercalate with radially elongate, epidermal cells of the foot (Crandall-Stotler & Guerke 1980, Ligrone *et al.* 1993). Further studies of foot structure and placental organization are needed to decipher the phylogenetic signal that may be implicit in their diversity.

Setae can be either chlorophyllose or hyaline when young, but are always white when mature. They are basically cylindrical, but are often tapered or constricted just above the foot. Seta anatomy varies from a massive, generalized type consisting of ten or more cells in width, as in many Pelliidae, to highly reduced types with obvious quadrant organization, e.g. Cephaloziaceae, Cephaloziellaceae, and Lejeuneaceae (Schuster 1966, p. 584). In some taxa the epidermal cells are larger than those of the interior; in most Lejeuneaceae the setae are articulate, meaning that the rectangular epidermal cells are arranged in regular tiers. In almost all liverworts, after sporogenesis is completed, the fragile parenchymatous cells of the seta elongate up to 20 times their original length, thus elevating the capsule up and out of the enclosing gametophytic tissues. This elongation process involves substantial uptake of water, is auxin-mediated, and can involve the synthesis of additional wall materials (Thomas & Doyle 1976). In most Marchantiidae, seta elongation is abbreviated or absent, but the structure of the unelongated seta is comparable to that of other liverworts. Capsule dehiscence and spore release occur shortly after seta elongation ceases, often within a few hours of capsule emergence. The seta collapses soon thereafter due to loss of cell turgor.

Variations in capsule shape, capsule wall structure, and dehiscence properties are of considerable taxonomic importance. Shapes vary from the generalized ovoidal type to spheroidal, ellipsoidal or long cylindric forms. In all Jungermanniopsida, as well as *Treubia* and Blasiales, the capsule wall consists of two or more layers of cells, each of which typically displays a specific pattern of darkly pigmented wall thickenings (Fig. 1.8C–E). The thickenings are secondary wall deposits, laid down after expansion of the capsule wall cells is complete, during the late stages of sporogenesis and elater differentiation. Commonly, in the outer layer of wall cells, the thickenings are deposited as scattered I- or J-shaped bands on the longitudinal, radial walls to produce a nodular pattern in surface view (Fig. 1.8D). Cells in the inner wall layers, in contrast, deposit annular or semiannular, U-shaped thickenings that extend from the radial walls across the inner tangential wall; these impart a banded appearance in surface view (Fig. 1.8E). In the vast majority of liverworts, capsule dehiscence occurs along differentiated sutures. Usually two such sutures extend longitudinally from near the capsule base, over the capsule apex and down to the base on the other side, thereby dividing the capsule wall into four sectors, or valves. With drying, the cell walls between the two rows of suture cells tear along the middle lamella. The transverse thickening bands of the inner wall layers make them more rigid than the outer, and consequently when the sutures tear, the separated valves bend outwards (Ingold 1939), releasing the mass of spores and elaters. In several taxa, the valves are very long and spirally twisted (e.g. Balantiopsidaceae). Neither the chemical nature of the thickening bands nor the mechanisms regulating their deposition are known. The fact that they are autofluorescent and have a homogeneous, osmiophilic appearance in TEM micrographs, however, suggests that they are composed of polyphenolics.

In *Haplomitrium* and the Marchantiidae, capsule walls are unistratose and dehiscence rarely occurs along four valves. In *Haplomitrium* each capsule wall cell bears a single longitudinal, annular thickening band and the capsule opens along one to four slits (Bartholomew-Began 1991). In many Marchantiidae the capsule breaks apart into irregular plates of cells, a phenomenon also seen in some of the Pelliidae, e.g. *Fossombronia*, whereas in others, e.g. the Aytoniaceae, dehiscence involves an apical operculum. In the Ricciaceae, the capsule wall actually deteriorates before the spores are mature, leaving them in a cavity lined by the calyptra; spore dispersal in this group requires thallus degeneration.

An almost universal feature of liverwort sporophytes is the presence of elaters in the capsule (Fig. 1.8C). The elaters are always unicellular and dead at maturity, and are usually thin and elongate with spiral thickening bands in their walls. In some Marchantiopsida, e.g. the Aytoniaceae, the division that produces the elater initials is followed directly by sporocyte meiosis and elater differentiation, resulting in a 4:1 spore:elater ratio in the mature capsule. In most hepatics, however, the sporocyte initial divides several times prior to meiosis, while the elater initial remains undivided. As a consequence, in most hepatics, including many genera of the Marchantiopsida, spore:elater ratios are 8:1 or greater. Among liverworts, *Pellia epiphylla* and *Conocephalum conicum*, both of which display endosporic, precocious spore germination, are reported to have spore:elater ratios less than 4:1 (Bischler 1998). In Pelliaceae many of the elaters arise from a basal pad of sterile tissue, or elaterophore; these are not homologous to the elaters that are sister cells to the sporocytes. In other groups, e.g. Aneuraceae, an elaterophore is formed at the capsule apex; in the Jubulaceae, Frullaniaceae, and Lejeuneaceae, the elaters are dispersed, but remain attached to the apices of the capsule valves after dehiscence.

Hygroscopically induced movements of the elaters help to break up the spore mass after the capsule opens (Ingold 1939). In developing capsules, however, these sterile cells may serve as a dispersed tapetum (Crandall-Stotler 1984). Bartholomew-Began (1991) has shown that the immature elaters of *Haplomitrium* contain lipids and starch bodies, both of which disappear as elater thickenings are deposited. At the same time, the capsule lumen itself also contains numerous lipid droplets (Crandall-Stotler 1984). A similar nutritive function has often been postulated for the nurse cells of the Sphaerocarpales (Parihar 1961). Schuster (1992b, p. 799) suggested that the nurse cells are not homologous to elaters because they are not formed as a consequence of a fixed spore/elater division, but instead seem to be sporocytes that fail to undergo meiosis. The studies of Doyle (1962) and Kelley & Doyle (1975), while demonstrating that the nurse cells are tapetal, do not resolve the question of origin. Although variation in form occurs, the production of dispersed sterile unicells in the archesporium appears to be a significant defining character of the Marchantiophyta as suggested by Mishler & Churchill (1984).

Prophase sporocytes in the Haplomitriopsida and Jungermanniopsida are deeply quadrilobed (Fig. 1.8F), whereas in the Marchantiidae they are spheroidal and unlobed. This difference in sporocyte morphology has historically been cited to support the basic dichotomy between marchantioids and other liverworts (Schuster 1984b). Recently, Brown & Lemmon (2006) have shown that quadrilobed-shaping involves two intersecting girdling bands of microtubules that develop in very early prophase to establish the planes of meiotic cytokinesis. Polar organizers differentiated in each of the lobes then generate the quadripolar microtubular system (QMS) that is involved in karyokinesis.

Comparable premeiotic girdling bands have not been observed in the Marchantiidae, but only *Conocephalum* and *Dumortiera* have been studied (Brown & Lemmon 1988, Shimamura *et al.* 2004). In most liverworts, sporocytes bear multiple plastids and meiosis is polyplastidic, but in *Haplomitrium blumii*, Blasiales, and several genera of the Marchantiidae (*Monoclea, Dumortiera, Wiesnerella, Lunularia*, and *Marchantia*), plastid number is reduced to one in the sporocyte and meiosis is monoplastidic (Renzaglia *et al.* 1994, Shimamura *et al.* 2003). *Conocephalum* has polyplastidic sporocytes, but produces spores in rhomboidal or linear rather than tetrahedral arrays through a unique process of cytoplasmic partitioning (Brown & Lemmon 1988, 1990).

Spores vary greatly in size, shape and ornamentation, tending to be smaller and less highly ornamented in the Jungermanniidae and Metzgeriidae than in the Pelliidae and Marchantiopsida (Schuster 1984a). Spore wall ornamentation is due to sculpturing of the exine, which in several taxa is patterned by callosic deposits in the preprophase sporocyte (Brown *et al.* 1986, Brown & Lemmon 1987). Variations in spore wall architecture are informative in the systematics of some taxa, e.g. *Fossombronia, Riccia*, and the Aytoniaceae. All liverworts are isosporous and most disperse their spores as monads. Spores that remain in tetrads after dispersal are, however, diagnostic of some species of *Sphaerocarpos* and *Riccia*, and may occasionally also be found in other taxa, e.g. *Haplomitrium, Fossombronia*, and *Aneura*.

1.2.5 *Spore germination and sporeling patterns*

Spore germination is initiated with swelling and division of the spore protoplast to form a multicellular protonema. In exosporic germination, the swollen protoplast ruptures the spore wall before it divides, usually after release from the capsule. In many epiphytic taxa, including all members of the Porellales, in contrast, the protonema is formed inside the stretched spore wall, i.e. germination is endosporic. Endosporic germination is usually precocious, i.e. germination occurs prior to capsule dehiscence, but in a few taxa, e.g. *Radula* and *Trichocoleopsis*, endosporic germination occurs after spore release. In the Jungermanniopsida, the most common protonema is a multicellular, globose to cylindrical structure, from which a single gametophore develops. Other expressions include a filamentous protonema, as in Cephaloziaceae, a plate-like protonema, as in Radulaceae and Metzgeriaceae, and a biphasic protonema, as in some Lejeuneaceae. The exosporic protonema of *Haplomitrium* initially comprises two tiers of quadrants from which a cell mass and then a system of highly branched cylindrical, leafless axes, or stolons, arise (Bartholomew-Began 1991).

Among the Marchantiopsida only the Blasiales and *Conocephalum* exhibit endosporic germination. In most taxa, an elongate hyaline germ tube emerges from the ruptured spore, often bearing a germ rhizoid at its base. Initially, transverse divisions at the apex of the germ tube form a short filamentous protonema, the terminal cell of which is soon partitioned into quadrants. The quadrants may continue to produce tiers of cells basally, forming a quadriseriate cylindrical phase, e.g. *Sphaerocarpos* (Nehira 1983, p. 369), or the quadrant stage may be very abbreviated, e.g., *Marchantia*. With additional vertical divisions, the terminal quadrant may be converted to a discoid plate. Ultimately, the apical cell of the young thallus is formed in one of the cells of either the quadrant stage or the plate. This very distinctive pattern of sporeling development, known as the "golf-tee" type, is restricted to the Marchantiidae. Although it has been suggested that the fine details of sporeling ontogeny are phylogenetically informative (Fulford 1956, Nehira 1983), this view has yet to be critically tested.

1.3 Morphology, molecules, and classification

The application of molecular methods to unraveling the evolutionary history of liverworts has resulted in revolutionary changes in our concepts of liverwort phylogeny (see, for example, Heinrichs *et al.* 2005, Forrest *et al.* 2006, He-Nygrén *et al.* 2006). Analyses of character evolution have demonstrated that there is substantial homoplasy in many of the characters previously used to define genera, families and even suborders (Crandall-Stotler *et al.* 2005). That does not mean, however, that morphology cannot provide phylogenetically informative characters. In fact, many of the novel relationships resolved in molecular analyses are supported by morphological signals, e.g. the relationships between *Haplomitrium* and *Treubia*, *Blasia* and the Marchantiopsida, and *Pleurozia* and the Metzgeriales (for discussion see Crandall-Stotler *et al.* 2005, Forrest *et al.* 2006, Renzaglia *et al.* 2007). Resolving incongruency between molecule-based phylogenies and traditional schemes derived intuitively from morphology requires critical re-evaluation of morphological characters to correct faulty interpretations of homology, as well as re-assessment of specimen identity to eliminate erroneous DNA sequences (see discussion in Forrest *et al.* 2006). Total evidence analyses that incorporate ontogenetic and ultrastructural data are essential to future efforts to clarify the evolution of structural characters (Renzaglia *et al.* 2007).

To accommodate the many changes arising from molecular phylogenetic studies, several authors have proposed modifications to the taxonomic hierarchy of liverworts above the family level (Frey & Stech 2005, Heinrichs *et al.* 2005, Forrest *et al.* 2006, He-Nygrén *et al.* 2006). The classification scheme presented below, which circumscribes families as well as higher ranks, integrates morphology with these hypotheses and others generated from molecular analyses, including but not limited to the following: Schill *et al.* (2004), Yatsentyuk *et al.* (2004), Heinrichs *et al.* (2006, 2007), Hentschel *et al.* (2006), de Roo *et al.* (2007), Hendry *et al.* (2007), Heselwood & Brown (2007), Liu *et al.* (2008), and Wilson *et al.* (2007). The resolution of liverwort phylogeny is very much a work in progress, with fewer than 30% of liverwort genera (< 5% of species) sampled for molecular analyses. Consequently, the relationships of many lineages are unresolved. For example, in the Marchantiidae the branching order of Sphaerocarpales and the recently established Neohodgsoniales and Lunulariales (Long 2006b) are equivocal, and the hierarchial relationships of the paraphyletic assemblage of families in the crown group of the Marchantiales are unresolved. For this reason, we have not recognized any subordinal rankings in the Marchantiales. There are substantial changes at all hierarchial levels from our previously published classification (Crandall-Stotler & Stotler 2000). A discussion of these changes and a complete classification that provides author citations, place of publication and diagnoses for ranks of family and above are presented in Crandall-Stotler *et al.* (2008). The classification scheme that follows has been extracted from this publication and reflects our current state of understanding of liverwort phylogeny. Certainly, as new ontogenetic, ultrastructural and molecular data are generated, the scheme presented herein will be scrutinized and refined.

PHYLUM MARCHANTIOPHYTA Stotler & Crand.-Stotl.

CLASS HAPLOMITRIOPSIDA Stotler & Crand.-Stotl.

Plants bearing foliar appendages at discrete nodes; axes (stems) secreting copious mucilage from epidermal cells, forming unique associations with glomeromycotean fungi; apical cells tetrahedral; androecia and gynoecia loosely organized (apical disks in some species of *Haplomitrium*); early antheridial ontogeny forming one primary androgonial initial; spermatids with a massive spline; anacrogynous (acrogynous in *Haplomitrium* subg. *Calobryum*); sporophytes large, enclosed by a fleshy shoot calyptra or coelocaule.

SUBCLASS TREUBIIDAE Stotler & Crand.-Stotl.

Plants prostrate, dorsiventrally flattened; leaves in two rows, unequally divided into a small dorsal lobule and large ventral lobe, with the lobe fleshy, confluent with the stem, longitudinal or slightly succubous, polystratose except near the margins; rhizoids ventral, scattered; oil bodies large, in specialized cells; gemmae multicellular, not in receptacles; gametangia protected by dorsal lobules; capsules ovoidal, wall 3- to 5-stratose; dehiscence 4-valved.

ORDER TREUBIALES Schljakov
Treubiaceae Verd: *Apotreubia* S. Hatt. & Mizut., *Treubia* K. I. Goebel

SUBCLASS HAPLOMITRIIDAE Stotler & Crand.-Stotl.

Plants differentiated into highly branched leafless stolons and erect leafy shoots, isophyllous or anisophyllous; leaves in three rows, with the third row of leaves dorsal, transverse or weakly succubous, undivided, polystratose only near the base; stems with a central strand of thin-walled, hydrolyzed cells; rhizoids absent; oil bodies small, homogeneous, in all cells; gemmae absent; antheridia and archegonia scattered on the stem, in leaf axils, or on apical discs; capsules cylindrical, wall unistratose, dehiscence along 1, 2 or 4 sutures, non-valvate.

ORDER CALOBRYALES Hamlin
Haplomitriaceae Dědeček: *Haplomitrium* Nees

CLASS MARCHANTIOPSIDA Gonquist, Takht & W. Zimm.

Plants thalloid, or rarely leafy; apical cell cuneate with four cutting faces; thallus often differentiated into assimilatory and storage tissues, usually with persistent ventral scales with appendages and dimorphic rhizoids; oil-bodies single in specialized cells or lacking; gametangia on specialized branches or embedded dorsally in the thallus; early antheridial ontogeny forming four primary androgonial initials; archegonial neck of six cell rows; embryos often octamerous; sporophytes with seta usually short or absent; capsule wall usually unistratose; sporocytes unlobed, spores usually polar and highly ornamented.

SUBCLASS BLASIIDAE He-Nygrén, Juslén, Ahonen, Glenny & Piippo

Thallus simple, lacking dorsiventral differentiation; wing margins scarcely (*Cavicularia*) to deeply lobed and "leaf-like" (*Blasia*), with the lobes longitudinal in insertion; midrib bearing a strand of calcium oxalate deposits (*Blasia*) or with three strands of elongate, hydrolyzed cells (*Cavicularia*); air chambers and air pores absent; ventral scales without appendages, in two rows on midrib, with a row of *Nostoc*-containing auricles (domatia) to the outside of each row of scales; rhizoids all smooth; oil bodies absent or few in unspecialized cells; multicellular gemmae present (in flasks or crescent-shaped cups); dioicous; antheridia partially embedded dorsally on thallus, arranged in two rows; sporophytes dorsal at thallus apex; pseudoperianth absent; involucre tubular; seta elongate, massive; elaters present; capsule wall 2- to 4-stratose, dehiscence by four valves.

ORDER BLASIALES Stotler & Crand.-Stotl.
Blasiaceae H. Klinggr.: *Blasia* L., *Cavicularia* Steph.

SUBCLASS MARCHANTIIDAE Engl.

Thallus differentiated into layers or not; air chambers and air pores present or absent; ventral scales present or absent, appendaged or not; rhizoids smooth or smooth and pegged; specialized oil cells usually present; multicellular gemmae present in specialized structures or absent; antheridia embedded in dorsal part of thallus, or in cushions on thallus, or in stalked receptacles; sporophytes on stalked receptacles or borne dorsally on thallus or embedded in thallus; involucre present, rarely absent; seta usually very short or absent, rarely elongate; elaters usually present; capsule dehiscence by longitudinal valves or slits, or by a lid, sometimes cleistocarpous.

ORDER SPHAEROCARPALES Cavers
Plants delicate, with stems bearing longitudinally inserted leaves (Sphaerocarpaceae) or small ventral scales and a large dorsal wing (Riellaceae), sometimes sexually dimorphic (*Sphaerocarpos*); leaves and dorsal wing unistratose; air chambers and air pores absent; ventral scales absent (Sphaerocarpaceae) or present (*Riella*); rhizoids all smooth; specialized oil cells absent or present (*Riella*); specialized asexual structures absent (gemmae in *Riella*); dioicous, rarely monoicous; antheridia in flask-shaped dorsal perigonial involucres (Sphaerocarpaceae) or embedded in pockets on margin of wing (*Riella*); each archegonium and sporophyte enclosed in dorsal or ventral flask-shaped pseudoperianth; seta very short; elaters absent; capsule cleistocarpous; spores shed singly or in tetrads.

Sphaerocarpaceae Heeg: *Sphaerocarpos* Boehm., *Geothallus* Campb.
Riellaceae Engl.: *Riella* Mont.

ORDER NEOHODGSONIALES D. G. Long
Thallus differentiated into layers, with compound air pores; ventral scales in two rows, without appendages; rhizoids all smooth; specialized oil cells present; specialized asexual structures present (gemma cups); monoicous; antheridia on unbranched stalked receptacle; archegonia and young sporophytes enclosed in campanulate pseudoperianths; sporophytes on branched stalked receptacle; involucre bivalved; seta not elongated; elaters present; capsule dehiscence by irregular valves.

Neohodgsoniaceae D. G. Long: *Neohodgsonia* Perss.

ORDER LUNULARIALES D. G. Long

Thallus differentiated into layers, upper layer with air chambers, dorsal surface with simple air pores; ventral scales in two rows, with single appendage; rhizoids smooth and pegged; specialized oil cells present; specialized asexual structures present (crescent-shaped gemma cups); dioicous; antheridia in terminal cushions on thallus; sporophytes on stalked deeply 4-lobed receptacle; pseudoperianth absent; involucre tubular; seta elongate, massive; elaters present; capsule dehiscence by lid and four valves.

Lunulariaceae H. Klinggr.: *Lunularia* Adans.

ORDER MARCHANTIALES Limpr.

Thallus usually differentiated into layers, upper layer with air chambers, dorsal surface with simple or compound air pores (rarely absent); ventral scales in 2–10 rows, sometimes absent, usually with 1–3(6) appendages; rhizoids usually smooth and pegged, sometimes only smooth; specialized oil cells usually present; specialized asexual structures absent or present; monoicous or dioicous; antheridia embedded dorsally in thallus or on stalked receptacles; sporophytes on stalked receptacle, or terminal or dorsal on thallus or embedded in thallus; pseudoperianth absent or present; involucre bivalved, cup-shaped, scale- or flap-like or tubular, sometimes absent; seta usually very short or absent, rarely elongate; elaters present or absent; capsule dehiscence by longitudinal valves, longitudinal slit or lid, sometimes cleistocarpous.

Marchantiaceae Lindl.: *Bucegia* Radian, *Marchantia* L., *Preissia* Corda

Aytoniaceae Cavers: *Asterella* P. Beauv., *Cryptomitrium* Austin ex Underw., *Mannia* Opiz, *Plagiochasma* Lehm. & Lindenb., *Reboulia* Raddi

Cleveaceae Cavers: *Athalamia* Falconer, *Sauteria* Nees, *Peltolepis* Lindb.

Monosoleniaceae Inoue: *Monosolenium* Griff.

Conocephalaceae Müll. Frib. ex Grolle: *Conocephalum* Hill

Cyathodiaceae Stotler & Crand.-Stotl.: *Cyathodium* Kunze

Exormothecaceae Müll. Frib. ex Grolle: *Aitchisoniella* Kashyap, *Exormotheca* Mitt., *Stephensoniella* Kashyap

Corsiniaceae Engl.: *Corsinia* Raddi; *Cronisia* Berk.

Monocarpaceae D. J. Carr ex Schelpe: *Monocarpus* D. J. Carr

Oxymitraceae Müll. Frib. ex Grolle: *Oxymitra* Bisch. ex Lindenb.

Ricciaceae Rchb.: *Riccia* L., *Ricciocarpos* Corda

Wiesnerellaceae Inoue: *Wiesnerella* Schiffn.

Targioniaceae Dumort.: *Targionia* L.

Monocleaceae A. B. Frank: *Monoclea* Hook.
Dumortieraceae D. G. Long: *Dumortiera* Nees

CLASS JUNGERMANNIOPSIDA Stotler & Crand.-Stotl.

Plants thalloid or leafy; oil bodies usually present in all cells (absent in a few taxa); rhizoids monomorphic, smooth-walled; early antheridial ontogeny forming two primary androgonial cells; archegonial neck usually of five cell rows; embryos filamentous; sporophytes with seta elongation pronounced; capsule wall 2- or more stratose; sporocytes lobed, spores cryptopolar to apolar, rarely polar.

SUBCLASS PELLIIDAE He-Nygrén, Juslén, Ahonen, Glenny & Piippo

Plants mostly thalloid without air chambers, if leafy, leaves developing from one primary initial, never lobed, arranged in two ranks; branches exogenous in origin, terminal or intercalary, lateral or ventral; antheridia on dorsal surface of midrib or stem, with or without perigonia (on abbreviated ventral branches in *Hymenophyton*); gynoecia usually anacrogynous, on dorsal surface of midrib or stem (acrogynous on thallus in *Pellia*, on abbreviated branches in *Hymenophyton* and *Podomitrium*).

ORDER PELLIALES He-Nygrén, Juslén, Ahonen, Glenny & Piippo
Plants thalloid or leafy with the leaves succubous; apical cell tetrahedral, cuneate, or hemidiscoid; stalked papillae or uniseriate hairs dispersed or in two rows on ventral surface; rhizoids hyaline or brownish to pale reddish brown; ventral branches rare; antheridia arranged in two rows, or scattered to weakly clustered on the thallus, each in a conical or flask-shaped chamber with an apical ostiole; archegonia naked and arranged in two rows along the midrib (*Noteroclada*), or in an acrogynous cluster, protected by a perichaetial flap or sheath (*Pellia*); sporophytes enclosed by a shoot calyptra and caulocalyx (*Noteroclada*) or perichaetial pseudoperianth (*Pellia*); capsules spheroidal, with conspicuous basal elaterophore, dehiscing into four valves; spore germination precocious and endosporic.

Pelliaceae H. Klinggr.: *Noteroclada* Taylor ex Hook. & Wilson, *Pellia* Raddi

ORDER FOSSOMBRONIALES Schljakov
Plants thalloid or leafy; foliose scales, uniseriate hairs, or stalked papillae, arranged in two rows on the ventral surface of the midrib or stem; oil bodies of the *Massula* type; ventral branches rare; gynoecia anacrogynous; capsules

usually spheroidal (cylindrical in *Makinoa*); dehiscence not valvate, irregular (by a single slit in *Makinoa*); spore germination exosporic.

SUBORDER CALYCULARIINEAE He-Nygrén, Juslén, Ahonen, Glenny & Piippo
Plants thalloid with well defined midrib; apical cell hemidiscoid; foliose ventral scales; rhizoids hyaline; antheridia in several rows on the midrib, with laciniate perigonial scales; gynoecia anacrogynous, with archegonia and perichaetial scales clustered; sporophytes enclosed by a shoot-calyptra and caulocalyx; capsules spheroidal, with a basal elaterophore, dehiscing irregularly into 5 to 7 unequal segments.

Calyculariaceae He-Nygrén, Juslén, Ahonen, Glenny & Piippo: *Calycularia* Mitt.

SUBORDER MAKINOINEAE He-Nygrén, Juslén, Ahonen, Glenny & Piippo
Plants thalloid with an inconspicuous midrib; apical cell cuneate; 3- to 6-celled ventral hairs; rhizoids reddish brown; androecia large, up to 80 antheridia sunken in thallus depressions, protected by a posterior lunulate ridge of thallus tissue; archegonia in small dorsal clusters protected by a posterior flap of thallus tissue; sporophytes enclosed by a coelocaule; capsules cylindrical, with rudimentary basal elaterophore, dehiscing along one slit.

Makinoaceae Nakai: *Makinoa* Miyake

SUBORDER FOSSOMBRONIINEAE R. M. Schust. ex Stotler & Crand.-Stotl.
Plants thalloid or leafy; apical cell tetrahedral (Petalophyllaceae), lenticular (Fossombroniaceae) or cuneate (Allisoniaceae); ventral appendages foliose scales or filamentous hairs or stalked slime papillae; rhizoids purplish or brownish (hyaline in Petalophyllaceae); antheridia scattered (Fossombroniaceae) or in clusters on the midrib, with or without perigonial scales; archegonia scattered or clustered, with or without perichaetial scales; sporophytes protected by a shoot calyptra and either a caulocalyx or perichaetial pseudoperianth (only a true calyptra in *Allisonia*); capsules spheroidal, lacking an elaterophore, dehiscence irregular or in 5–7 unequal segments.

Petalophyllaceae Stotler & Crand.-Stotl.: *Petalophyllum* Nees & Gottsche ex Lehm., *Sewardiella* Kashyap
Allisoniaceae Schljakov: *Allisonia* Herzog
Fossombroniaceae Hazsl.: *Austrofossombronia* R. M. Schust., *Fossombronia* Raddi

ORDER PALLAVICINIALES W. Frey & M. Stech
Plants thalloid (leafy in *Phyllothallia*), midrib usually well defined; apical cells cuneate, lenticular or hemidiscoid; ventral appendages stalked papillae or hairs, dispersed or in two rows; ventral branches common; antheridia

associated with perigonial scales, in clusters or rows on midrib; archegonia associated with perichaetial scales, clustered; sporophytes enclosed by a coelocaule or by a shoot calyptra and perichaetial pseudoperianth or caulocalyx; capsules ellipsoidal to cylindrical, with a multistratose apical cap (except *Phyllothallia*), dehiscence usually 2- or 4-valved, valves apically coherent (irregular in *Phyllothallia*).

SUBORDER PHYLLOTHALLIINEAE R. M. Schust.

Plants leafy, with the leaves opposite, distant to contiguous, with well defined internodes; apical cell cuneate; ventral stalked papillae dispersed; antheridia and perigonial scales in clusters at nodes; archegonia and perichaetial scales in clusters at nodes; sporophytes enclosed by a coelocaule; capsules spheroidal, wall multistratose, dehiscing into 12–14 irregular segments.

Phyllothalliaceae E. A. Hodgs.: *Phyllothallia* E. A. Hodgs.

SUBORDER PALLAVICINIINEAE R. M. Schust.

Plants thalloid, with wings sometimes deeply lobed, midrib with 1 or 2(4) strands of elongate, hydrolyzed conducting cells (strands lacking in *Sandeothallus* and some species of *Moerckia*).

Sandeothallaceae R. M. Schust.: *Sandeothallus* R. M. Schust.
Moerckiaceae Stotler & Crand.-Stotl.: *Hattorianthus* R. M. Schust. & Inoue, *Moerckia* Gottsche
Hymenophytaceae R. M. Schust.: *Hymenophyton* Dumort.
Pallaviciniaceae Mig.: *Greeneothallus* Hässel, *Jensenia* Lindb., *Pallavicinia* Gray, *Podomitrium* Mitt., *Seppeltia* Grolle, *Symphyogyna* Nees & Mont., *Symphyogynopsis* Grolle, *Xenothallus* R. M. Schust.

SUBCLASS METZGERIIDAE Barthol.-Began

Plants mostly thalloid, without air chambers, if leafy, leaves developing from three primary initials, arranged in two ranks; apical cells lenticular; branches exogenous or endogenous in origin, terminal or intercalary, lateral or ventral; androecia on abbreviated lateral or ventral branches (except *Verdoornia*); gynoecia acrogynous, on abbreviated lateral or ventral branches (except *Verdoornia*); capsule dehiscence 4-valved.

ORDER PLEUROZIALES Schljakov

Plants leafy; leaves succubous, unequally complicate-bilobed, with the larger lobe shallowly bifid and the small lobule usually forming a complex, valvate water sac (leaves simple in *P. paradoxa*); underleaves and ventral slime papillae lacking; branches endogenous, lateral (*Plagiochila*-type); androecia on abbreviated branches, with antheridia solitary in the axils of reduced perigonial leaves; gynoecia on abbreviated branches, with archegonia enclosed by a perianth and 2 to 5 series of modified perichaetial leaves; sporophytes enclosed by a shoot calyptra and perianth; capsules ovoid to subspheroidal, wall 8- to 10-stratose, epidermal cells with 2-phase ontogeny, walls of inner cells with complex reticulate thickenings; spore germination endosporic.

Pleuroziaceae Müll. Frib.: *Pleurozia* Dumort. [including *Eopleurozia* R. M. Schust.]

ORDER METZGERIALES Chalaud
Plants thalloid; 1- or 2-celled ventral slime papillae dispersed or in two rows; archegonial neck reduced, only weakly differentiated from the venter; sporophytes enclosed by a fleshy shoot calyptra or coelocaule; capsules ovoid, ellipsoid or cylindric, with an apical elaterophore, capsule wall 2-stratose, cells in both layers with wall thickenings; spore germination exosporic; asexual reproduction by gemmae common.

Metzgeriaceae H. Klinggr.: *Apometzgeria* Kuwah., *Austrometzgeria* Kuwah, *Metzgeria* Raddi, *Steereella* Kuwah.
Aneuraceae H. Klinggr.: *Aneura* Dumort. [including *Cryptothallus* Malmb.], *Riccardia* Gray, *Lobatiriccardia* (Mizut. & S. Hatt.) Furuki, *Verdoornia* R. M. Schust.
Mizutaniaceae Furuki & Z. Iwats.: *Mizutania* Furuki & Z. Iwats.
Vandiemeniaceae Hewson: *Vandiemenia* Hewson

SUBCLASS JUNGERMANNIIDAE Engl.

Plants leafy, very rarely thalloid (e.g., *Pteropsiella*); leaves developing from two primary leaf initials, frequently divided into two or more lobes, arranged in two or three rows, with the third row ventral; isophyllous, or anisophyllous with the ventral leaves (underleaves or amphigastria) smaller and/or morphologically different from the lateral leaves; apical cell tetrahedral; antheridia in axils of modified leaves, rarely underleaves (male bracts and bracteoles); archegonia acrogynous, usually surrounded by a perianth and modified leaves and underleaves (female bracts and bracteoles); capsules variable in shape, wall 2- to 10-stratose, dehiscence 4-valved.

ORDER PORELLALES Schljakov
Leaves incubous, complicate, unequally 2- or 3-lobed, with the smaller lobe(s) or lobules, ventral; lobules commonly forming inflated water sacs; underleaves present or absent, sometimes with water sacs, morphologically different from the leaves; rhizoids fascicled, from the underleaf base; branches exogenous, lateral; spore germination precocious and endosporic (unknown in *Goebeliella*).

SUBORDER PORELLINEAE R. M. Schust.

Plants robust, highly branched, pinnate or bipinnate; leaves 3-lobed (2-lobed in Porellaceae); water sacs when present of the *Frullania*-type; branches normally of the *Frullania*-type; underleaves present; gynoecia with multiple archegonia and several series of bracts and bracteoles; sporophytes enclosed by a calyptra and perianth (coelocaule in Lepidolaenaceae); perianths 3-keeled; elaters free and randomly dispersed in the capsule.

Porellaceae Cavers: *Ascidiota* C. Massal., *Macvicaria* W. E. Nicholson, *Porella* L.

Goebeliellaceae Verd.: *Goebeliella* Steph.

Lepidolaenaceae Nakai: *Gackstroemia* Trevis., *Lepidogyna* R. M. Schust., *Lepidolaena* Dumort., *Jubulopsis* R. M. Schust.

SUBORDER RADULINEAE R. M. Schust.

Plants irregularly pinnate to bipinnate, with branches of the *Radula*-type; leaves 2-lobed, with the ventral lobule slightly inflated near the keel; underleaves absent; rhizoids in fascicles from leaf lobules; androecia on abbreviated branches; gynoecia usually on a leading axis, with 2 to 4 archegonia; bracts in a single series; bracteoles absent; sporophytes enclosed by a shoot calyptra or stem perigynium and perianth; perianths 2-keeled, dorsiventrally compressed, with the mouth truncate; capsules cylindric, wall 2-stratose, both epidermal and inner cells with wall thickenings; multicellular discoid gemmae in some species.

Radulaceae Müll. Frib.: *Radula* Dumort.

SUBORDER: JUBULINEAE Müll. Frib.

Plants usually with underleaves (absent in a few Lejeuneaceae); leaves 2- or 3-lobed; water sacs of the *Frullania*- or *Lejeunea*-types; rhizoids fascicled from the underleaf base; sporophytes enclosed by a stalked, true calyptra and perianth; perianths beaked; capsules spheroidal, wall 2-stratose; elaters vertically aligned, attached to the valve apices; spores with rosette markings (absent in *Jubula*).

Frullaniaceae Lorch: *Frullania* Raddi [including *Amphijubula* R. M. Schust., *Neohattoria* Kamim., *Schusterella* S. Hatt., Sharp & Mizut., and *Steerea* S. Hatt. & Kamim.]

Jubulaceae H. Klinggr.: *Jubula* Dumort., *Nipponolejeunea* S. Hatt.

Lejeuneaceae Cavers: *Acanthocoleus* R. M. Schust., *Acantholejeunea* (R. M. Schust.) R. M. Schust., *Acrolejeunea* (Spruce) Schiffn., *Amblyolejeunea* Ast, *Anoplolejeunea* (Spruce) Schiffn., *Aphanolejeunea* A. Evans, *Aphanotropis* Herzog, *Archilejeunea* (Spruce) Schiffn., *Aureolejeunea* R. M. Schust., *Austrolejeunea* (R. M. Schust.) R. M. Schust., *Blepharolejeunea* S. W. Arnell, *Brachiolejeunea* (Spruce) Schiffn., *Bromeliophila* R. M. Schust., *Bryopteris*

(Nees) Lindenb. *Calatholejeunea* K. I. Goebel, *Caudalejeunea* (Steph.) Schiffn., *Cephalantholejeunea* (R. M. Schust. & Kachroo) R. M. Schust., *Cephalolejeunea* Mizut., *Ceratolejeunea* (Spruce) J. B. Jack & Steph., *Cheilolejeunea* (Spruce) Schiffn. [including *Cyrtolejeunea* A. Evans], *Chondriolejeunea* (Benedix) Kis & Pócs, *Cladolejeunea* Zwick., *Cololejeunea* (Spruce) Schiffn. [including *Metzgeriopsis* K. I. Goebel], *Colura* (Dumort.) Dumort., *Cyclolejeunea* A. Evans, *Cystolejeunea* A. Evans, *Dactylolejeunea* R. M. Schust., *Dactylophorella* R. M. Schust., *Dendrolejeunea* (Spruce) Lacout., *Dicranolejeunea* (Spruce) Schiffn., *Diplasiolejeunea* (Spruce) Schiffn., *Drepanolejeunea* (Spruce) Schiffn. [including *Capillolejeunea* S. W. Arnell and *Rhaphidolejeunea* Herzog], *Echinocolea* R. M. Schust., *Echinolejeunea* R. M. Schust., *Evansiolejeunea* Vanden Berghen, *Frullanoides* Raddi, *Fulfordianthus* Gradst., *Haplolejeunea* Grolle, *Harpalejeunea* (Spruce) Schiffn., *Hattoriolejeunea* Mizut., *Kymatolejeunea* Grolle, *Leiolejeunea* A. Evans, *Lejeunea* Lib. [including *Amphilejeunea* R. M. Schust., *Crossotolejeunea* (Spruce) Schiffn., *Cryptogynolejeunea* R. M. Schust. and *Dicladolejeunea* R. M. Schust.], *Lepidolejeunea* R. M. Schust., *Leptolejeunea* (Spruce) Schiffn., *Leucolejeunea* A. Evans, *Lindigianthus* Kruijt & Gradst., *Lopholejeunea* (Spruce) Schiffn., *Luteolejeunea* Piippo, *Macrocolura* R. M. Schust., *Macrolejeunea* (Spruce) Schiffn., *Marchesinia* Gray, *Mastigolejeunea* (Spruce) Schiffn., *Metalejeunea* Grolle, *Microlejeunea* Steph., *Myriocolea* Spruce, *Myriocoleopsis* Schiffn., *Neopotamolejeunea* E. Reiner, *Nephelolejeunea* Grolle, *Neurolejeunea* (Spruce) Schiffn., *Odontolejeunea* (Spruce) Schiffn., *Omphalanthus* Lindenb. & Nees, *Oryzolejeunea* (R. M. Schust.) R. M. Schust., *Otolejeunea* Grolle & Tixier, *Phaeolejeunea* Mizut., *Physantholejeunea* R. M. Schust., *Pictolejeunea* Grolle, *Pluvianthus* R. M. Schust. & Schäf.-Verw., *Prionolejeunea* (Spruce) Schiffn., *Ptychanthus* Nees, *Pycnolejeunea* (Spruce) Schiffn., *Rectolejeunea* A. Evans, *Schiffneriolejeunea* Verd., *Schusterolejeunea* Grolle, *Siphonolejeunea* Herzog, *Sphaerolejeunea* Herzog, *Spruceanthus* Verd., *Stenolejeunea* R. M. Schust., *Stictolejeunea* (Spruce) Schiffn., *Symbiezidium* Trevis., *Taxilejeunea* (Spruce) Schiffn., *Thysananthus* Lindenb., *Trachylejeunea* (Spruce) Schiffn. [including *Potamolejeunea* (Spruce) Lacout.], *Trocholejeunea* Schiffn., *Tuyamaella* S. Hatt., *Tuzibeanthus* S. Hatt., *Verdoornianthus* Gradst., *Vitalianthus* R. M. Schust. & Giancotti, *Xylolejeunea* X-L. He & Grolle

ORDER PTILIDIALES Schljakov

Plants regularly pinnate to bipinnnate; leaves asymmetrically 3-lobed, with the dorsal lobe largest; lobes with marginal cilia, plane, or with the ventralmost lobe forming a water sac of either the *Frullania*-type (*Neotrichocolea*, branch leaves only) or the *Lejeunea*-type (*Trichocoleopsis*); leaf insertion transverse to weakly incubous or succubous (*Herzogianthus*); underleaves bifid or quadrifid; rhizoids in fascicles from the underleaf base; branches of the *Frullania*-type; androecia on leading axes; gynoecia on leading axes; capsules ovoid to ellipsoidal, walls 4- to 7-stratose; spore germination exosporic or endosporic (*Trichocoleopsis*); gemmae absent.

SUBORDER PTILIDIINEAE R. M. Schust.

Ptilidiaceae H. Klinggr.: *Ptilidium* Nees

Neotrichocoleaceae Inoue: *Neotrichocolea* S. Hatt., *Trichocoleopsis* S. Okamura

Herzogianthaceae Stotler & Crand.-Stotl.: *Herzogianthus* R. M. Schust. See p. 54.

ORDER JUNGERMANNIALES H. Klinggr.

Leaves succubous, incubous, or transverse, undivided or variously lobed, sometimes complicate, but then usually with the smaller lobe(s), or lobules, dorsal, rarely with inflated water sacs of the *Lejeunea*-type; underleaves present or absent; rhizoids fascicled from the underleaf base or scattered along the ventral side of the stem; branches exogenous or endogenous, lateral or ventral; spore germination usually exosporic.

SUBORDER PERSSONIELLINEAE R. M. Schust.

Plants large, anisophyllous or distichous (isophyllous in *Pleurocladopsis*); leaves complicate-bilobed, with the lobes symmetric or if unequal, usually with the smaller lobe dorsal, with the keel often winged; leaf insertion transverse, but with dorsal lobes incubously shingled; rhizoids scattered (fascicled in *Pachyschistochila*), magenta to purple (hyaline in *Pachyschistochila*), with the apices highly branched and sometimes septate; branches lateral, of the *Plagiochila*-, *Frullania*-, and *Radula*-type; androecia dispersed on leading axes, with the bracts scarcely differentiated, with the antheridia long-stalked; perianths absent; sporophytes enclosed in a coelocaule; gemmae absent.

Perssoniellaceae R. M. Schust. ex Grolle: *Perssoniella* Herzog

Schistochilaceae H. Buch: *Gottschea* Nees ex Mont. [including *Paraschistochila* R. M. Schust.], *Pachyschistochila* R. M. Schust. & J. J. Engel, *Pleurocladopsis* R. M. Schust., *Schistochila* Dumort.

SUBORDER LOPHOCOLEINEAE Schljakov

Leaves transverse, succubous, or incubous, divided into 2 to 4 lobes or undivided; underleaves usually conspicuous (reduced or lacking in *Phycolepidozia*, *Brevianthus*, *Chonocolea*, and Plagiochilaceae); perianths, when present, often with three broad keels; capsule wall usually polystratose, only rarely 2-stratose; spore germination exosporic, mostly of the *Nardia*-type; gemmae rare.

Pseudolepicoleaceae Fulford & J. Taylor: *Archeophylla* R. M. Schust., *Blepharostoma* (Dumort.) Dumort., *Chaetocolea* Spruce, *Herzogiaria* Fulford ex Hässel, *Isophyllaria* E. A. Hodgs. & Allison, *Pseudolepicolea* Fulford & J. Taylor [including *Archeochaete* R. M. Schust. and *Lophochaete* R. M. Schust.], *Temnoma* Mitt.

Trichocoleaceae Nakai: *Eotrichocolea* R. M. Schust., *Leiomitra* Lindb., *Trichocolea* Dumort.

Grolleaceae Solari ex R. M. Schust.: *Grollea* R. M. Schust.

Mastigophoraceae R. M. Schust.: *Dendromastigophora* R. M. Schust., *Mastigophora* Nees

Herbertaceae Müll. Frib. ex Fulford & Hatcher: *Herbertus* Gray, *Olgantha* R. M. Schust., *Triandrophyllum* Fulford & Hatcher

Vetaformataceae Fulford & J. Taylor: *Vetaforma* Fulford & J. Taylor

Lepicoleaceae R. M. Schust.: *Lepicolea* Dumort.

Phycolepidoziaceae R. M. Schust.: *Phycolepidozia* R. M. Schust.

Lepidoziaceae Limpr.: *Acromastigum* A. Evans, *Amazoopsis* J. J. Engel & G. L. S. Merr., *Arachniopsis* Spruce, *Bazzania* Gray, *Chloranthelia* R. M. Schust., *Dendrobazzania* R. M. Schust. & W. B. Schofield, *Drucella* E. A. Hodgs., *Hyalolepidozia* S. W. Arnell ex Grolle, *Hygrolembidium* R. M. Schust., *Isolembidium* R. M. Schust., *Kurzia* G. Martens, *Lembidium* Mitt., *Lepidozia* (Dumort.) Dumort., *Mastigopelma* Mitt., *Megalembidium* R. M. Schust., *Micropterygium* Lindenb., Nees & Gottsche, *Monodactylopsis* (R. M. Schust.) R. M. Schust., *Mytilopsis* Spruce, *Neogrollea* E. A. Hodgs., *Odontoseries* Fulford, *Paracromastigum* Fulford & J. Taylor, *Protocephalozia* (Spruce) K. I. Goebel, *Pseudocephalozia* R. M. Schust., *Psiloclada* Mitt., *Pteropsiella* Spruce, *Sprucella* Steph., *Telaranea* Spruce ex Schiffn., *Zoopsidella* R. M. Schust., *Zoopsis* Hook. f. ex Gottsche, Lindenb. & Nees

Lophocoleaceae Vanden Berghen: *Amphilophocolea* R. M. Schust., *Chiloscyphus* Corda [including *Campanocolea* R. M. Schust.], *Clasmatocolea* Spruce, *Conoscyphus* Mitt., *Cyanolophocolea* R. M. Schust., *Evansianthus* R. M. Schust. & J. J. Engel [including *Austrolembidium* Hässel], *Hepatostolonophora* J. J. Engel & R. M. Schust., *Heteroscyphus* Schiffn., *Lamellocolea* J. J. Engel, *Leptophyllopsis* R. M. Schust., *Leptoscyphopsis* R. M. Schust., *Leptoscyphus* Mitt., *Lophocolea* (Dumort.) Dumort.], *Pachyglossa* Herzog & Grolle, *Perdusenia* Hässel, *Physotheca* J. J. Engel & Gradst., *Pigafettoa* C. Massal., *Platycaulis* R. M. Schust., *Pseudolophocolea* R. M. Schust. & J. J. Engel, *Stolonivector* J. J. Engel, *Tetracymbaliella* Grolle, *Xenocephalozia* R. M. Schust.

Brevianthaceae J. J. Engel & R. M. Schust.: *Brevianthus* J. J. Engel & R. M. Schust.
Chonecoleaceae R. M. Schust. ex Grolle: *Chonecolea* Grolle
Plagiochilaceae Müll. Frib. & Herzog: *Acrochila* R. M. Schust., *Chiastocaulon* Carl, *Pedinophyllopsis* R. M. Schust. & Inoue, *Pedinophyllum* (Lindb.) Lindb., *Plagiochila* (Dumort.) Dumort. [including *Rhodoplagiochila* R. M. Schust., *Steereochila* Inoue, *Szweykowskia* Gradst. & E. Reiner], *Plagiochilidium* Herzog, *Plagiochilion* S. Hatt., *Proskauera* Heinrichs & J. J. Engel, *Xenochila* R. M. Schust.

SUBORDER CEPHALOZIINEAE Schljakov

Leaves usually succubous (transverse in Cephaloziellaceae), undivided or 2-lobed, with the margins entire or with small teeth; underleaves absent or very small; rhizoids scattered; branches of the ventral *Bazzania*-type common; sporophytes usually enclosed by a calyptra and perianth; gemmae common.

Adelanthaceae Grolle: *Adelanthus* Mitt. [including *Pseudomarsupidium* Herzog], *Calyptrocolea* R. M. Schust., *Wettsteinia* Schiffn.
Jamesoniellaceae He-Nygrén, Juslén, Ahonen, Glenny & Piippo: *Anomacaulis* (R. M. Schust.) Grolle, *Cryptochila* R. M. Schust., *Cuspidatula* Steph., *Denotarisia* Grolle, *Jamesoniella* (Spruce) Carrington, *Nothostrepta* R. M. Schust., *Pisanoa* Hässel, *Protosyzygiella* (Inoue) R. M. Schust., *Syzygiella* Spruce, *Vanaea* (Inoue & Gradst.) Inoue & Gradst.
Cephaloziaceae Mig.: *Alobiella* (Spruce) Schiffn., *Alobiellopsis* R. M. Schust., *Anomoclada* Spruce, *Apotomanthus* (Spruce) Schiffn., *Cephalozia* (Dumort.) Dumort., *Cladopodiella* H. Buch, *Fuscocephaloziopsis* Fulford, *Haesselia* Grolle & Gradst., *Hygrobiella* Spruce, *Iwatsukia* N. Kitag., *Metahygrobiella* R. M. Schust., *Nowellia* Mitt., *Odontoschisma* (Dumort.) Dumort., *Pleurocladula* Grolle, *Schiffneria* Steph., *Schofieldia* J. D. Godfrey, *Trabacellula* Fulford
Cephaloziellaceae Douin: *Allisoniella* E. A. Hodgs. [including *Protomarsupella* R. M. Schust.], *Amphicephalozia* R. M. Schust., *Cephalojonesia* Grolle, *Cephalomitrion* R. M. Schust., *Cephaloziella* (Spruce) Schiffn., *Cephaloziopsis* (Spruce) Schiffn., *Cylindrocolea* R. M. Schust., *Gymnocoleopsis* (R. M. Schust.) R. M. Schust., *Kymatocalyx* Herzog, *Stenorrhipis* Herzog
Scapaniaceae Mig. [including Chaetophyllopsidaceae R. M. Schust. and Lophoziacae Cavers]: *Anastrepta* (Lindb.) Schiffn., *Anastrophyllum* (Spruce) Steph., *Andrewsianthus* R. M. Schust. [including *Cephalolobus* R. M. Schust.], *Barbilophozia* Loeske, *Chaetophyllopsis* R. M. Schust., *Chandonanthus* Mitt., *Diplophyllum* (Dumort.) Dumort., *Douinia* (C. N. Jensen) H. Buch, *Gerhildiella* Grolle, *Gymnocolea* (Dumort.) Dumort., *Hattoria* R. M. Schust., *Isopaches* H. Buch, *Krunodiplophyllum* Grolle, *Lophozia* (Dumort.) Dumort.,

Macrodiplophyllum (H. Buch) Perss., *Plicanthus* R. M. Schust., *Pseudocephaloziella* R. M. Schust., *Roivainenia* Perss., *Scapania* (Dumort.) Dumort., *Scapaniella* H. Buch, *Schistochilopsis* (N. Kitag.) Konst., *Sphenolobopsis* R. M. Schust. & N. Kitag., *Sphenolobus* (Lindb.) Berggr., *Tetralophozia* (R. M. Schust.) Schljakov, *Tritomaria* Schiffn. ex Loeske

SUBORDER JUNGERMANNIINEAE R. M. Schust. ex Stotler & Crand.-Stotl.
Leaves succubous, rarely transverse (incubous in *Isotachis* and Calypogeiaceae), undivided or 2(4)-lobed; anisophyllous or distichous (isophyllous in Antheliaceae); stem perigynia, hollow marsupia of the *Calypogeia*-type or solid marsupia of the *Tylimanthus*-type common; perianths often absent; capsules ellipsoidal to cylindric, with the wall often 2-stratose; gemmae present in some taxa.

Myliaceae Schljakov: *Leiomylia* J. J. Engel & Braggins, *Mylia* Gray
Trichotemnomataceae R. M. Schust.: *Trichotemnoma* R. M. Schust.
Balantiopsidaceae H. Buch: *Anisotachis* R. M. Schust., *Acroscyphella* N. Kitag. & Grolle [= *Austroscyphus* R. M. Schust., nom. illeg.], *Balantiopsis* Mitt., *Eoisotachis* R. M. Schust., *Hypoisotachis* (R. M. Schust.) J. J. Engel & G. L. S. Merr., *Isotachis* Mitt., *Neesioscyphus* Grolle, *Ruizanthus* R. M. Schust.
Blepharidophyllaceae R. M. Schust.: *Blepharidophyllum* Ångstr., *Clandarium* (Grolle) R. M. Schust.
Acrobolbaceae E. A. Hodgs.: *Acrobolbus* Nees, *Austrolophozia* R. M. Schust., *Enigmella* G. A. M. Scott & K. G. Beckm., *Goebelobryum* Grolle, *Lethocolea* Mitt., *Marsupidium* Mitt., *Tylimanthus* Mitt.
Arnelliaceae Nakai: *Arnellia* Lindb., *Gongylanthus* Nees, *Southbya* Spruce, *Stephaniella* J. B. Jack, *Stephaniellidium* S. Winkl. ex Grolle
Jackiellaceae R. M. Schust.: *Jackiella* Schiffn.
Calypogeiaceae Arnell: *Calypogeia* Raddi, *Eocalypogeia* (R. M. Schust.) R. M. Schust., *Metacalypogeia* (S. Hatt.) Inoue, *Mnioloma* Herzog.
Delavayellaceae R. M. Schust.: *Delavayella* Steph.
Mesoptychiaceae Inoue & Steere: *Hattoriella* (Inoue) Inoue, *Leiocolea* (Müll. Frib.) H. Buch, *Liochlaena* Nees, *Mesoptychia* (Lindb.) A. Evans
Jungermanniaceae Rchb.: *Arctoscyphus* Hässel, *Bragginsella* R. M. Schust., *Cryptocolea* R. M. Schust., *Cryptocoleopsis* Amak., *Cryptostipula* R. M. Schust., *Diplocolea* Amak., *Gottschelia* Grolle, *Horikawaella* S. Hatt. & Amakawa [*Invisocaulis* R. M. Schust. nom. nud.], *Jungermannia* L., *Nardia* Gray, *Notoscyphus* Mitt., *Scaphophyllum* Inoue, *Solenostoma* Mitt. [including *Plectocolea* (Mitt.) Mitt.]
Geocalycaceae H. Klinggr.: *Geocalyx* Nees, *Harpanthus* Nees, *Saccogyna* Dumort., *Saccogynidium* Grolle

Gyrothyraceae R. M. Schust.: *Gyrothyra* M. Howe
Antheliaceae R. M. Schust.: *Anthelia* (Dumort.) Dumort.
Gymnomitriaceae H. Klinggr.: *Acrolophozia* R. M. Schust., *Apomarsupella* R. M. Schust., *Eremonotus* Lindb. & Kaal. ex Pearson [including *Anomomarsupella* R. M. Schust.], *Gymnomitrion* Corda, *Herzogobryum* Grolle, *Lophonardia* R. M. Schust., *Marsupella* Dumort., *Nanomarsupella* (R. M. Schust.) R. M. Schust., *Nothogymnomitrion* R. M. Schust., *Paramomitrion* R. M. Schust., *Poeltia* Grolle, *Prasanthus* Lindb.

Acknowledgments

The financial support of NSF grant EF-0531750 is gratefully acknowledged. We also thank John Engel and Matt von Konrat for providing us with critical field specimens from New Zealand, and Christine Davis for sharing unpublished sequence data with us.

References

Apostolakos, P. & Galatis, B. (1998). Microtubules and gametophyte morphogenesis in the liverwort *Marchantia paleacea* Bert. In *Bryology for the Twenty-first Century*, ed. J. W. Bates, N. W. Ashton & J. G. Duckett, pp. 205–21. Leeds: Maney and British Bryological Society.

Bartholomew-Began, S. E. (1991). A morphogenetic re-evaluation of *Haplomitrium* Nees (Hepatophyta). *Bryophytorum Bibliotheca*, **41**, 1–297.

Bischler, H. (1998). Systematics and evolution of the genera of the Marchantiales. *Bryophytorum Bibliotheca*, **51**, 1–201.

Blomquist, H. L. (1929). The relation of capillary cavities in the Jungermanniaceae to water absorption and storage. *Ecology*, **10**, 556–7.

Bopp, M. & Capesius, I. (1998). A molecular approach to bryophyte systematics. In *Bryology for the Twenty-first Century*, ed. J. W. Bates, N. W. Ashton & J. G. Duckett, pp. 79–88. Leeds: Maney and British Bryological Society.

Bopp, M. & Feger, F. (1960[1961]). Das Grundschema der Blattentwicklung bei Lebermoosen. *Revue Bryologique et Lichénologique*, **29**, 256–73.

Brown, R. C. & Lemmon, B. E. (1987). Involvement of callose in determination of exine patterning in three hepatics of the subclass Jungermanniidae. *Memoirs of the New York Botanical Garden*, **45**, 111–21.

Brown, R. C. & Lemmon, B. E. (1988). Cytokinesis occurs at boundaries of domains delimited by nuclear-based microtubules in sporocytes of *Conocephalum conicum* (Bryophyta). *Cell Motility and the Cytoskeleton*, **11**, 139–46.

Brown, R. C. & Lemmon, B. E. (1990). Sporogenesis in bryophytes. In *Microspores, Evolution and Ontogeny*, ed. S. Blackmore & R. B. Knox, pp. 55–94. London: Academic Press.

Brown, R. C. & Lemmon, B. E. (2006). Polar organizers and girdling bands of microtubules are associated with γ-tubulin and act in establishment of meiotic quadripolarity in the hepatic *Aneura pinguis* (Bryophyta). *Protoplasma*, **227**, 77–85.

Brown, R. C., Lemmon, B. E. & Renzaglia, K. S. (1986). Sporocytic control of spore wall pattern in liverworts. *American Journal of Botany*, **73**, 593–6.

Buch, H. (1911). *Über die Brutorgane der Lebermoose*. Helsinfors: Kaiserliche Alexanders-Universität in Finland.

Buch, H. (1930). Über die Entstehung der verschiedenen Blattflächenstellungen bei den Lebermoosen. *Annales Bryologici*, **3**, 25–40.

Buchloh, G. (1951). Symmetrie und Verzweigung der Lebermoose. Ein Beitrag zur Kenntnis ihrer Wuchsformen. *Sitzungsberichte der Heidelberger Akademie der Wissenschaften, Mathematisch-naturwissenschaftliche Klasse*, **1951**, 211–79.

Campbell, D. H. (1913). The morphology and systematic position of *Calycularia radiculosa* (Steph.). In *Leland Stanford Junior University. 1913. Dudley Memorial Volume*. Stanford, pp. 43–61. [*Leland Stanford Junior University Publications, University Series*, 11.]

Campbell, E. O. (1954). The structure and development of *Monoclea forsteri* Hook. *Transactions of the Royal Society of New Zealand*, **82**, 237–48.

Capesius, I. & Bopp, M. (1997). New classification of liverworts based on molecular and morphological data. *Plant Systematics and Evolution*, **207**, 87–97.

Chalaud, G. (1928). *Le cycle évolutif de* Fossombronia pusilla *Dum*. Paris: Librairie Générale de l'Enseignement.

Clausen, E. (1964). The tolerance of hepatics to desiccation and temperature. *Bryologist*, **67**, 411–17.

Clee, D. A. (1937). Leaf arrangement in relation to water conduction in the foliose Hepaticae. *Annals of Botany (London)*, n.s., **1**, 325–8.

Cook, M. E. & Graham, L. E. (1998). Structural similarities between surface layers of selected Charophycean algae and bryophytes and the cuticles of vascular plants. *International Journal of Plant Sciences*, **159**, 780–7.

Crandall, B. J. (1969). Morphology and development of branches in the leafy Hepaticae. *Beihefte zur Nova Hedwigia*, **30**, 1–261.

Crandall-Stotler, B. (1972). Morphogenetic patterns of branch formation in the leafy Hepaticae – a résumé. *Bryologist*, **75**, 381–403.

Crandall-Stotler, B. (1976). The apical cell and early development of *Pleurozia purpurea* Lindb. *Lindbergia*, **3**, 197–208.

Crandall-Stotler, B. (1981). Morphology/anatomy of hepatics and anthocerotes. *Advances in Bryology*, **1**, 315–98.

Crandall-Stotler, B. (1984). Musci, hepatics and anthocerotes – an essay on analogues. In *New Manual of Bryology*, vol. 2, ed. R. M. Schuster, pp. 1093–129. Nichinan: Hattori Botanical Laboratory.

Crandall-Stotler, B. & Guerke, W. R. (1980). Developmental anatomy of *Jubula* Dum. (Hepaticae). *Bryologist*, **83**, 179–201.

Crandall-Stotler, B. & Stotler, R. (2000). Morphology and classification of the Marchantiophyta. In *Bryophyte Biology*, ed. A. J. Shaw & B. Goffinet, pp. 21–70. Cambridge: Cambridge University Press.

Crandall-Stotler, B., Furuki, T. & Iwatsuki, Z. (1994). The developmental anatomy of *Mizutania riccardioides* Furuki & Iwatsuki, an exotic liverwort from southeast Asia. *Journal of the Hattori Botanical Laboratory*, **75**, 243–55.

Crandall-Stotler, B. J., Stotler, R. E. & Ford, C. H. (2002). Contributions toward a monograph of *Petalophyllum* (Marchantiophyta). *Novon*, **12**, 334–7.

Crandall-Stotler, B. J., Forrest, L. L. & Stotler, R. E. (2005). Evolutionary trends in the simple thalloid liverworts (Marchantiophyta, Jungermanniopsida subclass Metzgeriidae). *Taxon*, **54**, 299–316.

Crandall-Stotler, B. J., Stotler, R. E. & Long, D. G. (2008). Phylogeny and classification of the Marchantiophyta. *Edinburgh Journal of Botany*, **65**, in press.

Davis, C. (2004). A molecular phylogeny of leafy liverworts (Jungermanniidae, Marchantiophyta). *Monographs in Systematic Botany from the Missouri Botanical Garden*, **98**, 61–86.

de Roo, R. T., Hedderson, T. A. & Söderström, L. (2007). Molecular insights into the phylogeny of the leafy liverwort family Lophoziaceae Cavers. *Taxon* **56**, 301–14.

Douin, R. (1912). Le sporophyte chez les hépatiques. *Revue Générale de Botanique*, **24**, 5–27 [of reprint].

Doyle, W. T. (1962). The morphology and affinities of the liverwort *Geothallus*. *University of California Publications in Botany*, **33**, 185–267.

Duckett, J. G. & Ligrone, R. (1995). The formation of catenate foliar gemmae and the origin of the oil bodies in the liverwort *Odontoschisma denudatum* (Mart.) Dum. (Jungermanniales): a light and electron microscope study. *Annals of Botany*, **76**, 406–19.

Duckett, J. G. & Soni, S. L. (1972). Scanning electron microscope studies on the leaves of Hepaticae. I. Ptilidiaceae, Lepidoziaceae, Calypogeiaceae, Jungermanniaceae, and Marsupellaceae. *Bryologist*, **75**, 536–49.

Duckett, J. G., Renzaglia, K. S. & Pell, K. (1991). A light and electron microscope study of rhizoid-ascomycete associations and flagelliform axes in British hepatics with observations on the effects of the fungi on host morphology. *New Phytologist*, **118**, 233–57.

Evans, A. W. (1912). Branching in the leafy Hepaticae. *Annals of Botany (London)*, **26**, 1–37.

Evans, A. W. (1918). The air chambers of *Grimaldia fragrans*. *Bulletin of the Torrey Botanical Club*, **45**, 235–51.

Evans, A. W. (1939). The classification of the Hepaticae. *Botanical Review*, **5**, 49–96.

Flegel, M. & Becker, H. (2000). Characterization of the contents of oil bodies from the liverwort *Radula complanata*. *Plant Biology*, **2**, 208–10.

Forrest, L. L., Davis, E. C., Long, D. G., Crandall-Stotler, B. J., Clark, A. & Hollingsworth, M. L. (2006). Unraveling the evolutionary history of the liverworts (Marchantiophyta): multiple taxa, genomes and analysis. *Bryologist*, **109**, 303–34.

Frey, W. & Stech, M. (2005). A morpho-molecular classification of the liverworts (Hepaticophytina, Bryophyta). *Nova Hedwigia*, **81**, 55–78.

Fulford, M. H. (1956). The young stages of the leafy Hepaticae: a résumé. *Phytomorphology*, **6**, 199–235.

Galatis, B., Apostolakos, P. & Katsaros, C. (1978). Ultrastructural studies on the oil bodies of *Marchantia paleacea* Bert. I. Early stages of oil-body cell differentiation: origination of the oil body. *Canadian Journal of Botany*, **56**, 2252–67.

Goebel, K. (1893). Archegoniatenstudien. V. Die Blattbildung der Lebermoose und ihre biologische Bedeutung. *Flora*, **77**, 423–59.

Goebel, K. (1895). Über Function und Anlegung der Lebermoos-Elateren. *Flora*, **80**, 1–37.

Goebel, K. (1912). Archegoniatenstudien. XV. Die Homologie der Antheridien- und der Archegonienhüllen bei den Lebermoosen. *Flora*, **105**, 53–70.

Goebel, K. (1930). *Organographie der Pflanzen*, 3rd edn. Jena: G. Fischer.

Gradstein, S. R., Cleef, A. M. & Fulford, M. H. (1977). Studies on Colombian cryptogams IIA. Hepaticae – oil body structure and ecological distribution of selected species of tropical Andean Jungermanniales. *Proceedings of the Koninklijke Nederlandse Akademie van Wetenschappen*, Series C, **80**, 377–420.

Grolle, R. (1964). Eine neue *Echinocolea* auf Celebes. *Botanical Magazine*, **77**, 333–5.

Groth-Malonek, M., Pruchner, D., Grewe, F. & Knoop, V. (2005). Ancestors of trans-splicing mitochondrial introns support serial sister group relationships of hornworts and mosses with vascular plants. *Molecular Biology and Evolution*, **22**, 117–25.

Hallet, J.-N. (1978). Le cycle cellulaire de l'apicale muscinale: données nouvelles et caractéres originaux. *Bryophytorum Bibliotheca*, **13**, 1–20.

Hébant, C. (1977). The conducting tissues of bryophytes. *Bryophytorum Bibliotheca*, **10**, 1–157.

Heinrichs, J., Gradstein, S. R., Wilson, R. & Schneider, H. (2005). Towards a natural classification of liverworts (Marchantiophyta) based on the chloroplast gene *rbc*L. *Cryptogamie, Bryologie*, **26**, 131–50.

Heinrichs, J., Lindner, M., Groth, H. *et al.* (2006). Goodbye or welcome Gondwana? – insights into the phylogenetic biogeography of the leafy liverwort *Plagiochila* with a description of *Proskauera*, gen. nov. (Plagiochilaceae, Jungermanniales). *Plant Systematics and Evolution*, **258**, 227–50.

Heinrichs, J., Hentschel, J., Wilson, R., Feldberg, K. & Schneider, H. (2007). Evolution of leafy liverworts (Jungermannniidae, Marchantiophyta): estimating divergence times from chloroplast DNA sequences using penalized likelihood with integrated fossil evidence. *Taxon*, **56**, 31–44.

Hendry, T. A., Yang, Y., Davis, E. C. *et al.* (2007). Evaluating phylogenetic positions of four liverworts from New Zealand, *Neogrollea notabilis*, *Jackiella curvata*, *Goebelobryum unguiculatum*, and *Herzogianthus vaginatus*, using three chloroplast genes. *Bryologist*, **110**, 738–51.

Hentschel, J., Wilson, R., Burghardt, M. *et al.* (2006). Reinstatement of Lophocoleaceae (Jungermanniopsida) based on chloroplast gene *rbc*L data: exploring the importance of female involucres for the systematics of Jungermanniales. *Plant Systematics and Evolution*, **258**, 211–26.

He-Nygrén, X., Ahonen, I., Juslén, A., Glenny, D. & Piippo, S. (2004). Phylogeny of liverworts - beyond a leaf and a thallus. *Monographs in Systematic Botany from the Missouri Botanical Garden*, **98**, 87–118.

He-Nygrén, X., Juslén, A., Ahonen, I., Glenny, D. & Piippo, S. (2006). Illuminating the evolutionary history of liverworts (Marchantiophyta) - towards a natural classification. *Cladistics*, **22**, 1–31.

Heselwood, M. M. & Brown, E. A. (2007). A molecular phylogeny of the liverwort family Lepidoziaceae Limpr. in Australasia. *Plant Systematics and Evolution*, **265**, 193–219.

Hess, S., Frahm, J.-P. & Theisen, I. (2005). Evidence of zoophagy in a second liverwort species, *Pleurozia purpurea*. *Bryologist*, **108**, 212–18.

Hofmeister, W. (1851). *Vergleichende Untersuchungen der Keimung, Entfaltung und Fruchtbildung höherer Kryptogamen*. Leipzig: Friedrich Hofmeister.

Hollensen, R. H. (1973). A new type of branching in *Blepharostoma trichophyllum* (L.) Dum. *Journal of the Hattori Botanical Laboratory*, **37**, 205–9.

Howe, M. A. (1894). Chapters in the early history of hepaticology - I. *Erythea*, **2**, 130–5.

Hutchinson, A. H. (1915). Gametophyte of *Pellia epiphylla*. *Botanical Gazette*, **60**, 134–43.

Ingold, C. T. (1939). *Spore Discharge in Land Plants*. Oxford: Clarendon Press.

Kelley, C. B. & Doyle, W. T. (1975). Differentiation of intracapsular cells in the sporophyte of *Sphaerocarpos donnellii*. *American Journal of Botany*, **62**, 547–59.

Knapp, E. (1930). Untersuchungen über die Hüllorgane um Archegonien und Sporogonien der akrogynen Jungermanniaceen. *Botanische Abhandlungen*, **16**, 1–168.

Kobiyama, Y. (2003). Comparative development and ultrastructure of the specialized parenchyma cells and/or hydrolyzed cells in select liverworts and hornworts. Unpublished Ph.D. dissertation, Southern Illinois University, Carbondale, Illinois.

Leitgeb, H. (1871a). Beiträge zur Entwicklungsgeschichte der Pflanzenorgane. IV. Wachstumsgeschichte von *Radula complanata*. *Sitzungsberichte der Kaiserlichen Academie der Wissenschaften. Wein. Mathematisch-naturwissenschaftliche Classe*, **63**, 13–60.

Leitgeb, H. (1871b). Über die Verzweignug der Lebermoose. *Botanische Zeitung, Berlin*, **29**, 557–65.

Leitgeb, H. (1872). Über die endogene Sprossbildung bei Lebermoosen. *Botanische Zeitung, Berlin*, **30**, 33–41.

Leitgeb, H. (1874–1881). *Untersuchungen über die Lebermoose*. I–VI; I. *Blasia pusilla*, 1874; II. Die Foliosen Jungermannieen, 1875; III Die Frondosen Jungermannieen, 1877; IV. Die Riccieen, 1879; V. Die Anthoceroteen, 1879; VI. Die Marchantieen, 1881. Vols. I–III, Jena: O. Deistung's Buchhandlung; vols. IV–VI, Graz: Leuschner & Lubensky.

Ligrone, R. & Duckett, J. G. (1996). Development of water-conducting cells in the antipodal liverwort *Symphyogyna brasiliensis* (Metzgeriales). *New Phytologist*, **132**, 603–15.

Ligrone, R., Duckett, J. G. & Renzaglia, K. S. (1993). The gametophyte-sporophyte junction in land plants. *Advances in Botanical Research*, **19**, 232–317.

Liu, Y., Jia, Y., Wang, W. *et al.* (2008). Phylogenetic relationships of two endemic genera *Trichocoleopsis* and *Neotrichocolea* (Hepaticae) from east Asia. *Annals of the Missouri Botanical Garden*, **95**, in press.

Long, D. G. (2006a). Revision of the genus *Asterella* P. Beauv. in Eurasia. *Bryophytorum Bibliotheca*, **63**, 1–299.

Long, D. G. (2006b). New higher taxa of complex thalloid liverworts (Marchantiophyta – Marchantiopsida). *Edinburgh Journal of Botany*, **63**, 257–62.

Longton, R. E. & Schuster, R. M. (1983). Reproductive biology. In *New Manual of Bryology*, vol. 1, ed. R. M. Schuster, pp. 386–462. Nichinan: Hattori Botanical Laboratory.

Magill, R. E. (ed.). (1990). Glossarium Polyglottum Bryologiae. *Monographs in Systematic Botany from the Missouri Botanical Garden*, **33**, 1–297.

Mishler, B. D. & Churchill, S. P. (1984). A cladistic approach to the phylogeny of the "bryophytes." *Brittonia*, **36**, 406–24.

Mishler, B. D. & DeLuna, E. (1991). The use of ontogenetic data in phylogenetic analyses of mosses. *Advances in Bryology*, **4**, 121–67.

Müller, K. (1939). Untersuchungen über die Ölkörper der Lebermoose. *Berichte der Deutschen Botanischen Gesellschaft*, **57**, 325–70.

Müller, K. (1948). Morphologische und anatomische Untersuchungen an Antheridien beblätter Jungermannien. *Botaniska Notiser*, **1948**, 71–80.

Müller, K. (1954[1952]). Marchantiineae. In *Die Lebermoose Europas, Dr. L. Rabenhorst's Kryptogamen-Flora von Deutschland, Österreich und der Schweiz*, 3rd edn, vol. 6, pp. 320–409. Leipzig: Eduard Kummer.

Murray, R. V. & Crandall-Stotler, B. J. (2005). SEM survey of hydrolyzed strand cell diversity in the simple thalloid liverworts (Jungermanniopsida, subclass Metzgeriidae). *Botany 2005 Abstracts. Scientific Meeting [American Bryological and Lichenological Society], August 13–17, 2005, Austin, Texas*, p. 33 [abstract].

Nehira, K. (1983). Spore germination, protonema development and sporeling development. In *New Manual of Bryology*, vol. 1, ed. R. M. Schuster, pp. 343–85. Nichinan: Hattori Botanical Laboratory.

Oostendorp, C. (1987). The bryophytes of the Palaeozoic and the Mesozoic. *Bryophytorum Bibliotheca*, **34**, 1–112.

Parihar, N. S. (1961). *An Introduction to Embryophyta*, vol. I, *Bryophyta*, 3rd edn. Allahabad: Central Book Depot.

Pfeffer, W. (1874). Die Ölkörper der Lebermoose. *Flora*, **57**, 2–6, 17–27, 33–43.

Pihakaski, K. (1972). Histochemical studies on the oil bodies of two liverworts, *Pellia epiphylla* and *Bazzania trilobata*. *Acta Botanica Fennica*, **9**, 65–76.

Proctor, M. C. F. (1979[1980]). Structure and eco-physiological adaptation in bryophytes. In *Bryophyte Systematics*, ed. G. C. S. Clarke & J. G. Duckett, pp. 479–509. London: Academic Press.

Qiu, Y.-L., Li, L., Wang, B. *et al.* (2006). The deepest divergences in land plants inferred from phylogenomic evidence. *Proceedings of the National Academy of Sciences, U.S.A.*, **103**, 15511–16.

Renzaglia, K. S. (1982). A comparative developmental investigation of the gametophyte generation in the Metzgeriales (Hepatophyta). *Bryophytorum Bibliotheca*, **24**, 1–253.

Renzaglia, K. S., Brown, R. C., Lemmon, B. E., Duckett, J. G. & Ligrone, R. (1994). Occurrence and phylogenetic significance of monoplastidic meiosis in liverworts. *Canadian Journal of Botany*, **72**, 65–72.

Renzaglia, K. S., Schuette, S., Duff, R. J. *et al.* (2007). Bryophyte phylogeny: advancing the molecular and morphological frontiers. *Bryologist* **110**, 179–213.

Schill, D. B., Long, D. G., Moeller, M. & Squirrell, J. (2004). Phylogenetic relationships between Lophoziaceae and Scapaniaceae based on chloroplast sequences. *Monographs in Systematic Botany from the Missouri Botanical Garden*, **98**, 141–9.

Schuster, R. M. (1966). *The Hepaticae and Anthocerotae of North America, East of the Hundredth Meridian*, vol. I. New York: Columbia University Press.

Schuster, R. M. (1984a). Comparative anatomy and morphology of the Hepaticae. In *New Manual of Bryology*, vol. 2, ed. R. M. Schuster, pp. 760–891. Nichinan: Hattori Botanical Laboratory.

Schuster, R. M. (1984b). Evolution, phylogeny and classification of the Hepaticae. In *New Manual of Bryology*, vol. 2, ed. R. M. Schuster, pp. 892–1070. Nichinan: Hattori Botanical Laboratory.

Schuster, R. M. (1992a). The oil-bodies of the Hepaticae. I. Introduction. *Journal of the Hattori Botanical Laboratory*, **72**, 151–62.

Schuster, R. M. (1992b). *The Hepaticae and Anthocerotae of North America, East of the Hundredth Meridian*, vol. V. Chicago, IL: Field Museum of Natural History.

Schuster, R. M. (1992c). *The Hepaticae and Anthocerotae of North America, East of the Hundredth Meridian*, vol. VI. Chicago, IL: Field Museum of Natural History.

Shimamura, M., Mineyuki, Y. & Deguchi, H. (2003). A review of the occurrence of monoplastic meiosis in liverworts. *Journal of the Hattori Botanical Laboratory*, **94**, 179–86.

Shimamura, M., Brown, R. C., Lemmon, B. E. *et al.* (2004). γ-Tubulin in basal land plants: characterization, localization and implication in the evolution of acentriolar microtubule organizing centers. *Plant Cell*, **16**, 45–59.

Stotler, R. E. (1969 [1970]). The genus *Frullania* subgenus *Frullania* in Latin America. *Nova Hedwigia* **18**, 397–555.

Stotler, R. & Crandall-Stotler, B. (1977). A checklist of the liverworts and hornworts of North America. *Bryologist*, **80**, 405–28.

Suire, C. (2000). A comparative transmission electron microscopic study on the formation of oil-bodies in liverworts. *Journal of the Hattori Botanical Laboratory*, **89**, 209–32.

Suire, C., Bouvier, F., Backhaus, R. *et al.* (2000). Cellular localization of isoprenoid biosynthetic enzymes in *Marchantia polymorpha*. Uncovering a new role of oil bodies. *Plant Physiology*, **124**, 971–8.

Thiers, B. M. (1982). Branching in the Lejeuneaceae I: A comparison of branch development in *Aphanolejeunea* and *Cololejeunea*. *Bryologist*, **85**, 104–9.

Thomas, R. T. & Doyle, W. T. (1976). Changes in the carbohydrate constituents of elongating *Lophocolea heterophylla* setae (Hepaticae). *American Journal of Botany*, **63**, 1054–9.

Thomas, R. J., Stanton, D. S. & Grusak, N. A. (1979). Radioactive tracer study of sporophyte nutrition in hepatics. *American Journal of Botany*, **66**, 398–403.

Verdoorn, F. (1930). Die Frullaniaceae der Indomalesischen Inseln (De Frullaniaceis VII). *Annales Bryologici*, suppl., **1**, 1-187.

von Konrat, M. J. & Braggins, J. E. (2001). A taxonomic assessment of the initial branching appendages in the liverwort genus *Frullania* Raddi. *Nova Hedwigia*, **72**, 283-310.

Wellman, C. H., Osterloff, P. & Mohluddin, U. (2003). Fragments of the earliest land plants. *Nature*, **425**, 282–5.

Wilson, R., Gradstein, S. R., Schneider, H. & Heinrichs, J. (2007). Unraveling the phylogeny of Lejeuneaceae (Jungermanniopsida): evidence of four main lineages. *Molecular Phylogenetics and Evolution*, **43**, 270–82.

Wood, A. J. (2007). The nature and distribution of vegetative desiccation-tolerance in hornworts, liverworts and mosses. *Bryologist*, **110**, 163–77.

Yatsentyuk, S. P., Konstantinova, N. A., Ignatov, M. S., Hyvönen, J. & Troitsky, A. V. (2004). On phylogeny of Lophoziaceae and related families (Hepaticae: Jungermanniales) based on *trn*L-*trn*F intron-spacer sequences of chloroplast DNA. *Monographs in Systematic Botany from the Missouri Botanical Garden*, **98**, 150–65.

Zwickel, W. (1932). Studien über die Ocellen der Lebermoose. *Beihefte zum Botanischen Centralblatt*, **49**, 569–648.

Note added in proof.

Herzogianthaceae Stotler & Crand.-Stotl., fam. nov. Type genus: *Herzogianthus* R. M. Schust. (*Familia haec a Ptilidiaceis similis sed differt foliis dimorphis, succubis vel subtransversalibus, cum foliis ramorum vaginatis et connatis dorsaliter, cum ciliis foliorum setosis, unicellulis; amphigastriis quadrifidis in caulibus robustioribus; sporis* > *60* μ*m.*)

2

Morphology, anatomy, and classification of the Bryophyta

BERNARD GOFFINET, WILLIAM R. BUCK AND
A. JONATHAN SHAW

2.1 Introduction

With approximately 13 000 species, the Bryophyta compose the second most diverse phylum of land plants. Mosses share with the Marchantiophyta and Anthocerotophyta a haplodiplobiontic life cycle that marks the shift from the haploid-dominated life cycle of the algal ancestors of embryophytes to the sporophyte-dominated life cycle of vascular plants. The gametophyte is free-living, autotrophic, and almost always composed of a leafy stem. Following fertilization a sporophyte develops into an unbranched axis bearing a terminal spore-bearing capsule. The sporophyte remains physically attached to the gametophyte and is at least partially physiologically dependent on the maternal plant. Recent phylogenetic reconstructions suggest that three lineages of early land plants compose an evolutionary grade that spans the transition to land and the origin of plants with branched sporophytes (see Chapter 4). The Bryophyta seem to occupy an intermediate position: their origin predates the divergence of the ancestor to the hornworts and vascular plants but evolved from a common ancestor with liverworts (Qiu *et al.* 2006). The origin of the earliest land plants can be traced back to the Ordovician and maybe the Cambrian (Strother *et al.* 2004). Although unambiguous fossils of mosses have only been recovered from sediments dating from younger geological periods (Upper Carboniferous), divergence time estimates based on molecular phylogenies suggest that the origin of mosses dates back to the Ordovician (Newton *et al.* 2007) and thus that their unique evolutionary history spans at least 400 million years. During this time, the lineage has undergone

Bryophyte Biology: Second Edition, ed. B. Goffinet & A. J. Shaw. Published by Cambridge University Press.

multiple radiations that have resulted in a broad spectrum of morphological, ontogenetic, anatomical, and cytological diversity.

In this chapter, we describe the features that unite the Bryophyta and characterize their main lineages. Detailed descriptions of moss morphology are provided by Ruhland (1924), Goebel (1898a), and Campbell (1895). The morphological diversity of the Bryophyta is exceptionally well illustrated in the second edition of the illustrated glossary of bryophytes by Malcolm & Malcolm (2006).

2.2 Modular architecture of the vegetative plant body

The vegetative plants correspond to the gametophyte: a haploid multicellular body whose function is to develop sex organs or gametangia. The architecture of the gametophyte follows a modular pattern: the meristematic activity of the apical cell yields cells undergoing divisions to form building blocks or metamers, which are assembled into modules (Mishler & DeLuna 1991). A hierarchical arrangement of modules forms a branch system, which may be reiterated. The shape of the apical cell typically approximates an inverted tetrahedron, as seen in some liverworts, with three oblique triangular cutting faces and a convex outer surface (Crandall-Stotler 1980). Only members of *Fissidens* possess a lenticular apical cell with two sides, but even here the apical cell starts out as a tetrahedral cell early in stem ontogeny (Chamberlin 1980). The apical cell gives rise to derivatives in three (two) directions in a clockwise sequence. Each derivative follows a precise pattern of divisions that leads to a building block or metamer. The first division in the derivative cell isolates an inner cell from which cortical and conducting tissues will be formed. The outer cell develops into the epidermis, including the leaf and branch initials. The branch initial occurs always below the leaf initial. All metamers formed by an apical cell compose a module. Longitudinal growth of the module is accomplished through division, enlargement and elongation of cells composing each metamer. In *Takakia lepidozioides* the meristematic activity is accounted for by a ring of cells surrounding a rather quiescent central cell (Crandall-Stotler 1986). In *Sphagnum*, the activity of the tetrahedral apical cell is complemented by that of a secondary subapical meristem composed of cells of the primary metamer that undergo several anticlinal divisions resulting in lines of about nine cells, which each dramatically elongate by a factor of nearly ten (Ligrone & Duckett 1998).

Vegetative growth of mosses results from the accumulation of cell lines, and all cells of one module have an origin that can be traced to a single apical cell. The apical cell of each branch started out as an initial on the epidermis of the axis (stem or branch) onto which it is attached.

2.3 Organography of the gametophyte

Macroscopically the vegetative body of mosses can be divided into rhizoids, stems and branches, and leaves.

2.3.1 *Rhizoids*

The filaments that function in anchoring the plant to the substrate, and may be involved in water conduction, are analogous to roots but differ in their very simple architecture. Each rhizoid is in fact a uniseriate (rarely multiseriate) filament of elongate and smooth or roughened cells separated by oblique crosswalls. The multicellular rhizoids of mosses, except for those of *Sphagnum*, *Andreaea*, and *Andreaeobryum*, are thigmotropic, winding tightly around solid objects (Newton *et al.* 2000). This ability is best expressed in some Polytrichaceae, where rhizoids may form rope-like bundles composed of narrow rhizoids coiling around a central larger rhizoid (Whigglesworth 1947). Rhizoids grow from epidermal cells either at the base of the stem, along the ventral side of the stem and branches, or the costa. Rhizoids rarely emerge from specialized cells, nematogens, at the apical portion of the lamina. The stems of various terricolous mosses are sheathed in a more or less extensive coat of white or reddish brown rhizoids, which may serve in the external conduction of water (Schofield 1981; see Chapter 6), although a clear pattern between rhizoid abundance and water availability is not evident (Crundwell 1979). In some mosses, large, highly branched rhizoids originate from large cells lining the branch initials, and may have a protective function like pseudoparaphyllia (Schofield 1985). In *Sphagnum* rhizoids occur only at the base of the thalloid protonemata and are lacking on mature leafy plants, except for a single species from New Caledonia (Iwatsuki 1986). *Takakia* is characterized by the complete absence of rhizoids (Schuster 1997), as are some pleurocarpous mosses, such as species of *Scorpidium* (Koponen 1982). In most cases, the lack of rhizoids in nature contrasts with their presence *in vitro*, suggesting that their development may be environmentally controlled in the wild (Duckett 1994a) as it is in culture (Duckett *et al.* 1998). Furthermore, the density, length, and branching of rhizoids produced by pleurocarpous mosses is very much influenced by the nature of the substrate colonized (Odu 1978). The environmental factors that stimulate rhizoid production continue to elude bryologists.

Some mosses develop perennating structures called gemmae or tubers on their rhizoids (Imura & Iwatsuki 1990). Tubers, which lack an abscission cell diagnostic of gemmae (Duckett *et al.* 1998) may be uni- or multicellular, uniseriate or spherical. Their development may be triggered by drought (Arts 1990), which may support the view that tubers offer a means to resist prolonged dry periods (Arts 1986) but their adaptive value to dry environments remains to be critically tested (Newton & Mishler 1994). Rhizoidal tubers occur in various lineages of acrocarpous mosses, including the Polytrichaceae, Funariaceae, Bryaceae, Pottiaceae, and Dicranaceae. They seem, by contrast, to be absent in pleurocarpous mosses. Some rhizoidal appendages similar to tubers result from modification of rhizoidal cells following fungal infections (Martínez-Abaigar *et al.* 2005). Indeed the development of zoosporangia within the moss cells leads to swelling of the rhizoid tip cells or the side branch initials. Both cells are characterized by thin walls, and hence are vulnerable to infection by oomycetes. Points of entry of the fungus in the rhizoid are often revealed by pegs or ingrowths of cell wall material deposited by the cell as a likely response to the fungal aggression (Martínez-Abaigar *et al.* 2005).

2.3.2 *Stems*

Vegetative axes display an architecture that follows a very simple Bauplan: epidermal cells surround a cortex of large parenchyma cells, which may surround a central strand of narrow, putative structural and water-conducting cells. Some authors (e.g. Zander 1993) refer to the outer two layers as cortex and central cylinder, respectively, and others such as Malcolm & Malcolm (2006) equate "central cylinder" with "central strand". The transition between epidermis and cortex can be either abrupt (Fig. 2.1K) or gradual. The pigmentation of the outer cells is sometimes shared with cortical cells, and both tissues can have incrassate cell walls. The epidermis is uni- to multistratose and its cells retain their cytoplasm and organelles. Stomata are always lacking, but a cuticle, even if thin, covers the surface. In various mosses, the epidermal cells are thin-walled and inflated, and the epidermis is then referred to as a hyalodermis. This tissue is conspicuously developed in *Sphagnum*, where the hyalocysts, which may have one or more conspicuous pores on their surface, function in external water movement. In many pleurocarpous mosses the stems and branches are clothed with paraphyllia, slightly branched epidermal outgrowths that differ from rhizoids in their green color and shorter cells. Paraphyllia likely serve in external water conduction, but are also photosynthetically active given the abundance of chloroplasts.

Juvenile leaves hide delicate filaments that line the insertion of the leaf. These hairs originate from the leaf initial and secrete a mucilage of polysaccharides (Ligrone 1986) that may be essential in preventing the delicate growing apices from dehydrating (Schofield & Hébant, 1984). In various pleurocarpous mosses, axillary hairs may develop elsewhere on the stem (Ignatov & Hedenäs 2007); in these cases their function remains ambiguous. Axillary hairs typically consist of a single unbranched row of several short to elongate hyaline cells

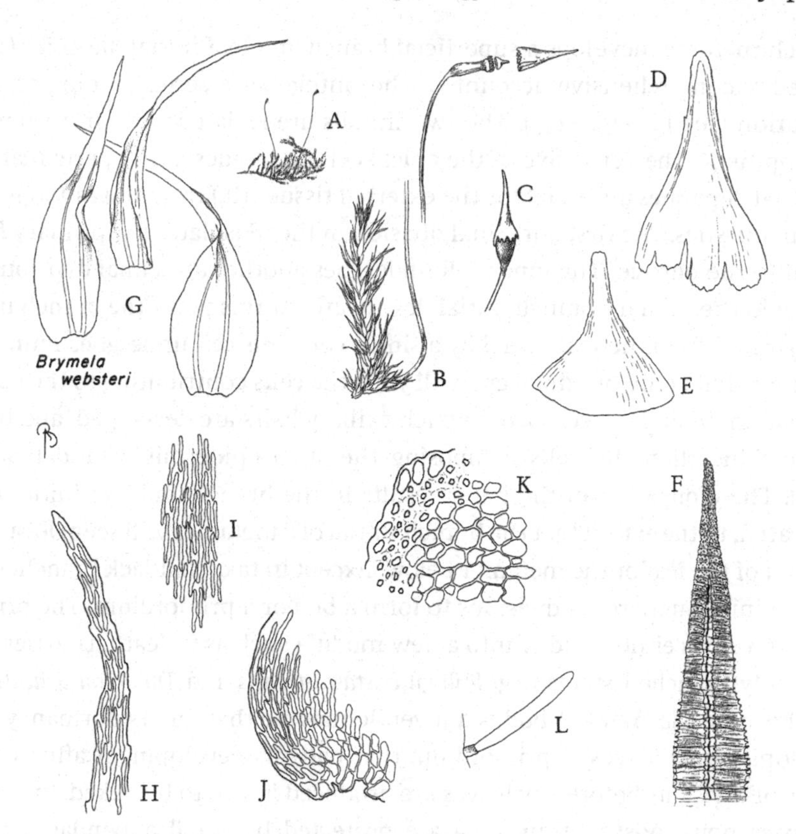

Fig. 2.1. *Brymela websteri* (Pilotrichaceae), as an example of a moss. (A) Aspect showing plagiotropic habit with somewhat erect branches. (B) Detail of branch system showing sporophyte borne on a lateral branch; the sporophyte is composed of a seta, capsule and operculum with the urn covered by a calyptra. (C) Young capsule with calyptra. (D) Calyptra. (E) Operculum. (F) Exostome tooth, composed of two columns of cells. (G) Leaves with strong double costa. (H) Leaf apex. (I) Laminal cells at mid-leaf. (J) Cells at base of leaf. (K) Portion of stem cross-section. (L) Axillary hair.

(rarely only one such cell) that often are subtended by one or more brown cells (Fig. 2.1L). The apical cell is typically longer and club-shaped. They vary greatly in size, number and shape, and although generally overlooked, seem to be taxonomically and hence phylogenetically informative (Hedenäs 1989a, Kruijer 2002), although infraspecific variation is possible (Zander 1993). Similar hairs are occasionally found associated with branches, as in *Dicranum* or *Encalypta*. These trichomes, like ordinary axillary hairs, exude mucilage, but because they originate from a metamer distinct from that of the leaf that subtends them they are in fact best seen as mucilaginous leaves (Berthier *et al.* 1971). *Andreaeobryum* and *Takakia* carry beaked mucilage or slime papillae (Murray 1988, Schuster 1984).

Each metamer develops a superficial branch and leaf initial. Berthier (1973) offered a comprehensive account of the ontogenetic series leading to their formation (see his Fig. $8Z_{1C}$). The two initials are isolated early in metameric development. The derivative of the apical cell undergoes a first periclinal division that separates an initial for the external tissues (IE) from the initial for the internal tissues. The first anticlinal division in the IE isolates the primary foliar initial to the outside. The inner cell undergoes another anticlinal division that will yield the primary branch initial downward. At this point the branch initial is separated from the leaf initial by a single cell. This cell undergoes numerous divisions that at the proximal end will yield the cells contributing to the base of the leaf, including the cells from which axillary hairs are developed, and below the leaf insertion the cells composing the stem epidermis (Crandall-Stotler 1980). The elongation of the latter results in the branch and leaf initial to be separated, to the extent that the branch initial of a metamer will seem located in the axil of the leaf of the metamer below. Except in taxa that lack branches, the branch initial undergoes divisions to form a bud or a primordium. The primordium develops either readily into a new module such as in feathermosses with pinnately branched stems (e.g. *Ptilium crista-castrensis* and *Thuidium delicatulum*) or it becomes dormant. A bud is a juvenile module that enters dormancy after developing tiny leaves. A primordium that halts its development after a short series of divisions before any leaves are produced is said to be naked. In various pleurocarpous mosses, primordia are protected by small appendages called pseudoparaphyllia. The term has traditionally been reserved for specialized structures restricted to the immediate vicinity of a branch primordium or branch bud. The shape of pseudoparaphyllia varies from filamentous to foliose; most species seem to produce only one type, but exceptions exist (Akiyama 1986). Akiyama & Nishimura (1993) distinguish "true" pseudoparaphyllia from scaly leaves based on ontogenetic grounds: the former arises from the stem, the latter from the branch bud. Ignatov & Hedenäs (2007) reject such distinction and broaden the concept of pseudoparaphyllia to include all appendages produced near leaf decurrencies and corners, and around primordia, and even those scattered along the stem, that have traditionally been called paraphyllia. They restrict the latter term to those structures developed in longitudinal rows on the stem. The phylogenetic significance of these appendages remains ambiguous, in part due to the controversy about their homology across lineages.

The anatomical complexity of the stem varies among mosses (Hébant 1977), with the Polytrichopsida exhibiting the greatest internal differentiation, whereas the Andreaeopsida show the least cytological variation (Kawai 1989). The parenchyma cells composing the cortex are typically somewhat larger than the epidermal cells, and in many peristomate mosses, outer and inner parenchyma cells may be morphologically distinct (Kawai 1989 and references therein). The cortex may serve as a structural or a storage tissue, and the thickness of the cell wall varies accordingly. A photosynthetic function may be restricted to the outer layers or to the young portions of the stem. Stereids are rather narrow prosenchymatous cells (i.e. long-tapered) with incrassate walls impregnated with a polyphenolic compound (other than lignin; Schofield & Hébant 1984). Stereids typically retain their protoplast at maturity. They occur in the central axis along with hydroids (see below) or below the epidermis in the cortex. The remainder of the cortex typically consists of more or less large parenchymatous cells, with flat or somewhat rounded ends. These cells may accumulate lipids or starch, which can be hydrolyzed and redistributed throughout the plant.

Transport within the plant is accomplished in part by undifferentiated parenchyma cells of the cortex, or by specialized conducting cells, the hydroids (Hébant 1977). These hydroids, with the associated parenchyma cells and stereids compose the water conducting tissue or hydrome. It is best developed in members of the Polytrichopsida, and is reduced to completely lacking in various lineages of the Bryopsida (Hébant 1977), most notably in aquatic mosses (Haberlandt 1886, Vitt & Glime 1984). Hydroids resemble stereids in their prosenchymatous shape, but lack protoplastic content at maturity. They occur in *Takakia* and peristomate mosses but are lacking in the Sphagnopsida, Andreaeopsida, and Andreaeobryopsida (Ligrone *et al.* 2000). Their walls are impregnated with polyphenols other than lignin, which is diagnostic of tracheids of vascular plants (Miksche & Yasuda 1978). Hydroids also lack spiral or annular secondary wall depositions. Furthermore, xylans that link strands of cellulose in secondary cell walls in vascular plants, and are considered essential for the evolution of vascular and supportive tissues, are lacking in mosses (Carafa *et al.* 2005). Immunocytological techniques Ligrone *et al.* (2002) revealed that water-conducting cells in particular exhibit great diversity in cell wall chemistry. *Takakia* differed from other bryophytes in the composition of its cell walls, suggesting that their water-conducting cells are not homologous to those of peristomate mosses. Furthermore, in *Takakia* the cells are short with the end walls nearly perpendicular to the axis. The contact surface between two consecutive cells is thus reduced but flow is facilitated by the presence of small perforations derived from plasmodesmata. By contrast, the hydroids of the Polytrichopsida and Bryopsida are slightly to highly elongate and lack small pores, at least at maturity. The thickness of the hydroid walls varies: thin all around to thickened and heterogeneous lateral walls and thin end-walls across which much of the transport takes place (Scheirer 1980, Ligrone *et al.* 2000).

Actively dividing cells are by definition undifferentiated and hence unlikely to engage in photosynthetic activities to sustain their energetic needs. Similarly, cell elongation and differentiation requires large amounts of energy to fuel anabolic reactions. Developing moss metamers depend on supplies of photosynthates and likely other organic compounds from other metamers and even modules. Long-distance symplastic transport from old to young tissues has been demonstrated in *Sphagnum* and the Polytrichaceae (Raven 2003). Although the speed of such transport seems hardly indicative of optimal specialization, cells involved in it share a series of attributes that are reminiscent of those exhibited by the sieve cells of tracheophytes, such as the lack of a vacuole and nuclear degeneration (Ligrone *et al.* 2000). Food-conducting cells are elongate and their end-walls contain many pores derived from plasmodesmata. Differentiation of these cells is most pronounced in the Polytrichales, and the term leptoid is restricted to these. For the food-conducting cells of other mosses the term "conducting parenchyma cells" is preferred (Ligrone *et al.* 2000). Food-conducting cells are characterized primarily by cytological aspects of their organization. Such features are found in other cells throughout the moss plant, suggesting that food transport occurs not just in the stem (e.g. rhizoids; see tables in Ligrone *et al.* 2000). Hence conducting parenchyma cells, even more than leptoids, are not distinguished with ease in light microscopy. In the stem, leptoids and their analogous parenchyma cells can be found throughout the cortex, with maybe a preferential location around hydroid strands.

The transverse section of the stem of various mosses reveals satellite bundles of water and food-conducting cells around the main axial strand. In longitudinal section these strands would connect at their distal ends with the conducting tissues in the leaf nerve, and hence are called leaf traces. In most mosses, these traces do not join with the main axial conducting tissues, and instead disappear within the cortex. True leaf traces reach the central strand. The thickness of the leaf trace (i.e. the number of hydroids composing it) and whether or not leaf traces join the axial hydrome varies among and even within species, although true leaf traces are commonly developed in the Polytrichaceae, whereas false traces seem to characterize the Bryopsida. It is worth noting that both the strength of the leaf trace and its contact with the main strand, as well as the width of the axial hydrome itself, may weaken along a moisture gradient (Hébant 1977), for example when a species is transferred to and grown in aquatic conditions (Zastrow 1934).

The moss stem is typically a solid organ, except for a central cavity resulting from the collapse of the axial hydrome. Only in *Canalohypopterygium* and *Catharomnion* (Hypopterygiaceae) are axillary cavities present in the cortex, extending from the stolon through the stipe and rachis to the branches

(Kruijer 2002). Whether these cavities form via schizogeny or lysogeny is not clear. These cavities are filled with lipids, which may serve to store energy, or they may be useful in deterring herbivores or infectious bacteria and fungi, but such function remains to be tested (Pelser *et al.* 2002).

2.3.3 *Modifications of the stem*

In mosses the female gametangia are developed at the apex of a module. The transition from vegetative to sexual module results in the cessation of growth of that module, except in the Sphagnopsida and the Andreaeopsida. In these lineages, the gametophytic tissue resumes growth after fertilization to elevate the sporangium above the gametangial leaves, a role restricted to the sporophytic seta in other mosses. The gametophytic stalk elevating the sporophyte is called a pseudopodium. In *Andreaea* the pseudopodium develops through a meristematic activity of cells from the archegonial stalk (Roth 1969). In some taxa, it retains a stem-like appearance with scattered reduced leaves, axillary hairs and even archegonia (Murray 1988), which suggests that in these species the cauline tissue (i.e. the receptacle) participates in pseudopodium formation. The pseudopodium of peatmosses develops solely from the receptacle beneath the vaginula (Roth 1969), as evidenced by the presence of scattered aborted archegonia. Murray (1988) speculated that the Andreaeopsida and Sphagnopsida acquired a pseudopodium independently. All other mosses except for some species of *Neckeropsis* (Neckeraceae; Touw 1962) lack a pseudopodium, at least one elevating the capsule. In *Aulacomnium*, the term pseudopodium is used differently and represents a defoliated extension of the stem that carries brood bodies typically arranged in a terminal crown. Whether the apical or a new subapical meristematic region provides the metamers to the pseudopodium formation is not known.

One last series of modifications in the stem anatomy occurs at the gametophyte–sporophyte junction. The sporophyte in mosses as in all bryophytes is permanently attached to and nutritionally dependent on the female gametophyte. Such maternal care or matrotrophy is considered a critical innovation preceding the origin of embryophytes and thus essential to the evolution of land plants (Graham & Wilcox 2000). Matrotrophy is facilitated by cytological and ultrastructural modifications on one or both sides of the generational junction, a region called the placenta (Ligrone *et al.* 1993). Here we address only changes pertaining to the gametophytic tissues (see below for characteristics of the sporophytic foot). A space resulting from the lysis of gametophytic placental cells exists in all mosses (Frey *et al.* 2001). The degree of ingrowths varies from short to labyrinth-like and long and from fine to coarse. Such cell-wall modifications are lacking in *Takakia*, *Sphagnum*, *Andreaea*, and the

Polytrichaceae, as well as in *Dicranum* among the Bryopsida. Although labyrinth walls are lacking, specialization may occur in the thickness and texture of the wall. All other mosses examined, from *Tetraphis* to *Brachythecium*, develop gametophytic transfer cells early and in one to three layers depending on the lineage. Ligrone *et al.* (1993) reported that in *Diphyscium* and *Bryum* the gametophytic tissues are penetrated by tubular outgrowths of elongate epidermal cells of the foot. Frey *et al.* (2001) could not confirm such haustoria, which Roth (1969) described as developing late in the ontogeny of the sporophyte. Specialization of the placental cells pertains further to their cytological and ultrastructural characteristics: the cytoplasm is often dense and rich in lipids, the vacuole is typically reduced but large in *Sphagnum*, the endoplasmic reticulum extensive, mitochondria numerous and large, chloroplasts numerous, often less differentiated, rich in lipid-filled globuli and sometimes filled with starch (Ligrone *et al.* 1993).

Stems have a broad range of morphological and anatomical diversity, with the most complex anatomy displayed by the Polytrichaceae (Hébant 1977). One member of this family holds the record for tallest terricolous moss: *Dawsonia superba* can reach 70 cm in height. Pendent epiphytes in moist tropical forests or aquatic species that benefit from buoyancy may have their gametophyte grow to one meter or more in length. Alternatively, reductionary trends lead to virtually invisible shoots: in *Buxbaumia* the female stem is reduced to a tiny axis with a few leaves, and the sole antheridium is protected by one leaf, sessile on the protonema (Goebel, *fide* Ruhland 1924).

2.3.4 *Leaves*

The sole unifying character of moss leaves is that they are always sessile on the stem, inserted along their entire base, and hence are never petiolate. In some species of Calymperaceae, the leaf lamina is contracted to the costa between the distal end of the sheathing base and the lower end of the green lamina; here the "petiole" is intralaminar rather than supporting the whole leaf (Reese & Tan 1983). Leaves develop from the single leaf initial present in each metamer. Their arrangement on the stem or phyllotaxy is shaped first by the spatial arrangement of metamers, and thus the shape of the apical cell, and further dictated by ecophysiological constraints. In most mosses, a spiral arrangement of leaves, especially on orthotropic shoots minimizes shade and maximizes light interception. The apical cell produces two successive derivatives at an angle closer to 137° (Crandall-Stotler 1984). If the angle were 120° (as would be predicted if the apical cell were perfectly triangular in section), every fourth leaf would be aligned vertically with the first one. Any deviation from 120° results in more spiral turns needed for any two leaves to be aligned. In some cases, the leaves form five conspicuous ranks (e.g. *Conostomum*); in others the phyllotaxy is such that the alignment of leaves is obscure at best. In *Fissidens* leaves are developed in two opposing rows. Such distichous arrangement is due to the lenticular shape of the apical cell, which thus has two rather than three cutting faces. Distichous leaves also characterize other genera, such as *Bryoxiphium* or *Distichum*, but their apical cells are tetrahedral. In *Schistostega*, the distichous arrangement results from torsion of the stem, and consequently characterizes only leaves below the apex. Apical leaves and all those on fertile stems are radially inserted.

Mosses creeping over the substrate in low light environments seem to adopt a complanate posture of their leaves, which, although inserted in more than two ranks, lie at maturity in a single plane (e.g. *Plagiothecium*). The shift to complanate leaves is likely dependent on an oblique versus transverse insertion of the leaves. In many taxa, the base of the leaf is differentiated to fulfill a function other than or in addition to photosynthesis. In various Polytrichaceae or Bartramiaceae, the base of the leaf clasps the stem, providing additional robustness to the insertion and thus support to the leaf. Clasping leaf-bases may also create capillary spaces essential for the external conduction of water.

The leaves developing on young stems or branches often lack the characters of mature leaves. A sharp morphological contrast between juvenile leaves at the base of the module and mature leaves apically (*sensu* Mishler 1988), often coupled with a gradual transition between them, is referred to as a heteroblastic leaf series. This series refers to transformations in morphology of leaves along the length of the shoot, and is thus different from the transformational series of a given leaf during its maturation: an immature or "young" mature leaf may not resemble a juvenile leaf at any of its developmental stages. Heteroblastic series are common in mosses, although rarely very conspicuous. Juvenile leaves may offer important phylogenetic clues (Mishler 1988). For one, it seems that even distantly related mosses exhibiting strikingly different mature leaf morphologies exhibit highly similar juvenile leaf morphologies (e.g. *Tortula* and *Funaria*, Mishler 1986).

In many branched plagiotropic mosses, stem and branch leaves are morphologically distinct. In *Macromitrium* the leaves on the creeping stems are much reduced compared to those of the erect branches, which are similar to the stem leaves of the close relative *Orthotrichum*, whose stems grow upright. In mosses with erect stems composed of a vertical stipe with a horizontal branched rachis, the stipe bears reduced, often scale-like and ecostate leaves whereas the rachis produces well-developed leaves with a strong midrib. Such variations in leaf morphology between modules is referred to as anisophylly (Newton 2007).

Leaf dimorphism also occurs around the circumference of a module (heterophylly), with dorsal and ventral leaves much reduced compared to the lateral ones. Reduced ventral leaves are often called amphigastria. Although common in liverworts, they occur only in few lineages of mosses, and most notably the Hypopterygiaceae. Such reduction may serve to maximize light exposure of lateral leaves in plagiotropic mosses (e.g. *Racopilum*), but remains to be explained in orthotropic taxa such as *Epipterygium*. Amphigastria may perform other functions such as create capillary spaces for external water or actually hold water as in *Cyathophorum tahitense*, where the amphigastrium base is inflated into a little pouch.

A functional basis for foliar dimorphism is evident when certain leaves are specialized for asexual reproduction, such as the stenophylls of some Calymperaceae with their reduced lamina and rod-like costa crowned by gemmae (Reese 2000). Splash-cups may be formed by dense rosettes of differentiated leaves surrounding discoid gemmae (*Tetraphis*) or clusters of antheridia (e.g. Polytrichaceae and various Mniaceae). In most mosses, foliar dimorphism occurs simply as a differentiation of vegetative and gametangial leaves, and between perigonial and perichaetial leaves. In autoicous species, the leaves surrounding antheridia often resemble juvenile leaves: they are much smaller than the vegetative or perichaetial leaves on the same plant. Perichaetial leaves are by contrast often larger and longer, sometimes following post-fertilization growth. The innermost perichaetial leaves can either be the smallest (e.g. Mniaceae) or more often the largest of the perichaetial leaves (e.g. Hypnales).

The interval between two successive leaves is determined by the elongation of the epidermal cells of each metamer. Most mosses seem to have a regular foliation of the axes. Stems that do not participate much in photosynthetic activity have widely spaced leaves. For example, in *Dendrohypopterygium arbuscula* the internodes on the erect stipe are long and the leaves distant from one another. Similarly, in *Rhodobryum roseum* long internodes separate the lower leaves, whereas the upper metamers hardly elongate, and as a result the upper leaves form a dense rosette.

Each metamer derived from the cauline apical cells yields an initial from which a leaf will develop. In most mosses, the apical cell of the leaf is two-sided, producing derivatives in two directions, alternating between the left and the right, but always in one plane. Only six derivatives may be produced in each direction, before the apical cells cease to divide (Frey 1970). Thus the lamina forms from building blocks resulting from the divisions of the derivative cells. The ontogenetic patterns differ between leaves composed of a parenchymatous network of five- to six-sided cells common to the Polytrichopsida and most acrocarps, and those leaves composed of elongate prosenchymatous cells, found in most pleurocarps (Frey 1970). The differences lay in the timing (delayed versus immediate) and the pattern of division of derivatives. Leaves with wide rhombic cells, such as those of the Bryaceae and Hookeriaceae, seem to be intermediate in their morphology, and hence may mark the transition from a parenchymatous to prosenchymatous cell network (Frey 1970). Detailed ontogenetic studies of these taxa are lacking. In *Andreaea*, the apical cell exhibits two or one cutting face(s). In *Buxbaumia*, the highly reduced leaves resemble those of jungermannioid liverworts, in that they lack a single apical cell (Goebel 1898a). Whether their development involves a basal meristem is not clear.

The lamina is unistratose in most mosses, except for the costa and in some cases the marginal region. Stomata are always absent. A cuticle may be present but if present then thin and largely ineffective in preventing water loss (Proctor 1979, 1984). The costa is derived from a variable number of so-called "Grundzellen" or fundamental cells. A series of periclinal and anticlinal divisions early in leaf ontogeny, results in multiple median layers of cells, present even in young leaves at the apex of the stem. In *Leucobryum*, whose leaves are 3–6-stratose for much of their width (except for a narrow winglike margin), periclinal divisions occur in juvenile leaves, suggesting that this region corresponds to the costa. Other multistratose streaks in the leaf, including the margin, are formed later in development.

Except for the deeply segmented leaves of *Takakia*, and the perigonial leaves of *Buxbaumia*, moss leaves are never lobed: the lamina is a single blade. In fertile stems of *Schistostega*, consecutive leaves may be connected by their lamina at their base, giving the impression of a deeply incised or pinnately lobed ribbon, but the leaves themselves are entire. In all other respects the leaves of mosses exhibit a tremendous diversity, clearly too vast to fully describe here. The leaf is typically composed of a single chlorophyllose blade. The most notable exception is the leaf of *Fissidens*, which, at least in the lower half, appears Y-shaped in transverse section. The two arms of the Y correspond to the vaginant laminae that embrace the stem and the leaf above it. The vaginant lamina makes up half to 3/4 of the leaf length. The lower portion of the Y represents the dorsal lamina. At the intersection of the three blades lies the costa. The dorsal lamina extends down the costa but most often does not contribute to the insertion of the leaf, although in some rare cases it is decurrent (Robinson 1970). Where the vaginant lamina ends in the upper half of the leaf, a single ventral laminal blade faces the dorsal wing. The dorsal and ventral blades are complanate with the stem, and are best seen as outgrowth of the costa. The vaginant laminae that are transversely inserted onto the stem constitute the true leaf (Salmon 1899).

The general outline of the lamina varies between orbicular, to lingulate truncate to linear. The margin of the lamina is sometimes entire but various degrees of dentation and serration are common. Long multicellular marginal cilia are rare (e.g. *Thelia*), and sometimes restricted to perichaetial leaves (e.g. *Hedwigia ciliata*). Teeth and cilia may be evenly distributed but they often characterize only a particular segment of the margin. The teeth may occur in pairs or singly. Depending on the length of the protrusion, we distinguish laciniae composed of multiple cells, serrations composed of one whole cell, serrulations due to protruding cell apices only and dentations due to lateral conical projection of the cell wall. All these, except for laciniae, are common in mosses and scattered across the phylogenetic tree. In *Sphagnum* walls of the marginal and some laminal cells may be partially resorbed and the affected area appears fringed.

The thickness of the lamina (excluding the costa, see below) varies from uni- to multistratose. In some mosses, only the margin is composed of multiple layers (Mniaceae), whereas in others, the leaf is several cell-layers thick across its width or only in discrete longitudinal strands (*Vittia*, *Orthotrichum* spp., or *Grimmia* spp.). Among pleurocarpous mosses, pachydermous leaves have been acquired independently in several lineages (Hedenäs 1993, Vanderpoorten *et al.* 2003), seemingly as a response to a transition to an aquatic habitat.

Whereas all mosses predating the origin of *Oedipodium* lack a midrib in their leaves, such costae characterize many taxa of the Bryopsida. The costa is typically single, and unbranched, vanishing in the upper lamina, or extending well beyond the leaf border (excurrent), forming a smooth or densely toothed awn. In many pleurocarpous taxa and very few acrocarps, the costa is double and V-shaped (see Fig. 2.1G), with the arms of the V varying from barely visible and short to conspicuous and long. Only in a few taxa, does the costa appear truly branched (e.g. *Antitrichia*). In some mosses (e.g. *Eurhynchium* and *Pohlia* spp.), the costa ends below the leaf apex and forms a spine projecting from the dorsal surface of the leaf. In *Pilotrichum* the costa emerges as a crest for part of its length, and may even produce propagula (Buck 1998). Although the appearance of the costa is generally fixed for a given species, its development can be environmentally altered: growing costate species in an aquatic medium results in the loss of expression of the costa in newly formed leaves (Zastrow 1934). This may be indicative of the costa fulfilling a role in water and nutrient transport. The costa may contain water- (hydroids) and food-conducting cells. The latter are less specialized than leptoids per se, and the term deuter or guide cell may be preferred (Hébant 1977). Bands of stereids may cover the guide cells on one or both sides and the surfaces of the costa may be covered by laminal cells. Thus in transverse section the costa may appear homo- or heterogeneous. In fact the costal anatomy varies significantly among mosses (Kawai 1968) and offers diagnostic features and hence phylogenetic information. The true pleurocarpous mosses share a homogeneous costa, i.e., a costa lacking cell differentiation

(Hedenäs 1994). In the Leucobryaceae, the costa is extremely broad, occupying more than half the leaf width. The costa of the Polytrichaceae is typically mounted by lamellae of green cells that run from the apex of the green lamina to the transition of the sheathing base. The margin of the lamellae is covered in wax (Proctor 1992) preventing water from filling the intralamellar space, thereby allowing for CO_2 absorption over a much larger surface (Proctor 2005). In the Dicranaceae, ridges characterize the dorsal surface of the costa of various taxa, and in *Bryoxiphium* and *Sorapilla* a single chlorophyllose dorsal extension may run partially down the costa.

Laminal cells are far from monomorphic along the dimensions of the leaf. In many acrocarpous mosses, the basal cells are rectangular, compared with short isodiametric upper cells. In the angles of the leaf insertion, primarily in pleurocarps, the alar cells may form groups of small, dense, quadrate or of inflated and thin-walled cells, as is diagnostic of the Sematophyllaceae. Large hyaline cells seem designed for rapid absorption of excessive water that may be essential for delaying dehydration of the leaf, as their differentiation is weakened when the shoots develop under submerged conditions (Zastrow 1934). However, the absorption of excess water would also result in increased turgor and the resulting forces may lead to changes in leaf posture, for example by pushing the leaf away from the stem. In *Ulota* the basal marginal cells differ from inner laminal cells in their quadrate shape and incrassate transverse walls. The differentiation of marginal cells extends in many mosses further up the blade. In *Mnium* linear cells line the whole lamina, forming a differentiated margin two or more cells wide and thick. Such a margin of elongate cells may be essential for conducting water within the leaf. Alternatively, a strong border may provide a structural reinforcement to the strength of the leaf against abrasion in rheophilous taxa (Vitt & Glime 1984). An intramarginal band of elongate cells, or teniolae, diagnoses various species of Calymperaceae, but not necessarily those with the longest leaves. A structural role is thus not obvious.

Cells may vary in morphological attributes across the leaf blade, but walls of a given cell may also vary in thickness and ornamentation. *Racomitrium* is diagnosed by wavy longitudinal anticlinal walls, and in *Dicranum* the thick axial walls are often porose. In *Steyermarkiella* the transverse walls of the hyalocysts (see below) are perforate. External ornamentations of the periclinal walls determines, along with the thickness of the cuticle, the shine of the leaf surface. Mammillae are protrusions of the cell lumina above the surface of the cell, whereas papillae are solid cell-wall protuberances. Papillae vary from small and knobby to tall and antler-like, from cylindrical to crescent-shaped, from single to numerous per cell. The presence of papillae (or that of a cuticle) is revealed by the matte appearance of the leaf, versus a shininess when cells are smooth. Whether the role of papillae is to merely increase light absorption is doubtful. Their presence dramatically increases the surface of the cell through which water or gases needed for photosynthesis are exchanged. Papillae also create capillary spaces essential for holding excessive water needed to delay loss of turgor in photosynthetic cells (Proctor 2000a). Papillae may thus be a necessary evil for some plants: they need to hold on to water when available, but by doing so they cover the cells with a CO_2-impermeable layer. A solution may be to impregnate the apex of the papilla with a water-repellent cuticle, and hence to use papillae as snorkels emerging from the water surface (Proctor 2000b).

The pattern of cell wall thickening within cells and across the lamina determines the three-dimensional shape of the leaf and the changes in habit of the leaf when dry. Differences in width and thickness of dorsal and ventral cell walls in unistratose leaves shape the general curvature of the leaf: concave or convex. When such differentiation is more localized, discrete shifts in curvature occur along the laminar surface, such as longitudinal plications (plicate leaves), or transverse or random undulations (undulate or rugose leaves). The adaptive value of such modifications of the leaf are obscure, but it is worth noting that in *Tomentypnum nitens* plications are lacking in leaves grown in submerged conditions (Zastrow 1934). Many mosses exhibit hygroscopic movements of the leaf. For example, the leaves of *Helicophyllum* always roll inward from the apex down. In *Ulota* the leaves are curled or crisped when dry, but spreading when moist. The basis for these movements must be accounted for by differences in length or thickness of adjacent walls of individual cells. Van Zanten (1974) studied the role that is played by the marginal swelling tissue at the transition between the clasping base and the green lamina in *Dawsonia* (Polytrichaceae). He showed that the movement of the lamina is controlled by the vertical lamellae in the joint area, which are composed of cells with differentially thickened walls: thicker walls swell upon hydration and pressure is exerted, resulting in the leaf bending. The swelling tissue composed of transversely elongate incrassate cells serves not in directing the movement but rather in preventing tearing under the torsion of the blade upon wetting.

The most conspicuous dimorphism in laminal cells involves the juxtaposition of chlorophyllose or assimilative cells and hyaline leucocysts either in a single stratum or across the thickness of the leaf. Metabolic activities in mosses may be constantly limited by the scarcity of water, and in the absence of stringent mechanisms to control water balance mosses have acquired means to hold excess water to delay dehydration and thus prolong periods suitable for photosynthetic activities (Proctor 2000a,b; Chapter 6). Leuco- or hyalocysts are cells devoid of their cytoplasm that function as reservoirs to temporarily store water. In *Sphagnum*, chlorophyllose assimilative cells alternate with leucocysts within the unistratose lamina. The walls of the hyalocysts often bear internal thickenings in the form of fibrils spun anticlinally. Dying the walls reveals pores on exposed surfaces of the leucocysts. This diagnostic cell dimorphism is lacking in juvenile leaves of *Sphagnum* (Mishler 1988), in late leaves of species such as *S. ehyalinum* (Shaw & Goffinet 2000), and can be lost in newly developed shoots when a typical species is grown in a water-saturated low-light environment (BG, pers. obs.).

Leucocysts are also conspicuously developed in some Bryopsida. In *Encalypta* and some Pottiaceae, for example, the typically hyaline basal cells are elongate and perforate on their outer (periclinal) and inner (anticlinal) walls, and likely involved in water storage or conduction. In the Leucobryaceae, leucocysts form multiple layers with one or more rows of chlorophyllose cells embedded in between. The walls of the leucocysts are pored, but unlike in *Sphagnum* the pores connect adjacent hyaline cells, rather than opening the cell to the atmosphere; the outer wall may be broken but actual pores are lacking (Robinson 1985). A water-holding function for the leucocysts seems intuitive, but would lead to the chlorophyllose cells being sealed from the gaseous atmosphere and hence deprived of essential carbon dioxide. Robinson (1985) argued that the hyaline cells of older mature leaves down the stem serve primarily in excess water storage, and that in young, photosynthetically active leaves the leucocysts are filled with air needed for gas exchange. A leucobryaceous leaf architecture is otherwise known only from some Calymperaceae and Dicranaceae.

Leaves offer many of the features essential for the identification of bryophytes. None of the mosses has entirely lost leaves, although some species rely on a persistent protonema to form their vegetative body, but even here a few leaves are produced to protect sex organs. By contrast, the leaves of *Syrrhopodon prolifer* var. *tenuifolius* reach nearly 6 cm in length!

2.4 Branching in mosses

The vegetative body of mosses is typically composed of modules of either identical or distinct hierarchical ranks as a result of branching. Every metamer derived from the apical cell comprises a single branch initial below the leaf initial, with the two separated by several epidermal cells. Although every metamer thus has the potential to develop into a branch, not all initials develop into primordia and branches, and if they do, their relative location on the module may vary among species or higher rank lineages. Depending on the distribution of branches, the polarity of their maturation along the stem, their density and their function in the hierarchical system, mosses exhibit fairly distinct life forms (LaFarge-England 1996).

Fig. 2.2. Close-up of branching in *Thuidium delicatulum*. Note the monopodial type of branching with modules of secondary and tertiary rank, giving the branch system a feathery outline.

The development of branch initials is triggered in some mosses by the cessation of meristematic activity at the apex of the module, due, for example, to the consumption of the apical cell in the formation of a sex organ. Such putative apical dominance, likely mediated by auxin (Cooke *et al.* 2002) is much weaker and maybe non-existent in mosses that branch freely. Determinate growth characterizes the primary module of many acrocarps and subsequent hierarchical modules of many pleurocarps (Newton 2007). Branch development is initiated either at the base of a module (basitonous branching) or at the apex (acrotonous branching), and older branches are thus located above or below younger branches, respectively. Based on the function of the new module, two main branching patterns are recognized. Sympodial branching refers to the development of innovations of the same hierarchical rank. Thus, a stem that ceases to grow because of determinate growth produces a new axis, which acts as a stem. Sympodial branching basically refers to a preservation of function between two successive modules, in other words the reiteration of a module or branch system. Most mosses with plagiotropic shoots develop branches continuously. Production of gametangia is transferred to the branches and the main module remains vegetative, capable of virtually indeterminate growth. Such monopodial branching pattern results in a somewhat feathery outline of the moss (Fig. 2.2), depending on the frequency of branch production. The branches themselves may repeat the pattern. As a result, beside the primary module (i.e. the stem), and secondary modules (the first set of branches), the plant body may further be composed of tertiary and more rarely quaternary modules. A function or even morphological differentiation of successive modules is not always evident, and in such cases, all modules are best regarded as primary modules (Newton 2007). The spatial distribution of the branches is dependent on (a) the elongation of each metamer, in particular of the cells that separate two consecutive branch initials, and (b) the dormancy of branch initials or bud primordia, since not all primordia necessarily develop into branches. In *Dendrohypopterygium arbusculum* the umbrella-like disposition of branches likely results from a lack of elongation of internodes, and in *Hypnum imponens* the pinnate branches that line the stem at regular intervals alternate with dormant buds. In *Sphagnum*, only every fourth (Ruhland 1924; but third according to Crum 2001) branch initial develops into a branch. Furthermore, each branch undergoes a series of immediate branching events, giving rise to fascicles of two to seven branches. At first the fascicles are tightly arranged at the apex of the stem into a compact capitulum. Further down the stem, branches differentiate and become either spreading or pendent. Such fascicular arrangement of branches is lacking in *Ambuchanania*, the sister taxon of *Sphagnum*, and some *Sphagna*.

Although the distribution of the sex organs (i.e., carpy, see below) influences the mode of branching, it only shapes the potential of individual modules. The function of the module ultimately determines which ability is expressed. For example, in *Macromitrium* the stem gives rise to erect monopodial branches, whose function it is to develop terminal sex organs. These branches may indeed produce a sympodial innovation to carry on the function of the sexual module. However, under some circumstances, the branch continues to grow and becomes plagiotropic, ultimately contributing to the clonal growth of the plant through indeterminate growth and monopodial branches. Thus, in this case the branch has reverted to a stem function and thereby adopted the alternative branching mode. In essence, the type of branching pattern may vary among modules of distinct hierarchical rank, and within a module over time. Rarely does sexual dimorphism pertain to branching, but in phyllodioicous species with genetically determined dwarf males, the latter lack the ability to branch whereas the female plant may produce abundant innovations.

Branching is not merely a process whereby new foliate shoots are added to the vegetative body. Some innovations are functionally more specialized. In most acrocarpous mosses, sympodial innovations are the only means for the plant to persist and engage in more than one sexual reproductive cycle. Branching contributes significantly to clonal growth either above or below ground. Indeed, many Polytrichaceae develop subterranean stolons or rhizomes from which aerial branches develop. In the Gigaspermaceae, the rhizomes are perennial and offer a means of survival from destruction of the delicate above-ground shoots. In *Canalohypopterygium* rudimentary bristle-like branches may serve as storage organs for oils. A few mosses rely on highly specialized branches for asexual reproduction. In *Orthodicranum flagellare*, stiff branches with minute leaves occur in clusters at the apex of the stem. Similar brood branches are crowded at the apices of primary branches of *Platygyrium repens*. Acrocarpous mosses producing both male and female sex organs on the same plant but in distinct clusters (i.e. autoicous) produce the female sex organs at the apex of stems, and rely on branches to host the perigonia (male inflorescences), or vice versa.

Branching patterns in mosses are diverse and complex (Meusel 1935, LaFarge-England 1996). Branching results in branch systems that are reiterated as the moss grows. In its simplest form, the branch system is composed of a primary module that is repeated by sympodial branching; most primordia remain dormant. Complex vegetative bodies arise from the reiteration of a branch system composed of multiple modules of distinct hierarchical rank. Variations in termination, origin, and orientation of modules, combined with differences in modularity and reiteration, allow for virtually endless combinations of architectural patterns in mosses, and most notably among the pleurocarpous mosses (Newton 2007). Assessing the phylogenetic significance of characters associated with branching has been plagued by ambiguities regarding the homology of modules within and between plant bodies, in part due to the functional plasticity of stems and branches in some species.

2.5 Sex organs: distribution, development and dehiscence

The sex of a plant is likely genetically determined. Heteromorphic bivalents, reminiscent of the XY sex chromosomes, occur in various mosses (Anderson 1980), and their distribution among plants seem to correlate with the sex in at least some of them, although the mechanisms by which they function is not understood (Newton 1984). However, not every plant produces gametangia, and sex expression may vary over short distances with patterns of expressions varying between the sexes (Stark *et al.* 2005). Sex organs are thus genetically determined but their development is triggered by environmental parameters (Bopp & Bhatla 1990).

As in all embryophytes, the sex organs are multicellular, offering some type of protection to the developing gametes, the sperms and egg. Gametangia are often accompanied by sterile unbranched filaments (i.e. paraphyses), and are surrounded by specialized leaves to form the perichaetium and perigonium. Sex organs are always developed superficially and at the apex of a module. In some cases the gametangium is the module. Female and male sex organs may be borne on one bisexual individual (monoicy) or two unisexual plants (dioicy). Plants capable of producing both sex organs may do so within a single cluster with either antheridia surrounding (paroicy) or mixed with archegonia (synoicy), or in distinct clusters on distinct modules (autoicy). Rarely is the production of different sex organs spread in time, resulting in only one sex present at any given time (pseudodioicy; Ramsay 1979). Sexual dimorphism in dioicous species is rare in mosses (Vitt 1968), at least in terms of morphological differentiation (Stark 2005). The most striking dimorphism between male and female plants characterizes phyllodioicous or pseudoautoicous species, wherein the male plants are reduced to a single bud emerging from the protonema growing epiphytically on the female plant (Ramsay 1979).

All mosses develop their archegonia at the apex of a module, and hence all bryophytes are acrogynous (Goebel 1898a). Acrocarpy is defined by the terminal cauline position of the perichaetium. Cladocarpy and pleurocarpy refer to the apical location of the female sex organ on branches that are well developed and bearing a heteroblastic series of leaves or that are highly reduced, respectively (LaFarge-England 1996), although exceptions exist (Newton 2007). The distinction between these two modes of perichaetial position may be somewhat ambiguous given the continuum in branch development, and this ambiguity leads to conflicting interpretations of the phylogenetic significance of pleurocarpy (e.g., Buck & Vitt 1986, Hedenäs 1994, LaFarge-England 1996, Bell & Newton 2007, O'Brian 2007).

Archegonia are rarely developed as single organs, such as in *Takakia* (Schuster 1997), some *Sphagna* (Crum 2001), or *Splachnobryum obtusum* (Arts 1996). The sole or first archegonium is always derived from the apical cell. When multiple archegonia occur in a perichaetium, the additional sex organs seem to originate from the branch initial of segments below, suggesting that perichaetia are condensed branching systems with highly reduced branches.

At maturity, the archegonium consists of three main parts: a solid, more or less elongate stalk, a venter with a multistratose jacket and a single egg, and a neck. The first division of the archegonial initial yields a lower and an upper cell. The latter undergoes a couple of vertically oblique divisions that demarcate a two-side apical cell, whose divisions may lead to a biseriate column that will mostly compose the stalk of the archegonium. Unlike in liverworts and hornworts, the upper cell will then be reshaped into a tetrahedral apical cell with three cutting faces. Among the Bryophyta, only *Sphagnum* lacks this newly derived apical cell (Ruhland 1924). The derivatives of the new apical cell form the venter. The apical cell then divides transversely to yield a cover cell and a central cell, the latter undergoing a series of transverse divisions giving rise to the canal cells and the egg. Further longitudinal growth involves divisions of existing cells and derivatives from the apical cell with now four cutting faces (three obliquely lateral ones and one facing downward), although in some species this cell ceases to divide soon after the development of the central initials.

Six cells compose the circumference of the neck. The number of cells composing the vertical axis of the canal varies among, but seemingly not within species. Fertilization is made possible following the disintegration of the neck canal cells and opening of the distal end of the neck. According to Zielinski (1909), the "dehiscence" of the archegonium is due to the swelling of the mucilage in the apical cells and not in the canal. The mucilage accumulating in the distal cells swells upon hydration, resulting in an increase of the internal pressure. This pressure acts on the superficial cuticle, which breaks and rolls back, taking with it the adhering cell walls.

Male gametangia are typically formed at the apex of a module and in that sense all mosses are also acrandrous. The first and in some cases the sole antheridium is always derived from the actual apical cell. Consequently, a module developing antheridia will cease to grow. Additional antheridia are formed from segments below. The next antheridium occupies the position of a leaf (Ruhland 1924). Additional antheridia can be developed from either basal cells of the primary antheridium, or epidermal cells of the segments; whether this ambiguity has been resolved since Ruhland (1924) is not clear. In *Polytrichum piliferum*, antheridia are developed from cells below the leaf initial (Frey 1970). Similarly, in *Sphagnum*, antheridia occur singly below a leaf, and hence seem to be derived from the branch-initial (Leitgeb 1882). *Polytrichum* modules retain their vegetative apical cell, which enables them to resume growth following sexual maturation. Here the antheridia occur in terminal splash-cups.

At maturity the antheridium is typically elongate-cylindrical, but rarely subspherical as in *Sphagnum* and *Buxbaumia*. Their development in bryophytes is described and contrasted to that of other land plants by Renzaglia & Garbary (2001). The antheridium consists of a stalked spermatogenous cylinder. Their development involves at first a two-sided apical cell that forms a short biseriate nascent antheridium. A set of two oblique divisions cuts each upper cell into three derivatives, and each of the central cells undergoes one additional division. At this stage, the apex of the developing antheridium is composed of four inner cells, and four outer cells. The former will compose the spermatogenous tissue and the latter, the protective jacket. The inner cell of each segment undergoes a series of divisions to yield a vast number of sperm mother cells. With the final division a pair of spermatids are formed. A spectacular transformation (Renzaglia & Garbary 2001) results in two sperm cells each surrounded by a thick wall and bathed in a medium rich in lipid droplets. The lipid, which is lacking in *Sphagnum*, may be essential to the dispersal of the sperm, favoring their spreading on the surface of the water by lowering the surface tension (Muggoch & Walton 1942). Sperm carry the bare necessities in terms of organelles (e.g., one large mitochondrion and one plastid) and cytoplasm (Miller & Duckett 1985). At syngamy only the nucleus is transferred to the egg, and organelles are only maternally inherited by the zygote (Natcheva & Cronberg 2007).

The antheridial jacket remains unistratose, and grows primarily through cell elongation. Bryophyte antheridia differ from those of liverworts and hornworts in the mode of dehiscence that involves specialized opercular cells (Renzaglia & Garbary 2001). At maturity the antheridium is more than half filled by a pressurized fluid, and hence the sperm mass only occupies a fraction of the inner volume (Paolillo 1975, Hausmann & Paolillo 1977). At the tip thick-walled cells compose an operculum (Goebel 1898b) which ruptures from the jacket below as the internal pressure increases. According to Ruhland (1924), the cuticle covering the antheridium prevents the tear from spreading, and thus insures that the dehiscence is narrow. *Sphagnum* lacks differentiated opercular cells, and its antheridia dehisce "by irregular valvelike tears rolling down from the apex" (Crum 2001).

Paraphyses are likely homologues of axillary hairs. They are typically uniseriate, chlorophyllose at first, typically becoming hyaline or brownish at maturity. Paraphyses are lacking in all peatmosses, and in some species they are absent from either the perichaetium or the perigonium. The role of the perigonial paraphysis is not fully elucidated. Beside their potential role in protecting the developing gametangia from dehydration either by their globose terminal cell sheltering the antheridia, or through mucilage production (which seems to have been shown only in *Diphyscium*), Goebel (1898b) suggested that paraphyses filled with water may, when in dense formation in the perigonium, favor the expulsion of the sperm mass by exerting pressure on the mature antheridium. Finally, these filaments may be involved in attracting vectors recruited to disperse sperm (Harvey-Gibson & Miller-Brown 1927), but such a hypothesis remains to be tested, although it has recently been demonstrated that microarthropods can mediate sperm dispersal (Cronberg *et al.* 2006). In the absence of a biotic vector, sperm of most mosses are dispersed by water and in a most spectacular fashion in antheridia arranged in splash-cups (Andersson 2002).

2.6 Asexual reproduction

Sexual reproduction in mosses as in other bryophytes is strictly dependent on the availability of water, for the motile sperm to reach the apex of the archegonial neck and then move down the neck to fertilize the egg. Even following syngamy, successful reproduction may be compromised by sporophyte abortion due to desiccation stress (Stark 2001). Given this constraint, it is not surprising that mosses developed means of reproduction independent of water availability. Reproduction without sex is known from most if not all families of mosses (Correns 1899), and various mosses engage in both forms of reproduction. Individual plant fragments offer a common means of reproduction, but leaf fragments alone rarely regenerate a plant (Correns 1899). Despite their delicate nature, leafy shoots remain viable even after passing through the digestive tracts of bats (Parson *et al.* 2007). With a few exceptions, the specialized diaspores are formed exogenously, on stems, leaves, rhizoids and protonemata (Imura & Iwatsuki 1990). They vary greatly in shape and size, germination type and even mode of abscission, and two or three diaspore types may be produced by a single species and even a single specimen, at least in culture (Correns 1899, Duckett & Ligrone 1992). Newton & Mishler (1994) recognized 15 groups of diaspores, from protonematal gemmae, modified shoots and leaves, to cauline and foliar gemmae. Laaka-Lindberg *et al.* (2003) followed a similar classification, but recognized only 13 classes. The main inconsistency among classifications relates to the usage of the terms asexual versus vegetative, and propagule versus gemmae. Newton & Mishler (1994) reserved the term "asexual" in a strict sense to spores produced via selfing or sex reproduction involving two clones or siblings. All other diaspores are referred to as vegetative. They further follow Imura & Iwatsuki (1990) in defining propagules by the presence of an apical cell from which the new gametophore will arise, but restrict the term gemmae to small structures derived from a secondary protonema. Laaka-Lindberg *et al.* (2003), however, considered only those diaspores such as gemmae, whose germination will recapitulate the ontogeny of the whole plant as truly asexual. Consequently, propagules, which are defined by the presence of an apical cell, are treated as vegetative diaspores. Leaves and rhizoids may occur on propagules but are lacking in asexual diaspores. Whereas Imura & Iwatsuki (1990) emphasized that germination of gemmae always leads to a protonemal phase before a gametophore is formed, Laaka-Lindberg *et al.* (2003) saw this more as a trend rather than as a diagnostic feature. In fact, Duckett & Ligrone (1992) argued that the vast majority of diaspores produce filaments first. These strands emanate from specific cells that can be recognized prior to germination (Correns 1899). The liberation mechanisms for moss diaspores are similar to those described for other bryophytes or even sporic vascular plants, except in some taxa, where breakage occurs through a differentiated abscission or tmema cell, unlike in any other group of land plants (Duckett & Ligrone 1992, Ligrone *et al.* 1996).

2.7 Components of the sporophyte

With more than one archegonium maturing within a single perichaetium, multiple fertilization events are likely and may even be common, unless unfertilized eggs and their archegonium abort rapidly. Should syngamy have occurred in two or more archegonia, typically only one embryo will pursue its development. Polysety is indeed rare but has evolved independently in various lineages.

2.7.1 *Early embryogenesis*

Upon sexual reproduction a zygote, the first diploid cell, is formed, marking the beginning of the sporophytic phase. The zygote always undergoes a transverse division. The lower or hypobasal cell never develops into a conspicuous part of the mature sporophyte and carries no significance for the development of the mature sporophyte (Roth 1969). The epibasal cell divides transversely again, forming a small uniseriate filament. Soon an apical cell with two cutting faces is differentiated. The derivative cells form two lines, and each cell will develop into a segment. About 20 segments are formed by the time the apical region ceases to grow. Each segment undergoes further divisions, in all three planes, and the embryo soon has a three-dimensional architecture (Wenderoth 1931). The frequency of divisions in each segment decreases gradually downward and rather abruptly upward. The transition between the more or less median segments and the upper ones marks the location of the new meristem. The segments above this region form the capsule. Those below will develop into the seta and the foot. The seta meristem contributes cells only acropetally, and hence only to the seta. Apical growth thus ceases early and when the embryo is less than a millimeter long the presumptive tissues yielding the capsule and the seta are in place. All subsequent growth of the sporophyte is thus of an intercalary nature, by means of a seta meristem (Roth 1969, French & Paolillo 1975a). The Sphagnopsida and the Andreaeopsida lack a seta meristem and hence lack a seta altogether. Ligrone & Duckett (1998) viewed the intercalary meristems as a residual primary meristem. The ontogenetic origin of the seta from an intercalary meristem rather than an apical cell led Kato & Akiyama (2005) to question the homology of the seta and the branched sporophytic axes of polysporangiophytes. They preferred to view the moss sporophyte as a sporogonium, homologous to a vascular plant sporangium. Under this hypothesis the seta would be homologous to a sporangial stalk. Polysporangiophytes would have acquired the ability to branch by delaying the "inception of the sporangial phase" through interpolation of a more complex vegetative phase. As a consequence the intercalary meristem would likely be an autapomorphy for the Bryophyta or a fraction thereof.

2.7.2 *The sporophytic placenta*

The base of the sporophyte lacks an apical meristem, and hence the foot is formed by segments derived from the intercalary meristem. The foot is broadly defined as the portion of the sporophyte that is enveloped by the gametophyte, and thus includes the portion of the seta sheathed by the vaginula and the base of the sporophyte that is anchored in the actual cauline tissue (Roth 1969). The basalmost portion of the foot is typically elongate and tapered, except in *Sphagnum* and *Andreaea* that have a bulbous or short conical foot, respectively. In many lineages conspicuous cytological modifications insure the stability of the gametophyte/sporophyte junction and favor transfer of nutrients and water from the haploid to the diploid generation (Roth 1969, Ligrone *et al.* 1993).

Matrotrophy characterizes all embryophytes, and takes place at the junction of the maternal tissues and the developing embryo (see also features of the vaginula below). The placenta is composed of tissues of both generations. Ultrastructural and cytological modifications on the gametophyte-side (see above) are always matched by similar changes in adjacent sporophytic cells, but in many cases the differentiation is unbalanced and conspicuous only on the sporophytic side of the junction (Ligrone *et al.* 1993). Only in *Sphagnum* is specialization absent on both sides of the junction. In *Takakia*, *Andreaea*, *Andreaeobryum*, and all other mosses studied, transfer cells occur in the sporophyte. The morphology of the ingrowth varies among lineages (Ligrone *et al.* 1993). In *Diphyscium* and *Bryum*, cells of the sporophytes penetrate the gametophytic tissue in a manner that resembles a haustorium. Roth (1969) argued against the ingrowth of the placental cells being essential to the transfer of organic nutrients to the developing embryo since, in all cases, their development continues at a time when the capsule begins enlarging, and when the sporophyte builds its own assimilative capabilities. Although sugars are transferred from the gametophyte to the sporophyte (Renault *et al.* 1992), sporophytes past a certain age are able to complete their development autonomously (Haberlandt 1886, Bopp 1954). Roth (1969) considered that the differentiation of the foot cells (and those of the vaginula) fulfill primarily mechanical functions, strengthening the connection between the sporophyte and gametophyte, even with regard to environmental stresses such as periodic drought. Certainly in the case of *Diphyscium*, with its massive sporophyte inserted by a short stalk on a short female plant, such a function would seem essential.

2.7.3 *Protection of the developing embryo*

Following sexual reproduction the apex of the gametophyte undergoes a series of transformations designed to nurture and protect the embryo. The urn enclosing the developing embryo develops from tissues of the archegonium, the receptacle and even the perichaetial leaves, with contributions varying among lineages (Roth 1969). The development of the epigonium (i.e., post-fertilization archegonium) parallels that of the embryo, and in most mosses ruptures well before sporogenesis into a calyptra (Fig. 2.1C, D) and a vaginula. The calyptra is derived basipetally (i.e. with the youngest tissues at the base) nearly exclusively from the archegonial stalk. The calyptra of *Andreaea* is composed of a membranous body topped by an archegonial neck. By contrast, in *Andreaeobryum* the calyptra is multistratose and massive, and persistent to maturity (Murray 1988). The multistratose epigonium of *Sphagnum* degenerates rapidly to a delicate unistratose membrane.

The role of the calyptra continues to elude bryologists. Although widely considered as essential for the normal development of the capsule, the mechanisms of control are ambiguous. Developmental phenotypes of the sporophyte obtained after removing the calyptrae include: swelling of the seta (Bopp 1956); erect and actinomorphic capsule, and stronger and faster negative geotropic response in *Funaria hygrometrica* (Herzfelder 1923); decreased spore development and viability based on isolated capsules of the same species (Bopp 1954); and acceleration of capsule swelling in the Polytrichaceae (Bopp 1956). French & Paolillo (1975b) showed that the effect of calyptra removal decreased with the age of the sporophyte: the development was more dramatically altered when the calyptra was removed from young vs. old sporophytes. Similar phenotypes were described from mutants of *Funaria hygrometrica* and *Physcomitrium pyriforme* (Oehlkers & Bopp 1957). Apogamous sporophytes (see below) grow without a calyptra and always exhibit a deviant morphology, which may further indicate the critical role of the calyptra in the development of the sporangium. Bopp (1961) summarized the role of the calyptra as follows: the calyptra inhibits the development of the capsule (i.e. its swelling) until the internal differentiation within the sporophyte apex is complete. The premature loss of the calyptra inevitably results in abnormal, incomplete and non-functional capsules. The nature of the underlying mechanisms is still poorly if at all understood. Calyptrae boiled in solvents and refitted onto the developing sporophytes retain their ability to dictate capsule ontogeny, suggesting that the control is not hormonal but rather mechanical in nature (Bopp 1961).

At maturity the calyptrae, and to a lesser extent vaginulae, display a broad spectrum of morphologies (Janzen 1917). In some mosses, the calyptra clasps around the mature sporangium and remains attached to the sporophyte (e.g. *Pyramidula*), although even in this case tearing occurs to allow for spores to be dispersed. Unless the calyptra covers only the operculum (e.g. *Physcomitrella*) it must be shed along with or before sporangial dehiscence. Mosses lack active mechanisms to free themselves of calyptrae, and can only facilitate the removal by wind by loosening the fit of the calyptra on the urn. Mitrate calyptrae sit like a cap on the capsule. In *Orthotrichum*, the rostrum of the operculum elevates the calyptra, which is then easily blown off. A cucullate calyptra is characterized by a single long slit extending nearly to the apex. Its loss is facilitated either by the asymmetric growth of the capsule (e.g. *Funaria*) or by an oblique rostrum (e.g. *Zygodon*). The base of the calyptra is entire, broadly (e.g. *Schlotheimia*) or deeply lobed (e.g. *Macromitrium* sp.), or fringed (e.g. *Daltonia*). The surface is commonly smooth, but ridges or pleats (e.g. *Orthotrichum*), papillae (e.g. *Leratia*) or hairs (e.g. *Racopilum* sp.) occur in various lineages.

The hairiness of the vaginula often matches that of the calyptra (e.g. *Zygodon*, Malta 1926). In other aspects, the vaginula is morphologically fairly uniform, except for its size, across mosses, although a systematic survey is lacking. The inner layer of the vaginula may also exhibit wall ingrowths similar to those seen in the placenta, except that they are developed even later with respect to the ontogeny of the sporophyte (Roth 1969). The vaginula forms a tight but still independent cylinder around the sporophyte (the upper portion of the foot). Its function is likely structural, by solidifying the anchor of the sporophyte.

2.7.4 *Architecture of the mature sporophyte*

The function of the sporophyte is to produce and disperse spores. The sporangial tissue is enclosed in an urn that can be elevated onto a stalk, or seta (Fig. 2.3). Although some species mature their sporangia among the perichaetial leaves, favoring establishment of offspring *in situ*, most mosses raise the spore-bearing capsules above the vegetative leaves. Like the vegetative axes, the seta is composed of an epidermis, a cortex and a central strand of conducting cells, except if the seta is highly reduced in size. The epidermal and sometimes the cortical cells are pigmented and the color of the seta varies from yellow, bright red, brown to rarely black. Stomata are always lacking in the seta. The surface is rarely roughened by projecting cell ends (e.g. *Brachythecium* spp.), cilia (e.g. *Calyptrochaeta*) or warts (e.g. Buxbaumiaceae). The seta is often strongly twisted upward either clock- or counterclockwise (dextrorse or sinistrorse, respectively). In some species of *Campylopus*, the seta is cygneous or flexuose when dry and unwinds upon moistening (Frahm & Frey 1987). In *Rhachitheciopsis tisserantii* the seta is curved downward when moist, hiding the capsule among the vegetative leaves and elevates the capsules when dry (Goffinet 1997a). Rapid hygroscopic movements of the seta most likely promote the dispersal of spores. They are accounted for in both cases by asymmetrically thickened cortical cells of the seta.

Fig. 2.3. *Funaria hygrometrica.* This typical acrocarpous moss is characterized by a ruderal habitat, an annual life cycle, small gametophytes and an asymmetric capsule borne on a long seta. Note the line on the capsule, marking the line of dehiscence. The sporogenous mass occupies the upper half of the capsule.

An axial strand of hydroids is typically present in the seta of mosses, including in species lacking such conducting cells in the gametophyte (e.g. *Orthotrichum* spp.). The surface of the seta is covered in a waxy cuticle and lacks appendages essential for external water conduction (Ligrone *et al.* 2000). The anatomical complexity of the seta parallels that of the stem in the Polytrichaceae, which exhibit the most highly developed strands of hydroids surrounded by leptoids (Hébant 1977). In other mosses, organic compounds are transferred between specialized parenchyma cells (Ligrone *et al.* 2000), even if such cells are lacking in the gametophyte (e.g. *Funaria*; Hébant 1977). A hydroid strand may be lacking in taxa with immersed capsules (e.g. *Stoneobryum*; Goffinet 1997b), in which capillary forces may suffice to supply the capsule with water. When present, the hydrome rarely extends far into the central axis (i.e. the columella) of the capsule (Hébant 1977).

Bryophytes are characterized by a single sporangium born at the apex of an unbranched axis. Abnormal sporophytes bearing two sporangia have been reported for various mosses (e.g., Győrffy 1929, 1934), including *Sphagnum* (Győrffy 1931), and Leitgeb (1876) even considered such "doublefruits" not to be that uncommon. In most cases the branching occurs distally so that the two capsules are close to one another, even sheltered under a single calyptra. The sister capsules differ in shape and size at maturity, but both produce viable spores. Considering that the seta is developed by the intercalary meristem and that the tissues of the capsule are derived from the last set of divisions of the apical cell, the ultimate development of two capsules must be initiated at the earliest stages of embryogenesis (Leitgeb 1876). Alternatively, a dual fertilization, involving two eggs is also possible, although less likely considering that the anatomy of the seta suggests that a single axis with one axial hydrome of normal size is formed. Lal (1984) reported one or more capsules budding off laterally from parthenogenic sporophytes (i.e., developed from unfertilized eggs) in species of *Physcomitrium*. Similarly Tanahashi *et al.* (2005) observed occasionally branched sporophytes in cultures of mutant *Physcomitrella*; whether these are phenotypic expressions of the mutations or result from parthenogenesis is not clear, although the authors favored the latter explanation. If parthenogenesis is indeed demonstrated, and not merely hypothesized because cultures were not flooded to allow for fertilization, it may be an explanation for other occurrences of polysporangy in mosses. For example, in *Sphagnum* karyogamy between egg and ventral canal cells may result in a diploid cell that could develop into a sporophyte (Crum 2001). It is possible that in other mosses, too, the ventral canal cell fails to disintegrate (Ruhland 1924) and is involved in sporophyte formation.

Maturation of the sporophyte from fertilization to sporogenesis is likely a continuous process even if it is slowed down by environmental factors such as low temperatures (Greene 1960, Stark 2002). The phase of elongation of the sporophyte (the spear stage) culminates in the swelling of the capsule, which is composed of a sterile neck, the urn, and the operculum (see Fig. 2.3). The basal sterile tissues may be well developed and result in a distinct region below the urn tapered to the seta, or abruptly constricted to it (Győrffy 1917). In various Dicranales it forms a goiter-like protuberance, a struma. Several entomophilous Splachnaceae develop an inflated and brightly colored hypo- or apophysis designed to aid in attracting insects recruited for spore dispersal.

The spectrum of variation in capsule shape is seemingly endless. *Sphagnum*, *Physcomitrella*, and *Pleurophascum*, for example, have spherical capsules, whereas most other mosses have rather elongate and sometimes cylindrical ones, as in *Encalypta*. The capsule of *Buxbaumia* is conspicuously bilaterally symmetric, a feature shared with various taxa of the Funariaceae and Dicranaceae with curved capsules. Erect and radially symmetric capsules are common among epiphytic or saxicolous mosses, but rare among terricolous mosses, and the transformation between curved to erect capsules seems correlated with a shift to epiphytism in various lineages of the Hypnales (Buck *et al.* 2000). In the Mniaceae and many Bryaceae the radially symmetric capsule hangs from a strongly curved seta.

The capsule wall is the only tissue in mosses that may contain stomata, typically less than fifteen, rarely more than 100 (Paton & Pearce 1957). Heavily cuticularized guard cells typically occur in pairs (rarely four) but in the Funariaceae the cytokinesis is incomplete and the new wall only partially divides the two cells; the single guard cell is shaped like a tire inner tube with a central stoma (Sack & Paolillo 1983). The shape of the stoma is either round or elongate, with the former being more common. The shape may be correlated to the thickness of guard cell wall (Paton & Pearce 1957). The phylogenetic significance of the characteristics of the stomata is ambiguous (Hedenäs 1989b). The distribution of the stomata over the capsule surface is not random: often restricted to the sterile base, they are formed only rarely in the distal portion of the urn, and never in the operculum. The guard cells are generally exposed on the surface of the capsule (phaneroporous stomata) but in various lineages the guard cells are sunken below the surface and may even be overarched by subsidiary cells, creating a suprastomatal chamber. Cryptoporous stomata are homoplasious. Although these could likely reduce rates of transpiration, immersed stomata are not restricted to or common among xerophytic mosses.

In *Sphagnum* the stoma-like structures are not involved in gas exchange, but may be essential to the dehiscence of the capsule and dispersal of the spores (see Section 2.9 on spore dispersal), and hence are referred to as pseudostomata (Boudier 1988). Even in other mosses, the function of stomata is not clearly understood (Paton & Pearce 1957) although they behave like those of vascular plants, at least in young capsules (Garner & Paolillo 1973). In capsules nearing maturity, the stomata tend to remain open, possibly to favor dehydration of the tissues surrounding the spores, to prevent premature germination but also to allow for hygroscopic movements of the capsule (e.g., shrinkage of the wall) to favor spore dispersal. Although stomata and pseudostomata accomplish distinct functions it is not clear whether the latter are derived or not from the former. Given the absence of stomata in *Takakia*, *Andreaea*, and *Andreaeobryum* it is possible that stomata are not a defining feature of the Bryophyta (Cox *et al.* 2004), and hence they may not be considered homologous to those of hornworts and polysporangiophytes (as considered by Mishler & Churchill 1984).

A prerequisite to stomatal function is the presence of a cuticle and intercellular gas spaces (Raven 2002). The latter results in a dramatic increase in the surface area through which gases can be exchanged by the photosynthetic cells in multistratose tissues. A spongy tissue occurs in the capsule of all mosses bearing stomata (Crosby 1980), except maybe *Tetrodontium*, and is particularly conspicuous in *Buxbaumia*, where it lines much of the sporogenous tissue below the capsule wall (Fig. 769 in Brotherus 1924). The assimilative tissue either surrounds the sporogenous tissue in species lacking a distinct neck (e.g. *Buxbaumia*), extends into the sterile base (e.g. *Funaria*), or is restricted to the apophysis (e.g. *Splachnum*). A lacunose tissue is lacking in all lineages preceding *Oedipodium*, a pattern congruent with the view that air spaces are essential for stomata, which are lacking in all early lineages of the Bryophyta.

Dimorphism among exothecial cells is not restricted to the guard cells. In many species, the wall of the urn shrinks or contracts upon drying. As in all other hygroscopic movements in mosses, changes in capsule shape with atmospheric moisture are determined by patterns of cell wall thickness. In the Orthotrichaceae, the capsule is often ribbed when dry. The cells between the ribs are thin-walled and collapse when losing turgor. Those marking the ridges have thick walls and remain firm, even in the lack of cellular water. The constriction may affect only the upper portion of the capsule, thereby narrowing the opening, potentially regulating spore dispersal. In entomophilous Splachnaceae, the exothecial cells may have thinner longitudinal walls, and hence shrink vertically when losing water. Here the effect may be to push the spore mass closer to the capsule mouth to favor contact between insects and the sticky spores.

The axis of the capsule of mosses, unlike that of liverworts, is occupied at least partially by a columella; the only exception is *Archidium* (Snider 1975). In *Takakia*, *Sphagnum*, *Andreaea* and *Andreaeobryum*, the columella is dome-shaped and hence overarched by the sporogenous tissue. In all other mosses, the columella extends beyond the spore sac, in some cases remaining attached to the operculum upon dehiscence (systylious). The form of the columella varies among taxa and may be of taxonomic value. In all Bryopsida, the columella is of endothecial origin (Crum 2001). The spore sac is either of endothecial or amphithecial (Sphagnopsida only) origin.

Dehiscence of the capsule (see below) involves in most species of the Superclass V of the Bryophyta the loss of a lid or operculum at the apex of the capsule. Only in *Takakia*, *Andreaea*, *Acroschisma*, and *Andreaeobryum* are the lines of dehiscence vertical. In *Takakia* dehiscence begins in the center of the capsule and extends towards the poles in a spiral line (Smith & Davison 1993). *Andreaeobryum* and *Andreaea* (incl. *Acroschisma*) share valvate capsules, but in the former the valves vary in number and are formed irregularly along lines of least resistance, whereas in the latter the four valves are defined by distinct suture lines, composed of thin-walled cells, visible prior to the dehiscence (Murray 1988). The valves of *Andreaea* typically extend for much of the length of the capsule and remain connected at the apex. Only in *Acroschisma wilsonii* is the dehiscence restricted to the apical region, forming 4–8 valves.

The line of dehiscence in operculate mosses is sometimes defined by the presence of an annulus, a ring of cells at the capsule mouth composed of cells from the capsule wall or the underlying tissue. The annulus is often simple and not well differentiated. By contrast, in the Rhachitheciaceae or Funariaceae, the annulus is composed of 2–3 layers of cells and revoluble: it arches outward, unzipping the lid from the urn. Such movement is again indicated by patterns in cell-wall thickenings.

2.7.5 *Sporogenesis and spores*

Sporogenesis occurs relatively late in the development of the sporophyte, after seta elongation, which itself is often delayed by several months following fertilization. Brown & Lemmon (1990) broke down sporogenesis into five major stages: (1) differentiation of the spore mother cells (sporocytes), (2) nuclear divisions of meiosis, (3) cytoplasmic cleavage, (4) formation of the spore wall, and (5) dehydration and accumulation of storage compounds. The development of the sporogenous tissue coincides with the expansion of the capsule. Meiotic divisions within the sporangium are fairly synchronized. Each sporocyte contains, unlike other cells of the sporophyte or gametophyte, a single chloroplast that undergoes two consecutive divisions to yield four plastids equally distributed among the newly formed four lobes of the cytoplasm with each destined to belong to one of the future spores in the tetrad. The plastids are thus located at the poles of a tetrahedron, and are connected to one another by microtubules forming the quadripolar microtubule system that will later form the spindles essential to nuclear division.

The spores of extant bryophytes are always produced in tetrads, whereby every spore is in contact with the other three products of meiosis. However, a distinct trilete mark on the proximal pole is typically lacking. Patterns in spore wall development fall within three broad categories, according to the architecture of the wall: penta lamellate in *Sphagnum*, spongy exine and no tripartite lamellae in *Andreaea* and *Andreaeobryum*, and tripartite lamellate in the remainder of mosses, with a perine, a median exine, and an inner intine (Brown & Lemmon 1990). The exine is composed of sporopollenin or a sporopollenin-like compound that confers to it its highly mechanical and likely also physiological resistance to degradation and desiccation, as in other land plants. By contrast, polysaccharides compose the intine, which is laid down last by the spore. The intine of most mosses is thickened in one particular area where it faces a thinner exine. This area, called the aperture or leptoma, likely offers less resistance for the germ tube to emerge. The outermost layer, or perine, is contributed by the sporophytic tissues or their breakdown, rather than by the spores themselves. This layer, composed of pectin and callose-like compounds like the intine, contributes also to the resistance of the spore.

Since bryophytes are characterized by a single terminal sporangium, all spores are thus produced by the same sporogenous tissue, and hence cases of heterospory are by definition impossible. The sporangium yields a mass of spores whose size varies around single mean in most species (isospory). In some cases, a bimodal distribution of spore size reveals the presence of aborted spores (Mogensen 1978, 1983). Sexual dimorphism is rarely apparent at the spore stage. In many phyllodioicous species, each tetrad holds two large and two small spores. Upon germination the small spores yield small male gametophytes. The larger spores are assumed to develop into female gametophytes, although empirical observations are still lacking. Such production of two size classes of spores within a single sporangium is termed anisospory (Vitt 1968). Anisospory always leads to profound sexual dimorphism with males being dwarf; however, the reverse is not true. Hence many more mosses exhibit dwarf males than anisospory. Furthermore, not all mosses that develop dwarf males lack the ability to form regular-sized male plants. Une (1985) thus distinguished physiologically from genetically determined male dwarfism.

The number of sporocytes and hence ultimately the number of spores varies among species. Some species of *Archidium* may produce only four spores per indehiscent sporangium, several million spores are formed in each capsule of *Dawsonia lativaginata* (Crum 2001).

2.8 Fundamental peristome types

Early in their evolutionary history, mosses acquired a peristome, a set of teeth forming one or two rings lining the mouth of the sporangium. The peristome arises in virtually all cases from cells of the amphithecium. The peristome of the Polytrichopsida and the Tetraphidopsida is composed of teeth that are bundles of whole, thick-walled cells, hence the name nematodontous for the thread-like appearance of the teeth. As is revealed by its name, the urn of *Tetraphis* bears four teeth, which are massive and erect (Shaw & Anderson 1988). Most Polytrichaceae have 32 or 64 short teeth, protruding only slightly above the rim of the capsule. All other peristomate mosses (i.e. Bryopsida) have teeth composed solely of cell plates or cell wall remnants. Because the vertical plates within a column are jointed, permitting the tooth to bend in or outward, the peristome is called arthrodontous. Several architectural types of arthrodontous peristome can be recognized based on ontogenetic and morphological features: the *Timmia*-, the *Funaria*-, the *Dicranum*- and the *Bryum*-type peristome (Budke *et al.* 2007).

Three layers of the amphithecium contribute cells to the peristome: the inner, primary and outer peristomial layers, respectively referred to as the IPL, PPL and OPL (Fig. 2.4B). All peristomes share a developmental sequence that

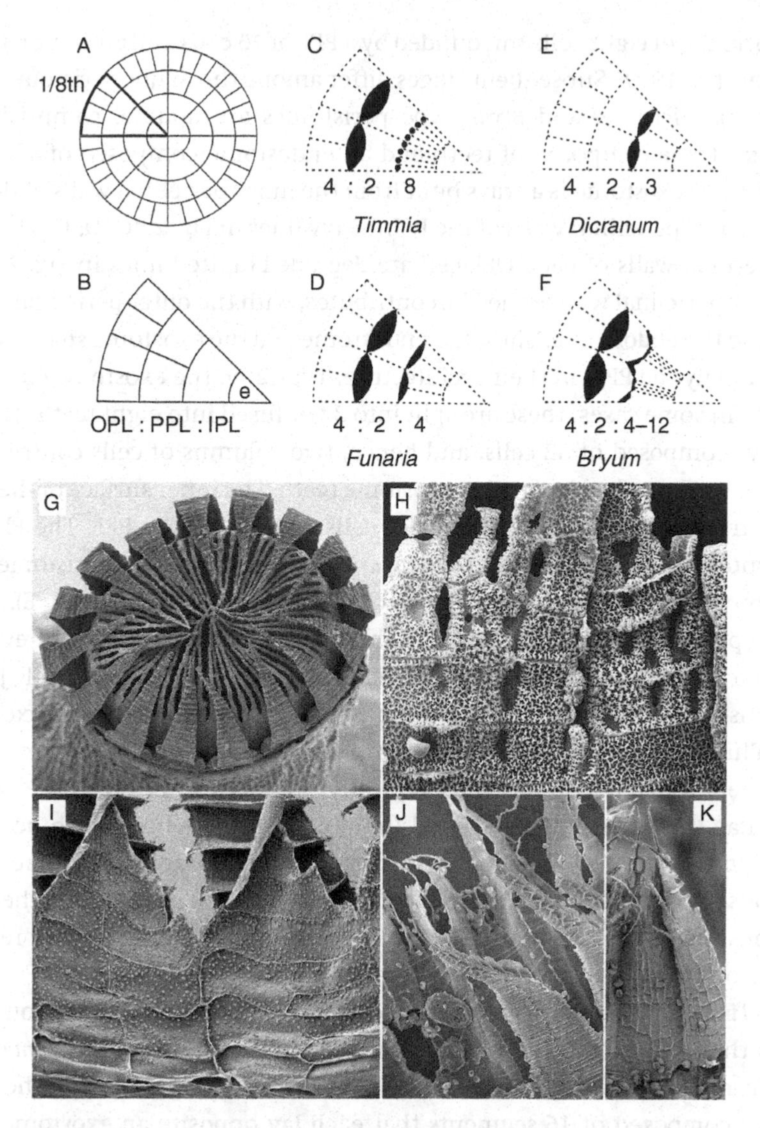

Fig. 2.4. Peristome architecture in mosses. (A) Diagram of a transverse section through the putative peristome forming region at the apex of an immature moss sporophyte. (B) Detail of $\frac{1}{8}$ of the section in (A), showing the endothecium (e) and the three innermost amphithecial layers that contribute to peristome formation: outer (OPL), primary (PPL) and inner (IPL) peristomial layer. (C–F) Diagram of $\frac{1}{8}$ of a *Timmia*-, *Funaria*-, *Dicranum*- and *Bryum*-type peristome. Black areas identify thickened cell walls composing the peristomes; dotted lines mark the walls of the IPL, PPL, and OPL cells that are resorbed and hence that are not contributing to the peristome. (A–F) Redrawn from Budke *et al.* (2007). (G) Diplolepideous peristome of *Timmia megapolitana*, showing the 64 filamentous appendages of the endostome. (H) Inner view of the peristome of *Funaria hygrometrica*, showing the four IPL cells composing the two segments, which lie opposite the two exostome teeth. (I) Outer view

leads to an IPL of eight cells surrounded by a PPL of 16 cells and an OPL of 32 cells (Goffinet *et al.* 1999). Subsequent stages differ among the major peristome types. The *Timmia*-, *Funaria*- and *Bryum*-type peristomes are double, comprising an outer exostome composed of teeth and an endostome composed of segments and cilia. The exostome is always built from the inner periclinal walls of the OPL and the outer periclinal walls of the PPL (heavy lines in Fig. 2.4C, D, F). The three other vertical walls of each OPL cell are degraded (dotted lines in Fig. 2.4C–F). The inner periclinal wall of the PPL contributes, with the outer periclinal wall of the IPL, to the endostome. Thus, the endostome and the exostome share one cell layer, namely the PPL, in their architecture. Typically, the exostome comprises 16 teeth. In some cases, these are split into 32 or fused into eight teeth. The OPL is always composed of 32 cells, and hence, two columns of cells contribute to the outer surface of each of the 16 exostome teeth. The inner surface of the tooth is built from one of the 16 columns of cells composing the PPL. The PPL also contributes to the endostome. In the *Funaria*- and *Dicranum*-type peristome, each endostome tooth is composed of one column of PPL cells (Fig. 2.4D, E). In the *Bryum*-type, each endostome segment shows a median vertical wall, revealing that each segment is composed of one half from two PPL cells (Fig. 2.4F, J).

Because two columns of cells compose the outer surface each exostome tooth, Philibert (1884) introduced the term diplolepideous for this architecture. A diplolepideous peristome thus typically comprises an exostome. In the Dicranidae the peristome is reduced to an endostome whose teeth bear only one column of cells on their outer surface: the peristome is said to be haplolepideous. By coincidence, the haplolepideous peristome is single and the diplolepideous one is double. Both peristome architectures can be further reduced and one or both rings of teeth lost completely (Vitt 1981).

The *Timmia*-, *Funaria*- and *Bryum*-types represent three diplolepidous peristomes that differ in the architecture of the endostome. The *Dicranum*-type is the sole model of a haplolepideous peristome. In the *Funaria*-type the endostome is composed of 16 segments that each lay opposite an exostome tooth (Fig. 2.4H). The *Bryum*-endostome is characterized by a basal membrane mounted with 16 segments alternate to exostome teeth and separated by

---

Caption for Fig. 2.4. (cont.)

of the peristome of *Tortula plinthobia*, each tooth is fenestrate along the vertical walls of the IPL, and hence one and half cells of the IPL face each PPL cell (outer cells in view here). (J) Diplolepideous peristome of *Pseudoscleropodium purum*, showing the keeled endostome segments alternating with the exostome teeth, and the cilia between two consecutive segments. (K) Inner view of the diplolepideous peristome of *Mnium thomsonii*, showing the numerous cells composing the IPL.

small, slender appendages called cilia (Fig. 2.4J). The cilia are thus opposite the teeth. This peristome is also referred to as diplolepideous alternate, in contrast to the diplolepideous opposite of *Funaria*. In *Timmia* the membrane bears only 64 filiform appendages similar to cilia (Fig. 2.4C, G). Their homology to the endostome segments of other diplolepideous mosses is ambiguous (Cox *et al.* 2004).

The conspicuous morphological differences between the main peristome types are paralleled by developmental divergences. In *Funaria* each of the eight IPL cells undergoes a set of three symmetric divisions that yield four identical cells for every two PPL cells (Fig. 2.4H; Schwartz 1994). In *Timmia*, the IPL cells undergo one additional round of symmetric divisions, leading to eight cells per eighth of the peristome (Budke *et al.* 2007). The first division in each of the eight IPL cells of the remaining diplolepideous mosses is strongly asymmetric (Shaw *et al.* 1989a). The number of subsequent divisions varies and yields between four and 12 cells per pair of PPL cells (Fig. 2.4K). Haplolepideous mosses, too, are characterized by an asymmetric division. Here it is followed by a single division, hence only three cells compose the IPL adjacent to two PPL cells (Shaw *et al.* 1989a,b). Thus, each segment of *Dicranum* bears one column of cells on the outer surface and one and one half columns on the inner surface (Fig. 2.4I). Patterns in cell division may be inferred from the arrangement of anticlinal walls in mature peristome. However, lateral displacements of IPL walls during the development may mislead assessments of symmetry of the division at maturity. Similarly, amphithecial cells immediately below the narrow presumptive peristome-forming zone lack the constraints that dictate the patterns of cell division in the IPL, and great care must be taken in ontogenetic studies to identify homologous layers (Budke *et al.* 2007).

These architectures and ontogenies characterize typical peristomes in the Timmiidae, Funariidae, Dicranidae, and Bryidae, but are by no means shared by all their species. For example, the peristome of the Orthotrichaceae is diplolepideous, with alternate segments but lacking cilia (Goffinet *et al.* 1999). The Encalyptaceae share with the Funariaceae a symmetric division, but at maturity their highly divided peristome is morphologically unique (Horton 1982, Vitt 1984). In the Rhachitheciaceae (Dicranidae), the endostome is composed of eight segments each built from a single column of PPL cells (Goffinet 1997a). In *Mittenia*, another haplolepideous moss, the peristome is derived from one endostomial layer (Shaw 1985). *Bruchia flexuosa* belongs to the Dicranales but its development prematurely stops, and the last divisions expected in the IPL are lacking (Shaw *et al.* 2000).

2.9 Spore dispersal

The dehiscence of the sporangium (stegocarpy) exposes the spore mass and allows for the dispersal of the spores. In the Sphagnopsida the line of dehiscence is horizontal, and the release of spores is explosive. The mechanism relies on the presence of pseudostomata along the equatorial line. Pseudostomata are essentially cavities resulting from the collapse of the outer periclinal wall (Boudier 1988). The base of the depression is formed by thick cell walls. Upon dehydration the neighboring exothecial walls collapse toward the floor of the cavity pushing it inward. As the mature capsule emerges from the calyptra and is exposed to drying winds, the cells lose water; the capsule constricts along the equator. At the same time the columella degenerates and is replaced by gases, which are compressed by the shrinking capsule. The internal pressure and thus the tension on the exothecial cell walls increases. The thin walls of the subapical cells tear and the operculum is projected at once, and a cloud of spores is released (Ingold 1965). In *Andreaea*, the capsule dehisces along vertical lines but the valves remain connected at their apex. As the atmospheric moisture decreases, the exothecial cells lose their water content and the valves arch outward, thereby exposing the spores. As the humidity increases, the cells swell and a reverse movement occurs. This closes the sporangium and protects the spore mass from water, which would trigger the premature germination of spores, but also agglutinate the spores and inhibit their effective dispersal by wind. In *Takakia*, a similar movement of the capsule wall is likely to control the release of spores. In all other mosses, except of course for the indehiscent or cleistocarpous ones, the line of dehiscence is equatorial or (typically) subapical, and the mouth of the capsule is typically lined with peristome teeth.

In nematodontous and arthrodontous mosses, the peristome may control the release of spores (Ingold 1959). In the Polytrichopsida, the peristome teeth are short. The mouth of the capsule is closed by a thin membrane, the epiphragm, that expands from the apex of the columella to the inner surface of the teeth. Between the teeth, the epiphragm is free and hence small holes persist. The capsule resembles a salt-shaker, as the small spores are released through the marginal pores between the teeth. *Dawsonia* lacks an epiphragm, but the teeth are long and twisted, forming a mesh-like tissue over the capsule mouth. The massive teeth of *Tetraphis* move only slightly as the ambient humidity changes, but sufficiently so to open tiny gaps between them for the spores to escape when the air dries out.

The movement of teeth is most spectacular in arthrodontous mosses. Here the teeth may bend from an overarching position all the way back whereby the teeth are recurved over the capsule wall. Such dramatic movements are made possible by (a) the fundamental architecture of the teeth and (b) much thicker outer versus inner surfaces of the teeth. Species contrasting in their habitat preferences may favor spore release under different conditions. Many terricolous species have peristomes that close the capsule mouth under moist conditions and expose the spore mass when the air is dry (xerocastique peristome). Other mosses, especially epiphytes, favor dispersal under moist conditions (hygrocastique peristome). In some aquatic mosses, such as *Cinclidotus* and *Fontinalis*, the inner peristome forms a trellis or a solid dome that prevents water from entering the capsule and all spores from leaving the urn at once.

Wind is the primary and in the vast majority of mosses the sole dispersal vector. Insects are only recruited as dispersal agents by the Splachnaceae, and only those that grow on substrates of animal origin. Olfactory and visual cues attract flies foraging or looking for these substrates to lay their eggs (Koponen 1990). The chemical adaptation is complemented by morphological innovations such as sticky spores, a pseudocolumella which acts as a piston to elevate the spore mass to the mouth of the capsule or, more strikingly, the expansion of the sterile base of the urn either to amplify the visual cue or to provide a suitable landing platform for insects (Koponen 1990). Recent phylogenetic investigation suggests that entomophily, insect-mediated spore dispersal, arose early in the evolutionary history of the Splachnaceae, and was subsequently lost, maybe due to the severe biotic constraints shaping this system (Goffinet & Buck 2004).

A few mosses lack specific dehiscence mechanisms, and the sporangium remains closed. Such cleistocarpy characterizes various taxa such as *Kleioweisiopsis* and other Ditrichaceae, *Pleurophascum*, *Gigaspermum*, and *Bryobartramia*, among others, but is not known among the Hypnanae (Shaw *et al.* 2000). Spores are dispersed following the disintegration of the capsule wall, either *in situ* if the plant is an annual, or away from the maternal plant if the whole capsule is dispersed. Only for *Voitia* have birds been invoked as a dispersal agent, after ingesting the capsule. Dispersal following trampling of the colonies by herds of caribou is also a possibility. In all cases, cleistocarpy seems to be a result of reduction from peristomate ancestors and in a few cases peristomial fragments remain inside the indehiscent capsule (e.g. *Tetraplodon paradoxus*).

2.10 Early gametophyte development

The vegetative phase of the life cycle begins with the germination of the spore, typically a single cell protected by a wall with a complex ultrastructure and impregnated with sporopollenin, conferring physical and physiological protection as well as resistance against decay on the meiotic product. The aperture is a specialized area in the spore wall through which a germ tube typically emerges (Brown & Lemmon 1981). It is not prominent because it is beneath the exine ornamentation. Germination is fuelled by the combustion of protein, lipid, and starch reserves accumulated in the spore during its maturation in the capsule (Stetler & DeMaggio 1976).

Upon germination, a sporeling is formed. Based on the timing of the first divisions, and the architecture of the sporeling, Nehira (1983) distinguished 14 sporeling types in mosses. The series of studies focusing on protonematal morphogenesis in mosses by Duckett and his coworkers (see references in Duckett *et al.* 2004) sheds further light on the diversity of structures and patterns involved in the earliest stages of vegetative growth. Cell divisions may precede actual germination of the spore. Such endosporic development occurs in several unrelated mosses. A more extreme head-start is provided to the sporeling of *Brachymenium leptophyllum* wherein spores often germinate within the protective confines of the capsule and with protonemata emerging from capsules bent down to the ground (Kürschner 2004). In this case, like in the majority of mosses, the development of the spore is exosporic: the spore germinates and all divisions add cells to the emerging germ tube.

The architecture of the protonema (i.e. the first multicellular stage in the life cycle of most mosses) in exosporic mosses is typically filamentous, with three components (chloronema, caulonema, and rhizoids) that are rather distinct but morphogenetically connected, in the sense that transformations between any two of them are not unidirectional or irreversible (Duckett *et al.* 1998). In exemplars of most major lineages preceding the Bryopsida, a thalloid protonema emerges from the filaments. In *Sphagnum* the rosette-shaped thallus dominates the protonematal stage. Similarly in *Tetraphis* and *Oedipodium*, leaf-like assimilative appendages are developed from the green filaments. At the base of each such thalloid structure in *Tetraphis*, a leafy gametophore arises. In *Andreaea* small appendages unlike any other protonemata (Duckett *et al.* 2004) may form, but are not essential to the development of gametophores (Murray 1988). In *Diphyscium* the protonema comprises filaments, but clavate branches and funnel-shaped structures are also formed, which Duckett (1994b) regarded as caulonematal derivatives. Germination of spores of *Takakia* has not yet succeeded, but vegetative regeneration lacks a filamentous stage, suggesting that spores also do not form a uniseriate protonema. Among the Bryopsida, and also *Buxbaumia*, thalloid appendages are lacking, and the protonemata are typically entirely filamentous in nature. *Buxbaumia* develops a highly reduced protonema composed of rather short rhizoids and an erect chloronema.

The chloronema is always filamentous, composed of green cells, and characterized by intercalary growth. It lacks buds and its function is essentially assimilative in nature. The caulonema, dark-pigmented filaments of cells separated by oblique crosswalls with numerous plasmadesmata, emanates from the chloronema and in most bryopsid mosses develops the actual buds from which shoots will arise. Like rhizoids, caulonematal cells exhibit a cytoplasmic organization reminiscent of leptoids, suggesting that they are the site of cytoplasmic transport, essential to the growing gametophores. The caulonematal stage is lacking in some lineages (e.g. Orthotrichales) where buds are thus formed by the chloronema (Duckett *et al.* 1998). Protonematal cells are generally elongate, rarely short-cylindrical. In *Encalypta*, *Ptychomitrium*, and *Hedwigia* the primary protonemata are composed of globose cells.

The protonema of most mosses is ephemeral, but in *Pogonatum pensylvanicum* and some species (e.g. Ephemeraceae) it is perennial, producing short female shoots every year. In other mosses, the protonema contributes to the longevity of the population by producing asexual propagules. Indeed, much like rhizoids, protonematal filaments can bear gemmae that are filamentous (Duckett & Ligrone 1994), or spherical (Arts 1994), and in some cases even bulbils (i.e. highly undifferentiated shoot with a leaf primordium) may be formed (Mallón *et al.* 2006). Protonematal gemmae occur in an estimated 25% of all mosses (Duckett *et al.* 1998). Regardless of the germination or sporeling type, or the presence of differentiated asexual propagules, protonemata account for significant clonal reproduction since virtually all mosses, including *Sphagnum*, share the ability to form multiple gametophores from a single spore (Crum 2001). Indeed, monogametophytic protonemata characterize only a few species (Duckett *et al.* 1998).

2.11 Apogamy and apospory: a life cycle without sex and meiosis

The gametophyte is by definition the plant that bears the sex organs needed to develop sperms and eggs essential to sexual reproduction and thus sporophyte formation. The function of the sporophyte is to yield spores from which gametophytes can be regenerated. The apparent robust fundamental functional differentiation between the two alternating phases of the life cycle of sexually reproducing bryophytes is compromised by the observations of sporophytes emerging from gametophytic tissues and that of sporophytes "germinating", and protonemata developing leafy stems! Pringsheim (1876) and Stahl (1876) were the first to report cases of apospory or of "regeneration", that is the formation of protonemata from young sporophytic tissue and in particular the seta (Bryan 2001). Such observations have been extended to other taxa (e.g. Wettstein 1925), across much of the phylogenetic spectrum of mosses (from *Tetraphis* to *Hypnum*), but these refer exclusively to experimentally induced diploid gametophytes; *in vivo* observations seem to be lacking. Aposporous gametophytes differ from their haploid progenitors in the larger cells (Moutschen 1951), and larger gametangia but not in larger vegetative organs (Marchal & Marchal 1911). Furthermore, given their diploid nature, regenerants of dioicous species carry, as does the sporophyte, the loci defining both sexes. Aposporous leafy stems of the dioicous *Bryum caespiticium* produce primarily male inflorescences, which over time tend to acquire a single archegonium (rarely more), and hence become synoicous; few purely perichaetial plants are developed (Marchal & Marchal 1907). Great variation in the sex ratio among clonal bisexual plants suggests that sex expression is determined by external factors. Furthermore, sexual reproduction leading to a tetraploid sporophyte and the induction of a tetraploid aposporic gametophyte has been achieved primarily if not only with monoicous taxa (Marchal & Marchal 1911).

Apogamy, that is the formation of a sporophyte directly from gametophytic tissues rather than following sexual reproduction, was first described by Springer (1935) on aposporic gametophytes of *Phascum cuspidatum*. Marchal & Marchal (1911) observed club-shaped outgrowths on leaves and stems of diploid plants but interpreted these as asexual reproductive structures. Springer revealed that these so-called broodbodies actually contain spores. Although the sporophytes typically deviate in their morphology from normal sporophytes, in terms of shape and lack of stomata, they do produce viable spores that germinate into a protonema that in turn forms leafy stems, bearing gametangia and capable of sexual reproduction. The gametophytes derived from these spores exhibit much morphological variation, which Springer attributed to mutations due to faulty chromosomal reduction during sporogenesis, even though she was unable to demonstrate that meiosis actually occurred in the apogamous sporangium. Apogamy has been triggered in several species by Lazarenko (Crum 2001).

Only mosses whose genome is truly monoploid seem to lack the ability to develop apogamous sporogonia on vegetative plants (Chopra 1988). Bauer (1959) described apogamic behavior in wild diploid races of *Funaria hygrometrica*, a species that occurs in the wild with distinct ploidy levels (Fritsch 1991). *In vitro*, apogamous development of sporophytes is induced by various factors, such as low light intensity, increased sugar concentration in the medium, or growth hormones such as indol acetic acid (Chopra 1988), which are not exclusively artificial. It should be noted that apogamous sporophytes develop unprotected by a calyptra, which may explain their abnormal morphologies (see Springer 1935), and perhaps their incomplete ontogenies accounting for the generalized observation of "sterile sporophytes" (Chopra 1988). The significance of apospory as a mechanism of speciation and for the occurrence of ploidy races within some species (Fritsch 1991), remains unexplored (see Shaw, Chapter 11).

2.12 Origin and evolution of the Bryophyta

The fossil record of mosses is a poor indicator of absolute age of the phylum and its main lineages. Unequivocal records date from the Carboniferous

(Kenrick & Crane 1997) and *Sporogonites*, from the Lower Devonian, exhibits sporophytic characters reminiscent of mosses, but in the absence of a gametophyte its affinities remain ambiguous. Inferences from variation in chloroplast sequence data suggest that the transition to land occurred 425–490 mya, roughly during the Silurian or Ordovician period (Sanderson 2003), an estimate congruent with microfossil evidence (Edwards 2000, Wellman & Gray 2000, Wellman *et al.* 2003). Another estimate based on sequence data suggests, by contrast, an origin of the terrestrial flora at about 1000 mya, with a divergence between the mosses and polysporangiophytes as early as 700 mya (Heckman *et al.* 2001). This hypothesis is congruent with the report of a single bryophyte-like fossil from the Middle Cambrian (Yang *et al.* 2004). Recent reports of cryptospores from the Cambrian further point to an earlier origin of a land plant flora (Strother *et al.* 2004). Thus, although the relative relationships among land plants (Kenrick & Crane 1997) and particularly extant land plants (Qiu *et al.* 2006) are becoming increasingly resolved, the origin of the land plant flora as well as the timing of the major early radiation continue to elude plant biologists. It is, however, clear that bryophytes arose and diversified early with most orders and even various families established by the Cretaceous, as inferred from a two-plastid gene phylogeny (Newton *et al.* 2007) as well as actual fossil evidence (e.g. Konopka *et al.* 1997, 1998, Bell & York 2007).

In the early 1980s Crosby (1980) and Vitt (1984) proposed two distinct views of bryophyte phylogeny, and in particular the relationships among the Bryopsida *sensu* Vitt. Mishler and Churchill (1985) provided the first formal cladistic analysis of the mosses. Only over a decade later have these hypotheses been critically tested further, based on inferences from either DNA sequence data alone (e.g. Goffinet *et al.* 2001, Cox *et al.* 2004, Tsubota *et al.* 2004) or in combination with morphological characters (Newton *et al.* 2000; see Goffinet & Buck 2004 for review). Emerging from these analyses are the following hypotheses (Renzaglia *et al.* 2007): *Takakia* and *Sphagnum* compose the earliest divergence, but their relative branching order remains ambiguous. Similarly, the Andreaeopsida and Andreaeobryopsida may compose a grade or a clade. One major contribution of these recent studies is the resolution of *Oedipodium* as a sister-taxon to all peristomate mosses (Newton *et al.* 2000). Nematodontous mosses (Polytrichales and Tetraphidales) form a sister-group or more likely a basal grade to the true mosses, the Bryopsida. The Buxbaumiales and Diphysciales represent early evolutionary lines within the Bryopsida, with the latter sister to all true arthrodontous mosses. Within the Bryopsida the main relationships are: *Timmia* likely composes the earliest divergence within the Bryopsida; the Encalyptales share a common ancestor with the Funariales and Gigaspermales; the Dicranidae may compose a monophyletic lineage with either the Funariidae or Bryidae; the latter composes a grade leading to the pleurocarpous mosses or Hypnanae, with the Ptychomniales and Hookeriales composing a grade to the largest and ultimate clade of mosses, the Hypnales.

The ambiguity of the branching order in critical areas of the tree, whether near the root of all mosses or that of the pleurocarps, may indicate a rapid diversification, that is divergences over periods of time too short to allow for much fixation of characters in the ancestor of successive radiations. One such rapid and putative adaptive radiation may characterize the Hypnales (Shaw *et al.* 2003). If increase in lineages of the Hypnales over time was gradual, as suggested by inferences by Newton *et al.* (2007), the lack of resolution, and thus the lack of shared substitutions at critical nodes, may be explained by a dramatic reduction in the rate of molecular evolution.

Phylogenetic hypotheses provide the evolutionary history upon which character transformations can be reconstructed. Although much emphasis has been placed on reconstructing the relationships among lineages of mosses (e.g. Cox *et al.* 2004), including use of morphological characters (Newton *et al.* 2000), no critical or explicit attempt has been made to establish a phylogenetic pattern in character transformations. This shortcoming is explained by (a) the relative lack of robustness of critical nodes (see above), (b) difficulties in assessing homology (e.g. for early divisions in inner peristome formed by the amphithecial layer), (c) lack of pertinent studies elucidating the character-state for certain lineages (e.g. amphithecial development in basal lineages) and (d) the diversity of character-states encountered near a particular node (e.g. mode of dehiscence near root of the tree).

The morphological evolution of the Bryophyta is not a unidirectional trend, and hence the polarities in early character transformations are reversible: an acquired state can be lost, resulting in a putative plesiomorphy in an otherwise highly derived taxon. Reverse evolution (e.g. the loss of a costa, papillae, hydroids, stomata, peristome, or operculum, among others) is widespread in mosses, and may be associated with a shift in habitat (e.g. in epiphytic Hypnales; Huttunen *et al.* 2004) or in other life history traits (e.g. anemophilous Splachnaceae; Goffinet & Buck 2004). Such reduction significantly hampers testing phylogenetic affinities based on morphology. The problem is compounded by the possibility that even complex characters, such as peristomes, may be regained (Zander 2006).

2.13 Classification of the Bryophyta

Mosses offer a large array of structural diversity from which relationships can be inferred and hence lineages defined. Throughout the 200 year history of bryophyte systematics, which is well summarized by Vitt (1984) and Buck (2007), much weight has been placed on the complexity of the peristome and the distribution of sex organs for defining taxonomic units above the species rank. Modern classifications (e.g. Crosby 1980, Walther 1983, Vitt 1984) reflect major systematic concepts proposed by Fleischer (1920) and Brotherus (1924, 1925), wherein the peristomate mosses are divided into nematodontous and arthrodontous mosses (following Mitten 1859), with the latter subdivided into acrocarpous and pleurocarpous mosses based on the position of the perichaetia (following Bridel 1826–1827), and into haplolepideous and diplolepideous mosses based on the architecture of the outer ring of peristome teeth. *Sphagnum*, *Andreaea*, *Andreaeobryum*, and *Takakia* represent additional groupings distinguished by the presence of a pseudopodium and the mode of sporangial dehiscence.

The classification of the Bryophyta is undergoing constant revisions, particularly in the light of phylogenetic inferences. Most recent revisions (Buck & Goffinet 2000, Goffinet & Buck 2004) rest on results from phylogenetic reconstructions. Although some transfers have subsequently been reversed, and correctly so, as the original data were incomplete (e.g. *Pleurophascum* in Goffinet *et al.* 1999), based on misidentified vouchers (e.g. *Goniomitrium*; Goffinet & Cox 2000) or based on contaminant DNA, others have withstood critical testing (e.g. *Ephemerum*; Goffinet & Cox 2000). Other changes may be challenged for subjective reasons. Zander (2008) for example retains members of the Rhabdoweisiaceae in the Dicranaceae, on the grounds that the monophyly of the former family is not well supported, and that the Dicranaceae are not resolved as polyphyletic. However, nothing prevents two sister taxa from being treated as distinct taxa of the same rank. The lack of support for the recognition of a distinct Rhabdoweisiaceae cannot be translated as support for a broadly defined Dicranaceae. Nodal support is preferable but not necessary. For example, the sister-group relationship of the Gigaspermaceae to the remainder of the Funariales and the Encalyptales combined is weakly supported by nucleotide substitutions but consistent with the architecture of the chloroplast genome, which led Goffinet *et al.* (2007) to accommodate the Gigaspermaceae in their own order.

The classification proposed here builds on those presented by Buck & Goffinet (2000) and Goffinet & Buck (2004). The rank of superclass is adopted to unite all arthrodontous mosses in one taxon (i.e. Superclass V). Although we aim at accepting only monophyletic taxa, given the limited number of ranks available, paraphyletic taxa are inevitable (e.g. Bryanae). Much effort is currently dedicated to resolving the relationships among genera of the Hypnanae (Newton & Tangney 2007). Although current results reveal that many familial delimitations fail to reflect shared ancestry, it is premature to propose significant changes within the pleurocarpous mosses because of a lack of resolution between the major clades, and the lack of sequence data for many of the genera. Also, some new classifications are based solely on regional taxa (e.g. Ignatov & Ignatova 2004) and we have been unable to expand them to a global scale. Therefore, until such time as we can understand generic inclusions on a world basis, we are not following such examples.

Classification of mosses

BRYOPHYTA Schimp.

SUPERCLASS I

CLASS TAKAKIOPSIDA Stech & W. Frey: Leaves divided into terete filaments; capsules dehiscent by a single longitudinal spiral slit; stomata lacking.

ORDER TAKAKIALES Stech & W. Frey

Takakiaceae Stech & W. Frey Type: *Takakia* S. Hatt. & Inoue

SUPERCLASS II

CLASS SPHAGNOPSIDA Ochyra: Branches usually in fascicles; leaves composed of a network of chlorophyllose and hyaline cells; setae lacking; capsules elevated on a pseudopodium; stomata lacking.

ORDER SPHAGNALES Limpr.: Plants mostly branched, with branches in fascicles; stems with wood cylinder; leaves unistratose; antheridia subglobose; archegonia terminal on branches; capsules ovoid.

Sphagnaceae Dumort. Type: *Sphagnum* L.

ORDER AMBUCHANANIALES Seppelt & H.A. Crum: Plants sparsely branched, with branches not in fascicles; stems without wood cylinder; leaves partially bistratose; antheridia oblong-cylindric; archegonia terminal on stems; capsules cylindrical.

Ambuchananiaceae Seppelt & H.A. Crum. Type: *Ambuchanania* Seppelt & H.A. Crum

SUPERCLASS III

CLASS ANDREAEOPSIDA Rothm.: Plants on acidic rocks, generally autoicous; cauline central strand absent; calyptrae small; capsules valvate, with four valves attached at apex; seta absent, pseudopodium present; stomata lacking.

ORDER ANDREAEALES Limpr.

Andreaeaceae Dumort. Type: *Andreaea* Hedw.

Acroschisma (Hook.f. & Wilson) Lindl., *Andreaea* Hedw.

SUPERCLASS IV

CLASS ANDREAEOBRYOPSIDA Goffinet & W. R. Buck: Plants on calcareous rocks, dioicous; cauline central strand lacking; calyptrae large and covering whole capsule; capsules valvate, apex eroding and valves free when old; stomata lacking; seta present.

ORDER ANDREAEOBRYALES B. M. Murray

Andreaeobryaceae Steere & B. M. Murray. Type: *Andreaeobryum* Steere & B. M. Murray

SUPERCLASS V

CLASS OEDIPODIOPSIDA Goffinet & W. R. Buck: Leaves unicostate; calyptrae cucullate; capsule symmetric and erect, neck very long; stomata lacking; capsules gymnostomous.

ORDER OEDIPODIALES Goffinet & W. R. Buck

Oedipodiaceae Schimp. Type: *Oedipodium* Schwägr.

CLASS POLYTRICHOPSIDA Doweld: Plants typically robust, dioicous; cauline central strand present; stems typically rhizomatous; costa broad, with adaxial chlorophyllose lamellae; peristome nematodontous, mostly of (16)32–64 teeth.

ORDER POLYTRICHALES M. Fleisch.

Polytrichaceae Schwägr. Type: *Polytrichum* Hedw.

Alophozia Card., *Atrichopsis* Card., *Atrichum* P. Beauv., *Bartramiopsis* Kindb., *Dawsonia* R. Br., *Dendroligotrichum* (Müll. Hal.) Broth., *Hebantia* G. L. Sm. Merr., *Itatiella* G. L. Sm., *Lyellia* R. Br., *Meiotrichum* (G. L. Sm.) G. L. Sm. Merr., *Notoligotrichum* G. L. Sm., *Oligotrichum* Lam. & DC., *Plagioracelopus* G. L. Sm. Merr., *Pogonatum* P. Beauv., *Polytrichadelphus* (Müll. Hal.) Mitt., *Polytrichastrum* G. L. Sm., *Polytrichum* Hedw., *Pseudatrichum* Reimers, *Pseudoracelopus* Broth., *Psilopilum* Brid., *Racelopodopsis* Thér., *Racelopus* Dozy & Molk., *Stereobryon* G. L. Sm.

CLASS TETRAPHIDOPSIDA Goffinet & W. R. Buck: Leaves unicostate; calyptrae small conic; capsule symmetric and erect, neck short; peristome nematodontous, of four erect teeth.

ORDER TETRAPHIDALES M. Fleisch.

Tetraphidaceae Schimp. Type: *Tetraphis* Hedw.

Tetraphis Hedw., *Tetrodontium* Schwägr.

CLASS BRYOPSIDA Rothm.: Plants small to robust; leaves costate or not, typically lacking lamellae; capsules operculate; peristome at least partially arthrodontous.

Subclass Buxbaumiidae Ochyra: Leaves ecostate; calyptrae cucullate or mitrate; capsule strongly asymmetric and horizontal, neck short; peristome double.

Order Buxbaumiales M. Fleisch.

Buxbaumiaceae Schimp. Type: *Buxbaumia* Hedw.

Subclass Diphysciidae Ochyra: Gametophore small, perennial; leaves costate, often bistratose; capsules asymmetric, immersed among long perichaetial leaves; peristome double.

Order Diphysciales M. Fleisch.

Diphysciaceae M. Fleisch. Type: *Diphyscium* D. Mohr

Subclass Timmiidae Ochyra: Plants acrocarpous; leaves with sheathing bases; costa single, mostly with 2 stereid bands; laminal cells short, mammillose on upper surface; peristome double; endostome of 64 cilia from a high basal membrane; calyptrae cucullate, often adhering to the tip of the seta at maturity.

Order Timmiales Ochyra

Timmiaceae Schimp. Type: *Timmia* Hedw.

Subclass Funariidae Ochyra: Plants terricolous, acrocarpous; stem typically with central strand; annulus often well developed.

Order Gigaspermales Goffinet, Wickett, O. Werner, Ros, A. J. Shaw & C. J. Cox: Plants stoloniferous; capsules immersed, gymnostomous.

Gigaspermaceae Lindb. Type: *Gigaspermum* Lindb.

Chamaebryum Thér. & Dixon, *Costesia* Thér., *Gigaspermum* Lindb., *Lorentziella* Müll Hal., *Oedipodiella* Dixon

Order Encalyptales Dixon: Plants mostly of bare soil; upper laminal cells mostly pluripapillose, often with C-shaped papillae, basal laminal cells usually differentiated, smooth; calyptra completely covering the capsule.

Bryobartramiaceae Sainsb. Type: *Bryobartramia* Sainsb. Plants very small, acrocarpous; calyptrae remaining attached to vaginula, persisting as an epigonium; capsules cleistocarpous; stomata with two guard cells.

Bryobartramia Sainsb.

Encalyptaceae Schimp. Type: *Encalypta* Hedw. Plants very small to medium-size; laminal cells thick-walled, isodiametric above, rectangular and hyaline or reddish below; annulus not differentiated; calyptrae very large, enclosing the entire erect capsule.

Bryobrittonia R. S. Williams, *Encalypta* Hedw.

ORDER FUNARIALES M. Fleisch.: Peristome diplolepideous, opposite, endostome lacking cilia.

Funariaceae Schwägr. Type: *Funaria* Hedw. Protonema short-lived; costa well developed; laminal cells smooth and thin-walled; perigonial paraphyses with swollen apical cell; calyptrae smooth and naked; stomata with single guard cell; peristome opposite, following a 4:2:4 pattern, or lacking.

Aphanorhegma Sull., *Brachymeniopsis* Broth., *Bryobeckettia* Fife, *Clavitheca* O. Werner, Ros & Goffinet, *Cygnicollum* Fife & Magill, *Entosthodon* Schwägr., *Funaria* Hedw., *Funariella* Sérgio, *Goniomitrium* Hook.f. & Wilson, *Loiseaubryum* Bizot, *Nanomitriella* E. B. Bartram, *Physcomitrella* Bruch & Schimp., *Physcomitrellopsis* Broth. & Wager, *Physcomitrium* (Brid.) Brid., *Pyramidula* Brid.

Disceliaceae Schimp. Type: *Discelium* Brid. Protonemata persistent; costa weak to absent; calyptrae persistent below the urn; stomata lacking; stomata none, peristomes reduced.

Discelium Brid.

SUBCLASS DICRANIDAE Doweld: Plants typically acrocarpous; peristome haplolepideous, with a formula of (4):2:3; exostome typically absent; late state division in the IPL asymmetric.

ORDER SCOULERIALES Goffinet & W. R. Buck: Plants blackish, acro- or cladocarpous, saxicolous in riparian habitats; calyptrae mitrate, smooth; annulus not differentiated; capsules urceolate to globose.

Scouleriaceae S. P. Churchill *in* Funk & D. R. Brooks. Type: *Scouleria* Hook.

Scouleria Hook., *Tridontium* Hook.f.

Drummondiaceae Goffinet. Type: *Drummondia* Hook. Stem with central strand, cladocarpous; costa with differen-tiated adaxial stereids; laminal cells thick-walled; peristome reduced.

Drummondia Hook.

ORDER BRYOXIPHIALES H. A. Crum & L. E. Anderson: Leaves distichous with small dorsal extension along costa; capsules gymnostomous.

Bryoxiphiaceae Besch. Type: *Bryoxiphium* Mitt.

ORDER GRIMMIALES M. Fleisch.: Plants slender to robust, usually saxicolous; laminal cells with thick and often wavy walls; peristome of 16 entire or divided teeth.

Grimmiaceae Arn. Type: *Grimmia* Hedw. Plants typically of acidic rocks; leaves little different wet or dry, often terminated by hair-point; laminal cells mostly with sinuose walls.

Bucklandiella Roiv., *Codriophorus* P. Beauv., *Dryptodon* Brid., *Grimmia* Hedw., *Leucoperichaetium* Magill, *Niphotrichum* (Bednarek-Ochyra) Bednarek-Ochyra & Ochyra, *Racomitrium* Brid., *Schistidium* Bruch & Schimp.

Ptychomitriaceae Schimp. Type: *Ptychomitrium* Fürnr. Leaves often crispate when dry; laminal cells with straight walls, often bistratose; calyptrae cucullate.

Aligrimmia R. S. Williams, *Campylostelium* Bruch & Schimp., *Indusiella* Broth. & Müll. Hal., *Jaffueliobryum* Thér., *Ptychomitriopsis* Dixon, *Ptychomitrium* Fürnr.

Seligeriaceae Schimp. Type: *Seligeria* Bruch & Schimp.. Plants small, typically of calcareous rocks; alar cells differentiated; peristome mostly deeply inserted, relatively well developed.

Blindia Bruch & Schimp., *Brachydontium* Fürnr., *Hymenolomopsis* Thér., *Seligeria* Bruch & Schimp., *Trochobryum* Breidl. & Beck

Order Archidiales Limpr. Plants small, often with persistent protonemata; seta lacking; capsules cleistocarpous, with fewer than 200 large spores (often 4–60); columella lacking.

Archidiaceae Schimp. Type: *Archidium* Brid.

Order Dicranales H. Philib. *ex* M. Fleisch.: Plants small to large; laminal cells generally smooth; alar cells often differentiated; peristome single, lacking basal membrane, segments trabeculate and striate.

Fissidentaceae Schimp. Type: *Fissidens* Hedw. Leaves distichous and complanate, with vaginant lamellae; apical cell two-sided.

Fissidens Hedw.

Hypodontiaceae Stech & W. Frey Type: *Hypodontium* Müll. Hal. (*Plantae grandes, caulis filio centrali, folia limbata in sicco incurvatae, basi amplectenti, cellulis foliorum papillatis, costa turmis stereidarum duabus basi hyalina.*) Plants large, terricolous or saxicolous; central strand present; leaves incurled when dry, with clasping base; costa with 2 stereid bands; inner perichaetial leaves sheathing below but narrowly subulate or awned apically; calyptra cucullate; spores large.

Hypodontium Müll. Hal.

Eustichiaceae Broth. Type: *Eustichia* (Brid.) Brid. Leaves distichous; laminal cells quadrate and thick-walled; capsules ribbed; peristome of 16 teeth.

Eustichia (Brid.) Brid.

Ditrichaceae Limpr. Type: *Ditrichum* Hampe. Plants slender; alar cells not differentiated; peristome of 16 completely divided, terete teeth.

Astomiopsis Müll. Hal., *Austrophilibertiella* Ochyra, *Bryomanginia* Thér., *Ceratodon* Brid., *Cheilothela* Broth., *Chrysoblastella* R. S. Williams, *Cladastomum* Müll. Hal., *Cleistocarpidium* Ochyra & Bednarek-Ochyra, *Crumuscus* W. R. Buck & Snider, *Cygniella* H. A. Crum, *Distichium* Bruch & Schimp., *Ditrichopsis* Broth., *Ditrichum* Hampe, *Eccremidium* Hook.f. & Wilson, *Garckea* Müll. Hal., *Kleioweisopsis* Dixon, ×*Pleuriditrichum* A. L. Andrews & F. J. Herm., *Pleuridium* Rabenh., *Rhamphidium* Mitt., *Saelania* Lindb., *Skottsbergia* Cardot, *Strombulidens* W. R. Buck, *Trichodon* Schimp., *Tristichium* Müll. Hal., *Wilsoniella* Müll. Hal.

Bruchiaceae Schimp. Type: *Bruchia* Schwägr. Alar cells not differentiated; capsules with elongate necks; spores mostly with trilete markings, usually strongly ornamented.

Bruchia Schwägr., *Cladophascum* Sim, *Eobruchia* W. R. Buck, *Pringleella* Cardot, *Trematodon* Michx.

Rhachitheciaceae H. Rob. Type: *Rhachithecium* Le Jolis. Laminal cells rectangular in lower half, short to isodiametric above; alar cells not differentiated; perichaetial leaves differentiated; capsules ribbed, rarely smooth; endostome teeth fused or not; IPL of 8 or 16 cells only (peristome formula: (4):2:2 or (4):2:1).

Hypnodontopsis Z. Iwats. & Nog., *Jonesiobryum* B. H. Allen & Pursell, *Rhachitheciopsis* P. de la Varde, *Rhachithecium* Le Jolis, *Tisserantiella* P. de la Varde, *Uleastrum* W. R. Buck, *Zanderia* Goffinet

Erpodiaceae Broth. Type: *Erpodium* (Brid.) Brid. Plants cladocarpous; costa lacking; laminal cells often papillose; calyptrae mitrate.

Aulacopilum Wilson, *Erpodium* (Brid.) Brid., *Solmsiella* Müll. Hal., *Venturiella* Müll. Hal., *Wildia* Müll. Hal. & Broth.

Schistostegaceae Schimp. Type: *Schistostega* D. Mohr. Gametophores dimorphic, small, annual, arising from persistent luminescent protonemata; leaves ecostate, distichous or in five rows; capsules globose, gymnostomous, lacking stomata and annulus.

Schistostega D. Mohr

Viridivelleraceae I. G. Stone Type: *Viridivellus* I. G. Stone. Protonemata persistent; stems producing gametangia and associated leaves only; capsules gymnostomous.

Viridivellus I. G. Stone

Rhabdoweisiaceae Limpr. Type: *Rhabdoweisia* Bruch & Schimp. Plants small to medium size; stem lacking central strand; capsules ribbed, widest at mouth.

Amphidium Schimp., *Arctoa* Bruch & Schimp., *Cynodontium* Schimp., *Dichodontium* Schimp., *Dicranoweisia* Milde, *Glyphomitrium* Brid., *Holodontium* (Mitt.) Broth., *Hymenoloma* Dusén, *Kiaeria* I. Hagen, *Oncophorus* (Brid.) Brid., *Oreas* Brid., *Oreoweisia* (Bruch & Schimp.) De Not., *Pseudohyophila* Hilp., *Rhabdoweisia* Bruch & Schimp., *Symblepharis* Mont., *Verrucidens* Cardot

Dicranaceae Schimp. Type: *Dicranum* Hedw. Plants generally robust, acrocarpous or cladocarpous; cauline central strand present or not; leaves often with well differentiated alar cells; laminal cells elongate, thick-walled and porose; calyptra mitrate or cucullate; peristome of 16 flat teeth divided in upper two-thirds, typically with vertically pitted outer surface.

Anisothecium Mitt., *Aongstroemia* Bruch & Schimp., *Aongstroemiopsis* M. Fleisch., *Braunfelsia* Paris, *Brotherobryum* M. Fleisch., *Bryotestua* Thér. & P. de la Varde, *Camptodontium* Dusén, *Campylopodium* (Müll. Hal.) Besch., *Chorisodontium* (Mitt.) Broth., *Cnestrum* I. Hagen, *Cryptodicranum* E. B. Bartram, *Dicnemon* Schwägr., *Dicranella* (Müll. Hal.) Schimp., *Dicranoloma* (Renauld) Renauld, *Dicranum* Hedw., *Diobelonella* Ochyra, *Eucamptodon* Mont., *Eucamptodontopsis* Broth., *Holomitriopsis* H. Rob., *Holomitrium* Brid., *Hygrodicranum* Cardot, *Leptotrichella* (Müll. Hal.) Lindb., *Leucoloma* Brid., *Macrodictyum* (Broth.) E. H. Hegew., *Mesotus* Mitt., *Mitrobryum* H. Rob., *Muscoherzogia* Ochyra, *Orthodicranum* (Bruch & Schimp.) Loeske, *Paraleucobryum* (Limpr.) Loeske, *Parisia* Broth., *Platyneuron* (Cardot) Broth., *Pocsiella* Bizot, *Polymerodon* Herzog, *Pseudephemerum* (Lindb.) I. Hagen, *Pseudochorisodontium* (Broth.) C. H. Gao, Vitt, D. H. Fu & T. Cao, *Schliephackea* Müll. Hal., *Sclerodontium* Schwägr., *Sphaerothecium* Hampe, *Steyermarkiella* H. Rob., *Wardia* Harv. & Hook., *Werneriobryum* Herzog

Leucobryaceae Schimp. Type: *Leucobryum* Hampe. Plants robust, glaucous; cauline central strand lacking; costa broad, occupying most of the leaf, with median chlorophyllose cells and adaxial and abaxial layers of hyaline cells.

Atractylocarpus Mitt., *Brothera* Müll. Hal., *Bryohumbertia* P. de la Varde & Thér., *Campylopodiella* Cardot, *Campylopus* Brid., *Cladopodanthus* Dozy & Molk., *Dicranodontium* Bruch & Schimp., *Leucobryum* Hampe, *Microcampylopus* (Müll. Hal.) Fleisch., *Ochrobryum* Mitt., *Pilopogon* Brid., *Schistomitrium* Dozy & Molk.

Calymperaceae Kindb. Type: *Calymperes* Sw. Plants epiphytic; stem lacking central strand; leaves narrowly to broadly lanceolate; laminal cells papillose or smooth; often with hyaline cancellinae on either side of costa at leaf base; calyptrae persistent or not; peristome of 16 (rarely fused into 8) segments, smooth, papillose or vertically striate.

Arthrocormus Dozy & Molk., *Calymperes* Sw., *Exodictyon* Cardot, *Exostratum* L. T. Ellis, *Leucophanes* Brid., *Mitthyridium* H. Rob., *Octoblepharum* Hedw., *Syrrhopodon* Schwägr.

Order Pottiales M. Fleisch.: Plants minute to robust, generally orthotropic; upper laminal cells usually isodiametric and papillose; alar cells not differentiated; perichaetial leaves typically not differentiated; capsules erect; peristome typically papillose, not trabeculate.

Pottiaceae Schimp. Type: *Pottia* (Rchb.) Fürnr. Plants small to robust, primarily terrestrial; cauline central strand often present; leaves narrowly lanceolate to ligulate; laminal cells typically papillose; calyptrae cucullate, naked, smooth; peristome of 16 or 32 segments.

Acaulon Müll. Hal., *Aloinia* Kindb., *Aloinella* Cardot, *Anoectangium* Schwägr., *Aschisma* Lindb., *Barbula* Hedw., *Bellibarbula* P. C. Chen, *Bryoceuthospora* H. A. Crum & L. E. Anderson, *Bryoerythrophyllum* P. C. Chen, *Calymperastrum* I. G. Stone, *Calyptopogon* (Mitt.) Broth., *Chenia* R. H. Zander, *Chionoloma* Dixon, *Cinclidotus* P. Beauv., *Crossidium* Jur., *Crumia* W. B. Schofield, *Dialytrichia* (Schimp.) Limpr., *Didymodon* Hedw., *Dolotortula* R. H. Zander, *Ephemerum* Schimp., *Erythrophyllopsis* Broth., *Eucladium* Bruch & Schimp., *Ganguleea* R. H. Zander, *Gertrudiella* Broth., *Globulinella* Steere, *Gymnostomiella* M. Fleisch., *Gymnostomum* Nees & Hornsch., *Gyroweisia* Schimp., *Hennediella* Paris, *Hilpertia* R. H. Zander, *Hymenostyliella* E. B. Bartram, *Hymenostylium* Brid., *Hyophila* Brid., *Hyophiladelphus* (Müll. Hal.) R. H. Zander, *Leptobarbula* Schimp., *Leptodontiella*

R. H. Zander & E. H. Hegew., *Leptodontium* (Müll. Hal.) Lindb., *Luisierella* Thér. & P. de la Varde, *Microbryum* Schimp., *Micromitrium* Austin, *Mironia* R. H. Zander, *Molendoa* Lindb., *Nanomitriopsis* Cardot, *Neophoenix* R. H. Zander & During, *Pachyneuropsis* H. Mill., *Phascopsis* I. G. Stone, *Plaubelia* Brid., *Pleurochaete* Lindb., *Pottiopsis* Blockeel & A. J. E. Sm., *Pseudocrossidium* R. S. Williams, *Pseudosymblepharis* Broth., *Pterygoneurum* Jur., *Quaesticula* R. H. Zander, *Reimersia* P. C. Chen, *Rhexophyllum* Herzog, *Sagenotortula* R. H. Zander, *Saitobryum* R. H. Zander, *Sarconeurum* Bryhn, *Scopelophila* (Mitt.) Lindb., *Splachnobryum* Müll. Hal., *Stegonia* Venturi, *Stonea* R. H. Zander, *Streptocalypta* Müll. Hal., *Streptopogon* Mitt., *Streptotrichum* Herzog, *Syntrichia* Brid., *Teniolophora* W. D. Reese, *Tetracoscinodon* R. Br. ter, *Tetrapterum* A. Jaeger, *Timmiella* (De Not.) Schimp., *Tortella* (Lindb.) Limpr., *Tortula* Hedw., *Trachycarpidium* Broth., *Trachyodontium* Steere, *Trichostomum* Bruch, *Triquetrella* Müll. Hal., *Tuerckheimia* Broth., *Uleobryum* Broth., *Weisiopsis* Broth., *Weissia* Hedw., *Weissiodicranum* W. D. Reese, *Willia* Müll. Hal.

Pleurophascaceae Broth. Type: *Pleurophascum* Lindb. Plants robust; stems creeping with erect secondary stems; leaves concave, ecostate; cells short above, elongate below, smooth, strongly porose; setae elongate; capsules large, globose, cleistocarpous; calyptra cucullate.

Pleurophascum Lindb.

Serpotortellaceae W. D. Reese & R. H. Zander. Type: *Serpotortella* Dixon. Plants robust, cladocarpous, epiphytic; cauline central strand present; leaf margins entire and unistratose; perichaetial leaves differentiated; peristome well developed, reflexed when dry.

Serpotortella Dixon

Mitteniaceae Broth. Type: *Mittenia* Lindb. Plants small, with luminescent protonemata; cauline central strand lacking; leaves complanate, decurrent; perichaetia polysetous; peristome double, outer row homologous to bryoid endostome.

Mittenia Lindb.

Subclass Bryidae Engl.: Peristome double, of alternating teeth and segments; endostome ciliate; late stage division in the IPL asymmetric

Superorder Bryanae (Engl.) Goffinet & W. R. Buck: Plants acrocarpous, cladocarpous or pseudopleurocarpous; pseudoparaphyllia generally lacking; leaves erect to spreading, lanceolate to ovate, mostly costate, costal anatomy mostly heterogeneous; laminal cells generally short.

ORDER SPLACHNALES Ochyra: Laminal cells rhombic to elongate, typically smooth; capsules erect with differentiated neck; peristome single or double; cilia rudimentary or lacking.

Splachnaceae Grev. & Arn. Type: *Splachnum* Hedw. Plants mostly coprophilous; laminal cells thin-walled, rhomboidal; annulus not differentiated; capsules erect, neck often differentiated into broad hypophysis; endostome fused to exostome or lacking.

Aplodon R. Br., *Moseniella* Broth., *Splachnum* Hedw., *Tayloria* Hook., *Tetraplodon* Bruch & Schimp., *Voitia* Hornsch.

Meesiaceae Schimp. Type: *Meesia* Hedw. Plants acrocarpous, often of moist habitats; leaves often in rows; lower laminal cells often delicate and hyaline; setae elongate; capsules inclined to suberect but strongly curved and asymmetric, oblong-pyriform with a well-differentiated neck; peristome double with exostome teeth usually shorter than endostome segments; calyptra cucullate.

Amblyodon P. Beauv., *Leptobryum* (Bruch & Schimp.) Wilson, *Meesia* Hedw., *Neomeesia* Deguchi, *Paludella* Brid.

ORDER BRYALES Limpr.: Plants primarily terricolous; cauline central strand present; laminal cells rhombic to elongate, smooth; annulus differentiated; capsules pendent, neck differentiated; peristome double, typically well developed and ciliate; exostome incurved.

Catoscopiaceae Broth. Type: *Catoscopium* Brid. Plants small, slender; leaves in three ranks; laminal cells quadrate and smooth; capsules black, asymmetric, horizontal; peristome double and reduced.

Catoscopium Brid.

Pulchrinodaceae D. Quandt, N. E. Bell & Stech. Type: *Pulchrinodus* B. H. Allen. Plants with foliose pseudoparaphyllia; stems with central strand; leaves ecostate; laminal cells smooth, strongly porose, bistratose at base; alar cells strongly differentiated; perigonia terminal, discoid; perigonia stalked.

Pulchrinodus B. H. Allen

Bryaceae Schwägr. Type: *Bryum* Hedw. Plants erect, mostly unbranched, acrocarpous; laminal cells mostly rhomboidal, smooth, thin-walled; costa single, strong; capsules inclined to pendulous, smooth, with differentiated neck.

Acidodontium Schwägr., *Anomobryum* Schimp., *Brachymenium* Schwägr., *Bryum* Hedw., *Leptostomopsis* (Müll. Hal.) J. R. Spence & H. P. Ramsay, *Mniobryoides* Hörmann, *Osculatia* De Not., *Perssonia* Bizot, *Ptychostomum* Hornsch., *Rhodobryum* (Schimp.) Limpr., *Roellia* Kindb., *Rosulabryum* J. R. Spence

Phyllodrepaniaceae Crosby. Type: *Phyllodrepanium* Crosby. Plants small; leaves complanate, in four rows; peristome single, of 16 segments.

Mniomalia Müll. Hal., *Phyllodrepanium* Crosby

Pseudoditrichaceae Steere & Z. Iwats. Type: *Pseudoditrichum* Steere & Z. Iwats. Plants very small; leaves ovate lanceolate; laminal cells thick-walled; capsules erect; peristome double; cilia lacking.

Pseudoditrichum Steere & Z. Iwats.

Mniaceae Schwägr. Type: *Mnium* Hedw. Plants acro- or cladocarpous; leaves often bordered and often toothed; laminal cells thin-walled, rhomboidal to elongate.

Cinclidium Sw., *Cyrtomnium* Holmen, *Epipterygium* Lindb., *Leucolepis* Lindb., *Mielichhoferia* Nees & Hornsch., *Mnium* Hedw., *Ochiobryum* J. R. Spence & H. P. Ramsay, *Orthomnion* Wilson, *Plagiomnium* T. J. Kop., *Pohlia* Hedw., *Pseudobryum* (Kindb.) T. J. Kop., *Pseudopohlia* R. S. Williams, *Rhizomnium* (Broth.) T. J. Kop., *Schizymenium* Harv., *Synthetodontium* Cardot, *Trachycystis* T. J. Kop.

Leptostomataceae Schwägr. Type: *Leptostomum* R. Br. Plants forming dense mats; stems heavily tomentose; leaf margins entire, unbordered; annulus poorly differentiated to lacking; stomata cryptoporous; peristome strongly reduced.

Leptostomum R. Br.

Order Bartramiales D. Quandt, N. E. Bell & Stech: Plants often robust; laminal cells isodiametric, quadrate or rectangular, smooth or prorulose; annulus typically undifferentiated; capsules subglobose, erect or slightly curved, typically ribbed; neck undifferentiated.

Bartramiaceae Schwägr. Type: *Bartramia* Hedw.

Anacolia Schimp., *Bartramia* Hedw., *Breutelia* (Bruch & Schimp.) Schimp., *Conostomum* Sw., *Fleischerobryum* Loeske, *Flowersia* D. G. Griffin & W. R. Buck, *Leiomela* (Mitt.) Broth., *Neosharpiella* H. Rob. & Delgad., *Philonotis* Brid., *Plagiopus* Brid.

ORDER ORTHOTRICHALES Dixon: Plants medium-size, epiphytic or saxicolous; cauline central strand lacking; laminal cells typically papillose; capsules erect; peristome double or reduced; exostome recurved.

Orthotrichaceae Arn. Type: *Orthotrichum* Hedw. Plants acrocarpous or cladocarpous; laminal cells mostly isodiametric, thick-walled; calyptrae typically plicate and hairy; capsules erect, rarely immersed, often ribbed; OPL thick and teeth recurved when dry; cilia lacking.

Cardotiella Vitt, *Ceuthotheca* Lewinsky, *Codonoblepharon* Schwägr., *Desmotheca* Lindb., *Florschuetziella* Vitt, *Groutiella* Steere, *Leiomitrium* Mitt., *Leratia* Broth. & Paris, *Macrocoma* (Müll. Hal.) Grout, *Macromitrium* Brid., *Matteria* Goffinet, *Orthotrichum* Hedw., *Pentastichella* Müll. Hal., *Pleurorthotrichum* Broth., *Schlotheimia* Brid., *Sehnemobryum* Lewinsky-Haapasaari & Hedenäs, *Stoneobryum* D. H. Norris & H. Rob., *Ulota* D. Mohr, *Zygodon* Hook. & Taylor

ORDER HEDWIGIALES Ochyra: Plants medium to robust, plagiotropic, acrocarpous or cladocarpous; laminal cells thick-walled, papillose or smooth; capsules gymnostomous and immersed.

Hedwigiaceae Schimp. Type: *Hedwigia* P. Beauv. Protonemata globular; leaves ecostate; laminal cells pluripapillose; calyptrae smooth, naked.

Braunia Bruch & Schimp., *Bryowijkia* Nog., *Hedwigia* P. Beauv., *Hedwigidium* Bruch & Schimp., *Pseudobraunia* (Lesq. & James) Broth.

Helicophyllaceae Broth. Type: *Helicophyllum* Brid. Leaves unicostate, dimorphic, with lateral leaves strongly inrolled when dry, dorsal and ventral leaves reduced and appressed; laminal cells smooth.

Helicophyllum Brid.

Rhacocarpaceae Kindb. Type: *Rhacocarpus* Lindb. Leaves ecostate, bordered by narrow cells; laminal cells roughened; alar cells inflated.

Pararhacocarpus Frahm, *Rhacocarpus* Lindb.

ORDER RHIZOGONIALES Goffinet & W. R. Buck: Plants pseudo-pleurocarpous, often with basal sporophytes; leaves often complanate and asymmetric; laminal cells mostly short, smooth or unipapillose, basal cells not or weakly differentiated; setae elongate; capsules cylindric, often asymmetric; peristome often reduced.

Rhizogoniaceae Broth. Type: *Rhizogonium* Brid. Plants small to large; cauline central strand present; marginal laminal cells often differentiated, bi- or multistratose, often toothed; costa typically toothed above; annulus differentiated; capsules generally smooth; peristome typically well developed, ciliate, or reduced to endostome or exostome, or peristome absent.

Calomnion Hook.f. & Wilson, *Cryptopodium* Brid., *Goniobryum* Lindb., *Pyrrhobryum* Mitt., *Rhizogonium* Brid.

Aulacomniaceae Schimp. Type: *Aulacomnium* Schwägr. Plants acrocarpous but sporophytes sometimes lateral, often with leaf-like gemmae; leaves unicostate; capsules usually asymmetric, often furrowed.

Aulacomnium Schwägr., *Hymenodontopsis* Herzog, *Mesochaete* Lindb.

Orthodontiaceae Goffinet. Type: *Orthodontium* Wilson. Plants small to robust, acrocarpous but often with basal sporophytes; laminal cells short to elongate, lax; capsules often ribbed; annulus lacking; peristome often reduced, sometimes with exostome lacking, endostomial membrane reduced or lacking.

Hymenodon Hook.f. & Wilson, *Leptotheca* Schwägr., *Orthodontium* Wilson, *Orthodontopsis* Ignatov & B. C. Tan

Superorder Hypnanae W. R. Buck, Goffinet & A. J. Shaw: Plants pleurocarpous, typically freely branching; pseudoparaphyllia usually present; leaves mostly ovate, costate or not, costal anatomy usually homogeneous; laminal cells generally elongate.

ORDER HYPNODENDRALES N. E. Bell, Ang. Newton & D. Quandt: Plants often stipitate; costae single, with anatomy heterogeneous; laminal cells mostly short; setae elongate; peristome double.

Braithwaiteaceae N. E. Bell, Ang. Newton & D. Quandt. Type: *Braithwaitea* Lindb. Leaves cymbiform, obtuse; costa strong, excurrent; peristome reduced, exostome teeth with vestigial trabeculae, endostome segments linear from a low basal membrane.

Braithwaitea Lindb.

Racopilaceae Kindb. Type: *Racopilum* P. Beauv. Stems plagiotropic; leaves dimorphic with dorsal ones reduced; costa excurrent; capsules long-exserted; peristome double, well developed.

Powellia Mitt., *Racopilum* P. Beauv.

Pterobryellaceae W. R. Buck & Vitt. Type: *Pterobryella* (Müll. Hal.) A. Jaeger. Plants robust and large, stipitate from rhizomatous stem, frondose to dendroid; cauline central strand lacking; capsules short oval; annulus differentiated; peristome double, with long teeth and segments but reduced cilia.

Cyrtopodendron M. Fleisch., *Pterobryella* (Müll. Hal.) A. Jaeger, *Sciadocladus* Lindb. *ex* Kindb.

Hypnodendraceae Broth. Type: *Hypnodendron* (Müll. Hal.) Mitt. Plants robust, rhizomatous and stipitate; secondary stems erect, frondose to dendroid; laminal marginal cells differentiated or not, unistratose, often toothed; capsules often ribbed when dry; annulus differentiated; peristome well developed, ciliate.

Bescherellia Duby, *Cyrtopus* (Brid.) Hook.f., *Dendro-hypnum* Hampe, *Franciella* Thér., *Hypnodendron* (Müll. Hal.) Mitt., *Mniodendron* Lindb. *ex* Dozy & Molk., *Spiridens* Nees, *Touwiodendron* N. E. Bell, Ang. Newton & D. Quandt

Order Ptychomniales W. R. Buck, C. J. Cox, A. J. Shaw & Goffinet. Plants usually robust and turgid, often phyllodioicous; stems sympodially branched, usually lacking a central strand; leaves usually plicate, often strongly toothed; costae short and double; laminal cells elongate, often thick-walled and porose; alar cells often colored; capsules mostly ribbed; endostomial segments lacking baffle-like cross walls; calyptrae often cucullate.

Ptychomniaceae M. Fleisch. Type: *Ptychomnion* (Hook.f. & Wilson) Mitt. Alar cells little or well differentiated, except for color; capsules long-exserted or immersed, smooth to strongly 8-ribbed, anisosporous or isosporous, calyptrae mitrate or cucullate.

Cladomnion Hook.f. & Wilson, *Cladomniopsis* M. Fleisch., *Dichelodontium* Broth., *Endotrichellopsis* During, *Euptychium* Schimp., *Garovaglia* Endl., *Glyphotheciopsis* Pedersen & Ang. Newton, *Glyphothecium* Hampe, *Hampeella* Müll. Hal., *Ombronesus* N. E. Bell, Pederson & Ang. Newton, *Ptychomniella* (Broth.) W. R. Buck, C. J. Cox, A. J. Shaw & Goffinet, *Ptychomnion* (Hook.f. & Wilson) Mitt., *Tetraphidopsis* Broth. & Dixon

Order Hookeriales M. Fleisch.: Laminal cells mostly thin-walled, often short; alar cells mostly not differentiated; exothecial cells mostly collenchymatous; opercula mostly rostrate; exostome teeth often furrowed, endostomial segments with baffle-like cross walls; calyptrae often mitrate.

Hypopterygiaceae Mitt. Type: *Hypopterygium* Brid. Plants dendroid; amphigastria differentiated; leaves often limbate;

costa single; laminal cells short, mostly smooth; alar cells not differentiated; exostome teeth not furrowed; endostomial segments lacking baffle-like cross walls.

Arbusculohypopterygium Stech, T. Pfeiffer & W. Frey, *Canalohypopterygium* W. Frey & Schaepe, *Catharomnion* Hook.f. & Wilson, *Cyathophorum* P. Beauv., *Dendrocyathophorum* Dixon, *Dendrohypopterygium* Kruijer, *Hypopterygium* Brid., *Lopidium* Hook.f. & Wilson

Saulomataceae W. R. Buck, C. J. Cox, A. J. Shaw & Goffinet Type: *Sauloma* (Hook.f. & Wils.) Mitt. Plants slender, usually erect; leaves ecostate; laminal cells short, firm-walled; capsules erect, symmetric; exostome teeth usually furrowed.

Ancistrodes Hampe, *Sauloma* (Hook.f. & Wilson) Mitt., *Vesiculariopsis* Broth.

Daltoniaceae Schimp. Type: *Daltonia* Hook. & Taylor. Stems lacking central strand; pseudoparaphyllia absent or rarely filamentous; laminal cells oval to long-hexagonal, differentiated at leaf margins or rarely not; costa single; calyptrae unistratose at middle, fringed at base or not, usually naked but rarely densely hairy.

Achrophyllum Vitt & Crosby, *Adelothecium* Mitt., *Benitotania* H. Akiyama, Yamaguchi & Suleiman, *Bryobrothera* Thér., *Calyptrochaeta* Desv., *Crosbya* Vitt, *Beeveria* Fife, *Daltonia* Hook. & Taylor, *Distichophyllidium* M. Fleisch., *Distichophyllum* Dozy & Molk., *Ephemeropsis* K. I. Goebel, *Leskeodon* Broth., *Leskeodontopsis* Zanten, *Metadistichophyllum* Nog. & Z. Iwats.

Schimperobryaceae W. R. Buck, C. J. Cox, A. J. Shaw & Goffinet. Type: *Schimperobryum* Margad. Plants robust, epiphytic; leaves complanate with short, double costa; laminal cells hexagonal, porose; setae short; capsules erect; exostome teeth cross-striolate, not furrowed; cilia absent; calyptra mitrate, not fringed.

Schimperobryum Margad.

Hookeriaceae Schimp. Type: *Hookeria* Sm. Stems with central strand; pseudoparaphyllia filamentous or absent; gemmae often on rhizoids; laminal cells large and lax; costa usually short and double; calyptrae multistratose at middle, naked.

Crossomitrium Müll. Hal., *Hookeria* Sm.

Leucomiaceae Broth. Type: *Leucomium* Mitt. Stems lacking central strand; pseudoparaphyllia absent; laminal cells linear, lax; costa lacking; calyptrae cucullate.

Leucomium Mitt., *Rhynchostegiopsis* Müll. Hal., *Tetrastichium* (Mitt.) Cardot

Pilotrichaceae Kindb. Type: *Pilotrichum* P. Beauv. Stems lacking central strand; pseudoparaphyllia none or foliose; laminal cells various; costa strong and double, or short and double; calyptrae unistratose at middle, usually hairy.

Actinodontium Schwägr., *Amblytropis* (Mitt.) Broth., *Brymela* Crosby & B. H. Allen, *Callicostella* (Müll. Hal.) Mitt., *Callicostellopsis* Broth., *Cyclodictyon* Mitt., *Diploneuron* E. B. Bartram, *Helicoblepharum* (Mitt.) Broth., *Hemiragis* (Brid.) Besch., *Hookeriopsis* (Besch.) A. Jaeger, *Hypnella* (Müll. Hal.) A. Jaeger, *Lepidopilidium* (Müll. Hal.) Broth., *Lepidopilum* (Brid.) Brid., *Neohypnella* E. B. Bartram, *Philophyllum* Müll. Hal., *Pilotrichidium* Besch., *Pilotrichum* P. Beauv., *Stenodesmus* (Mitt.) A. Jaeger, *Stenodictyon* (Mitt.) A. Jaeger, *Thamniopsis* (Mitt.) M. Fleisch., *Trachyxiphium* W. R. Buck

ORDER HYPNALES (M. Fleisch.) W. R. Buck & Vitt: Stems monopodially or sympodially branched; alar cells often differentiated; opercula various, mostly not rostrate; exostome seldom furrowed; calyptrae mostly cucullate, naked.

Rutenbergiaceae M. Fleisch. Type: *Rutenbergia* Besch. Stems sympodially branched, lacking a central strand; secondary stems little branched; costa single; laminal cells prorulose; alar cells well differentiated; capsules immersed; calyptrae mitrate, hairy.

Neorutenbergia Bizot & Pócs, *Pseudocryphaea* Broth., *Rutenbergia* Besch.

Trachylomataceae W. R. Buck & Vitt. Type: *Trachyloma* Brid. Stems sympodially branched; secondary stems stipitate frondose, complanate-foliate; alar cells weakly differentiated; asexual propagula of stem-borne, filamentous gemmae; exostome teeth pale, densely papillose.

Trachyloma Brid.

Fontinalaceae Schimp. Type: *Fontinalis* Hedw. Plants aquatic; stems sympodially branched; costa single or short and double (and then the leaves concave to carinate); capsules immersed or short-exserted; endostome forming a trellis; calyptrae mitrate or cucullate.

Brachelyma Cardot, *Dichelyma* Myrin, *Fontinalis* Hedw.

Climaciaceae Kindb. Type: *Climacium* F. Weber & D. Mohr. Plants dendroid; stems sympodially branched, with paraphyllia or longitudinal lamellae on stipe; leaves decurrent or not; costa single; laminal cells relatively short, smooth.

Climacium F. Weber & D. Mohr, *Pleuroziopsis* E. Britton

Amblystegiaceae G. Roth. Type: *Amblystegium* Schimp. Plants typically growing in moist areas; stems monopodially branched; paraphyllia sometimes present; costa mostly single but often variable; laminal cells mostly short, sometimes elongate, smooth or rarely prorulose; alar cells not to strongly differentiated; setae often relatively long in comparison to size of plants; capsules strongly curved and asymmetric; exostome teeth yellow-brown, cross-striolate.

Amblystegium Schimp., *Anacamptodon* Brid., *Bryostreimannia* Ochyra, *Campyliadelphus* (Kindb.) R. S. Chopra, *Campylium* (Sull.) Mitt., *Conardia* H. Rob., *Cratoneuron* (Sull.) Spruce, *Cratoneuropsis* (Broth.) M. Fleisch., *Drepanocladus* (Müll. Hal.) G. Roth, *Gradsteinia* Ochyra, *Hygroamblystegium* Loeske, *Hygrohypnella* Ignatov & Ignatova, *Hygrohypnum* Lindb., *Hypnobartlettia* Ochyra, *Koponenia* Ochyra, *Leptodictyum* (Schimp.) Warnst., *Limbella* (Müll. Hal.) Müll. Hal., *Limprichtia* Loeske, *Ochyraea* Váňa, *Palustriella* Ochyra, *Pictus* C. C. Towns., *Pseudocalliergon* (Limpr.) Loeske, *Pseudohygrohypnum* Kanda, *Sanionia* Loeske, *Sasaokaea* Broth., *Sciaromiella* Ochyra, *Sciaromiopsis* Broth., *Scorpidium* (Schimp.) Limpr., *Sinocalliergon* Sakurai, *Serpoleskea* (Limpr.) Loeske, *Vittia* Ochyra

Calliergonaceae Vanderpoorten, Hedenäs, C. J. Cox & A. J. Shaw. Type: *Calliergon* (Sull.) Kindb. Plants typically growing in moist areas; stems monopodially branched; costa single; laminal cells mostly elongate, smooth; alar cells often enlarged and inflated; setae elongate; capsules mostly asymmetric; peristome hypnoid.

Calliergon (Sull.) Kindb., *Hamatocaulis* Hedenäs, *Loeskypnum* H. K. G. Paul, *Straminergon* Hedenäs, *Warnstorfia* Loeske

Helodiaceae Ochyra. Type: *Helodium* Warnst. Stems monopodially branched; paraphyllia present, filamentous to narrowly foliose, the cells elongate, not papillose; costa single; laminal cells mostly prorulose; alar cells often well differentiated; exostome teeth cross-striolate.

Actinothuidium (Besch.) Broth., *Bryochenea* C. H. Gao & K. C. Chang, *Helodium* Warnst.

Rigodiaceae H. A. Crum. Type: *Rigodium* Schwägr. Plants terrestrial or weakly epiphytic, more or less stipitate; stems monopodially branched; paraphyllia absent; stipe, stem and branch leaves differentiated; costa single; laminal cells short, smooth; alar cells not or weakly differentiated; setae smooth; capsules curved and asymmetric; exostome teeth densely cross-striolate.

Rigodium Schwägr.

Leskeaceae Schimp. Type: *Leskea* Hedw. Plants terrestrial or epiphytic; stems monopodially branched, often terete-foliate; paraphyllia non-papillose; leaves mostly short-acuminate; costa mostly single; laminal cells short, usually unipapillose; alar cells weakly differentiated; capsules curved and asymmetric when plants terrestrial but in epiphytes often erect; exostome striate in terrestrial taxa but in epiphytes often pale, weakly ornamented; endostome often reduced.

Claopodium (Lesq. & James) Renauld & Cardot, *Fabronidium* Müll. Hal., *Haplocladium* (Müll. Hal.) Müll. Hal., *Hylocomiopsis* Cardot, *Leptocladium* Broth., *Leptopterigynandrum* Müll. Hal., *Lescuraea* Bruch & Schimp., *Leskea* Hedw., *Leskeadelphus* Herzog, *Leskeella* (Limpr.) Loeske, *Lindbergia* Kindb., *Mamillariella* Laz., *Miyabea* Broth., *Orthoamblystegium* Dixon & Sakurai, *Platylomella* A. L. Andrews, *Pseudoleskea* Bruch & Schimp., *Pseudoleskeella* Kindb., *Pseudoleskeopsis* Broth., *Ptychodium* Schimp., *Rigodiadelphus* Dixon, *Rozea* Besch., *Schwetschkea* Müll. Hal.

Thuidiaceae Schimp. Type: *Thuidium* Bruch & Schimp. Plants terrestrial; stems monopodially branched; paraphyllia present, the cells papillose; stem and branch leaves differentiated; costa single; laminal cells short, papillose; alar cells not or weakly differentiated; setae often roughened; capsules typically curved and asymmetric; exostome teeth densely cross-striolate; calyptrae naked or sparsely hairy.

Abietinella Müll. Hal., *Boulaya* Cardot, *Cyrto-hypnum* (Hampe) Hampe & Lorentz, *Fauriella* Besch., *Pelekium* Mitt., *Rauiella* Reimers, *Thuidiopsis* (Broth.) M. Fleisch., *Thuidium* Bruch & Schimp.

Regmatodontaceae Broth. Type: *Regmatodon* Brid. Plants epiphytic; stems monopodially branched; paraphyllia absent; costa single; laminal cells short, smooth; alar cells not or weakly, differentiated; capsules erect; exostome teeth much shorter than endostome segments.

Regmatodon Brid.

Stereophyllaceae W. R. Buck & Ireland. Type: *Stereophyllum* Mitt. Plants terrestrial or epiphytic; stems monopodially branched; costa typically single; laminal cells elongate, mostly smooth, rarely unipapillose; alar cells differentiated, collenchymatous, extending across base of costa; setae smooth; capsules inclined to erect; exostome teeth cross-striolate to papillose.

Catagoniopsis Broth., *Entodontopsis* Broth., *Eulacophyllum* W. R. Buck & Ireland, *Juratzkaea* Lorentz, *Pilosium* (Müll. Hal.) M. Fleisch., *Sciuroleskea* Broth., *Stenocarpidium* Müll. Hal., *Stereophyllum* Mitt.

Brachytheciaceae G. Roth. Type: *Brachythecium* Schimp. Plants mostly growing in mesic woodlands, terrestrial; stems monopodially branched; leaves often plicate; costa single, often projecting as a small spine; laminal cells elongate; alar cells mostly weakly differentiated; setae sometimes roughened; capsules often relatively short, curved, asymmetric; opercula conic to rostrate; exostome teeth mostly red-brown; calyptrae mostly naked.

Aerobryum Dozy & Molk., *Aerolindigia* M. Menzel, *Brachytheciastrum* Ignatov & Huttunen, *Brachythecium* Schimp., *Bryhnia* Kaurin, *Bryoandersonia* H. Rob., *Cirriphyllum* Grout, *Clasmatodon* Hook.f. & Wilson, *Donrichardsia* H. A. Crum & L. E. Anderson, *Eriodon* Mont., *Eurhynchiadelphus* Ignatov & Huttunen, *Eurhynchiastrum* Ignatov & Huttunen, *Eurhynchiella* M. Fleisch., *Eurhynchium* Bruch & Schimp., *Flabellidium* Herzog, *Helicodontium* Schwägr., *Homalotheciella* (Cardot) Broth., *Homalothecium* Schimp., *Juratzkaeella* W. R. Buck, *Kindbergia* Ochyra, *Lindigia* Hampe, *Mandoniella* Herzog, *Meteoridium* (Müll. Hal.) Manuel, *Myuroclada* Besch., *Nobregaea* Hedenäs, *Okamuraea* Broth., *Oxyrrhynchium* (Schimp.) Warnst., *Palamocladium* Müll. Hal., *Plasteurhynchium* Broth., *Platyhypnidium* M. Fleisch., *Pseudopleuropus* Takaki, *Pseudoscleropodium* (Limpr.) M. Fleisch., *Remyella* Müll. Hal., *Rhynchostegiella* (Schimp.) Limpr., *Rhynchostegium* Bruch & Schimp., *Schimperella* Thér., *Sciuro-hypnum* (Hampe) Hampe, *Scleropodium* Bruch & Schimp., *Scorpiurium* Schimp., *Squamidium* (Müll. Hal.) Broth., *Stenocarpidiopsis* M. Fleisch., *Tomentypnum* Loeske, *Zelometeorium* Manuel

Meteoriaceae Kindb. Type: *Meteorium* (Brid.) Dozy & Molk. Plants epiphytic, often pendent; stems monopodially branched, often very elongate; costa short and double or single; laminal cells mostly elongate, sometimes short, often variously papillose; alar cells not or weakly differentiated; setae often short, roughened; capsules often immersed, erect, symmetric; exostome teeth cross-striolate to papillose; endostome often reduced; calyptrae mitrate or cucullate, often hairy.

Aerobryidium M. Fleisch., *Aerobryopsis* M. Fleisch., *Barbella* M. Fleisch., *Barbellopsis* Broth., *Chrysocladium* M. Fleisch., *Cryptopapillaria* M. Menzel, *Diaphanodon* Renauld & Cardot, *Duthiella* Renauld, *Floribundaria* M. Fleisch., *Lepyrodontopsis* Broth., *Meteoriopsis* Broth., *Meteorium* (Brid.) Dozy & Molk., *Neodicladiella* W. R. Buck, *Neonoguchia* S. H. Lin, *Pseudospiridentopsis* (Broth.) M. Fleisch., *Pseudotrachypus* P. de la Varde & Thér., *Sinskea* W. R. Buck, *Toloxis* W. R. Buck, *Trachycladiella* (M. Fleisch.) M. Menzel & W. Schultze-Motel, *Trachypodopsis* M. Fleisch., *Trachypus* Reinw. & Hornsch.

Myriniaceae Schimp. Type: *Myrinia* Schimp. Plants often epiphytic, small; stems monopodially branched; costa single, often slender; laminal cells elongate, smooth; alar cells weakly differentiated; capsules often erect; peristomes mostly variously reduced; calyptrae rarely hairy.

Austinia Müll. Hal., *Macgregorella* E. B. Bartram, *Merrilliobryum* Broth., *Myrinia* Schimp., *Nematocladia* W. R. Buck

Fabroniaceae Schimp. Type: *Fabronia* Raddi. Plants epiphytic, often small; stems monopodially branched, sometimes fragile; leaves mostly acuminate; costa single, slender; laminal cells short, smooth; alar cells mostly weakly differentiated; capsules typically erect; peristome often reduced; exostome teeth often paired.

Dimerodontium Mitt., *Fabronia* Raddi, *Ischyrodon* Müll. Hal., *Levierella* Müll. Hal., *Rhizofabronia* (Broth.) M. Fleisch.

Hypnaceae Schimp. Type: *Hypnum* Hedw. Stems monopodially branched; pseudoparaphyllia foliose or rarely filamentous; paraphyllia none; leaves often falcate or homomallous; costa short and double (or absent); laminal cells mostly linear; capsules mostly inclined and asymmetric; exothecial cells usually not collenchymatous; opercula apiculate to short-rostrate; exostome teeth mostly cross-striolate; calyptrae mostly naked.

Acritodon H. Rob., *Andoa* Ochyra, *Bardunovia* Ignatov & Ochyra, *Breidleria* Loeske, *Bryocrumia* L. E. Anderson, *Buckiella* Ireland, *Callicladium* H. A. Crum, *Calliergonella* Loeske, *Campylophyllopsis* W. R. Buck nom. nov. (*Campylidium* (Kindb.) Ochyra, nom. inval. [Art. 20.2], *Biodiversity Poland* 3: 182. 2003; *Campylium* [unranked] *Campylidium* Kindb., Eur. N. Amer. Bryin. 2: 119. 1896), *Campylophyllum* (Schimp.) M. Fleisch., *Caribaeohypnum* Ando & Higuchi, *Chryso-hypnum* (Hampe) Hampe, *Crepidophyllum* Herzog,

Ctenidiadelphus M. Fleisch., *Cyathothecium* Dixon, *Ectropotheciella* M. Fleisch., *Ectropotheciopsis* (Broth.) M. Fleisch., *Ectropothecium* Mitt., *Elharveya* H. A. Crum, *Elmeriobryum* Broth., *Entodontella* M. Fleisch., *Eurohypnum* Ando, *Foreauella* Dixon & P. de la Varde, *Gammiella* Broth., *Giraldiella* Müll. Hal., *Gollania* Broth., *Hageniella* Broth., *Herzogiella* Broth., *Homomallium* (Schimp.) Loeske, *Hondaella* Dixon & Sakurai, *Horridohypnum* W. R. Buck, *Hyocomium* Bruch & Schimp., *Hypnum* Hedw., *Irelandia* W. R. Buck, *Isopterygiopsis* Z. Iwats., *Leiodontium* Broth., *Leptoischyrodon* Dixon, *Macrothamniella* M. Fleisch., *Mahua* W. R. Buck, *Microctenidium* M. Fleisch., *Mittenothamnium* Henn., *Nanothecium* Dixon & P. de la Varde, *Orthothecium* Bruch & Schimp., *Phyllodon* Bruch & Schimp., *Plagiotheciopsis* Broth., *Platydictya* Berk., *Platygyriella* Cardot, *Podperaea* Z. Iwats. & Glime, *Pseudohypnella* (M. Fleisch.) Broth., *Pseudotaxiphyllum* Z. Iwats., *Ptilium* De Not., *Pylaisia* Schimp., *Rhacopilopsis* Renauld & Cardot, *Rhizohypnella* M. Fleisch., *Sclerohypnum* Dixon, *Stenotheciopsis* Broth., *Stereodon* (Brid.) Mitt., *Stereodontopsis* R. S. Williams, *Syringothecium* Mitt., *Taxiphyllopsis* Higuchi & Deguchi, *Taxiphyllum* M. Fleisch., *Tripterocladium* (Müll. Hal.) A. Jaeger, *Vesicularia* (Müll. Hal.) Müll. Hal., *Wijkiella* Bizot & Lewinsky

Catagoniaceae W. R. Buck & Ireland. Type: *Catagonium* Broth. Stems monopodially branched; pseudoparaphyllia filamentous; leaves conduplicate; costa short and double or absent; laminal cells linear, smooth; alar cells not differentiated; exostome teeth cross-striolate.

Catagonium Broth.

Pterigynandraceae Schimp. Type: *Pterigynandrum* Hedw. Plants terrestrial or epiphytic, mostly relatively small; stems monopodially branched, mostly terete-foliate; paraphyllia absent; costa short and double; laminal cells short, prorulose; alar cells weakly differentiated; gemmae stem-borne; setae smooth; capsules often erect; peristome often reduced.

Habrodon Schimp., *Heterocladium* Bruch & Schimp., *Iwatsukiella* W. R. Buck & H. A. Crum, *Myurella* Bruch & Schimp., *Pterigynandrum* Hedw., *Trachyphyllum* A. Gepp

Hylocomiaceae M. Fleisch. Type: *Hylocomium* Bruch & Schimp. Plants mostly robust; stems monopodially or sympodially branched; paraphyllia often present; leaves often strongly toothed; costae often strong and double; laminal cells elongate, smooth or prorulose; alar cells weakly differentiated; setae very elongate; capsules typically curved and asymmetric; exostome teeth yellow- to red-brown, often with reticulate pattern.

Ctenidium (Schimp.) Mitt., *Hylocomiastrum* Broth., *Hylocomium* Bruch & Schimp., *Leptocladiella* M. Fleisch., *Leptohymenium* Schwägr., *Loeskeobryum* Broth., *Macrothamnium* M. Fleisch., *Meteoriella* S. Okamura, *Neodolichomitra* Nog., *Orontobryum* M. Fleisch., *Pleurozium* Mitt., *Puiggariopsis* M. Menzel, *Rhytidiadelphus* (Limpr.) Warnst., *Rhytidiopsis* Broth., *Schofieldiella* W. R. Buck

Rhytidiaceae Broth. Type: *Rhytidium* (Sull.) Kindb. Plants robust; stems monopodially branched; paraphyllia none; leaves plicate, rugose; costa single; laminal cells linear, strongly porose, prorulose; alar cells well differentiated; exostome teeth yellow-brown, cross-striolate.

Rhytidium (Sull.) Kindb.

Symphyodontaceae M. Fleisch. Type: *Symphyodon* Mont. Stems monopodially branched; laminal cells mostly prorulose; alar cells not or weakly differentiated; setae mostly roughened; capsules symmetric, typically spinose; exostome teeth papillose to cross-striolate; calyptrae cucullate or mitrate.

Chaetomitriopsis M. Fleisch., *Chaetomitrium* Dozy & Molk., *Dimorphocladon* Dixon, *Symphyodon* Mont., *Trachythecium* M. Fleisch., *Unclejackia* Ignatov, T. Kop. & D. Norris

Plagiotheciaceae (Broth.) M. Fleisch. Type: *Plagiothecium* Bruch & Schimp. Plants terrestrial; stems monopodially branched, mostly complanate-foliate; leaves decurrent; costa short and double or absent; laminal cells elongate, often strongly chlorophyllose; alar cells differentiated into the decurrencies; setae smooth; capsules often curved and asymmetric; peristome teeth mostly pale yellow; exostome typically cross-striolate below; endostome well developed.

Plagiothecium Bruch & Schimp., *Struckia* Müll. Hal.

Entodontaceae Kindb. Type: *Entodon* Müll. Hal. Plants often epiphytic; stems monopodially branched; costa short and double or absent; laminal cells linear, smooth; alar cells subquadrate, numerous; capsules erect and symmetric, long-exserted; columella often exserted; peristome inserted below mouth of capsule; endostome mostly strongly reduced.

Entodon Müll. Hal., *Erythrodontium* Hampe, *Mesonodon* Hampe, *Pylaisiobryum* Broth.

Pylaisiadelphaceae Goffinet & W. R. Buck. Type: *Pylaisiadelpha* Cardot. Stems monopodially branched; leaves usually not falcate; costa short and double or none; laminal cells mostly linear, mostly smooth, sometimes papillose; alar cells quadrate, few; exothecial cells not collenchymatous; opercula often straight-rostrate; exostome teeth not furrowed.

Aptychella (Broth.) Herzog, *Brotherella* M. Fleisch., *Clastobryopsis* M. Fleisch., *Clastobryum* Dozy & Molk., *Heterophyllium* (Schimp.) Kindb., *Isocladiella* Dixon, *Isopterygium* Mitt., *Mastopoma* Cardot, *Platygyrium* Bruch & Schimp., *Pterogonidium* Broth., *Pseudotrismegistia* H. Akiyama & Tsubota, *Pylaisiadelpha* Cardot, *Taxitheliella* Dixon, *Taxithelium* Mitt., *Trismegistia* (Müll. Hal.) Müll. Hal., *Wijkia* H. A. Crum

Sematophyllaceae Broth. Type: *Sematophyllum* Mitt. Stems monopodially branched; leaves often golden green, often falcate; costa short and double or none; laminal cells mostly linear, smooth or papillose; alar cells well differentiated; exothecial cells collenchymatous; opercula mostly obliquely rostrate; exostome teeth often furrowed, cross-striolate.

Acanthorrhynchium M. Fleisch., *Acroporium* Mitt., *Allionellopsis* Ochyra, *Aptychopsis* (Broth.) M. Fleisch., *Chinostomum* Müll. Hal., *Clastobryella* M. Fleisch., *Clastobryophilum* M. Fleisch., *Colobodontium* Herzog, *Donnellia* Austin, *Hydropogon* Brid., *Hydropogonella* Cardot, *Macrohymenium* Müll. Hal., *Meiotheciella* B. C. Tan, W. B. Schofield & H. P. Ramsay, *Meiothecium* Mitt., *Papillidiopsis* (Broth.) W. R. Buck & B. C. Tan, *Paranapiacabaea* W. R. Buck & Vital, *Potamium* Mitt., *Pterogoniopsis* Müll. Hal., *Piloecium* (Müll. Hal.) Broth., *Radulina* W. R. Buck & B. C. Tan, *Rhaphidostichum* M. Fleisch., *Schraderella* Müll. Hal., *Schroeterella* Herzog, *Sematophyllum* Mitt., *Timotimius* W. R. Buck, *Trichosteleum* Mitt., *Trolliella* Herzog, *Warburgiella* Müll. Hal.

Cryphaeaceae Schimp. Type: *Cryphaea* D. Mohr. Stems sympodially branched; secondary stems little or not branched; costa single; laminal cells short, smooth or sometimes prorulose; alar cells numerous; capsules immersed or seldom emergent; exostome teeth pale, papillose; endostome rudimentary to absent; calyptrae mitrate.

Cryphaea D. Mohr, *Cryphaeophilium* M. Fleisch., *Cryphidium* (Mitt.) A. Jaeger, *Cyptodon* (Broth.) M. Fleisch., *Cyptodontopsis* Dixon, *Dendroalsia* E. Britton, *Dendrocryphaea* Broth., *Dendropogonella* E. Britton, *Pilotrichopsis* Besch., *Schoenobryum* Dozy & Molk., *Sphaerotheciella* M. Fleisch.

Prionodontaceae Broth. Type: *Prionodon* Müll. Hal. Plants epiphytic; stems sympodially branched; axillary hairs as in *Breutelia* (Bartramiaceae); leaves usually plicate and with strongly toothed margins; costa single; laminal cells short, papillose; alar cells differentiated in large areas; capsules immersed to emergent; annulus revoluble; exostome teeth papillose; endostome segments united into a reticulum.

Prionodon Müll. Hal.

Leucodontaceae Schimp. Type: *Leucodon* Schwägr. Plants mostly epiphytic; stems sympodially branched; secondary stems often not or little branched, mostly curled when dry; leaves rapidly spreading when moist, mostly plicate; costa short and double or none; laminal cells oval to linear, mostly smooth, rarely prorulose; alar cells numerous; capsules usually exserted, often anisosporous; annulus not differentiated; exostome teeth pale, papillose; endostome mostly rudimentary; spores often large.

Antitrichia Brid., *Dozya* Sande Lac., *Eoleucodon* H.A. Mill. & H. Whittier, *Felipponea* Broth., *Leucodon* Schwägr., *Pterogonium* Sw., *Scabridens* E. B. Bartram

Pterobryaceae Kindb. Type: *Pterobryon* Hornsch. Plants mostly epiphytic; stems sympodially branched; secondary stems often well branched, and thus stipitate; pseudoparaphyllia filamentous; stem and branch leaves often differentiated, branch leaves sometimes 5-seriate; costa mostly single, sometimes short and double or absent; laminal cells mostly linear, mostly smooth, sometimes prorulose; alar cells usually differentiated, often thick-walled and colored; capsules mostly immersed; exostome teeth pale, often smooth; endostome most rudimentary; calyptrae cucullate or mitrate, often hairy.

Calyptothecium Mitt., *Cryptogonium* (Müll. Hal.) Hampe, *Henicodium* (Müll. Hal.) Kindb., *Hildebrandtiella* Müll. Hal., *Horikawaea* Nog., *Jaegerina* Müll. Hal., *Micralsopsis* W.R. Buck, *Muellerobryum* M. Fleisch., *Neolindbergia* M. Fleisch., *Orthorrhynchidium* Renauld & Cardot, *Orthostichidium* Dusén, *Orthostichopsis* Broth., *Osterwaldiella* Broth., *Penzigiella* M. Fleisch., *Pireella* Cardot, *Pseudopterobryum* Broth., *Pterobryidium* Broth. & Watts, *Pterobryon* Hornsch., *Pterobryopsis* M. Fleisch., *Renauldia* Müll. Hal., *Rhabdodontium* Broth., *Spriridentopsis* Broth., *Symphysodon* Dozy & Molk., *Symphysodontella* M. Fleisch.

Phyllogoniaceae Kindb. Type: *Phyllogonium* Brid. Plants epiphytic; stems sympodially branched; secondary stems irregularly branched, strongly complanate-foliate; leaves conduplicate, cucullate, auriculate; costa short and double or absent; laminal cells linear, smooth; alar cells differentiated in small groups; capsules immersed or shortly exserted; exostome teeth pale, not or scarcely ornamented; endostome rudimentary or absent; calyptrae cucullate or mitrate, naked or hairy.

Phyllogonium Brid.

Orthorrhynchiaceae S.H. Lin. Type: *Orthorrhynchium* Reichardt. Plants terrestrial; stems monopodially branched; leaves conduplicate, cucullate; costa short and double or absent; laminal cells linear, smooth; alar cells undifferentiated; capsules short-exserted, erect; exostome teeth pale, unornamented; endostome none; calyptrae mitrate, hairy.

Orthorrhynchium Reichardt

Lepyrodontaceae Broth. Type: *Lepyrodon* Hampe. Plants terrestrial or epiphytic; stems sympodially branched; secondary stems not or little branched; leaves sometimes plicate; costa single and weak or short and double to absent; laminal cells linear, smooth, thick-walled and porose; alar cells few or scarcely differentiated; capsules long-exserted; peristome usually only endostomial.

Lepyrodon Hampe

Neckeraceae Schimp. Type: *Neckera* Hedw. Plants terrestrial or epiphytic; stems mostly sympodially branched, sometimes monopodial; stipes sometimes differentiated and plants then frondose; leaves mostly complanately arranged; costa typically single, sometimes short and double; laminal cells fusiform to linear, rarely shorter, mostly smooth, rarely prorulose or papillose; alar cells mostly few or weakly differentiated; capsules immersed (mostly in epiphytes) to long-exserted (mostly in terrestrial taxa); exostome teeth often pale, usually cross-striolate at least at extreme base, papillose above; endostome often reduced; calyptrae mostly cucullate.

Baldwiniella M. Fleisch., *Bissetia* Broth., *Bryolawtonia* D.H. Norris & Enroth, *Caduciella* Enroth, *Crassiphyllum* Ochyra, *Cryptoleptodon* Renauld & Cardot, *Curvicladium* Enroth, *Dixonia* Horik. & Ando, *Dolichomitra* Broth., *Handeliobryum* Broth., *Himantocladium* (Mitt.) M. Fleisch., *Homalia* (Brid.) Bruch & Schimp., *Homaliadelphus* Dixon & P. de la Varde, *Homaliodendron* M. Fleisch., *Hydrocryphaea* Dixon, *Isodrepanium* (Mitt.) E. Britton, *Metaneckera* Steere, *Neckera* Hedw., *Neckeropsis* Reichardt, *Neomacounia* Ireland, *Noguchiodendron*

Ninh & Pócs, *Pendulothecium* Enroth & S. He, *Pinnatella* M. Fleisch., *Porotrichodendron* M. Fleisch., *Porotrichopsis* Broth. & Herzog, *Porotrichum* (Brid.) Hampe, *Thamnobryum* Nieuwl., *Touwia* Ochyra

Echinodiaceae Broth. Type: *Echinodium* Jur. Plants epipetric or less often on soil or bases of trees; stems sympodially branched, wiry; secondary stems irregularly branched; leaves mostly subulate; costa single, mostly excurrent; laminal cells short, smooth; alar cells weakly differentiated; capsules long-exserted, inclined to horizontal; exostome teeth reddish, cross-striolate; endostome well developed.

Echinodium Jur.

Leptodontaceae Schimp. Type: *Leptodon* D. Mohr. Plants mostly epiphytic; stems sympodially branched, often curled when dry; secondary stems irregularly branched to bipinnate; costa typically single; laminal cells isodiametric to long-hexagonal, smooth, unipapillose or prorulose; alar cells numerous; capsules immersed to short-exserted; exostome teeth pale, unornamented to spiculose; endostome rudimentary; calyptrae hairy.

Alsia Sull., *Forsstroemia* Lindb., *Leptodon* D. Mohr, *Taiwanobryum* Nog.

Lembophyllaceae Broth. Type: *Lembophyllum* Lindb. Plants often turgid; stems monopodially branched; leaves mostly strongly concave; costa mostly short and double (rarely single); laminal cells elongate, smooth; alar cells often somewhat differentiated; capsules mostly erect and immersed to short-exserted; endostome mostly reduced; calyptrae rarely mitrate, naked or hairy.

Acrocladium Mitt., *Bestia* Broth., *Camptochaete* Reichardt, *Dolichomitriopsis* S. Okamura, *Fallaciella* H.A. Crum, *Fifea* H.A. Crum, *Isothecium* Brid., *Lembophyllum* Lindb., *Neobarbella* Nog., *Orthostichella* Müll. Hal., *Pilotrichella* (Müll. Hal.) Besch., *Weymouthia* Broth.

Myuriaceae M. Fleisch. Type: *Myurium* Schimp. Stems sympodially branched; secondary stems little or not branched; leaves mostly long-acuminate; costa short and double or none; laminal cells linear, smooth; alar cells well differentiated, mostly colored; capsules long-exserted, erect; exostome teeth reduced, smooth, often perforate; endostome rudimentary.

Eumyurium Nog., *Myurium* Schimp., *Oedicladium* Mitt., *Palisadula* Toyama

Anomodontaceae Kindb. Type: *Anomodon* Hook. & Taylor. Plants mostly epiphytic; stems sympodially or monopodially branched, secondary stems and/or branches often curled when dry, not complanate-foliate; paraphyllia none; leaves often acute to obtuse; costa single or short and double; laminal cells mostly short, papillose or prorulose; alar cells mostly poorly differentiated; capsules exserted, erect; exostome teeth pale to white, cross-striolate sometimes with overlying papillae to papillose; endostome often reduced.

Anomodon Hook. & Taylor, *Bryonorrisia* L. R. Stark & W. R. Buck, *Chileobryon* Enroth, *Curviramea* H. A. Crum, *Haplohymenium* Dozy & Molk., *Herpetineuron* (Müll. Hal.) Cardot, *Schwetschkeopsis* Broth.

Theliaceae M. Fleisch. Type: *Thelia* Sull. Plants terrestrial or on bases of trees; stems monopodially branched; paraphyllia present; leaves imbricate, little altered when moist, deltoid-ovate; costa single; laminal cells short, stoutly unipapillose; alar cells differentiated; capsules exserted, erect; exostome teeth white, smooth to papillose; endostome strongly reduced.

Thelia Sull.

Microtheciellaceae H. A. Mill. & A. J. Harr. Type: *Microtheciella* Dixon. Plants epiphytic; stems monopodially branched; costa single; laminal cells short, smooth; alar cells weakly differentiated;, capsules short-exserted, erect; exostome teeth truncate, reduced, weakly ornamented; endostome rudimentary.

Microtheciella Dixon

Sorapillaceae M. Fleisch. Type: *Sorapilla* Spruce & Mitt. Leaves distichous and complanate; capsules cladocarpous, immersed; peristome double, of 16 slender segments and 32 stout exostome knobs, cilia absent.

Sorapilla Spruce & Mitt.

Acknowledgments

The National Science Foundation is acknowledged for its financial support to A. J. Shaw through grant DEB-0529593.

References

Akiyama, H. (1986). Notes on, little known species of the genus *Leucodon* with immersed or laterally exserted capsules. *Acta Phytotaxonomica et Geobotanica*, **37**, 128–36.

Akiyama, H. & Nishimura, H. (1993). Further studies on branch buds in mosses; "pseudoparaphyllia" and "scaly leaves." *Journal of Plant Research*, **106**, 101–8.

Anderson, L. E. (1980). Cytology and reproductive biology of mosses. In *The Mosses of North America*, ed. R. J. Taylor & A. E. Leviton, pp. 37–76. San Francisco, CA: Pacific Division, AAAS.

Andersson, K. (2002). Dispersal of spermatozoids from splash-cups of the moss *Plagiomnium affine*. *Lindbergia*, **27**, 90–6.

Arts, T. (1986). Drought resistant rhizoidal tubers in *Fissidens cristatus* Wils. *ex* Mitt. *Lindbergia*, **12**, 119–20.

Arts, T. (1990). Moniliform rhizoidal tubers in *Archidium alternifolium* (Hedw.) Schimp. *Lindbergia*, **16**, 59–61.

Arts, T. (1994). Rhizoidal tubers and protonemal gemmae in European *Ditrichum* species. *Journal of Bryology*, **18**, 43–61.

Arts, T. (1996). The genus *Splachnobryum* in Africa, with new combinations in *Bryum* and *Gymnostomiella*. *Journal of Bryology*, **19**, 65–77.

Bauer, L. (1959). Auslösung apogamer Sporogonbildung am Regenerationsprotonema von Laubmoosen durch einen vom Muttersporogon abgegebenen Faktor. *Naturwissenschaften*, **46**, 154–5.

Bell, N. E. & Newton, A. E. (2007). Pleurocarpy in the rhizogoniaceous clade. In *Pleurocarpous Mosses: Systematics and Evolution*, ed. A. E. Newton & R. S. Tangney, pp. 41–64. London: Taylor & Francis.

Bell, N. E. & York, P. V. (2007). *Vetiplanaxis pyrrhobryoides*, a new fossil moss genus and species from Middle Cretaceous Burmese amber. *Bryologist*, **110**, 514–20.

Berthier J. (1973). Recherches sur la structure et le dévelopement de l'apex du gametophyte feuillé des mousses. *Revue Bryologique et Lichénologique*, **38**, 421–551.

Berthier, J., Bonnot, E. J. & Hébant, C. (1971). Analyse d'un exemple de développement foliare hétéroblastique chez les mousses: apparition de feuilles filamenteuses muscigènes au cours de l'ontogénèse des rameaux latéraux de certaines Dicranales et Encalyptales. *Comptes Rendus de l'Académie des Sciences*, **273**, 2232–5.

Bopp M. (1954). Untersuchungen über Wachstum und Kapselentwicklung normaler und isolierter Laubmoosporogone. *Zeitschrift für Botanik*, **42**, 331–52.

Bopp, M. (1956). Die Bedeutung der Kalyptra für die Entwicklung der Laubmoossporogone. *Berichte der Deutschen Botanischen Gesellschaft*, **69**, 455–68.

Bopp, M. (1961). Morphogenese der Laubmoose. *Biological Review*, **36**, 237–80.

Bopp, M. & Bhatla, S. C. (1990). Physiology of sexual reproduction in mosses. *Critical Reviews in Plant Sciences*, **9**, 317–27.

Boudier, P. (1988). Différenciation structurale de l'épiderme du sporogone chez *Sphagnum fimbriatum* Wilson. *Annales des Sciences Naturelles, Botanique, 13ème série*, **8**, 143–56.

Bridel, S. E. (1826–27). *Bryologia Universa*, seu, Systematica ad novam methodum dispositio, historia et descriptio omnium muscorum frondosorum huscusque cognitorum cum synonymia ex auctoribus probatissimis. Lipsiae (Leipzig): Sumtibus J.A. Barth.

Brotherus, V. F. (1924–25). Musci (Laubmoose). In *Die Natürlichen Pflanzenfamilien*, 2nd edn, vols. 10–11, ed. A. Engler. Leipzig: Wilhelm Engelmann.

Brown, R. C. & Lemmon, B. E. (1981). Aperture development in spores of the moss, *Trematodon longicollis* Mx. *Protoplasma*, **106**, 273–87.

Brown, R. C. & Lemmon, B. E. (1990). Sporogenesis in Bryophytes. In *Microspores: Evolution and Ontogeny*, ed. S. Blackmore & R. B. Knox, pp. 55–94. London: Academic Press.

Bryan, V. S. (2001). Apospory in mosses discovered by Nathanael Pringsheim in a brilliant epoch of botany. *Bryologist*, **104**, 40–6.

Buck, W. R. (1998). Pleurocarpous mosses of the West Indies. *Memoirs of The New York Botanical Garden*, **82**, 1–400.

Buck, W. R. (2007). The history of pleurocarp classification: Two steps forward, one step back. In *Pleurocarpous Mosses: Systematics and Evolution*, ed. A. E. Newton & R. S. Tangney, pp. 1–18. Boca Raton, FL: Taylor & Francis.

Buck, W. R. & Goffinet, B. (2000). Morphology and classification of mosses. In *Bryophyte Biology*, ed. A. J. Shaw & B. Goffinet, pp. 71–123. Cambridge: Cambridge University Press.

Buck, W. R. & Vitt, D. H. (1986). Suggestions for a new classification of pleurocarpous mosses. *Taxon*, **35**, 21–60.

Buck, W. R., Goffinet, B. & Shaw, A. J. (2000). Testing morphological concepts of orders of pleurocarpous mosses (Bryophyta) using phylogenetic reconstructions based on *trnL-trnF* and *rps4* sequences. *Molecular Phylogenetics and Evolution*, **16**, 180–98.

Budke, J. M., Jones, C. S. & Goffinet, B. (2007). Development of the enigmatic peristome of *Timmia megapolitana* (Timmiaceae; Bryophyta). *American Journal of Botany*, **94**, 460–7.

Campbell, D. H. (1895). *The Structure and Development of the Mosses and Ferns (Archegoniae)*. London: Macmillan.

Carafa, A., Duckett, J. G., Knox, J. P. & Ligrone, R. (2005). Distribution of cell-wall xylans in bryophytes and tracheophytes: new insights into basal interrelationships of land plants. *New Phytologist*, **168**, 231–40.

Chamberlin, M. A. (1980). The morphology and development of the gametophytes of *Fissidens* and *Bryoxiphium* (Bryophyta). M. A. thesis, Southern Illinois University, Carbondale.

Chopra, R. N. (1988). In vitro production of apogamy and apospory in bryophytes and their significance. *Journal of the Hattori Botanical Laboratory*, **64**, 169–75.

Cooke, T. J., Poli, D.-B., Sztein, A. E. & Cohen, J. D. (2002). Evolutionary pattern in auxin action. *Plant Molecular Biology*, **49**, 319–38.

Correns, C. (1899). *Untersuchungen über die Vermehrung der Laubmoose durch Brutorgane und Stecklinge*. Jena: Fischer.

Cox, C. J., Goffinet, B., Shaw, A. J. & Boles, S. B. (2004). Phylogenetic relationships among the mosses based on heterogeneous Bayesian analysis of multiple genes from multiple genomic compartments. *Systematic Botany*, **29**, 234–50.

Crandall-Stotler, B. (1980). Morphogenetic designs and a theory of bryophyte origins and divergence. *BioScience*, **30**, 580–5.

Crandall-Stotler, B. (1984). Musci, hepatics and anthocerotes – an essay on analogues. In *New Manual of Bryology*, vol. 2, ed. R. M. Schuster, pp. 1093–129. Nichinan: Hattori Botanical Laboratory.

Crandall-Stotler, B. (1986). Morphogenesis, developmental anatomy and bryophyte phylogenetics: contraindications of monophyly. *Journal of Bryology*, **14**, 1–23.

Cronberg, N., Natcheva, R. & Hedlund, K. (2006). Microarthropods mediate sperm transfer in mosses. *Science*, **313**, 1255.

Crosby, M. R. (1980). The diversity and relationships of mosses. In *The Mosses of North America*, ed. R. J. Taylor & A. E. Leviton, pp. 115–29. San Francisco, CA: Pacific Division, AAAS.

Crum, H. A. (2001). *Structural Diversity of Bryophytes*. University of Michigan Herbarium.

Crundwell, A. C. (1979). Rhizoids and moss taxonomy. In *Bryophyte Systematics*, ed. G. C. S Clarke & J. G. Duckett, pp. 347–63. London: Academic Press.

Duckett, J. G. (1994a). Studies of protonemal morphogenesis in mosses. VI. The foliar rhizoids of *Calliergon stramineum* (Brid.) Kindb. function as organs of attachment. *Journal of Bryology*, **18**, 239–52.

Duckett, J. G. (1994b). Studies of protonemal morphogenesis in mosses. V. *Diphyscium foliosum* (Hedw.) Mohr (Buxbaumiales). *Journal of Bryology*, **18**, 223–38.

Duckett, J. G. & Ligrone, R. (1992). A survey of diaspore liberation mechanisms and germination patterns in mosses. *Journal of Bryology*, **17**, 335–54.

Duckett, J. G. & Ligrone, R. (1994). Studies of protonemal morphogenesis in mosses. III. The perennial gemmiferous protonema of *Rhizomnium punctatum* (Hedw.) Kop. *Journal of Bryology*, **18**, 13–26.

Duckett J. G., Schmid, A. M. & Ligrone, R. (1998). Protonemal morphogenesis. In *Bryology for the Twenty-first Century*, ed. J. W. Bates, N. W. Ashton & J. G. Duckett, pp. 223–46. Leeds: Maney & British Bryological Society.

Duckett, J. G., Burch, J., Fletcher, P. W. *et al.* (2004). *In vitro* cultivation of bryophytes: a review of practicalities, problems, progress and promise. *Journal of Bryology*, **26**, 3–20.

Edwards, D. (2000). The role of Mid-Palaeozoic mesofossils in the detection of early bryophytes. *Philosophical Transaction of the Royal Society of London*, B**355**, 733–55.

Fleischer, M. (1904–23). *Die Musci der Flora von Buitenzorg (zugleich Laubmoosflora von Java)*, 4 vols. Leiden: Brill.

Frahm, J.-P. & Frey, W. (1987). The twist mechanism in the cygneous setae of the genus *Campylopus*. Morphology, structure and function. *Nova Hedwigia*, **44**, 291–304.

French, J. C. & Paolillo Jr., D. J. (1975a). Intercalary meristematic activity in the sporophyte of *Funaria* (Musci). *American Journal of Botany*, **62**, 86–96.

French, J. C. & Paolillo Jr., D. J. (1975b). On the role of the calyptra in permitting expansion of capsules in the moss *Funaria*. *Bryologist*, **78**, 438–46.

Frey, W. (1970). Blattentwicklung bei Moosen. *Nova Hedwigia*, **20**, 463-565.

Frey, W., Hofmann, M. & Hilger, H. H. (2001). The gametophyte-sporophyte junction: unequivocal hints for two evolutionary lines of archegoniate land plants. *Flora*, **196**, 431–45.

Fritsch, R. (1991). Index to chromosome counts. *Bryophytorum Bibliotheca*, **40**, 1–352.

Garner, D. L. B. & Paolillo Jr, D. J. (1973). On the functioning of stomata in *Funaria*. *Bryologist*, **76**, 423–7.

Goebel, K. (1898a). *Organographie der Pflanzen insbesondere der Archegoniaten und Saamenpflanzen*. Zweiter Teil. 1. Heft. Bryophyten. Jena, Germany.

Goebel, K. (1898b). Über den Öffnungsmechanismus der Moos-antheridien. *Annales du Jardin Botanique de Buitenzorg*, suppl. **2**, 65–72.

Goffinet, B. (1997a). The Rhachitheciaceae: revised circumscription and ordinal affinities. *Bryologist*, **100**, 425–39.

Goffinet, B. (1997b). Phylogeny of the Orthotrichales (Bryopsida). Doctoral dissertation, University of Alberta, Edmonton, Canada.

Goffinet, B. & Buck, W. R. (2004). Systematics of the Bryophyta (Mosses): from molecules to a new classification. *Monographs in Systematic Botany from the Missouri Botanical Garden*, **98**, 205–39.

Goffinet, B. & Cox, C. J. (2000). Phylogenetic relationships among basal-most arthrodontous mosses with special emphasis on the evolutionary significance of the Funariineae. *Bryologist*, **103**, 212–23.

Goffinet, B., Shaw, J., Anderson, L. E. & Mishler, B. D. (1999). Peristome development in mosses in relation to systematics and evolution. V. Diplolepideae: Orthotrichaceae. *Bryologist*, **102**, 581–94.

Goffinet, B., Cox, C. J., Shaw, A. J. & Hedderson, T. J. (2001). The Bryophyta (Mosses): Systematic and evolutionary inferences from an *rps4* gene (cpDNA) phylogeny. *Annals of Botany*, **87**, 191–208.

Goffinet, B., Wickett, N. J., Werner, O. *et al.* (2007). Distribution and phylogenetic significance of the 71-kb inversion in the plastid genome in Funariidae (Bryophyta). *Annals of Botany*, **99**, 747–53.

Graham, L. K. E. & Wilcox, L. W. (2000). The origin of alternation of generations in land plants: a focus on matrotrophy and hexose transport. *Philosophical Transaction of the Royal Society of London*, B**355**, 757–67.

Greene, S. W. (1960). The maturation cycle, or the stages of development of gametangia and capsules in mosses. *Transactions of the British Bryological Society*, **3**, 736–45.

Győrffy, I. (1917). Über die «Apophyse» der Moose. *Magyar Botanikai Lapok*, **16**, 131–5.

Győrffy, I. (1929). Monstruoses Sporophyton von *Tetraplodon bryoïdes* aus Suomi. *Annales Societatis Zoologicae Botanicae Fennicae Vanamo*, **9**(7), 299–319.

Győrffy, I. (1931). *Sphagnum*-Monstruositaten aus der Hohen-Tàtra. *Revue Bryologique et Lichénologique*, **4**, 191–3, pl. V.

Győrffy, I. (1934). Musci monstruosi transsilvanici. I. *Catharinea haussknechtii* torzok erdélyből. *Erdélyi Múzeum*, **39**, 341–8.

Haberlandt, G. (1886). Beiträge zur Anatomie und Physiologie der Laubmoose. *Jahrbücher für Wissenschaftliche Botanik*, **17**, 359–498, pl. 21–27.

Harvey-Gibson, R. J. & Miller-Brown, D. (1927). Fertilization of Bryophyta. *Polytrichum commune* (Preliminary note). *Annals of Botany*, **41**, 190–1.

Hausmann, M. K. & Paolillo Jr., D. J. (1977). On the development and maturation of antheridia in *Polytrichum*. *Bryologist*, **80**, 143–8.

Hébant, C. (1977). The conducting tissues of bryophytes. *Bryophytorum Bibliotheca*, **10**, i–xi, 1–157.

Heckman, D. S., Geiser, D. M., Eidell, B. R. *et al.* (2001). Molecular evidence for the early colonization of land by fungi and plants. *Science*, **293**, 1129–33.

Hedenäs, L. (1989a). Axillary hairs in pleurocarpous mosses – a comparative study. *Lindbergia*, **1**, 166–80.

Hedenäs, L. (1989b). Some neglected character distribution patterns among the pleurocarpous mosses. *Bryologist*, **92**, 157–63.

Hedenäs, L. (1993). Higher taxonomic level relationships among diplolepidous pleurocarpous mosses – a cladistic overview. *Journal of Bryology*, **18**, 723–81.

Hedenäs, L. (1994). Basal pleurocarpous diplolepideous mosses – a cladistic approach. *Bryologist*, **97**, 225–43.

Herzfelder, H. (1923). Experimente an Sporophyten von *Funaria hygrometrica*. *Flora*, **116**, 476–90.

Hofmeister, W. (1870). Ueber die Zellenfolge im Achsenscheitel der Laubmooses. *Botanische Zeitung*, **28**, 441–9, Taf. VII, 457–66, 473–8.

Horton, D. G. (1982). A revision of the Encalyptaceae (Musci), with particular reference to the North American taxa. Part I. *Journal of the Hattori Botanical Laboratory*, **53**, 365–418.

Hughes, J. G. (1969). Factors conditioning development of sexual and apogamous races of *Phascum cuspidatum* Hedw. *New Phytologist*, **68**, 883–900.

Huttunen, S., Ignatov, M. S., Müller, K. & Quandt, D. (2004). Phylogeny and evolution of epiphytism in the three moss families Meteoriaceae, Brachytheciaceae, and Lembophyllaceae. *Monographs in Systematic Botany from the Missouri Botanical Garden*, **98**, 328–61.

Ignatov, M. S. & Hedenäs, L. (2007). Homologies of stem structures in pleurocarpous mosses, especially of pseudoparaphyllia and similar structures. In *Pleurocarpous Mosses: Systematics and Evolution*, ed. A. E. Newton & R. S. Tangney, pp. 269–86. Boca Raton, FL: Taylor & Francis.

Ignatov, M. S. & Ignatova, E. A. (2004). Moss flora of the Middle European Russia. Vol. 2: Fontinalaceae-Amblystegiaceae [In Russian]. *Arctoa*, **11** (suppl. 2), 611–960.

Imura, S. & Iwatsuki, Z. (1990). Classification of vegetative diaspores on Japanese mosses. *Hikobia*, **10**, 435–43.

Ingold, C. T. (1959). Peristome teeth and spore discharge in mosses. *Transactions of the Botanical Society of Edinburgh*, **38**, 76–88.

Ingold, C. T. (1965). *Spore Liberation*. Oxford: Clarendon Press.

Iwatsuki, Z. (1986). A peculiar New Caledonian *Sphagnum* with rhizoids. *Bryologist*, **89**, 20–2.

Janzen, P. (1917). Die Haube der Laubmoose. *Hedwigia*, **58**, 158–280.

Kato, M. & Akiyama, H. (2005). Interpolation hypothesis for the origin of the vegetative sporophyte of land plants. *Taxon*, **54**, 443–50.

Kawai, I. (1968). Taxonomic studies on the midrib in Musci. (1) Significance of the midrib in systematic Botany. *The Science Reports of Kanazawa University*, **13**, 127–57.

Kawai, I. (1989). Systematic studies on the conducting tissues of the gametophyte in Musci: XVI. Relationships between the anatomical characteristics of the stem and the classification system. *Asian Journal of Plant Science*, **1**, 19–52.

Kenrick, P. & Crane, P. R. (1997). *The Origin and Early Diversification of Land Plants*. Washington, D.C.: Smithsonian Institution Press.

Konopka, A. S., Herendeen, P. S., Merrill, G. S. S. & Crane, P. R. (1997). Sporophytes and gametophytes of Polytrichaceae from the Capanian (Late Cretaceous) of Georgia, U.S.A. *International Journal of Plant Sciences*, **158**, 489–99.

Konopka, A. S., Herendeen, P. S. & Crane, P. R. (1998). Sporophytes and gametophytes of Dicranaceae from the Santonian (Late Cretaceous) of Georgia, USA. *American Journal of Botany*, **85**, 714–23.

Koponen, T. (1982). Rhizoid topography and branching patterns in mosses taxonomy. *Nova Hedwigia*, suppl. **71**, 95–9.

Koponen A. (1990). Entomophily in the Splachnaceae. *Botanical Journal of the Linnean Society*, **104**, 115–27.

Kruijer, J. D. H. (2002). Hypopterygiaceae of the world. *Blumea Supplement*, **13**, 1–388.

Kürschner, H. (2004). Intracapsular spore germination in *Brachymenium leptophyllum* (Müll. Hal.) A. Jaeger (Bryaceae, Bryopsida) – an achorous strategy. *Nova Hedwigia*, **78**, 447–51.

Laaka-Lindberg, S., Korpelainen, H. & Pohjamo, M. (2003). Dispersal of asexual propagules in bryophytes. *Journal of the Hattori Botanical Laboratory*, **93**, 319–30.

LaFarge-England, C. (1996). Growth-form, branching pattern, and perichaetial position in mosses: cladocarpy and pleurocarpy redefined. *Bryologist*, **99**, 170–86.

Lal, M. (1984). The culture of bryophytes including apogamy, apospory, parthenogenesis and protoplasts. In *The Experimental Biology of Bryophytes*, ed. A. F. Dyer & J. G. Duckett, pp. 97–115. London: Academic Press.

Leitgeb, H. (1876). Ueber verzweigte Moossporogonien. *Mitteilungen des Naturwissenschaftlichen Vereines für Steiermark*, **13**, 1–20.

Leitgeb, H. (1882). Die Antheridienstände der Laubmoose. *Flora*, **65**, 467–74.

Ligrone, R. (1986). Structure, development and cytochemistry of mucilage secreting hairs in the moss *Timmiella barbuloides* (Brid.) Moenk. *Annals of Botany*, **58**, 859–68.

Ligrone, R. & Duckett, J. G. (1998). Development of the leafy shoot in *Sphagnum* (Bryophyta) involves the activity of both apical and subapical meristems. *New Phytologist*, **140**, 581–95.

Ligrone, R., Duckett, J. G. & Renzaglia, K. S. (1993). The gametophyte-sporophyte junction in land plants. *Advances in Botanical Research*, **19**, 231–317.

Ligrone, R., Duckett, J. G. & Gambardella, G. (1996). Development and liberation of cauline gemmae in the moss *Aulacomnium androgynum* (Hedw.) Schwaegr. (Bryales): An ultrastructural study. *Annals of Botany*, **78**, 559–68.

Ligrone, R., Duckett, J. G. & Renzaglia, K. S. (2000). Conducting tissues and phyletic relationships of bryophytes. *Philosophical Transactions of the Royal Society of London*, B**355**, 795–813.

Ligrone, R., Vaughn, K. C., Renzaglia, K. S., Knox, J. P. & Duckett, J. G. (2002). Diversity in the distribution of polysaccharide and glycoprotein epitopes in the cell walls of bryophytes: new evidence for the multiple evolution of water-conducting cells. *New Phytologist*, **156**, 491–508.

Malcolm, B. & Malcolm, N. (2006). *Mosses and other Bryophytes, an Illustrated Glossary*. 2nd edn. Nelson, New Zealand: Microoptics Press.

Mallón, R., Reinoso, J., Rodríguez-Oubiña, J. & González, M. L. (2006). *In vitro* development of vegetative propagules in *Splachnum ampullaceum*: brood cells and chloronematal bulbils. *Bryologist*, **109**, 215–23.

Malta, N. (1926). Die Gattung *Zygodon* Hook. & Tayl. Eine monographische Studie. *Latvijas Universitates Botanika Darza Darbi*, **1**, 1–185.

Martínez-Abaigar, J., Núñez-Olivera, E., Matcham, H. W. & Duckett, J. G. (2005). Interactions between parasitic fungi and mosses: pegged and swollen-tipped rhizoids in *Funaria* and *Bryum*. *Journal of Bryology*, **27**, 47–53.

Marchal, É. & Marchal, É. (1907). Aposporie et sexualité chez les mousses. *Bulletin de l'Académie Royale de Belgique*, **7**, 766–89.

Marchal, E. & Marchal, E. (1911). Aposporie et sexualité chez les mousses. *Bulletin de l'Académie Royale de Belgique*, **9–10**, 750–78.

Meusel, H. (1935). Wuchsformen und Wuchstypen der europaischen Laubmoose. *Nova Acta Leopoldina (n.s.)*, **3**(12), 219–77.

Miksche, G. E. & Yasuda, S. (1978). Lignin of 'giant' mosses and some related species. *Phytochemistry*, **17**, 503–4.

Miller, C. C. J. & Duckett, J. G. (1985). Cytoplasmic deletion during spermatogenesis in mosses. *Gamete Research*, **13**, 253–70.

Mishler, B. D. (1986). Ontogeny and phylogeny in *Tortula* (Musci: Pottiaceae). *Systematic Botany*, **11**, 189–208.

Mishler, B. D. (1988). Relationships between ontogeny and phylogeny, with reference to bryophytes. In *Ontogeny and Systematics*, ed. C. J. Humphries, pp. 117–36. New York: Columbia University Press.

Mishler, B. D. & Churchill, S. P. (1984). A cladistic approach to the phylogeny of the "bryophytes." *Brittonia*, **36**, 406–24.

Mishler, B. D. & Churchill, S. P. (1985). Transition to a land flora: phylogenetic relationships of the green algae and bryophytes. *Cladistics*, **1**, 305–28.

Mishler, B. D. & DeLuna, E. (1991). The use of ontogenetic data in phylogenetic analyses of mosses. *Advances in Bryology*, **4**, 121–67.

Mitten, W. (1859). Musci Indiae Orientalis. An enumeration of the mosses of the East Indies. *Journal of the Proceedings of the Linnean Society*, Supplement to Botany, **1**, 1–171.

Mogensen, G. S. (1978). Spore development and germination in *Cinclidium* (Mniaceae, Bryophyta), with special reference to spore mortality and false anisospory. *Canadian Journal of Botany*, **56**, 1032–60.

Mogensen, G. S. (1983). The spore. In *New Manual of Bryology*, vol. 2, ed. R. M. Schuster, pp. 323–43. Nichinan: Hattori Botanical Laboratory.

Moutschen, J. (1951). Quelques cas nouveaux d'aposporie chez les mousses. *Lejeunia*, **15**, 41–50.

Muggoch, H. & Walton, J. (1942). On the dehiscence of the antheridium and the part played by surface tension in the dispersal of spermatocytes in Bryophyta. *Proceedings of the Royal Society of London*, B**130**, 448–61.

Murray, B. M. (1988). Systematics of the Andreaeopsida (Bryophyta): Two orders with links to *Takakia*. *Beiheft zur Nova Hedwigia*, **90**, 289–336.

Natcheva, R. & Cronberg, N. (2007). Maternal transmission of cytoplasmic DNA in interspecific hybrids of peat mosses, *Sphagnum* (Bryophyta). *Journal of Evolutionary Biology*, **20**, 1613–16.

Nehira, K. (1983). Spore germination, protonema development and sporeling development. In *New Manual of Bryology*, vol. 2, ed. R. M. Schuster, pp. 343–86. Nichinan: Hattori Botanical Laboratory.

Newton, A. E. (2007). Branching architecture in pleurocarpous mosses. In *Pleurocarpous Mosses: Systematics and Evolution*, ed. A. E. Newton & R. S. Tangney, pp. 287–307. Boca Raton, FL: Taylor & Francis.

Newton, A. E. & Mishler, B. D. (1994). The evolutionary significance of asexual reproduction in mosses. *Journal of the Hattori Botanical Laboratory*, **76**, 127–45.

Newton, A. E. & Tangney, R. S. (eds.) (2007). *Pleurocarpous Mosses: Systematics and Evolution*. Boca Raton, FL: Taylor & Francis.

Newton, A. E., Cox, C., Duckett, J. G. *et al.* (2000). Evolution of the major moss lineages. *Bryologist*, **103**, 187–211.

Newton, A. E., Wikström, N., Bell, N., Forrest, L. L. & Ignatov, M. S. (2007). Dating the diversification of the pleurocarpous mosses. In *Pleurocarpous Mosses: Systematics and Evolution*, ed. A. E. Newton & R. S. Tangney, pp. 337–66. Boca Raton, FL: Taylor & Francis.

Newton, M. E. (1984). The cytogenetics of bryophytes. In *The Mosses of North America*, ed. R. J. Taylor & A. E. Leviton, pp. 65–96. San Francisco, CA: Pacific Division, AAAS.

O'Brian, T. J. (2007). The phylogenetic distribution of pleurocarpous mosses. In *Pleurocarpous Mosses: Systematics and Evolution*, ed. A. E. Newton & R. S. Tangney, pp. 19–40. Boca Raton, FL: Taylor & Francis.

Odu, E. A. (1978). The adaptive importance of moss rhizoids for attachment to the substratum. *Journal of Bryology*, **10**, 163–81.

Oehlkers, F. & Bopp, M. (1957). Entwicklungsphysiologische Untersuchungen an Moosmutanten. II. Die Korrelation zwischen Sporogon und Kalyptra bei Mutanten von *Funaria* und *Physcomitrium*. *Zeitschrift für Induktive Abstammungs- und Vererbungslehre*, **88**, 608–18.

Paolillo Jr., D. F. (1975). The release of sperms from the antheridia of *Polytrichum formosum*. *New Phytologist*, **74**, 287–93.

Parsons, J. G., Cairns, A., Johnson, C. N. *et al.* (2007). Bryophyte dispersal by flying foxes: a novel discovery. *Oecologia*, **152**, 112–14.

Paton, J. A. & Pearce, J. V. (1957). The occurrence, structure and functions of the stomata in British Bryophytes. Part II. *Transactions of the British Bryological Society*, **3**, 242–59.

Pelser, P. B., Kruijer, H. J. D. & Verpoorte, R. (2002). What is the function of oil-containing rudimentary branches in the moss *Canalohypopterygium tamariscinum*? *New Zealand Journal of Botany*, **40**, 149–53.

Philibert, H. (1884). De l'importance du péristome pour les affinités naturelles des mousses. 2e article. *Revue Bryologique*, **11**, 65–72.

Pringsheim, N. (1876). Über vegetative Sprossung der Moosfrüchte. [Vorrlaüfige Mitteilung] *Monatsberichte der Königlichen Preussischen Akademie der Wissenschaft zu Berlin*, **1876**, 425–30.

Proctor, M. C. F. (1979). Surface wax on the leaves of some mosses. *Journal of Bryology*, **10**, 531–8.

Proctor, M. C. F. (1984). Structure and ecological adaptation. In *The Experimental Biology of Bryophytes*, ed. A. F. Dyer & J. G. Duckett, pp. 39–64. London: Academic Press.

Proctor, M. C. F. (1992). Scanning electron microscopy of lamella-margin characters and the phytogeography of the genus *Polytrichadelphus*. *Journal of Bryology*, **17**, 317–33.

Proctor, M. C. F. (2000a). The bryophyte paradox: tolerance of desiccation, evasion of drought. *Plant Ecology*, **151**, 14–49.

Proctor, M. C. F. (2000b). Physiological ecology. In *Bryophyte Biology*, ed. A. J. Shaw & B. Goffinet, pp. 225–47. Cambridge: Cambridge University Press.

Proctor, M. C. F. (2005). Why do Polytrichaceae have lamellae? *Journal of Bryology*, **27**, 221–9.

Qiu, Y. L., Li, L., Wang, B. *et al.* (2006). The deepest divergences in land plants inferred from phylogenomic evidence. *Proceedings of the National Academy of Sciences, U.S.A.*, **103**, 15511–16.

Ramsay, H. P. (1979). Anisospory and sexual dimorphism in the Musci. In *Bryophyte Systematics*, ed. G. C. S. Clarke & J. G. Duckett, pp. 479–509. London: Academic Press.

Raven, J. A. (2002). Selection pressures on stomatal evolution. *New Phytologist*, **153**, 371–86.

Raven, J. A. (2003). Long-distance transport in non-vascular plants. *Plant, Cell and Environment*, **26**, 73–85.

Reese, W. D. (2000). Extreme leaf dimorphism in Calymperaceae. *Bryologist*, **103**, 534–40.

Reese, W. D. & Tan, B. C. (1983). The "petiolate" Calymperaceae: A review with a new species. *Bulletin of the National Science Museum Series B*, **9**, 23–32.

Renault, S., Bonnemain, J. L., Faye, L. & Gaudillere, J. P. (1992). Physiological aspects of sugar exchange between the gametophyte and the sporophyte of *Polytrichum formosum*. *Plant Physiology*, **100**, 1815–22.

Renzaglia, K. S. & Garbary, D. J. (2001). Motile gametes of land plants: Diversity, development, and evolution. *Critical Reviews in Plant Sciences*, **20**, 107–213.

Renzaglia, K. S., Schuette, S., Duff, R. J. *et al.* (2007). Bryophyte phylogeny: advancing the molecular and morphological frontiers. *Bryologist*, **110**, 179–213.

Robinson, H. (1970). Observations on the origin of the specialized leaves of *Fissidens* and *Schistostega*. *Revue Bryologique et Lichénologique*, **37**, 941–7.

Robinson, H. (1985). The structure and significance of the Leucobryaceae leaf. *Monographs in Systematic Botany from the Missouri Botanical Garden*, **11**, 111–20.

Roth, D. (1969). Embryo und Embryotheca bei den Laubmoosen. Eine histogenetische und morphologische Untersuchung. *Bibliotheca Botanica*, **129**, 1–49.

Ruhland, W. (1924). Musci. Allgemeiner Teil. In *Die natürlichen Pflanzenfamilien*, 2nd edn, vol. 10, ed. A. Engler, pp. 1–100. Leipzig: Wilhelm Engelmann.

Sack, F. & Paolillo Jr., D. J. (1983). Structure and development of walls in *Funaria* stomata. *American Journal of Botany*, **70**, 1019–30.

Salmon, E. S. (1899). On the genus *Fissidens*. *Annals of Botany (Oxford)*, **13**, 103–30.

Sanderson, M. J. (2003). Molecular data from 23 proteins do not support a Precambrian origin of land plants. *American Journal of Botany*, **90**, 954–6.

Scheirer, D. C. (1980). Differentiation of bryophyte conducting tissues: structure and histochemistry. *Bulletin of the Torrey Botanical Club*, **107**, 298–307.

Schofield, W. B. (1981). Ecological significance of morphological characters in the moss gametophyte. *Bryologist*, **84**, 149–65.

Schofield, W. B. (1985). *Introduction to Bryology*. Caldwell, NJ: Blackburn Press.

Schofield, W. B. & Hébant, C. (1984). The morphology and anatomy of the moss gametophore. In *New Manual of Bryology*, vol. 2., ed. R. M. Schuster, pp. 627–57. Nichinan: Hattori Botanical Laboratory.

Schuster, R. M. (1997). On *Takakia* and the phylogenetic relationships of the Takakiales. *Nova Hedwigia*, **64**, 281–310.

Schuster, R. M. (1984). Comparative anatomy and morphology of the Hepaticae. In *New Manual of Bryology*, vol. 2, ed. R. M. Schuster, pp. 760–891. Nichinan: Hattori Botanical Laboratory.

Schwartz, O. M. (1994). The development of the peristome-forming layers in the Funariaceae. *International Journal of Plant Sciences*, **155**, 640–57.

Shaw, J. (1985). Peristome structure in the Mitteniales (ord. nov.: Musci), a neglected novelty. *Systematic Botany*, **10**, 224–33.

Shaw, J. & Anderson, L. E. (1988). Peristome development in mosses in relation to systematics and evolution. II. *Tetraphis pellucida* (Tetraphidaceae). *American Journal of Botany*, **75**, 1019–32.

Shaw, A. J. & Goffinet, B. (2000). Molecular evidence of reticulate evolution in the peatmosses (*Sphagnum*), including *S. ehyalinum* sp. nov. *Bryologist*, **103**, 357–74.

Shaw, A. J., Anderson, L. E. & Mishler, B. D. (2000). Paedomorphic sporophyte development in *Bruchia flexuosa* (Bruchiaceae). *Bryologist*, **103**, 147–55.

Shaw, J., Mishler, B. D. & Anderson, L. E. (1989a). Peristome development in mosses in relation to systematics and evolution. III. *Funaria hygrometrica*, *Bryum pseudocapillare*, and *B. bicolor*. *Systematic Botany*, **14**, 24–36.

Shaw, J., Mishler, B. D. & Anderson, L. E. (1989b). Peristome development in mosses in relation to systematics and evolution. IV. Haplolepideae: Ditrichaceae and Dicranaceae. *Bryologist*, **92**, 314–25.

Shaw, A. J., Cox, C. J., Goffinet, B., Buck, W. R. & Boles, S. B. (2003). Phylogenetic evidence of a rapid radiation of pleurocarpous mosses (Bryophyta). *Evolution*, **57**, 2226–41.

Smith, D. K. & Davison, P. G. (1993). Antheridia and sporophytes in *Takakia ceratophylla* (Mitt.) Grolle: Evidence for reclassification among the mosses. *Journal of the Hattori Botanical Laboratory*, **73**, 263–71.

Snider, J. A. (1975). Sporophyte development in the genus *Archidium* (Musci). *Journal of the Hattori Botanical Laboratory*, **39**, 85–104.

Springer, E. (1935). Über apogame (vegetativ enstandene) Sporogone an der bivalenten Rasse des Laubmooses *Phascum cuspidatum*. *Zeitschrift für Induktive Abstammungs- und Vererbungslehre*, **69**, 249–62.

Stahl, E. (1876). Über künstlich hervorgerufene Protonemabildung an dem Sporogonium der Laubmoose. *Botanische Zeitung*, **34**, 689–95.

Stark, L. R. (2001). Widespread sporophyte abortion following summer rains in Mojave Desert populations of *Grimmia orbicularis*. *Bryologist*, **104**, 115–25.

Stark, L. R. (2002). Phenology and its repercussions on the reproductive ecology of mosses. *Bryologist*, **105**, 204–18.

Stark, L. R. (2005). Do the sexes of the desert moss *Syntrichia caninervis* differ in desiccation tolerance? A leaf regeneration assay. *International Journal of Plant Sciences*, **166**, 21–9.

Stark, L. R., McLetchie, D. N. & Mishler, B. D. (2005). Sex expression, plant size, and spatial segregation of the sexes across a stress gradient in the desert moss *Syntrichia caninervis*. *Bryologist*, **108**, 183–93.

Stetler, D. A. & DeMaggio, A. E. (1976). Ultrastructural characteristics of spore germination in the moss *Dawsonia superba*. *American Journal of Botany*, **63**, 438–42.

Strother, P. K., Wood, G. D., Taylor, W. A. & Beck, J. H. (2004). Middle-Cambrian cryptospores and the origin of land plants. *Memoirs of the Association of Australian Palaeontologists* **29**, 99–113.

Tanahashi, T., Sumikawa, N., Kato, M. & Hasebe, M. (2005). Diversification of gene function: homologs of the floral regulator *FLO/LFY* control the first zygotic cell division in the moss *Physcomitrella patens*. *Development*, **132**, 1727-36.

Touw, A. (1962). Revision of the moss-genus *Neckeropsis* (Neckeraceae) I. Asiatic and Pacific species. *Blumea*, **11**, 373–425.

Tsubota, H., DeLuna, E., González, D., Ignatov, M. S. & Deguchi, H. (2004). Molecular phylogenetics and ordinal relationships based on analyses of a large-scale data set of 600 *rbc*L sequences of mosses. *Hikobia*, **14**, 149–70.

Une, K. (1985). Sexual dimorphism in the Japanese species of *Macromitrium* (Musci: Orthotrichaceae). *Journal of the Hattori Botanical Laboratory*, **59**, 487-513.

Vanderpoorten, A., Goffinet, B., Hedenäs, L., Cox, C. J. & Shaw, A. J. (2003). A taxonomic reassessment of the Vittiaceae (Hypnales, Bryopsida): evidence from phylogenetic analyses of combined chloroplast and nuclear sequence data. *Plant Systematics and Evolution*, **241**, 1–12.

Vitt, D. H. (1968). Sex determination in mosses. *Michigan Botanist*, **7**, 195–203.

Vitt, D. H. (1981). Adaptive modes of the sporophyte. *Bryologist*, **84**, 166–86.

Vitt, D. H. (1984). Classification of the Bryopsida. In *New Manual of Bryology*, vol. 2, ed. R. M. Schuster, pp. 696–759. Nichinan: Hattori Botanical Laboratory.

Vitt, D. H. & Glime, J. M. (1984). Structural adaptations of aquatic Musci. *Lindbergia*, **10**, 95–110.

Walther, K. (1983). Bryophytina. Laubmoose. In *A. Engler's Syllabus der Pflanzenfamilien. Aufl. 13*, vol. 2, ed. J. Gerloff & J. Poelt. Berlin: Gebrüder Bornträger.

Wellman, C. H. & Gray, J. (2000). The microfossil record of early land plants. *Philosophical Transactions of the Royal Society of London*, B**355**, 717–32.

Wellman, C. H., Osterloff, P. L. & Mohiuddin, U. (2003). Fragments of the earliest land plants. *Nature*, **425**, 282–4.

Wettstein, F. von (1925). Genetische Untersuchungen an Moosen (Musci und Hepaticae). *Bibliographia Genetica*, **1**, 1–38.

Wenderoth, H. (1931). Beiträge zur Kenntniss des Sporophyten von *Polytrichum juniperinum* Willdenow. *Planta*, **14**, 344–85.

Whigglesworth, G. (1947). Reproduction in *Polytrichum commune* L. and the significance of the rhizoid system. *Transactions of the British Bryological Society*, **1**, 4–13.

Whitehouse, H. L. K. (1966). The occurrence of tubers in European mosses. *Transactions of the British Bryological Society*, **5**, 103–16.

Wyatt, R. & Anderson, L. E. (1984). Breeding systems in bryophytes. In *The Experimental Biology of Bryophytes*, ed. A. F. Dyer & J. G. Duckett, pp. 39–64. London: Academic Press.

Yang R.-D., Mao, J.-R., Zhang, W.-H., Jiang, L.-J. & Gao, H. (2004). Bryophyte-like fossil (*Parafunaria sinensis*) from Early-Middle Cambrian Kaili formation in Guizhou Province, China. *Acta Botanica Sinica*, **46**, 180–5.

Zander, R. H. (1993). Genera of the Pottiaceae: mosses of harsh environments. *Bulletin of the Buffalo Society of Natural Sciences*, **32**, 1–378 + i–vi.

Zander, R. H. (2006). The Pottiaceae s. str. as an evolutionary Lazarus taxon. *Journal of the Hattori Botanical Laboratory*, **100**, 581–600.

Zander, R. H. (2008). Statistical evaluation of the clade "Rhabdoweisiaceae." *Bryologist*, **111**, 292–301.

Zanten, B. O. van (1974). The hygroscopic movement of the leaves of *Dawsonia* and some other Polytrichaceae. *Bulletin de la Société Botanique de France*, **121**, 63–6.

Zastrow, E. (1934). Experimentelle Studien über die Anpassung von Wasser – und Sumpfmoosen. *Pflanzenforschung*, **17**, 1–70.

Zielinski, F. (1909). Beiträge zur Biologie des Archegoniums und der Haube der Laubmoose. *Flora*, **100**, 1–36.

3

New insights into morphology, anatomy, and systematics of hornworts

KAREN S. RENZAGLIA, JUAN C. VILLARREAL AND R. JOEL DUFF

3.1 Introduction

Hornworts are a key lineage in unraveling the early diversification of land plants. An emerging, albeit surprising, consensus based on recent molecular phylogenies is that hornworts are the closest extant relatives of tracheophytes (Qiu *et al.* 2006). Prior to comprehensive molecular analyses, discrepant hypotheses positioned hornworts as either sister to all embryophytes except liverworts or the closest living relatives of green algae (Mishler *et al.* 1994, Qiu *et al.* 1998, Goffinet 2000, Renzaglia & Vaughn 2000). Morphological features are of little value in resolving the placement of hornworts within the green tree of life because this homogeneous group of approximately 150 species exhibits numerous developmental and structural peculiarities not found in any extant or fossil archegoniate. Until recently, hornworts were neglected at every level of study and thus even the diversity and the relationships within this group have remained obscure.

Virtually every aspect of hornwort evolution has been challenged and/or revised since the publication of the first edition of this book (Duff *et al.* 2004, 2007, Shaw & Renzaglia 2004, Cargill *et al.* 2005, Renzaglia *et al.* 2007). Phylogenetic hypotheses based on multigene sequences have revolutionized our concepts of interrelationships. New classification schemes have arisen from these analyses and continue to be fine-tuned as more taxa are sampled. Three new genera have been named, increasing the number of hornwort genera to 14, namely *Leiosporoceros, Anthoceros, Sphaerosporoceros, Folioceros, Hattorioceros, Mesoceros, Paraphymatoceros, Notothylas, Phaeoceros, Phymatoceros, Phaeomegaceros, Megaceros, Dendroceros*, and *Nothoceros* (Duff *et al.* 2007, Stotler *et al.* 2005). Developmental and ultrastructural studies have

Bryophyte Biology: Second Edition, ed. B. Goffinet & A. J. Shaw. Published by Cambridge University Press.

extended the morphological boundaries in the group and have revealed parallelisms and reversals in characters previously viewed as taxonomically informative. Coupled with robust molecular phylogenies, the newly acquired morphological data have provided a clearer picture of character transformations within the group.

The focus of this chapter is to synthesize and interweave newly gained insights on hornwort structure, phylogeny, and classification. We present a molecular phylogeny that provides the basis for the revised classification included herein. Classical morphological information is updated with more comprehensive studies of ultrastructure and development across a wide sampling of hornworts. We conclude with a brief discussion of inferences on the evolution of diagnostic hornwort characters, namely chloroplasts, stomata, antheridia, and spores.

3.2 Phylogeny

The past five years have witnessed both the advent and wide application of molecular systematic tools toward the development of a phylogeny and classification of hornworts. The first studies by Stech *et al.* (2003) and Duff *et al.* (2004) reported sequence-based phylogenies based on *trn*L–*trn*F and *rbc*L regions of the chloroplast genome, respectively. Though limited in taxon sampling, these studies revealed new and startling relationships among hornwort taxa. Duff *et al.* (2007) reported a more comprehensive molecular phylogeny utilizing three genes, one each from the nuclear, mitochondrial, and plastid genomes, and up to 62 hornwort samples, representing 12 of the 14 genera and one third of the recognized species. The results of this study are summarized in the phylogeny presented in Fig. 3.1.

Several major features of hornwort relationships are well supported both by these molecular phylogenies and through subsequent detailed morphological and ultrastructural analyses. The salient features are: (1) there is significant genetic distance between three lineages of hornworts: *Leiosporoceros*, *Anthoceros* s. lat., and the remaining hornworts; (2) taxa formerly recognized as belonging to *Phaeoceros* are polyphyletic and consequently, three new genera were segregated from this genus: *Phymatoceros* (Stotler *et al.* 2005), *Paraphymatoceros* (Hässel de Menéndez 2006), and *Phaeomegaceros* (Duff *et al.* 2007); (3) American species of *Megaceros* plus *Nothoceros* form a monophyletic clade sister to the Paleotropical *Megaceros* and *Dendroceros*, suggesting a new generic status to this *Nothoceros*/American *Megaceros* alliance; and (4) a sister relationship exists between *Phaeoceros* s. str. and *Notothylas*.

3.3 Classification

The classification scheme presented in Table 3.1 is based on the most current molecular data. There are few congruencies with any of the four

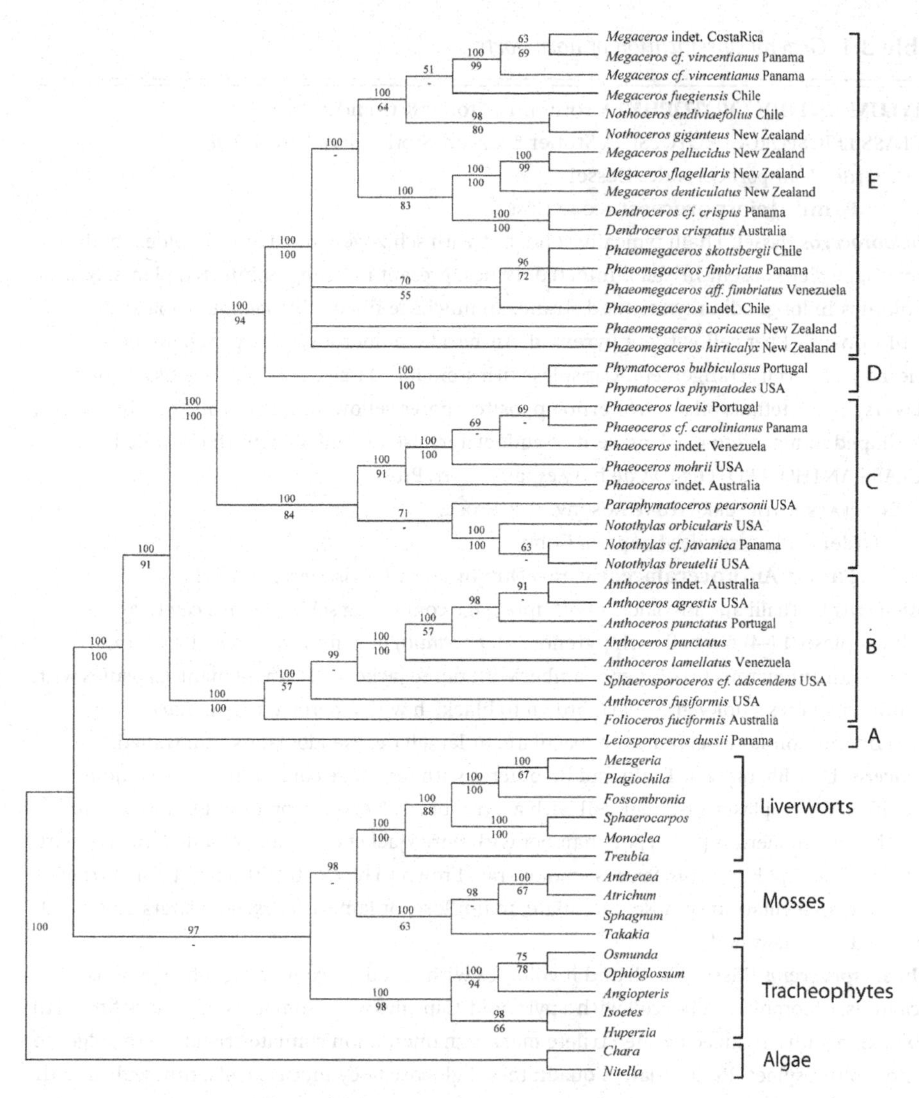

Fig. 3.1. Phylogenetic reconstruction of hornworts based on Bayesian analyses of three genomic regions; nuclear 18S, chloroplast *rbc*L, and mitochondrial *nad*5 (modified from Duff *et al.* 2007). Values on top of branches are Bayesian posterior probabilities and below branches are parsimony non-parametric bootstrap values. Hornwort clades discussed in the text are labeled A–E and represent the orders: A, Leiosporocerotales; B, Anthocerotales; C, Notothyladales; D, Phymatocerotales; E, Dendrocerotales.

classification schemes based on morphology that were highlighted in the first edition of this chapter (Schuster 1987, Hässel de Menéndez 1988, Hyvönen & Piippo 1993, Hasegawa 1994, Duff *et al.* 2007). One has only to look at the number and placement of genera in the revised classification scheme presented

Table 3.1 *General classification of hornworts*

PHYLUM ANTHOCEROTOPHYTA rothm. *ex* Stotler & Crand.-Stotler

CLASS LEIOSPOROCEROTOPSIDA Stotler & Crand.-Stotler *emend.* Duff *et al.*

Order Leiosporocerotales Hässel

Family Leiosporocerotaceae Hässel

Leiosporoceros Hässel: Thalli typically solid, but with schizogenous cavities in older thalli; mucilage clefts absent in *Nostoc*-infected tissues, present in young uninfected plants. *Nostoc* colonies in longitudinally oriented strands in mucilage-filled schizogenous canals. Chloroplasts 1 per cell without pyrenoid. Antheridia numerous (to 80 per chamber) with tiered jacket cell arrangement. Capsules with stomata. Massive sporogenous tissue (6–9 layers). Spore tetrads bilateral alterno-opposite. Spores yellow, minute, ovoid, nearly smooth; Y-shaped to monolete mark present. Pseudoelaters usually unicellular, thick-walled.

CLASS ANTHOCEROTOPSIDA de bary *ex* Jancz. *corr.* Prosk.

SUBCLASS ANTHOCEROTIDAE Rosenv. *corr.* Prosk.

Order Anthocerotales Limpr. *in* Cohn

Family Anthocerotaceae (Gray) Dumort. *corr.* Trevis. *emend.* Hässel

Anthoceros L.: Thalli and involucres with mucilage-containing schizogenous cavities. Chloroplasts 1 (–4) per cell with pyrenoid (*A. punctatus*) or starch-free area (*A. fusiformis*). Antheridia numerous (to 45) per chamber with tiered jacket cell arrangement. Capsules with stomata. Spores smoky gray, dark brown to blackish with a defined trilete mark; ornamentation spinose, punctate, baculate, or lamellate. Pseudoelaters thin-walled.

Folioceros D. C. Bharadwaj: Thalli and involucres with mucilage-containing schizogenous cavities. Chloroplasts 1 (–2) per cell with a pyrenoid (*F. fuciformis*) or absent (*F. assamicus*). Antheridia numerous (to 60) per chamber with tiered jacket cell arrangement. Capsules with stomata, except *F. incurvus*. Spores smoky gray, brown to blackish without a defined trilete mark; ornamentation spinose, reticulate, mamillose, or lamellate. Pseudoelaters elongated strongly, thick-walled.

Sphaerosporoceros Hässel: Thalli and involucres with mucilage-containing schizogenous cavities. Chloroplasts 1 per cell with a pyrenoid. Capsules with stomata. Spores dark brown to blackish with a reduced defined trilete mark; ornamentation connate-cristate with ridges to short blunt-spines. Pseudoelaters quadrate–subglobose to cylindrical cells, thin-walled with faint thickenings.

SUBCLASS NOTOTHYLATIDAE Duff *et al.*

Order Notothyladales Hyvönen & Piippo

Family Notothyladaceae (Milde) Müll. Frib. *ex* Prosk.

Subfamily Notothyladoideae Grolle

Notothylas Sull. *ex* A. Gray: Thalli solid. Chloroplasts 1 (–3) per cell with a pyrenoid (*N. orbicularis*) or absent (*N. nepalensis*). Antheridia 2–4 (–6) per chamber usually with non-tiered jacket cell arrangement. Sporophytes short, lying horizontally in the thallus, mostly or totally enclosed within the involucre. Stomata absent. Suture elaborate, rudimentary, or absent. Columella present (*N. dissecta*) or absent (*N. javanica*). Spores yellow to blackish with an equatorial girdle; ornamentation finely vermiculate, granulose to tuberculate. Pseudoelaters absent to subquadrate–elongated with or without annular thickenings.

Subfamily Phaeocerotoideae Hässel

Table 3.1 (*cont.*)

Phaeoceros Prosk.: Thalli solid. Marginal or short ventral tubers present or absent. Chloroplasts 1 (–2) per cell with pyrenoid present (*P. laevis*) or absent (*P. pearsonii*). Antheridia (1–) 2–6 (–8) per chamber with non-tiered jacket cell arrangement. Stomata present. Spores yellow to brownish when completely mature, with equatorial girdle; ornamentation spinose (*P. laevis–carolinianus* group) to distally covered by rounded protuberances (*P. himalayensis*). Pseudoelaters thin-walled.

Paraphymatoceros Hässel: Thalli solid. Apical flattened and disk-shaped tubers. Chloroplasts 1 (–2) per cell, without pyrenoid. Antheridia 2–5 per chamber with non-tiered jacket cell arrangement. Stomata present. Spores yellow to blackish-brownish when completely mature, with equatorial girdle; ornamentation of rounded protuberances in distal face. Pseudoelaters mostly unicellular (*P. hallii*), 4-celled in the other taxa (disintegrating in *P. diadematus*).

Hattorioceros (J. Haseg.) J. Haseg.: Thalli solid. Chloroplast and antheridium morphology unknown. Stomata present. Spores yellow to brownish. Spores small (usually less than 20 μm) without a triradiate mark, variable in shape, mostly ovoidal; ornamentation surface deeply canaliculate-striate. Pseudoelaters unevenly thick-walled.

Mesoceros Piippo: Thalli solid. Chloroplast morphology unknown. Antheridia 2–3 per chamber with a non-tiered jacket cell arrangement. Spores dark brown; ornamentation papillate to connate with reticulate ridges. Pseudoelaters thin-walled with faint thickenings.

Subclass Dendrocerotidae Duff *et al.*

Order Phymatocerales Duff *et al.*

Family Phymatocerotaceae Duff *et al.*

Phymatoceros Stotler *et al. emend.* Duff *et al.*: Thalli solid. Long-stalked ventral tubers. Chloroplasts 1 (–2) per cell with a pyrenoid (*P. bulbiculosus*) or absent (*P. phymatodes*). Antheridia 1–3 (–4) per chamber with non-tiered jacket cell arrangement (Schiffner 1937). Stomata present. Spores yellow to brownish when completely mature, with equatorial girdle; ornamentation finely vermiculate with a distal protuberance, distal spore ornamentation obscured by late spore wall deposition. Pseudoelaters thin-walled.

Order Dendrocerotales Hässel *emend.* Duff *et al.*

Family Dendrocerotaceae (Milde) Hässel

Subfamily Dendrocerotoideae R. M. Schust.

Dendroceros Nees: Epiphytic and epiphyllic. Thalli solid (subg. *Dendroceros*) or with mucilage-containing schizogenous cavities (subg. *Apoceros*), involucres solid in both subgenera. Thalli with a conspicuous midrib and perforated wings. *Nostoc* present as bulging globose colonies in the ventral and dorsal side of the thallus. Band or pit-field-like thickenings present in the thallus cell walls. Chloroplasts 1 per cell with a conspicuous pyrenoid with spherical inclusions. Antheridia 1 per chamber with a non-tiered jacket cell arrangement. Stomata absent. Spores multicellular owing to endosporic germination, colorless to pale yellow, appearing green in live spores owing to large chloroplasts and thin exine; ornamentation papillose to shortly tuberculate. Pseudoelaters with helical thickenings.

Megaceros Campb.: Thalli solid. Band or pit-field-like thickenings present in the thallus cell walls. Chloroplasts 1–8 (–12) per cell without pyrenoid. Antheridia 1 (–2) per chamber with non-tiered jacket cell arrangement. Stomata absent. Spores colorless to pale yellow, appearing green in live spores due to large chloroplasts and thin exine; ornamentation mamillose and/or tuberculate. Pseudoelaters with helical thickenings.

Table 3.1 (*cont.*)

Nothoceros (R. M. Schust.) J. Haseg.: Thalli solid, in rosette or pinnately branched with thin (less than 1 mm) branches resembling *Riccardia* or with a conspicuous midrib and imperforated wings. Band or pit-field-like thickenings present in the thallus cell walls. Chloroplasts 1–2 (–8) per cell. Pyrenoid absent (*N. endiviaefolius*) or present (*M. vincentianus*). Antheridia 1 (–2) per chamber with a non-tiered jacket cell arrangement. Stomata absent. Spores colorless to pale yellow, appearing green in live spores owing to large chloroplasts and thin exine; ornamentation mammillose and/or tuberculate, similar to that of *Megaceros*. Pseudoelaters with helical thickenings.

Subfamily Phaeomegacerotoideae Duff *et al.*

Phaeomegaceros Duff *et al.*: Thalli solid. Band or pit-field-like thickenings present in the thallus cell walls. Chloroplasts 1–2 per cell without a pyrenoid. Antheridia 1 (–rarely 8) per chamber with a non-tiered antheridial jacket cell arrangement. Stomata present. Spores yellow to brownish when completely mature, with equatorial girdle; ornamentation finely vermiculate with distal dimples. Pseudoelaters thin-walled to unevenly thick-walled.

Source: Based on Duff *et al.* (2007).

herein to understand the magnitude of change that has occurred in the past six years in regards to hornwort systematics. Hornworts are now accommodated in 14 genera compared with the previously widely recognized six, which were *Anthoceros, Phaeoceros, Folioceros, Notothylas, Megaceros, and Dendroceros*.

The sister relationship between *Leiosporoceros* and the remaining hornworts is reflected in the erection of a separate class, Leiosporocerotopsida, for this monospecific genus (Frey & Stech 2005, Stotler & Crandall-Stotler 2005). The Anthocerotopsida contains the remaining taxa and is divided into three subclasses: Anthocerotidae, Notothylatidae, and Dendrocerotidae. Within the Anthocerotidae, three morphologically similar genera, *Anthoceros*, *Sphaerosporoceros*, and *Folioceros*, are placed together in a single family and order. The Notothyladales comprises a single family and composes the only order of the Notothylatidae. The subfamily Notothyladoideae contains a single genus, while the remaining four genera, two of which have not yet been sampled for molecular analyses (*Mesoceros* and *Hattorioceros*), are accommodated in the Phaeocerotoideae. Sampling of *Paraphymatoceros* in our molecular analyses is restricted to one species, *P. hallii*. Analyses of *rbc*L sequences place *Phaeoceros pearsonii* sister to *P. hallii* and thus support transfer of *P. pearsonii* to this newly transcribed genus. The subclass Dendrocerotidae includes two genetically and morphologically distant orders: the monospecific Phymatocerales and the Dendrocerotales with four genera in a single family and two subfamilies.

3.4 Anatomy and development

The uniformity and uniqueness of morphological features within hornworts has been recognized for over a century (Campbell 1895, Goebel 1905, Bower 1935). Peculiarities in the structure and development of the sporophyte, chloroplasts, gametangia, and *Nostoc* colonies, among other traits, distinguish this small assemblage of bryophytes from all other land plants (Campbell 1895, Bartlett 1928, Renzaglia 1978, Renzaglia & Vaughn 2000). Based on the leafless habit of the gametophyte, hornworts were traditionally included within liverworts and were viewed as having an affinity with simple thalloids. With phylogenetic reconstructions pointing to a sister relationship between hornworts and tracheophytes, a thorough reappraisal of morphological transformations across cryptogams in both generations is warranted. This is beyond the scope of this chapter, and hence we consider here only the diversity and evolution of morphological features from the cell to the organ level within hornworts. This new synthesis is particularly timely owing to the recent contribution of significant new morphological knowledge across hornworts, especially within obscure tropical taxa (Villarreal & Renzaglia 2006a, b, Duff *et al.* 2007, Renzaglia *et al.* 2007). The recent morphological studies, in turn, were prompted by the surprising phylogenetic conclusions that emerged from molecular analyses focused solely on hornworts.

The vegetative gametophyte of hornworts is a flattened thallus, with or without a thickened midrib (Fig. 3.2A, B). Growing regions that contain solitary apical cells and immediate derivatives are located in thallus notches and are covered by mucilage that is secreted by epidermal cells (Fig. 3.2C, D). The apical cell and immediate derivatives contain well-developed chloroplasts that are intimately associated with the nucleus (Fig. 3.2C). Growth forms are correlated with apical cell geometry. The wedge-shaped apical cell of most taxa segments along four cutting faces: two lateral, one dorsal, and one ventral (Fig. 3.2E). The resulting growth forms tend to be orbicular and the thallus in cross-section gradually narrows from the center to lateral margins. In comparison, the hemidiscoid apical cell (Fig. 3.2D, F) of *Dendroceros* cuts along two lateral and one basal face and is responsible for producing a ribbon-shaped thallus with an enlarged midrib and monostromatic wings (Fig. 3.2B). A parallelism in this general habit is found in some *Nothoceros*, where the large thallus develops from a wedge-shaped apical cell but has a prominent midrib that tapers laterally to fragile wings (Duff *et al.* 2007, Villarreal *et al.* 2007). Aside from anthocerotes, wedge-shaped apical cells occur only in thalloid liverworts and some pteridophyte gametophytes (Crandall-Stotler 1980, Shaw & Renzaglia 2004). Hemidiscoid apical cells are rare and known only from *Dendroceros* and a few simple thalloid liverworts (Renzaglia 1982).

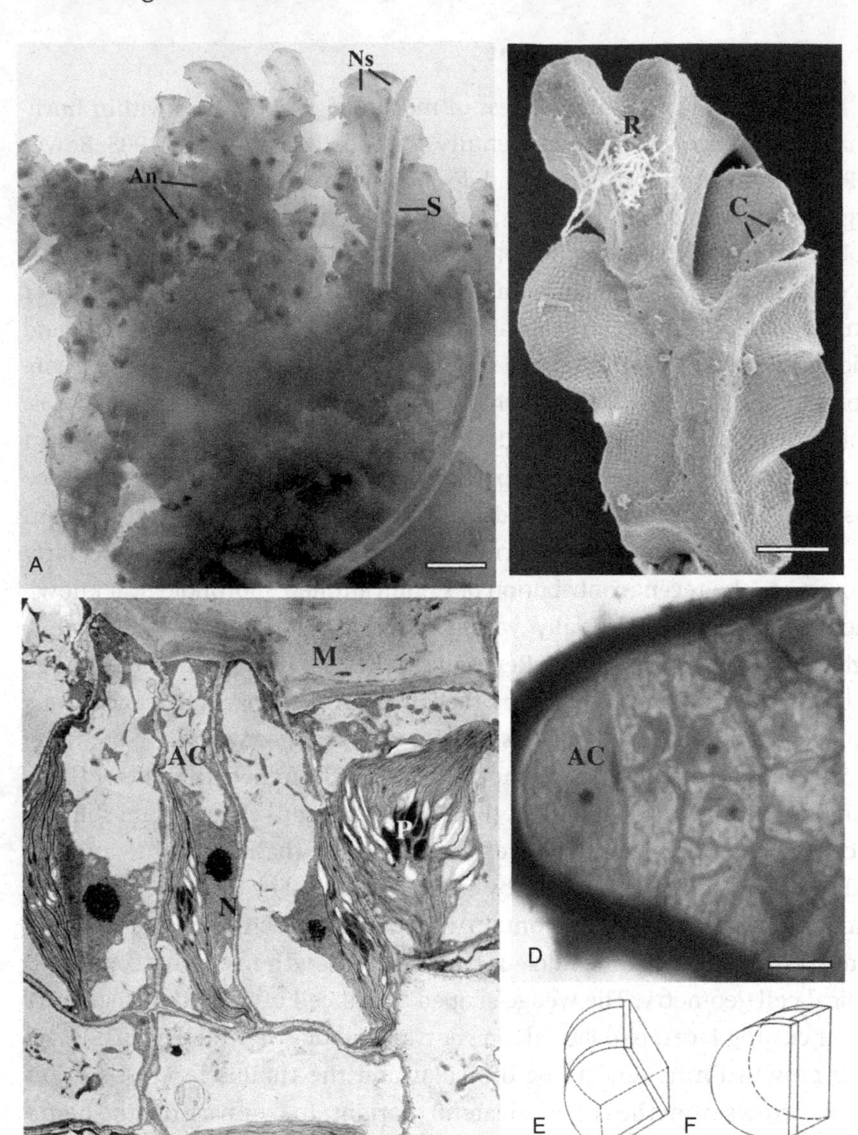

Fig. 3.2. (A) *Leiosporoceros dussii*. Light micrograph (LM) of female gametophyte with two mature sporophytes (S) and overlying male plant with abundant antheridial chambers (An) and longitudinal *Nostoc* canals (Ns). Bar = 1.0 mm. (B) *Dendroceros tubercularis*. Scanning electron micrograph (SEM) of ventral thallus showing swollen central midrib and monostromatic wings. Pore-like mucilage clefts (C) occur in two irregular rows on either side of the midrib and a tuft of rhizoids (R) is positioned below the terminal bifurcation. Bar = 0.25 mm. (C) *Phaeoceros carolinianus*. Transmission electron microscope (TEM) horizontal longitudinal section of growing notch overarched by mucilage (M). The rectangular apical cell (AC) and surrounding derivatives are highly vacuolated and contain a nucleus (N) associated with a well-developed chloroplast (P) containing a pyrenoid. Bar = 4.0 μm. (D) *Dendroceros*

At the cellular level, hornworts are known to contain solitary chloroplasts with central pyrenoids (or starch-free areas) and channel thylakoids, features shared with algae but found in no other land plants (Duckett & Renzaglia 1988, Vaughn *et al.* 1992) (Fig. 3.3A–D). Recent comparative studies, however, have revealed remarkable variability in chloroplast shape, number, and especially ultrastructure in hornworts (Duff *et al.* 2007, Renzaglia *et al.* 2007) (Fig. 3.3). For example, *Leiosporoceros* has plastids that lack a pyrenoid but often contain a central aggregation of large grana surrounded by starch (Fig. 3.3F). Channel thylakoids are abundant in these chloroplasts.

The classical hornwort pyrenoid is traversed by thylakoids, which separate lens-shaped to elongated subunits giving the appearance of "multiple pyrenoids" (Fig. 3.3A, B). The shape of pyrenoid subunits and the existence or location of pyrenoid inclusions have been considered to be of taxonomic value. For example, chloroplast structure in *Dendroceros* deviates from that of the typical hornworts in that the pyrenoid is spherical and contains irregularly shaped subunits with regularly spaced electron-opaque inclusions (Fig. 3.3G). Chloroplasts of *Megaceros* lack pyrenoids and may number as many as 14 per internal thallus cell (Fig. 3.3E) (Burr 1969). RUBISCO localizations in the pyrenoids and lack of grana end membranes (Fig. 3.3G) that characterize land plants may be viewed as plesiomorphies and suggest ties with charophytes (Vaughn *et al.* 1990, 1992). As in other land plants, RUBISCO is scattered among starch grains in the chloroplast stroma of *Megaceros*.

Cell division in all hornworts is monoplastidic and involves plastic division and morphogenetic migration that is tightly linked with nuclear division (Brown & Lemmon 1990, 1993). Spindle microtubules originate from an aggregation of electron-dense material at the poles, suggesting the vestige of algal-like centriolar centrosomes (Vaughn & Harper 1998). Further investigations into cell cycle and cytoskeletal proteins are required to clarify any homologies of this structure to the polar bodies of liverworts and to centrosomes of other eukaryotes.

Caption for Fig. 3.2. (cont.)
cavernosus. LM vertical longitudinal section of apical region. A hemidiscoid apical cell (AC) is overarched by mucilage. Bar = 0.5 μm. (E) Wedge-shaped apical cell characteristic of most hornworts. Two triangular lateral cutting-faces, one rectangular dorsal cutting face, and one rectangular ventral cutting face produce a total of four derivatives in spiraled rotation. Modified from Renzaglia (1978). (F) Hemisdiscoid apical cell of *Dendroceros* with two semicircular lateral cutting faces and a single rectangular basal cutting face. Modified from Renzaglia (1978).

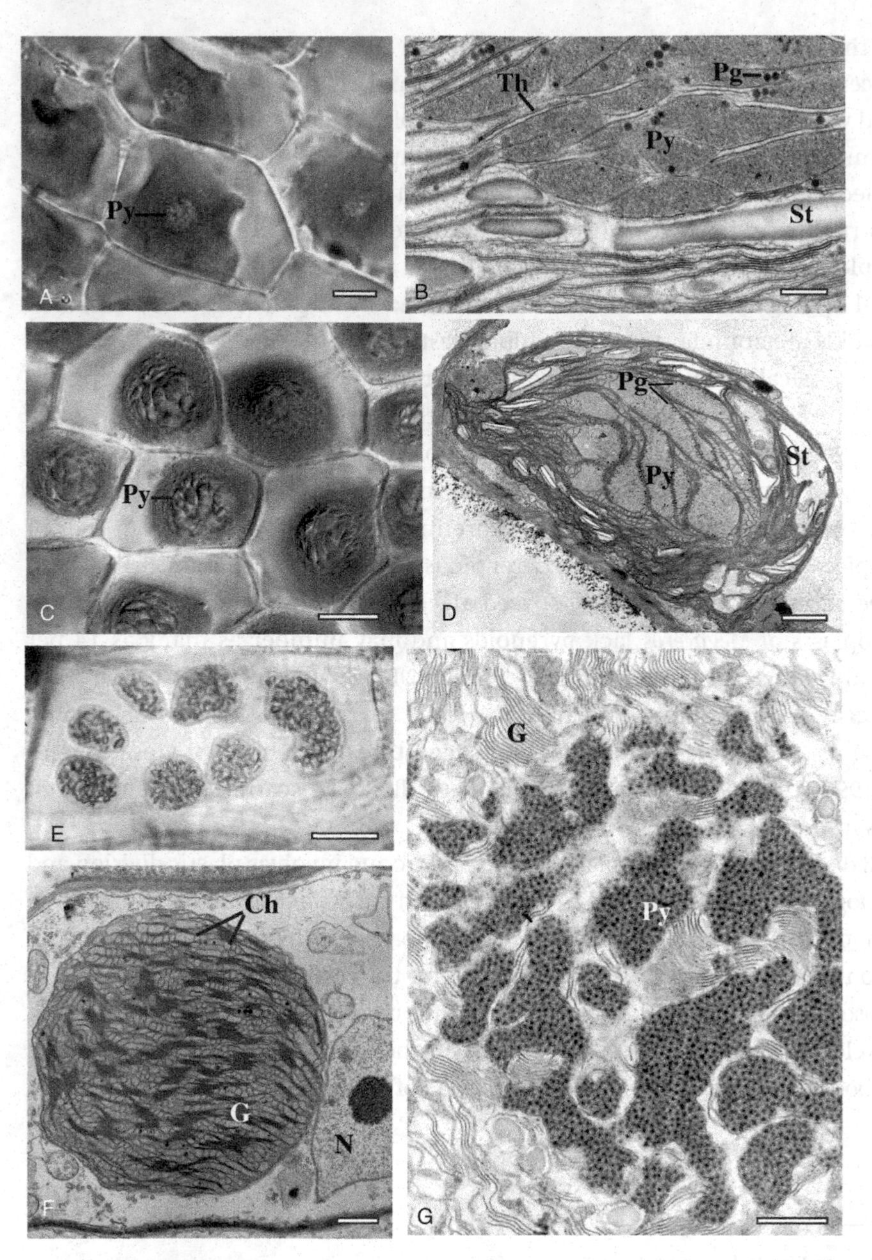

Fig. 3.3. (A) *Anthoceros agrestis*. LM of upper epidermis of gametophyte, each cell contains single plastid with central pyrenoid (Py). Bar = 10 μm. (B) *Folioceros fuciformis*. TEM of pyrenoid (Py) consisting of lens-shaped subunits delimited by thylakoids (Th) and scattered pyrenoglobuli (Pg). Starch (St) surrounds the pyrenoid and narrow grana traverse the plastid. Bar = 0.5 μm. (C) *Megaceros* cf. *vincentianus*. LM of upper epidermal cells of gametophyte, each with single spherical to lens-shaped plastid containing a modified central pyrenoid (Py) with abundant starch granules. Bar =10 μm. (D) *Anthoceros angustus*. TEM of chloroplast from thallus

The thickened thallus of the hornwort gametophyte lacks internal differentiation (Fig. 3.4A), except for the occurrence of rather extensive schizogenous mucilage canals in species of Anthocerotaceae (Fig. 3.4B) and *Dendroceros* (subgenus *Apoceros*). In some taxa, especially *Megaceros*, epidermal cells are smaller than in internal parenchyma cells (Fig. 3.4A). Unlike the sporophyte epidermis, all epidermal cells of the gametophyte contain chloroplasts (Fig. 3.4A, B, D). Mucilage-filled cells are abundant and scattered among photosynthetic parenchyma in most taxa (Fig. 3.4A). Band-like wall thickenings (Fig. 3.4C) and primary pit fields may occur in cells of the thallus that subtend archegonia and later the sporophyte foot (Leitgeb 1879, Proskauer 1960, Renzaglia 1978). Ultrastructural observations of the cells will enable an evaluation of the potential role in food transport. Vesicular–arbuscular endomycorrhizas are common in internal thallus cells of most taxa, and swollen, terminal tubers characterize some genera and species (Fig. 3.4E) (see below) (Renzaglia 1978, Ligrone 1988). Rhizoids are unicellular, smooth and may have branched tips (Hasegawa 1983). They are typically ventral in position and may develop from the outer cell derived from a periclinal division of an epidermal cell.

A distinctive feature of anthocerotes is the occurrence of apically derived mucilage clefts, primarily on the ventral thallus (Figs. 3.2B, 3.4D). Two cells that resemble stomatal guard cells surround a pore, which lacks the ability to close and open. Although considered by some to be homologous to the stomata in the sporophyte (Schuster 1992), this interpretation is dubious due to the function of these mucilage clefts as the site of entry for *Nostoc*, the colonial endosymbiont that is found in all hornworts. In most species, clefts are regularly produced from apical derivatives and each may attract the phycobiont. Once in the mucilage-filled internal chamber, the *Nostoc* increases in size and forms a

---

Caption for Fig. 3.3. (cont.)
epidermis. The modified pyrenoid ("starch-free area") is less dense than in other taxa and contains large subunits delimited by thylakoids. Abundant small pyrenoglobuli (Pg) line thylakoids and are scattered within the pyrenoid. Starch grains (St). Bar = 1.0 μm. (E) *Megaceros aenigmaticus*. LM of internal cells of thallus with seven starch-filled plastids that lack pyrenoids; the plastids on the right may be preparing for division. Bar = 10 μm. (F) *Leiosporoceros dussii*. TEM of chloroplast without a pyrenoid from thallus epidermis. Numerous channel thylakoids (Ch) run perpendicular to the main axis and interconnect short grana (G). Nucleus (N). Bar = 1.0 μm. (G) *Dendroceros tubercularis*. TEM of spherical pyrenoid (Py) with irregularly shaped subunits containing uniform electron-dense inclusions. Thylakoids, including grana, interrupt the pyrenoid. Grana (G) lack end membranes. Bar = 5.0 μm.

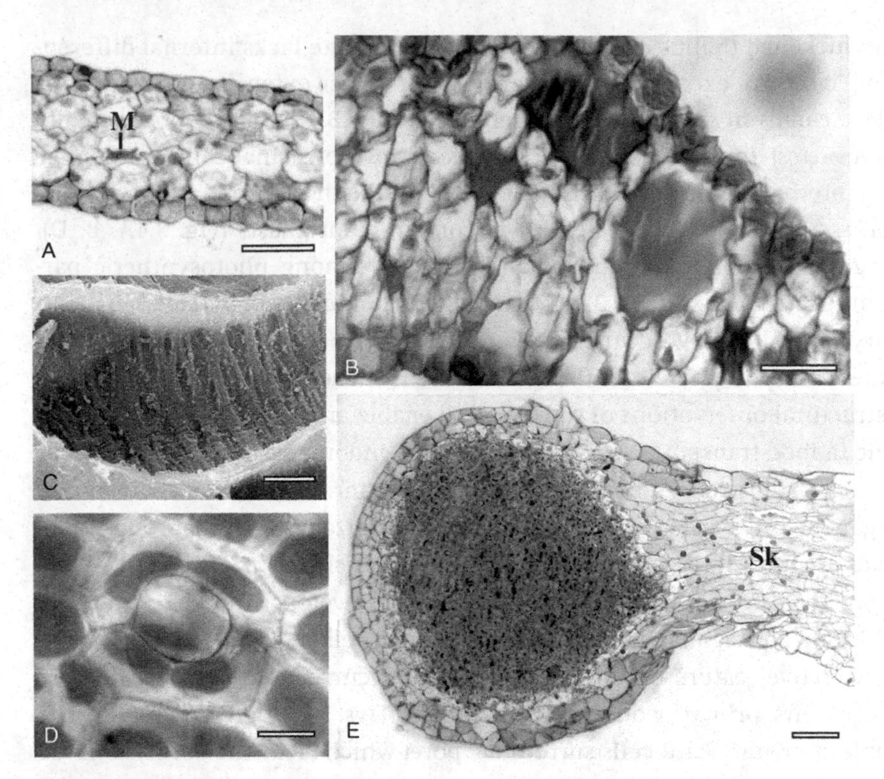

Fig. 3.4. (A) *Megaceros aenigmaticus*. LM transverse section of undifferentiated, simple thallus. Epidermal cells are smaller than internal cells, of which one is mucilage-filled (M). Bar = 25 µm. (B) *Anthoceros punctatus*. LM transverse section of gametophyte with numerous mucilage-containing schizogenous cavities near upper epidermis. Bar = 25 µm. (C) *Nothoceros giganteus*. SEM of internal gametophyte cell with band-like thickenings in cell wall as also occurs in *Megaceros*, *Phaeomegaceros* and *Dendroceros*. Bar = 15 µm. (D) *Megaceros aenigmaticus*. LM surface view of mucilage cleft in ventral epidermis of gametophyte. Both cleft cells contain recently divided plastids. Bar = 10 µm. (E) *Phymatoceros phymatodes*. LM longitudinal section of ventral spherical tuber, consisting of small oil-rich cells surrounded by 3–4 layers of cells. The tuber stalk (Sk) is 13–18 cells wide. Bar = 100 µm.

discrete spherical colony (Fig. 3.5A). Thallus outgrowths penetrate the algal colony (Fig. 3.5B). In *Leiosporoceros*, clefts are produced only in the sporeling; *Nostoc* enters and forms an intimate association directly behind the apical cell (Fig. 3.5C–E) (Villarreal & Renzaglia 2006a). As the thallus elongates through apical segmentation, so too does the *Nostoc* colony within an advancing narrow schizogenous canal (Fig. 3.5D, E). Unknown elsewhere in plants, these branching *Nostoc* canals run through the central thallus (Fig. 3.5C) and are visible to the naked eye as dark green strands.

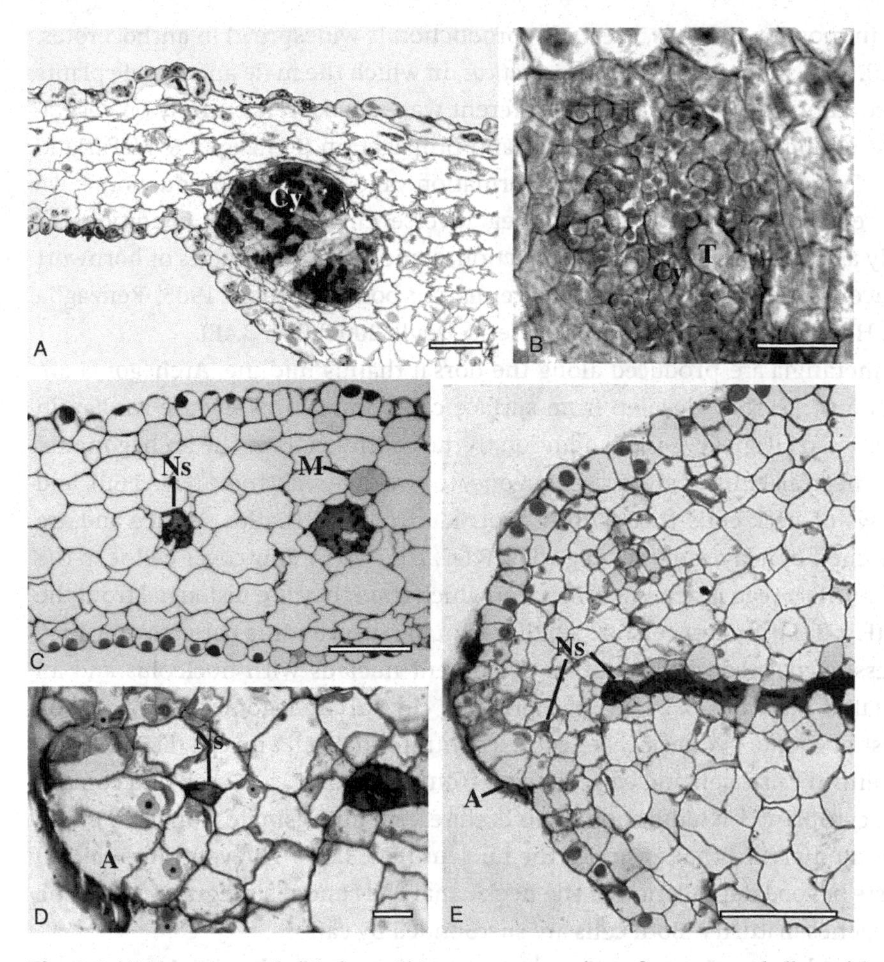

Fig. 3.5. (A) *Phymatoceros bulbiculosus.* LM transverse section of a mature thallus with embedded ventral cyanobacterium colony (Cy). Thallus cells (clear areas) interdigitate with and traverse *Nostoc* filaments. Bar = 100 μm. (B) *Phaeoceros carolinianus.* LM section through *Nostoc* colony. Thallus cells (T) penetrate the colony and are interspersed amongst small, spherical cells of cyanobacterium (Cy). Bar = 20 μm. (C) *Leiosporoceros dussii.* LM transverse section of a mature thallus with two central *Nostoc* canals (Ns) and scattered mucilage cells (M). Bar = 100 μm. (D) *Leiosporoceros dussii.* LM vertical longitudinal section of apical region showing schizogenous origin of *Nostoc* canal (Ns) between ventral and dorsal derivatives from the wedge-shaped apical cell (A). Bar = 15.0 μm. (E) *Leiosporoceros dussii.* Vertical longitudinal section of female plant showing central *Nostoc* canal (Ns) that originates behind the wedge-shaped apical cell (A) between ventral and dorsal derivatives. Bar = 100 μm.

As in most bryophytes, asexual reproduction is widespread in anthocerotes. Indeed, taxa such as *Megaceros aenigmaticus*, in which the male and female plants are geographically separated into different watersheds, rely entirely on vegetative reproduction for dissemination and propagation (Renzaglia & McFarland 1999). Fragmentation, regenerant formation, and gemmae production have been reported in various taxa. Under adverse environmental conditions or simply as a means of asexual reproduction, some genera or species of hornwort produce nutrient-filled tubers as perennating bodies (Goebel 1905, Renzaglia 1978, Hässel de Menéndez 2006, Stotler & Doyle 2006) (Fig. 3.4E).

Gametangia are produced along the dorsal thallus midline. Archegonia are exogenous, i.e. they develop from surface cells, and ultimately are sunken in thallus tissue (Fig. 3.6A–C). In addition to the central cells of the archegonium, the archegonial initial gives rise to a one- to two-layered venter (Fig. 3.6B), and six rows of neck cells that slightly protrude from the thallus surface and are overarched by a layer of mucilage (Fig. 3.6C, D). Two to four cover cells cap the canal until the egg reaches maturity, at which time they are dislodged from the neck (Fig. 3.6A). Venter cells are smaller than the surrounding parenchyma; they are less vacuolated and contain a prominent nucleus with nucleolus and an associated flattened plastid (Fig. 3.6B). The central archegonial cells typically consist of four to six neck canal cells, a ventral canal cell and egg (Fig. 3.6A, D). The ventral canal cell and egg originate from the venter canal cell and contain dense cytoplasm including abundant lipid reserve and a single elongated undifferentiated plastid that encircles the nucleus (Fig. 3.6D). The ventral canal cell persists beyond degradation of the neck canal cells and disintegrates when the egg reaches maturity. Both cells are surrounded by callose.

Antheridia are referred to as endogenous because they develop from subepidermal cells and ultimately are positioned within internal thallus chambers (Fig. 3.7A). One to 80 antheridia (all derived from the same subepidermal cell) are enclosed in each sunken chamber (Fig. 3.7A, B). In other embryophytes, antheridia develop from epidermal cells. In hornworts, the epidermal cell develops into the two-layered chamber roof. A schizogenous chamber forms below the roof and antheridial initials arise internally at the base of the chamber from epithelial (layer surrounding an internal space) cells. The designation of hornwort antheridia as endogenous refers only to the location of development and not to a developmental pathway inherently different from that in other bryophytes (Renzaglia *et al.* 2000). In fact, development of the antheridium proper in hornworts resembles that of other bryophytes, especially complex thalloid liverworts, in that the antheridial initial elongates without apical cell involvement and four primary spermatogones with eight peripheral jacket initials are produced in the formative stages of organogenesis. Thousands of minute spermatozoids are

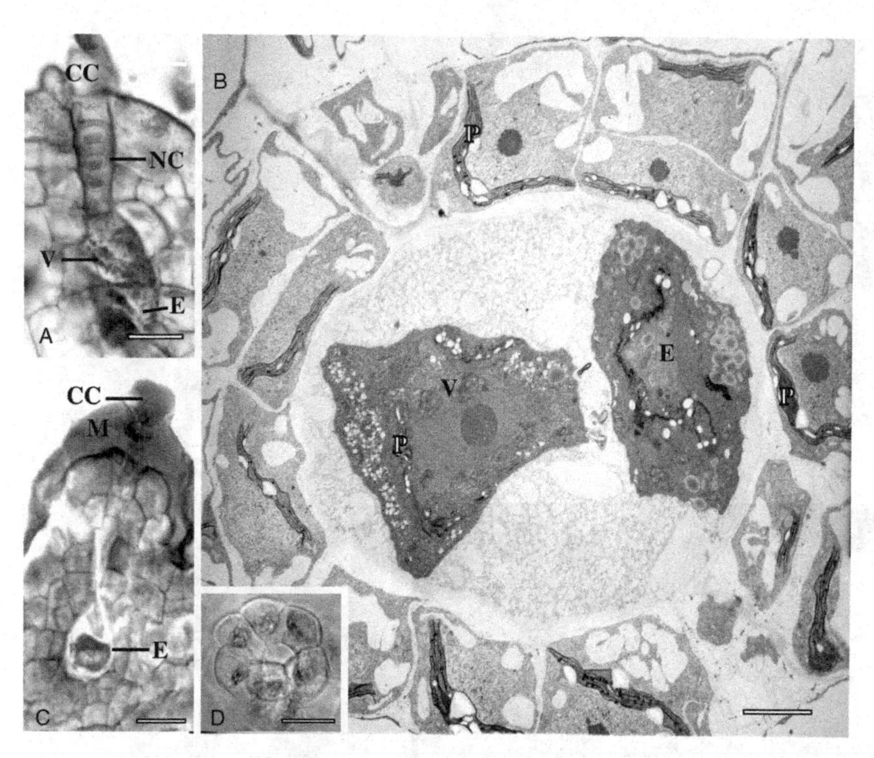

Fig. 3.6. (A) *Phaeoceros carolinianus.* LM longitudinal section of sunken archegonium with two cover cells (CC), six neck canal cells (NC), ventral canal cell (V), and egg (E)-containing nucleus. Bar = 20.0 μm. (B) *Phaeoceros carolinianus.* TEM oblique cross-section of venter of nearly mature archegonium containing ventral canal cell (V) and egg (E); both are embedded in a callosic matrix and contain an elongated plastid near a large central nucleus (visible in ventral canal cell) and dense lipid-filled cytoplasm. The surrounding venter is one- or two-layered. Venter cells are smaller than other thallus cells and contain less-dense cytoplasm with small vacuoles and an elongated plastid (P) adjacent to the nucleus. Bar = 4.0 μm. (C) *Dendroceros japonicus.* LM longitudinal section of mature archegonium that projects from the dorsal thallus, is overarched by mucilage (M), and has discharged the cover cells (CC). The venter contains an egg cell (E). Bar = 20.0 μm. (D) *Phaeoceros carolinianus.* LM surface view of mature archegonium containing six rows of neck cells, each with a single prominent chloroplast. Bar = 20.0 μm.

produced in each antheridium (Fig. 3.7A, C). When antheridia are mature, the plastids of the jacket layer typically have converted to orange-colored chromoplasts (Duckett 1975). The roof of each antheridial chamber ruptures and the jacket cells dissociate, thus liberating the spermatozoids (Fig. 3.7B).

Spermatogenesis provides clues to the phylogenetic history of hornworts (Renzaglia & Carothers 1986, Renzaglia & Duckett 1989, Garbary *et al.* 1993,

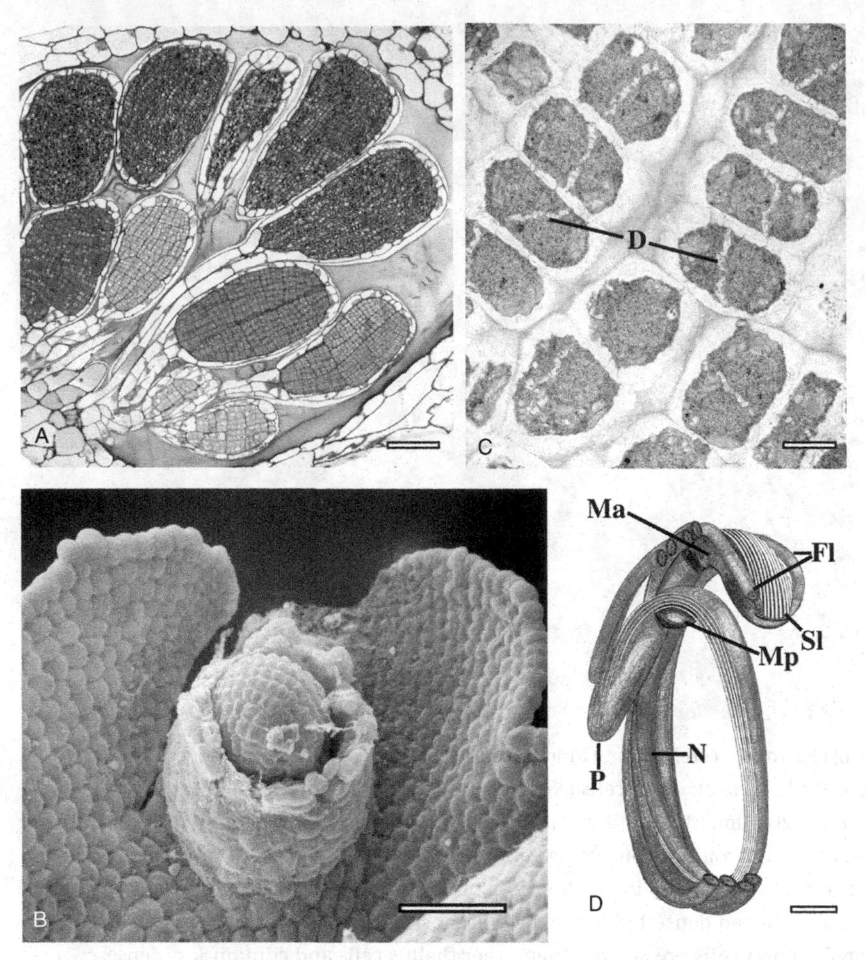

Fig. 3.7. (A) *Leiosporoceros dussii.* LM section of antheridial chamber showing eleven antheridia in different stages of development. Bar = 50 μm. (B) *Dendroceros tubercularis.* SEM of dorsal thallus with ruptured, projecting chamber containing a single antheridium. Bar = 0.1 mm. (C) *Notothylas orbicularis.* TEM of antheridium showing diagonal final mitotic division (D) that produces pairs of polygonal spermatids. Bar = 3.0 μm. (D) *Phaeomegaceros hirticalyx.* Three-dimensional reconstruction of biflagellated sperm cell. The locomotory apparatus consists of two flagella (Fl) that are inserted symmetrically into the cell anterior over a spline (Sl) of 12 microtubules and an underlying anterior mitochondrion (Ma). The cylindrical nucleus (N) with central constriction occupies most of the cell length, and a round posterior mitochondrion (Mp) is positioned in front of a plastid (P) with a single starch grain. Bar = 0.5 μm.

Graham 1993, Vaughn & Renzaglia 1998, Renzaglia & Garbary 2001). During spermatogenesis, pairs of bicentrioles arise *de novo* at the poles in the cell generation prior to the spermatid mother cell (Vaughn & Renzaglia 1998). Bicentrioles are diagnostic of archegoniates that produce biflagellated sperm cells but the timing of their origin in hornworts is earlier than in other taxa, where these organelles originate in the spermatid mother cell. Because green algal cells typically contain centrioles in all cell generations, this feature in hornworts was interpreted as a plesiomorphy (Vaughn & Harper 1998, Vaughn & Renzaglia 1998). As in *Coleochaete*, liverworts, and some pteridophytes, the final mitotic division in the spermatid mother cell is diagonal and spermatids develop in pairs (Fig. 3.7C). Sperm cell architecture varies little in the six genera (*Leiosporoceros, Anthoceros, Phaeoceros, Notothylas, Phaeomegaceros*, and *Megaceros*) that have been examined to date (Fig. 3.7D) (Renzaglia *et al.* 2007, K. S. Renzaglia, unpublished data). The mature spermatozoid is extremely small (approximately 3.0 μm in diameter), coiled, biflagellated, and symmetrical. Both flagella insert at the anterior extreme of the cell over a spline of 12 microtubules and are directed posteriorly. Spermatozoids contain an anterior mitochondrion, a cylindrical nucleus with mid-constriction, and a posterior mitochondrion associated with a plastid containing one starch grain. Unlike sperm cells in all other archegoniates which are sinistrally coiled, the hornwort cell is dextrally coiled (Fig. 3.7D).

Following fertilization, the first division of the zygote is longitudinal and the endothecium of the embryo gives rise to a central columella, if one exists (Renzaglia 1978). The amphithecium forms the sporogenous tissue, assimilative layer, and epidermis (Fig. 3.8A). This is in contrast to liverworts and most mosses in which the zygote undergoes a transverse first division and the endothecium gives rise to sporogenous tissue (the notable exception is *Sphagnum*) in addition to the columella. The foot matures before the remaining histogenic regions (Fig. 3.8A, B). The basal meristem is established early in development and is unifacial, producing cells above the foot that differentiate upwardly. Division patterns from this meristem upward mimic that of the embryo in the origin and differentiation of an amphithecium and endothecium (Fig. 3.8C). Growth of overarching gametophytic tissue occurs as the embryo develops, thus forming a protective involucre that in most taxa is ruptured with continued maturation of the sporophyte (see Fig. 3.2A). The involucre remains as a cylinder that surrounds the base of the sporophyte (Fig. 3.8A). Numerous archegonia are produced in an acropetal fashion and thus young plants will bear young sporophytes at different stages of development.

Although globose in general structure, the anatomical organization of the foot is highly variable among species (Renzaglia 1978, Renzaglia & Vaughn 2000). For example, palisade-like epidermal cells surround the relatively small foot of *Anthoceros* whereas the massive foot of *Megaceros* contains thousands of small undifferentiated cells (Renzaglia 1978). Typically, a parenchymatous inner foot is bordered by numerous smaller cells, including haustorial cells

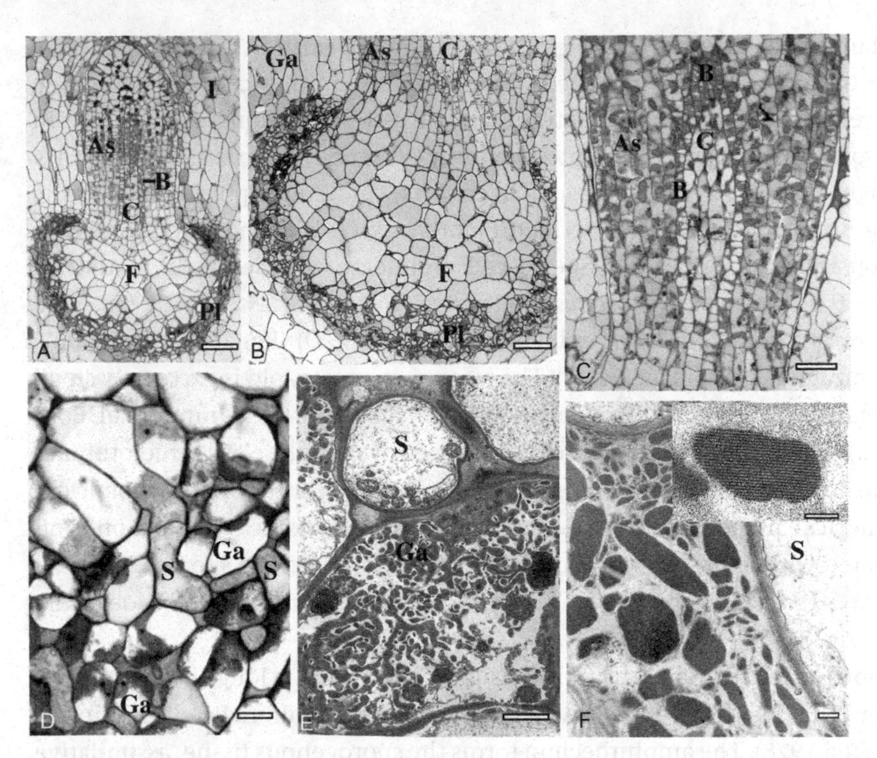

Fig. 3.8. (A) *Leiosporoceros dussii*. LM median longitudinal section of young sporophyte less than 1.0 mm long enclosed in the involucre (I). The prominent foot (F) consists of large central cells and smaller peripheral cells that interdigitate with gametophyte cells to form the placenta (Pl). A basal meristem (B) has begun to produce columella (C) and assimilative tissue (AS). Mature cells from embryonic divisions cap the sporophyte. Bar = 100.0 μm. (B) *Leiosporoceros dussii*. LM median longitudinal section of sporophyte more than 30 mm long showing placental region (Pl) at the interface between gametophyte (Ga) and foot. The slightly bulbous foot lacks a palisade layer as occurs in *Anthoceros*. Bar = 100 μm. (C) *Phaeomegaceros fimbriatus*. LM nearly median longitudinal section illustrating three histogenic regions: five or six layers of cells in assimilative region (As), archesporium with a single row of fertile tissue (B) and central columella (C). Bar = 50 μm. (D) *Phaeomegaceros fimbriatus*. LM of haustorial cells (S) intermingled with gametophytic cells (Ga) with wall ingrowths. Bar = 15 μm. (E) *Folioceros appendiculatus*. TEM of gametophyte cells (Ga) of the placenta with elaborate wall ingrowths adjacent to sporophyte cells (S) that lack ingrowths. Bar = 2.0 μm. (F) *Folioceros fuciformis*. TEM of protein crystals between sporophyte (S) and gametophyte (not visible) generations. Bar = 0.5 μm. Inset: Higher magnification showing substructure of protein crystal. Bar = 0.2 μm.

that penetrate and interdigitate with surrounding gametophytic cells (Villarreal & Renzaglia 2006b; Fig. 3.8A, B). Collectively, the cells at the interface between generations compose the placenta through which the sporophyte obtains nourishment. Transfer cells with elaborate wall labyrinths that facilitate intercellular transport are restricted to gametophyte cells (Fig. 3.8D, E), a feature that is shared with *Coleochaete* and rare in other bryophytes (Graham 1993, Ligrone *et al.* 1993). A distinctive feature of the hornwort placenta is the occurrence of abundant protein crystals between gametophyte and sporophyte cells in *Folioceros*, and some species of *Phaeoceros, Notothylas, Dendroceros*, and *Megaceros* (Ligrone *et al.* 1993, Vaughn & Hasegawa 1993) (Fig. 3.8F). These crystals likely derive from gametophytic cells and may be a source of amino acids for the developing sporophyte (Ligrone & Renzaglia 1990).

At maturity, the aerial sporophyte is an elongated cylindrical spore-bearing region which includes an epidermis, assimilative layer, sporogenous tissue, columella, and basal meristem (Fig. 3.9). Because of the programmed divisions from the basal meristem, spore production is continuous throughout the growing season, with spore maturation progressing from the base to the apex of the sporophyte. In *Notothylas*, the basal meristem functions for a limited period; the sporophyte remains small and is frequently retained within the protective tissue of the gametophyte. No parallels of the sporophyte developmental strategy of hornworts, which is essentially a process of elongating a sporangium from its base, are evident in any other embryophytes. Other monosporangiate archegoniates have determinate growth that produces a defined capsule and seta, whereas polysporangiate land plants exhibit apical growth of the sporophyte that produces repeating modules, some of which may bear discrete sporangia (Kenrick & Crane 1997). In all land plants except hornworts, spore maturation in a single sporangium is synchronized.

Stomata that resemble those of mosses and tracheophytes occur in the sporophyte of many hornworts. Guard cells are characterized by inner (ventral) wall thickenings and apparently they are the only epidermal cells that contain prominent plastids, especially amyloplasts (Fig. 3.9D, E). These features suggest homology with stomata of other embryophytes. Epidermal cells typically are elongated and less commonly isodiametric in some species that lack stomata (Fig. 3.9G). At the tip of the sporophyte, where spores are mature, walls of epidermal cells are thickened along the outer tangential and radial walls (Fig. 3.9A, C).

An assimilative (photosynthetic) layer of variable thickness (4–13 layers) underlies the epidermis (Fig. 3.9A–C). Substomatal cavities and prominent intercellular spaces characterize the outer assimilative layer in taxa with stomata (Fig. 3.9A). The inner region in species with thick assimilative layers is compacted, with smaller cells and chloroplasts than those in the spongy outer layer. In taxa without stomata such as *Megaceros* and *Dendroceros*, there is no spongy layer, i.e. the assimilate layer lacks intercellular spaces (Fig. 3.9C).

The sporogenous tissue is situated between the assimilative region and the columella. The columella usually comprises 16 cells in four rows of four in cross-section but may contain as many as 40 irregularly arranged cells

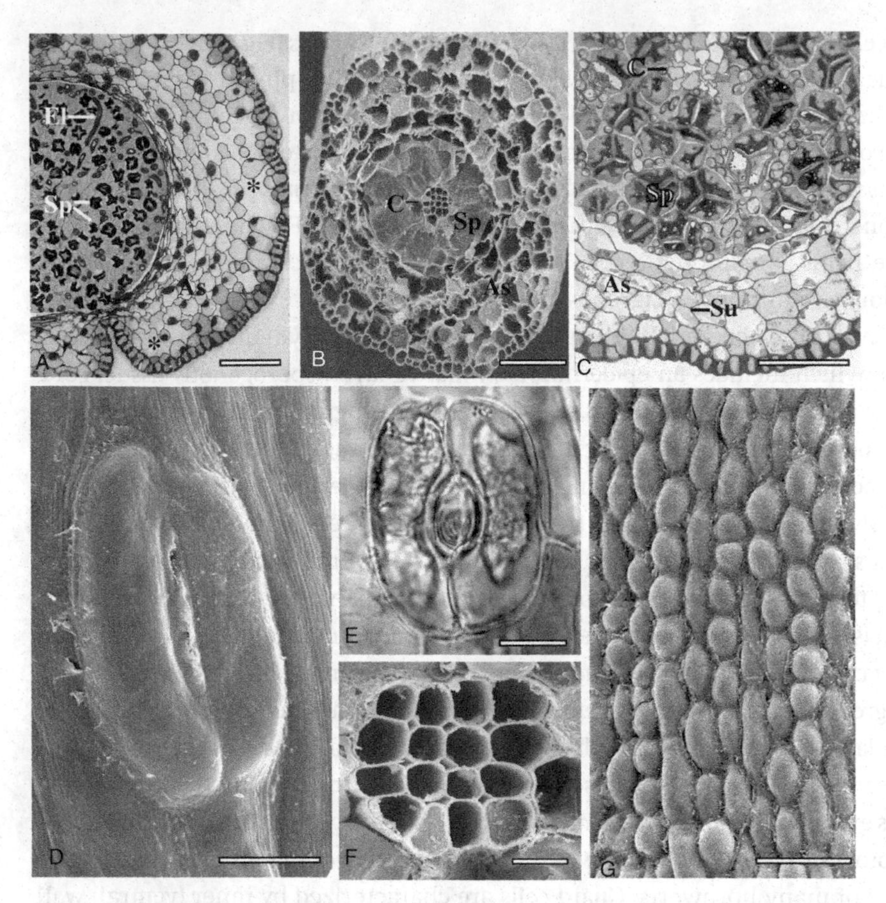

Fig. 3.9. (A) *Leiosporoceros dussii*. LM of sporophyte in transverse section. A single-layered epidermis with highly indented suture surrounds assimilative tissue (As) that consists of a spongy outer region with air spaces (*) open to stomata and compacted inner zone. Sporogenous tissue, with several layers of tetrads (Sp) intermixed with pseudoelaters (El), is bathed in mucilage. Bar = 100.0 μm. (B) *Phaeoceros carolinianus*. SEM of sporophyte in transverse section. A mostly compact assimilative tissue (As) surrounds a single layer of spore tetrads (Sp) and central columella (C). Bar ≈ 100.0 μm. (C) *Megaceros pellucidus*. LM of sporophyte in transverse section. Five or six layers of compact assimilative tissue (As) have slightly smaller and aligned cells along the suture (Su). Three or four layers of spore tetrads (Sp) are interspersed with longitudinally aligned pseudoelaters and bathed in mucilage around a central columella (C). Bar = 100 μm. (D) *Leiosporoceros dussii*. SEM of closed stoma in sporophyte epidermis. Bar = 20 μm. (E) *Phaeoceros carolinianus*. LM of open stoma in sporophyte epidermis showing massive starch-filled plastids in guard cells. Bar = 30.0 μm. (F) *Phaeoceros carolinianus*. SEM of sporophyte in transverse section showing 16-celled columella with small intercellular spaces. Bar = 15 μm. (G) *Dendroceros crispatus*. SEM of sporophyte epidermis with no stomata. Unlike those of most hornworts, epidermal cells in this species are not elongated. Bar = 40 μm.

(Fig. 3.9B, C, F). Sporogenous tissue is bathed in mucilage and consists of sporogenous cells or spores with pseudoelaters interspersed (Fig. 3.9A, C). Spore shape, wall ornamentation and pseudoelater architecture are variable across taxa and are widely used in taxonomy (Fig. 3.10A–F). Pseudoelaters are multicellular and range from thin-walled and isodiametric to elongated with tapering ends and evenly-thick or spirally thickened walls (Fig. 3.10B, C, E, F). These sterile cells do not undergo meiosis and are interspersed among sporogenous cells, thus separating sporocytes during differentiation. Sporophyte expansion further facilitates tetrad development that involves enlargement of nascent spores and the development of a sculptoderm.

Spore ornamentation and color offer the main characters to delimitate hornwort taxa. *Leiosporoceros* is the only known hornwort with nearly smooth, bean-shaped spores that are arranged in bilateral alterno-opposite tetrads (Fig. 3.10A) (Hässel de Menéndez 1986). Because of the arrangement, the proximal surface of these spores exhibits a modified Y-shaped mark. The remaining hornworts have tetrahedral, sometimes cruciate, tetrads (except *Hattorioceros*). Variability in distal wall ornamentation is seen in *Anthoceros* where it ranges from spinose and punctate (*A. punctatus* group) to lamellose (*A. angustus*). The sculptoderm on proximal faces is generally less ornate, but shows considerable variability, even within species of *Anthoceros* (e.g. hollows in *A. punctatus*, lamellae in *A. cavernosus*, and warts in *A. tuberculatus*). *Sphaerosporoceros* and *Folioceros* have rounded spores with inconspicuous trilete ridges (Asthana & Srivastava 1991) (Fig. 3.10B). In *Phaeoceros*, species of the *laevis-carolinianus* group have spinose spores with a conspicuous cingulum (Fig. 3.10C). Spores of *Hattorioceros* are strikingly different from other hornworts: they are small ($< 20\ \mu m$) with a canaliculate–striate surface and irregular shape (Fig. 3.10D). *Phymatoceros* spores are yellow-vermiculate spores with a distal bump (Fig. 3.10E). In *Phaeomegaceros* the yellow-vermiculate spores are differentiated by dimples on the distal surface (Fig. 3.10F). In the subfamily Dendrocerotoideae, *Dendroceros* has multicellular spores with finely spinulose (*D. crispus*) surfaces. Spores of *Megaceros* and *Nothoceros* are virtually identical with distal mammillae (Duff *et al.* 2007, Villarreal *et al.* 2007).

Although wall architecture has significant taxonomic value, we have discovered cases of convergent evolution, or perhaps hybridization, in our studies. For example, an unnamed species from Venezuela that nests within the *Phaeoceros* clade has spores similar to those of *Phymatoceros*, but the thallus and sporophyte anatomy are *Phaeoceros*-like.

Spore color is related to spore longevity. Yellow and brown spores are long-lived (up to 21 years in herbarium packets) because they have thicker walls and are filled with oils. The oil reserve performs a dual function of nutrient storage and protection against desiccation (Fig. 3.10G). Yellow spore color is plesiomorphic in

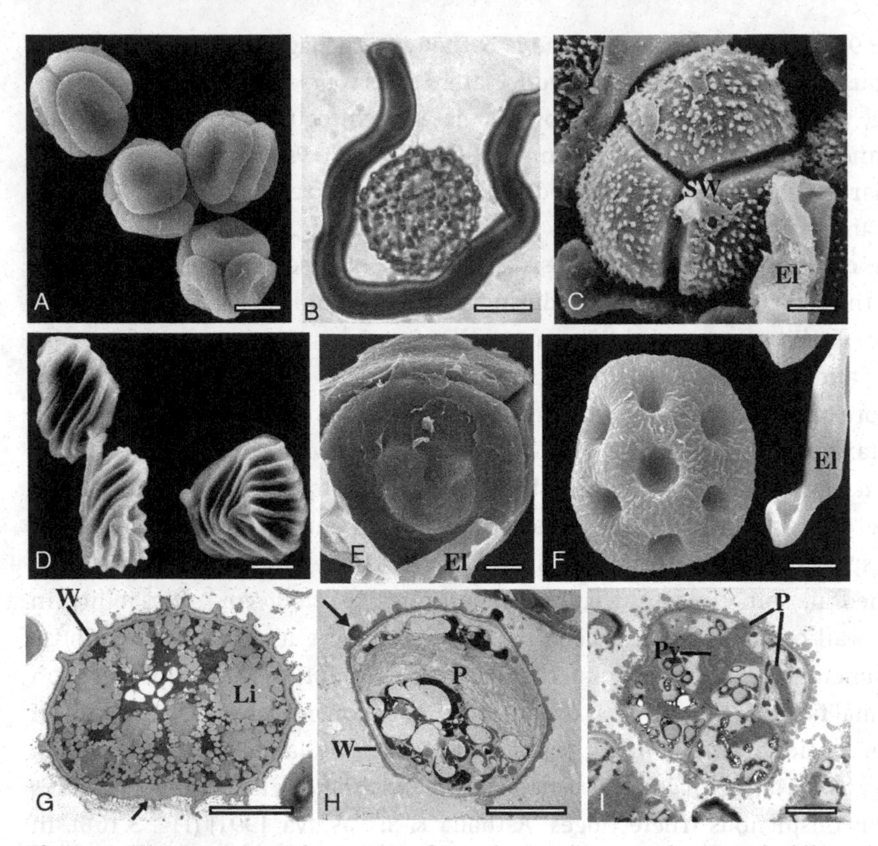

Fig. 3.10. (A) *Leiosporoceros dussii*. SEM of tetrads showing smooth spores in bilateral alterno-opposite arrangement. Bar = 10.0 μm. (B) *Folioceros appendiculatus*. LM of distal face of spore with thick-walled pseudoelater. Bar = 10.0 μm. (C) *Phaeoceros carolinianus*. SEM of spore tetrad with spinose spores surrounded by short smooth pseudoelater (El) and still enclosed in the sporophyte. Note remnant spore mother cell wall (SW) over spore surfaces. Bar = 10.0 μm. (D) *Hattorioceros striatisporus*. SEM of isolated minute spores of different sizes and shapes with striate-canaliculate ornamentation and no trilete mark. Bar = 5.0 μm. (E) *Phymatoceros phymatodes*. SEM of distal face of spore in a tetrad showing prominent mammilla and associated pseudoelater (El). Remnants of spore mother cell wall cover tetrad. Bar = 10.0 μm. (F) *Phaeomegaceros fimbriatus*. SEM of distal face of spore showing vermiculate surface with six depressions around a larger central one. El = pseudoelater. Bar = 8.0 μm. (G) *Folioceros fuciformis*. TEM of a mature spore full of lipids (Li) with a slightly thick spore wall (W). The aperture is arrowed. Bar = 10 μm. (H) *Megaceros gracilis*. TEM of a mature green spore with thin spore wall (W) and large chloroplast (P). External ornamentation is arrowed. Bar = 10 μm. (I) *Dendroceros granulatus*. TEM of precocious spore with multiple cells, each with a plastid (P) and conspicuous pyrenoid (Py).

hornworts and occurs in *Leiosporoceros, Phaeoceros, Paraphymatoceros*, some *Notothylas* taxa, *Hattorioceros, Phymatoceros*, and *Phaeomegaceros*. Dark spores are present in *Anthoceros*, (*Mesoceros*), *Folioceros, Sphaerosporoceros* and some species of *Notothylas*. "Green" spores, due to the presence of a chloroplast and a thin, colorless spore wall, are short-lived and restricted to tropical genera: *Megaceros, Nothoceros*, and the epiphytic *Dendroceros* (Fig. 3.10H, I).

Sporogenesis in hornworts resembles that in many other basal embryophytes in that meiosis is monoplastidic. The single plastid undergoes two series of division and the four resulting plastids define the meiotic poles (Fig. 3.11A). Associated with monoplastidy in achegoniates, but not in green algae, is a unique quadripolar microtubule system that is organized at the plastids and predicts polarity of the two meiotic divisions (Brown & Lemmon 1997).

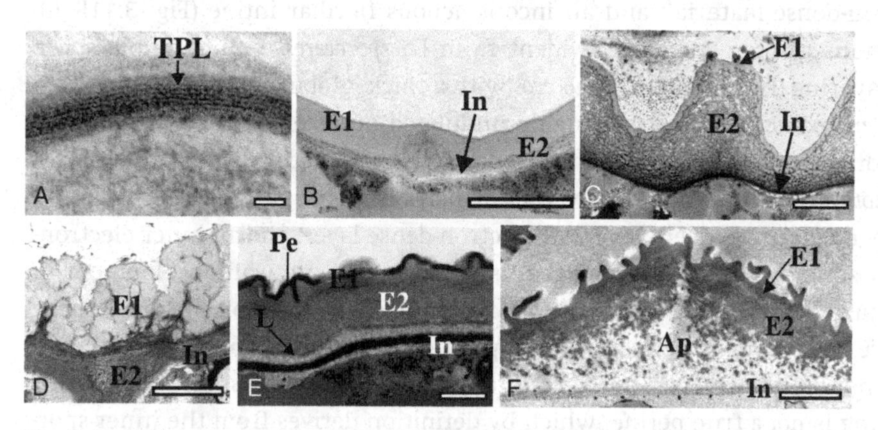

Fig. 3.11. (A) *Leiosporoceros dussii*. TEM of spore wall in nascent spore. A multilayered lamellae composed of tripartite lamellae (TPL) is seen just after meiosis; sporopollenin deposition obscures TPL in more mature spores (see B). Bar = 0.10 μm. (B) *Leiosporoceros dussii*. TEM of distal face of nearly mature spore. The three-layered wall is composed of a compacted and uniform exine 1 (E1), electron-translucent and granular exine 2 (E2), and electron-translucent intine (In). Bar = 1.0 μm. (C) *Anthoceros punctatus*. TEM of distal face of nearly mature spore. The three-layered wall is composed of a thin, compacted, homogeneous E1, a densely globular E2, and an electron-dense intine (In). Bar = 1.0 μm. (D) *Dendroceros granulatus*. TEM of distal face of multicellular spore. The three-layered wall is composed of an undulating homogeneous E1, a fibrillar electron-dense E2, and a thin translucent intine (In). Bar = 1.0 μm. (E) *Notothylas temperata*. TEM of nearly mature spore wall with complex wall. Dark-perine-like layer (Pe) deposited from spore mother cell wall covers outer exine (E1). One or two lamellae (L) lie between exine 2 (E2) and intine (In) of three layers: electron-lucent outer and inner layers, with an electron-dense layer between. Bar = 0.5 μm. (F) *Notothylas orbicularis*. TEM of proximal face at trilete mark where developing aperture (Ap) has greatly expanded exine 2 (E2) with globular sporopollenin deposition. Sporopollenin will eventually fill all but the mid-line of the aperture. Intine (In) and exine 1 (E1) layers are similar to remainder of spore. Bar = 1.0 μm.

Spore wall development in hornworts involves the presence of tripartite lamellae (TPL), not reported for hornworts prior to our studies. TPL are laid down immediately after meiosis and they coalesce to form a multilamellate layer (MLL) that delimits the outer exine (exine 1; Fig. 3.11A). Through precisely produced folds in the MLL, the spore wall sculpturing is determined in the initial stages of wall development. The TPL are reinforced with sporopollenin soon after ornamentation is established, and thus the fine lines are entirely obscured (Fig. 3.11B). To date, TPL have been found in early post-meiotic spore walls of *Leiosporoceros* (Fig. 3.11A), *Notothylas* (D. Long & K. S. Renzaglia, unpublished data) and *Megaceros*, but not in *Dendroceros* (S. Schuette & K. S. Renzaglia, unpublished data). The spore wall is deposited centripetally and consists of a thin outer layer (exine 1), a thick inner exine (exine 2) that forms by deposition of flocculent electron-dense material, and an inconspicuous fibrillar intine (Fig. 3.11B–E). Variations across genera are evident from *Leiosporoceros* with a simple spore wall layering (Fig. 3.11B) to *Anthoceros* with a thick, globular exine 2 (Fig. 3.11C) and *Dendroceros*, which has a highly undulated outer exine that stretches by unfolding as the cells divide during endosporic germination (Figs. 3.10I, 3.11D). In *Notothylas temperata*, the spore wall displays bands of varying opacity in the intine: an electron-lucent layer, an electron-dense layer, and an inner electron-lucent zone (Fig. 3.11E). During the final stages of spore wall development in most taxa, a thin dark band of fibrous material is laid down on the outer spore surface. This layer is derived from deposition of remnant sporocyte wall and intrasporal septum (Fig. 3.11E). Thus, although it is of extrasporic origin, this covering is not a true perine, which by definition derives from the inner sporangial wall. In a few species of yellow-spored genera, the remnant spore mother cell wall is responsible for secondary "browning" of the spores.

The trilete ridge serves as the site for spore germination and is differentiated into a simple aperture (Fig. 3.11F). In this aperture, exine 2 is greatly expanded whereas intine and exine 1 remain unchanged. Sporopollenin deposition is increased along the flanks of the aperture and thus fortifies the ridge, while the center of the aperture at maturity is nearly devoid of sporopollenin (Fig. 3.11F). An aperture-like region with similarly thickened exine 2 and a break in sporopollenin deposition occurs along the spore equator and forms the cingulum of many taxa (Villarreal & Renzaglia 2006b).

Dehiscence typically occurs along two longitudinal lines that originate near the sporophyte tip. Pseudoelaters and the columella facilitate spore separation and assist in dispersal. Spores remain in tetrads until nearly mature and are dispersed individually.

Spore germination results in the production of a single gametophyte. *Dendroceros* spores are precocious and initially endosporic. Multicellular "spores" are released from the capsule and develop upon contact with the substrate. In most taxa, germination is exosporic, resulting in a globose sporeling that produces an apical cell and flattens with continued development (Renzaglia 1978). Spores may overwinter or remain quiescent until favorable conditions for germination are encountered.

3.5 Evolution

Because the paleontological record for hornworts is depauperate it is difficult to assign dates to hornwort divergences. It is well accepted that the evolution of monosporangiate body plans preceded polysporangiate architectures and thus the hornwort clade was established prior to the appearance of tracheophytes in the Silurian. Yet no hornwort fossils have been found in any Paleozoic strata. It is particularly intriguing, therefore, that spores of *Leiosporoceros* (Fig. 3.10A) are remarkably similar in size, shape, and ornamentation to those of the earliest occurring land plants from the Ordovician (Kenrick 2003, Wellman *et al.* 2003). The primary difference is that these simple fossil spores are arranged in tetrahedral tetrads and not in the peculiar bilateral alterno-opposite arrangement that typifies *Leiosporoceros*.

The oldest confirmed hornwort fossils are sparse reports of spores and sporangia resembling the extant *Phaeoceros* and *Notothylas* from the Cretaceous (Jarzen 1979, Chitaley & Yawale 1980, Dettmann 1994). Cenozoic fossils of spores resembling *Phaeoceros* and *Phaeomegaceros* have been reported from Europe, Central America, and South America (Hooghiemstra 1984, Graham 1987, Ivanov 1997). Spore ornamentation of these fossils strongly resembles that of extant taxa, suggesting morphological stasis over the past hundred million years.

The fragile nature of the hornwort thallus explains the absence of fossil gametophytes. Without such tissue, morphological evolution within the group can only be inferred from phylogenetic analyses. It is plausible that hornworts were a highly diverse group in Pre-Cretaceous times and that they experienced episodes of extinctions. The nested phylogenetic position of epiphytic taxa (*Megaceros* and *Dendroceros*) supports the interpretation that diversification of these clades was correlated with angiosperm and fern radiation in the Cretaceous. A recent report of a preserved hornwort fossil from the Dominican Amber (Eocene–Oligocene) attributed to *Dendroceros* is consistent with this conjecture (Frahm 2005). With spore wall diversity and ultrastructure recently documented in hornworts (Duff *et al.* 2007), a critical assessment of spore fossils prior to the Cretaceous may reveal additional clues about hornwort diversification and provide valuable calibration points for molecular phylogenies.

The generic divergence of *Leiosporoceros* from the remaining hornworts prompted further morphological and molecular studies that reaffirm the distance between this and the other taxa (Duff & Moore 2005, Villarreal & Renzaglia 2006a). This robust taxon possesses features not seen in other hornworts, such as lack of RNA editing, small, smooth "monolete" spores, lack of ventral clefts in mature gametophytes, and central canals with *Nostoc* that run the length of the thallus. Unfortunately, it is impossible to determine whether these peculiarities are plesiomorphic in hornworts or simply features that evolved after *Leiosporoceros* diverged from the remaining taxa.

With only a few exceptions, morphological boundaries between hornwort genera remain ambiguous even though interrelationships have solidified as a result of phylogenetic inferences from DNA sequence data (Fig. 3.1). For example, based on appearance, several species within the genus *Nothoceros* could readily be placed within *Megaceros*. Moreover, genera such as *Sphaerosporoceros* and *Phymatoceros* are defined by single features pertaining to tubers, and spore shape and ornamentation; thus mature fertile plants are required for identification (Cargill & Scott 1997). *Notothylas* and *Dendroceros* are two of the 14 genera that are clearly demarcated by a suite of diagnostic characters. The placement of *Notothylas* within a paraphyletic *Phaeoceros* confirms the traditional view that the sporophyte of *Notothylas* is derived through extensive evolutionary reduction (Campbell 1895, Proskauer 1960, Renzaglia 1978, Schofield 1985, Schuster 1987). Nested within the *Megaceros–Nothoceros* assemblage, *Dendroceros* is defined by the midrib and monostromatic perforated wings, a hemidiscoid apical cell, a unique pyrenoid microanatomy, and precocious, endosporic germination that is associated with a peculiar spore wall structure. These features are viewed as adaptations to the epiphytic habit of this taxon.

With a solid backbone of relationships it is now possible to examine character evolution within hornworts (Fig. 3.12). Here we present brief descriptions of character transformations that demonstrate reductions, parallelisms, reversals, and niche-specific adaptations.

3.5.1 *Stomata*

Stomata are plesiomorphic in hornworts and were lost independently in two clades: *Notothylas* and the *Megaceros–Nothoceros–Dendroceros* assemblage. Genera that lack stomata have involucres that cover the epidermis in the region where guard cells differentiate, suggesting that developmental constraints may regulate stomata differentiation. In *Notothylas* and *Dendroceros*, most of the young sporophyte is covered by involucre, whereas in *Megaceros* and *Nothoceros* the involucre is markedly longer than those of other genera. It remains to be tested whether the capacity to produce stomata exists in these taxa but is inhibited by immersion within a mucilage-filled involucre. There are no examples of stomata, once lost, being regained in hornworts.

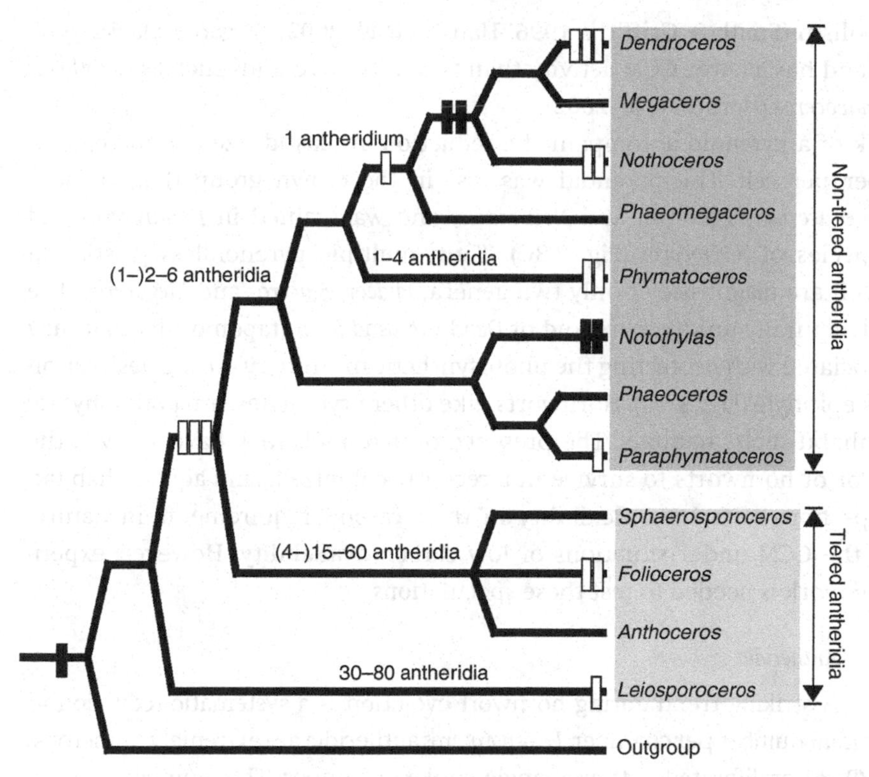

Fig. 3.12. Skeleton phylogeny of hornworts based on three genes (see Duff *et al.* 2007) with simplified inferences on the evolution of four characters. Chloroplast structure is shown with hollow rectangular bars: three bars = pyrenoid gain, two bars = partial loss of pyrenoid in some species of lineage, and one bar = no pyrenoid in entire lineage. Stomatal evolution is represented by solid black bars: = stomata present, = loss of stomata. Stomata are present in most lineages with two independent losses in *Notothylas* and Dendrocerotoideae. Antheridial number is shown above branches and presents a trend from abundant antheridia (30–80 in *Leiosporoceros*), to 15–60 in Anthocerotidae, to (1-) 2–6 in Notothyladaceae, 1–4 in *Phymatoceros*, to typically only one in the Dendrocerotaceae. Antheridial jacket cell arrangement is indicated with side arrows and transforms from tiered in the taxa with more than six antheridia per chamber to non-tiered in all other taxa.

3.5.2 *Chloroplast evolution*

The evolution of pyrenoids in hornworts involves parallelisms and reversals with multiple losses and gains. If a single chloroplast with a central pyrenoid is viewed as plesiomorphic, pyrenoid loss occurred independently at least six times. The pyrenoid is the site of occurrence of RUBISCO and is associated with a carbon concentration mechanism (CCM) and low $\Delta^{13}C$ values similar to those found in plants using C_4 photosynthesis and crassulacean acid metabolism (Smith & Griffiths 1996, Hanson *et al.* 2002). *Megaceros* lacks pyrenoids and has a lower CCM activity than taxa with pyrenoids such as *Notothylas* and *Phaeoceros* (Hanson *et al.* 2002).

Lack of a pyrenoid accompanied a reduction in plastid size and increase in number per cell. The pyrenoid was lost in the crown group that includes *Phaeomegaceros*, *Megaceros*, and *Nothoceros*, and was gained in *Dendroceros* and two species of *Nothoceros* (Fig. 3.3C). Thus, multiple pyrenoidless plastids in each cell are diagnostic of only two genera, *Phaeomegaceros* and *Megaceros*. The ultrastructurally unique pyrenoid of *Dendroceros* is an autapomorphy that may be associated with protecting the photosynthetic machinery during desiccation in this epiphyte (Fig. 3.3G). Hornworts, like other bryophytes, are poikilohydric and inhabit moist habitats. The presence of pyrenoids may have allowed the ancestor of hornworts to survive in a recently colonized semi-aquatic habitat. Perhaps hornworts have flexibility in their carbon requirements in nature, using the CCM under situations of low carbon availability. However, experimental work is needed to test these speculations.

3.5.3 *Antheridia*

A striking trend during hornwort evolution is a systematic reduction in antheridial number per chamber. *Leiosporoceros* antheridia are extremely numerous, with 30–80 proliferated within a single sunken chamber. This number progressively decreases from 15–60 in *Anthoceros–Folioceros–Sphaerosporoceros* to 2–6 in most genera to only one (rarely two) in the *Megaceros–Nothoceros–Dendroceros* crown group. A tiered arrangement of jacket cells around the antheridium is associated with large numbers of antheridia per chamber (Fig. 3.12). In chambers with six or fewer antheridia, the antheridial body expands to fill the chamber space. This process involves multiple divisions in the jacket, resulting in random cell arrangements.

3.5.4 *Spermatogenesis*

Ultrastructural examinations of sperm cell differentiation and mature architecture have revealed little variability among the five hornwort genera studied. A three-dimensional model that is characteristic of all species examined has been developed for the mature sperm cell of *Phaeomegaceros hirticalyx* (Fig. 3.7D). Hornwort spermatozoids are markedly different from those of other bryophytes and from those of all other land plants in a number of features. The two flagella are inserted side-by-side at the cell anterior and wrap twice around the cell. The nucleus is constricted in the middle and the plastid contains a single large starch grain. When viewed from the anterior end, the cell coils to the left, the opposite direction from other land plant sperm. Because these cells are bilaterally symmetrical, as opposed to asymmetrical as in other embryophytes, the direction of coiling may be inconsequential to swimming and thus was free to change during the evolutionary history of anthocerotes (Renzaglia *et al.* 2000).

3.6 Innovative morphology

Hornworts are unique among embryophytes in key morphogenetic characters. Diagnostic morphological features of the group includes chloroplast structure, endogenous antheridia, details of the microtubule-organizing center during mitosis, sperm cell architecture, sporophyte growth from a basal meristem, placental transfer cells restricted to the gametophyte generation, and non-synchronized sporogenesis. A striking contrast of hornworts when compared with mosses and liverworts is the lack of organized external appendages. No leaves, scales, slime papillae, or superficial gametangia occur in the group. Sex organs and *Nostoc* colonies are embedded within the thallus, and mucilage canals and cells are integrated into the undifferentiated thallus chlorenchyma. Vulnerable tissues such as the apical meristem and archegonia are protected externally by mucilage secreted by epidermal cells, and not by appendages. Small size, rapid life cycle, internal sequestration of structures, and mucilage proliferation may be the key to the persistence of this relatively isolated taxon through the millennia. Further insights into early land colonization strategies will be attain by continued investigation of the ultrastructure, morphogenesis, physiology, biochemistry, and phylogeny of this engaging plant group.

Acknowledgments

This study was supported by NSF grants DEB-9207626 and DEB-9527735. We thank Jeff Duckett, Scott Schuette, and one anonymous reviewer for comments on the manuscript, and Drs John Bozzola, Steven Schmitt, and H. Dee Gates at the Center for Electron Microscopy, Southern Illinois University, for assistance and use of the facility. Thanks to William Buck and the New York Botanical Garden (NYBG) for access to the holotype of *Hattorioceros striatisporus*.

References

Asthana, A. K. & Srivastava, S. C. (1991). Indian hornworts (a taxonomic study). *Bryophytorum Bibliotheca*, **42**, 1–158.

Bartlett, E. M. (1928). The comparative study of the development of the sporophyte in the Anthocerotaceae, with special reference to the genus *Anthoceros*. *Annals of Botany*, **42**, 409–30.

Bower, F. O. (1935). *Primitive Land Plants*. London: macmillan & Co.

Brown, R. C. & Lemmon, B. E. (1990). Monoplastidic cell division in lower land plants. *American Journal of Botany*, **77**, 559–71.

Brown, R. C. & Lemmon, B. E. (1993). Diversity of cell division in simple land plants hold clues to evolution of the mitotic and cytokinetic apparatus in higher plants. *Memoirs of the Torrey Botanical Club*, **25**, 45–62.

Brown, R. C. & Lemmon, B. E. (1997). The quadripolar microtubule system in lower land plants. *Journal of Plant Research*, **110**, 93–106.

Burr, F. A. (1969). Reduction in chloroplast number during gametophyte regeneration in *Megaceros flagellaris*. *Bryologist*, **72**, 200–9.

Campbell, D. H. (1895). *The Structure and Development of Mosses and Ferns (Archegoniatae)*. New York: Macmillan.

Cargill, D. C. & Scott, G. A. M. (1997). Taxonomic studies of the Australian Anthocerotales I. *Journal of the Hattori Botanical Laboratory*, **82**, 47–60.

Cargill, D. C., Duff, R. J., Villarreal, J. C. & Renzaglia, K. S. (2005). Generic concepts in hornworts: historical review, contemporary insights and future directions. *Australian Systematic Botany*, **18**, 7–16.

Chitaley, S. D. & Yawale, N. R. (1980). On *Notothylites nirulai* gen. et sp. nov. A petrified sporogonium from the Deccan-Intertrappean beds of Mohgaonkalan M.P. (India). *Botanique*, **9**, 111–18.

Crandall-Stotler, B. J. (1980). Morphogenetic designs and a theory of bryophyte origins and divergence. *BioScience*, **30**, 580–5.

Dettmann, M. E. (1994). Cretaceous vegetation: the microfossil record. In *History of the Australian Vegetation: Cretaceous to Recent*, ed. R. S. Hill, pp. 143–70. Cambridge: Cambridge University Press.

Duckett, J. G. (1975). An ultrastructural study of the differentiation of antheridial plastids in *Anthoceros laevis*. *Cytobiologie*, **10**, 432–48.

Duckett, J. G. & Renzaglia, K. S. (1988). Ultrastructure and development of plastids in bryophytes. *Advances in Bryology*, **3**, 33–93.

Duff, R. J. & Moore, F. (2005). Pervasive RNA editing inferred from *rbc*L transcripts among all hornworts except *Leiosporoceros*. *Journal of Molecular Evolution*, **61**, 571–8.

Duff, R. J., Cargill, D. C., Villarreal, J. C. & Renzaglia, K. S. (2004). Phylogenetic relationships of the hornworts based on *rbc*L sequence data: novel relationships and new insights. *Monographs in Systematic Botany, Missouri Botanical Garden*, **98**, 41–58.

Duff, R. J., Villarreal, J. C., Cargill, D. C. & Renzaglia, K. S. (2007). Progress and challenges toward developing a phylogeny and classification of the hornworts. *Bryologist*, **110**, 214–43.

Frahm, J.-P. (2005). The first record of a fossil hornwort (Anthocerotophyta) from Dominican Amber. *Bryologist*, **108**, 139–41.

Frey, W. & Stech, M. (2005). A morpho-molecular classification of the Anthocerotophyta (hornworts). *Nova Hedwigia*, **80**, 542–5.

Garbary, D. J., Renzaglia, K. S. & Duckett, J. G. (1993). The phylogeny of land plants: a cladistic analysis based on male gametogenesis. *Plant Systematics and Evolution*, **188**, 237–69.

Goebel, K. (1905). *Organography of Plants Especially of the Archegoniate and Spermatophyta.* Part II. *Special Organography.* Oxford: Clarendon Press.

Goffinet, B. (2000). Origin and phylogenetic relationships of bryophytes. In *Bryophyte Biology*, ed. A. J. Shaw & B. Goffinet, pp. 124–49. Cambridge: Cambridge University Press.

Graham, A. (1987). Miocene communities and paleoenvironments of Southern Costa Rica. *American Journal of Botany*, **74**, 1501–18.

Graham, L. E. (1993). *Origin of Land Plants.* New York: John Wiley.

Hanson, D., Andrews, T. J. & Badger, M. R. (2002). Variability of the pyrenoid-based CO_2 concentrating mechanism in hornworts (Anthocerotophyta). *Functional Plant Biology*, **29**, 407–16.

Hasegawa, J. (1983). Taxonomical studies on Asian Anthocerotae. III. Asian species of *Megaceros. Journal of the Hattori Botanical Laboratory*, **54**, 227–40.

Hasegawa, J. (1994). New classification of Anthocerotae. *Journal of the Hattori Botanical Laboratory*, **76**, 21–34.

Hässel de Menéndez, G. G. (1986). *Leiosporoceros* Hässel n. gen. and Leiosporocerotaceae Hässel n. fam. of Anthocerotopsida. *Journal of Bryology*, **14**, 255–9.

Hässel de Menéndez, G. G. (1988). A proposal for a new classification of the genera within the Anthocerotophyta. *Journal of the Hattori Botanical Laboratory*, **64**, 71–86.

Hässel de Menéndez, G. G. (2006). *Paraphymatoceros*, gen. nov. (Anthocerotophyta). *Phytologia*, **88**, 208–11.

Hooghiemstra, H. (1984). Vegetational and climatic history of the high plain of Bogotá, Colombia: a continuous record of the last 3,5 million years. *Dissertationes Botanicae*, **79**, 1–368.

Hyvönen, J. & Piippo, S. (1993). Cladistic analysis of the hornworts (Anthocerotophyta). *Journal of the Hattori Botanical Laboratory*, **74**, 105–19.

Ivanov, D. A. (1997). Miocene palynomorphs from the southern part of the Forecarpathian Basin (Northwest Bulgaria). *Flora Tertiaria Mediterranea*, **6** (4), 1–81.

Jarzen, D. H. (1979). Spore morphology of some Anthocerotaceae and the occurrence of *Phaeoceros* spores in the Cretaceous of North America. *Pollen et Spores*, **21**, 211–31.

Kenrick, P. (2003). Palaeobotany: fishing for the first plants. *Nature*, **425**, 248–9.

Kenrick, P. & Crane, P. R (1997). *The Origin and Early Diversification of Land Plants: A Cladistic Study.* Washington, D.C.: Smithsonian Institution Press.

Leitgeb, H. (1879). *Untersuchungen über die Lebermoose*, vol. 5, *Die Anthoceroteen.* Graz: Leuschner & Lubensky.

Ligrone, R. (1988). Ultrastructure of a fungal endophyte in *Phaeoceros laevis* (L.) Prosk. (Anthocerotophyta). *Botanical Gazette*, **149**, 92–100.

Ligrone, R. & Renzaglia, K. S. (1990). The sporophyte-gametophyte junction in the hornwort, *Dendroceros tubercularis* Hatt. (Anthocerotophyta). *New Phytologist*, **114**, 497–505.

Ligrone, R., Duckett, J. G. & Renzaglia, K. S. (1993). The gametophyte-sporophyte junction in land plants. *Advances in Botanical Research*, **19**, 231–317.

Mishler, B. D., Lewis, L. A., Buchheim, M. A. *et al.* (1994). Phylogenetic relationships of the "green algae" and "bryophytes." *Annals of the Missouri Botanical Garden*, **81**, 451–83.

Proskauer, J. (1960). Studies on Anthocerotales. VI. *Phytomorphology*, **10**, 1–19.

Qiu, Y. L., Choe, Y., Cox, J. C. & Palmer, J. D. (1998). The gain of three mitochondrial introns identifies liverworts as the earliest land plants. *Nature*, **394**, 671–4.

Qiu, Y. L., Li, L. B., Wang, B. *et al.* (2006). The deepest divergences in land plants inferred from phylogenomic evidence. *Proceedings of the National Academy of Sciences, U.S.A.*, **103**, 15511–16.

Renzaglia, K. S. (1978). A comparative morphology and developmental anatomy of the Anthocerotophyta. *Journal of the Hattori Botanical Laboratory*, **44**, 31–90.

Renzaglia, K. S. (1982). A comparative developmental investigation of the gametophyte generation in the Metzgeriales (Hepatophyta). *Bryophytorum Bibliotheca*, **24**, 1–205.

Renzaglia, K. S. & Carothers, Z. B. (1986). Ultrastructural studies of spermatogenesis in the Anthocerotales. IV. The blepharoplast and mid-stage spermatid of *Notothylas*. *Journal of the Hattori Botanical Laboratory*, **60**, 97–104.

Renzaglia, R. S. & Duckett, J. G. (1989). Ultrastructural studies of spermatogenesis in the Anthocerotophyta. V. Nuclear metamorphosis and the posterior mitochondrion of *Notothylas orbicularis* and *Phaeoceros laevis*. *Protoplasma*, **151**, 137–50.

Renzaglia, K. S. & Garbary, D. J. (2001). Motile male gametes of land plants: diversity, development and evolution. *Critical Review in Plant Sciences*, **20**, 107–213.

Renzaglia, K. S. & McFarland, K. D. (1999). Antheridial plants of *Megaceros aenigmaticus* in the Southern Appalachians: anatomy, ultrastructure and population distribution. *Haussknechtia Beiheft*, **9**, 307–16.

Renzaglia, K. S. & Vaughn, K. C. (2000). Anatomy, development and classification of hornworts. In *Bryophyte Biology*, ed. A. J. Shaw & B. Goffinet, pp. 1–35. Cambridge: Cambridge University Press.

Renzaglia, K. S., Duff, R. J., Nickrent, D. L. & Garbary, D. J. (2000). Vegetative and reproductive innovations of early land plants: implications for a unified phylogeny. *Philosophical Transactions of the Royal Society*, B**355**, 769–93.

Renzaglia, K. S., Schuette, S., Duff, R. J. *et al.* (2007). Bryophyte phylogeny: advancing the molecular and morphological frontiers. *Bryologist*, **110**, 179–213.

Schiffner, V. (1937). Kritische Bemerkungen über die Europäischen Lebermoose mit Bezug auf die Exemplare des Exsiccatenwerkes. *Hepaticae Europaeae Exsiccatae*. XII Serie. Wien.

Schofield, W. B. (1985). *Introduction to Bryology*. New York: Macmillan.

Schuster, R. M. (1987). Preliminary studies on Anthocerotae. *Phytologia*, **63**, 193–200.

Schuster, R. M. (1992). *The Hepaticae and Anthocerotae of North America*, vol. VI. Chicago: Field Museum of Natural History.

Shaw, A. J. & Renzaglia, K. S. (2004). Phylogeny and diversification of bryophytes. *American Journal of Botany*, **91**, 1557–81.

Smith, E. C. & Griffiths, H. (1996). A pyrenoid-based carbon-concentrating mechanism is present in terrestrial bryophytes of the class Anthocerotae. *Planta*, **200**, 203–12.

Stech, M., Quandt, D. & Frey, W. (2003). Molecular circumscription of the hornworts (Anthocerotophyta) based on the chloroplast DNA *trnL-trnF* region. *Journal of Plant Research*, **116**, 389–98.

Stotler, R. E. & Crandall-Stotler, B. (2005). A revised classification of the Anthocerotophyta and a checklist of the hornworts of North America, north of Mexico. *Bryologist*, **108**, 16–26.

Stotler, R. & Doyle, W. T. (2006). Contribution toward a bryoflora of California. III. Keys and annotated species catalogue for liverworts and hornworts. *Madroño*, **53**, 89–197.

Stotler, R. E., Doyle, W. T. & Crandall-Stotler, B. (2005). *Phymatoceros* Stotler, W. T. Doyle & Crand.-Stotler, gen. nov. (Anthocerotophyta). *Phytologia*, **87**, 113–16.

Vaughn, K. C. & Harper, D. L (1998). Microtubule-organizing centers and nucleating sites in land plants. *International Review of Cytology*, **181**, 75–149.

Vaughn, K. C. & Hasegawa, J. (1993). Ultrastructural characteristics of the placental region *of Folioceros* and their taxonomic significance. *Bryologist*, **96**, 112–21.

Vaughn, K. C. & Renzaglia, K. S. (1998). Origin of bicentrioles in Anthocerote spermatogenous cells. In *Bryology for the Twenty-first Century*, ed. J. W. Bates, N. W. Ashton & J. G. Duckett, pp. 189–203. Leeds: Maney and British Bryological Society.

Vaughn, K. C., Campbell, E. O., Hasegawa, J., Owen, H. A., & Renzaglia, K. S. (1990). The pyrenoid is the site of ribulose 1-5 bisphosphate carboxylase; oxygenase accumulation in the hornwort (Bryophyta: Anthocerotae) chloroplast. *Protoplasma*, **156**, 117–29.

Vaughn, K. C., Ligrone, R., Owen, H. A. *et al.* (1992). The anthocerote chloroplast: a review. *New Phytologist*, **120**, 169–90.

Villarreal, J. C. & Renzaglia, K. S. (2006a). Structure and development of *Nostoc* strands in *Leiosporoceros dussii* (Anthocerotophyta): a novel symbiosis in land plants. *American Journal of Botany*, **93**, 693–705.

Villarreal, J. C. & Renzaglia, K. S. (2006b). Sporophyte structure in the neotropical hornwort *Phaeomegaceros fimbriatus*: implications for phylogeny, taxonomy and character evolution. *International Journal of Plant Sciences*, **167**, 413-27.

Villarreal, J. C., Hässel de Menéndez, G. G. & Salazar Allen, N. (2007). *Nothoceros superbus* (Dendrocerotaceae), a new species of hornwort from Costa Rica. *Bryologist*, **110**, 279–85.

Wellman, C. H., Osterloff, P. L. & Mohiuddin, U. (2003). Fragments of the earliest land plants. *Nature*, **425**, 282–5.

4

Phylogenomics and early land plant evolution

BRENT D. MISHLER AND DEAN G. KELCH

4.1 Introduction

This is the era of whole-genome sequencing; molecular data are becoming available at a rate unanticipated even a few years ago. Sequencing projects in a number of countries have produced a growing number of fully sequenced organellar and nuclear genomes, providing computational biologists with tremendous opportunities, but also major challenges. The sheer amount of data is nearly overwhelming; comparative frameworks are needed. Comparative genomics was initially restricted to pairwise comparisons of genomes based on sequence similarity matching. The importance of taking a multispecies phylogenetic approach to systematically relating larger sets of genomes has only recently been realized.

Something can be learned about the function of genes by examining them in one organism, or by comparisons between two organisms. However, a much richer approach is to compare many organisms at once by using a phylogenetic approach, which lets us take advantage of the burgeoning number of phylogenetic comparative methods. A synthesis of phylogenetic systematics and molecular biology/genomics – two fields once estranged – is beginning to form a new field that could be called "phylogenomics" (Eisen 1998). We need to take advantage of the rich, multispecies approach provided by taking into account the history of life. Repeated, close sister-group comparisons between lineages differing in a critical phenotype (e.g. desiccation- or freezing-tolerance) can allow a quick narrowing of the search for genetic causes in a sort of natural experiment. Dissecting a complicated, evolutionarily advanced genotype-phenotype complex (e.g. development of the angiosperm flower), by tracing the components back through simpler ancestral reconstructions, can lead to

Bryophyte Biology: Second Edition, ed. B. Goffinet & A. J. Shaw. Published by Cambridge University Press.

quicker understanding via standard functional genomics approaches. Hence, phylogenomics allows one to go beyond the use of pairwise sequence similarities, and use phylogenic comparative methods to confirm and/or to establish gene function and interactions across many genomes at once.

Cross-genome phylogenetic approaches have the potential to provide insights into many open functional questions. A short list includes understanding the processes underlying genomic evolution, identifying key regulatory regions, understanding the complex relationship between phenotype and genomic changes, and understanding the evolution of complex physiological pathways in related organisms. Using such a comparative approach will aid in elucidating how these genes interact to perform specific biological processes. For example, Stuart *et al.* (2003) used microarray data from four completely sequenced genomes (yeast, nematode, insect, and human) to show co-expression relationships that have been conserved across a wide spectrum of animal evolution.

There are reciprocal benefits for the phylogeneticist as well, of course: the new comparative genomic data should also greatly increase the accuracy of reconstructions of the Tree of Life. Even though nucleotide sequence comparisons have become the workhorse of phylogenetic analysis at all levels, there are clearly phylogenetic problems for which nucleotide sequence data are poorly suited, because of their simple nature (having only four possible character states) and tendency to evolve in a regular, more or less clockwork fashion. In particular, "deep" branching questions (with relatively short internodes of interest mixed with long terminal branches) are notoriously difficult to resolve with DNA sequence data. Examples of such difficult cases in bryology range from the fundamental relationships of the major groups of bryophytes to embedded relationships within these groups such as the apparently rapid radiations of pleurocarpous mosses and leafy liverworts.

It is fortunate therefore, that fundamentally new kinds of structural genomic characters such as inversions, translocations, losses, duplications, and insertion/deletion of introns will be increasingly available in the future. These characters need to be evaluated by using much the same principles of character analysis that were originally developed for morphological characters. They must be looked at carefully to establish likely homology (e.g. examining the ends of breakpoints across genomes to see whether a single rearrangement event is likely to have occurred), independence, and discreteness of character states. It is also necessary to consider carefully the appropriate terminal units for comparative genomic analysis, especially since different parts of an organism's genome may or may not have exactly the same history. Thus close collaboration between systematists and molecular biologists will be required to code these genomic characters properly, and to assemble them into matrices with other data types.

The purpose of this chapter is to explore the relationships between genomics and phylogenetics in the land plants, in both directions, i.e. the uses of genomic characters in phylogenetic analysis and the uses of phylogenies in functional analysis of genes. We use examples involving bryophytes when possible; their position as the basal extant lineages in the land plants makes them especially important for comparative genomics.

4.2 The uses of comparative genomics in functional studies

Evolution by descent with modification is the most important organizing principle in biology. All living things are related to each other to a greater or lesser extent; thus similarities in the attributes they bear are dependent largely on their degree of relatedness. This was brought home to all biologists when the human genome was sequenced, and it became clear that very little of its structure has to do with being a human *per se*, but rather with being a great ape, a mammal, a metazoan, etc. For example, only 94 of 1278 protein families in the human genome appear to have arisen in vertebrates (Baltimore 2001)! The ubiquitous influence of the Tree of Life provides the key for exploring the full richness of biological data.

Realization of this interplay between phylogenetic history and functional/structural processes has ushered in a new era of scientific rigor in comparative biology, especially given the rapid development of explicit and testable hypotheses of phylogenetic relationships. Many advances have been made in improving evolutionary model building; "tree-thinking" is now central to all areas of systematics, ecology, and evolution (Donoghue 1989, Funk & Brooks 1990, Wanntorp *et al.* 1990, Brooks & McLennan 1991, Harvey & Pagel 1991, Martins 1996, Ackerly 1997, Weller & Sakai 1999).

Two main forms of phylogenetic reasoning are used in comparative genomics: sister-group comparisons and ancestor–descendant reconstructions (Fig. 4.1). The first approach, known as a *sister-group contrast* (also known as a phylogenetic contrast), involves the comparison of two closely related species that differ in some critical phenotype. It is much better to compare close relatives (as in Fig. 4.1A, right), than very distant relatives (as in Fig. 4.1A, left), because the background differences (i.e. biological differences not related to this particular phenotype) will be much less. Such contrasts are essentially natural experiments that can point to candidate genes likely to be causal for a particular trait. This is particularly true if one has several phylogenetically independent contrasts (i.e. pairs of species that are distantly related to other pairs) as this

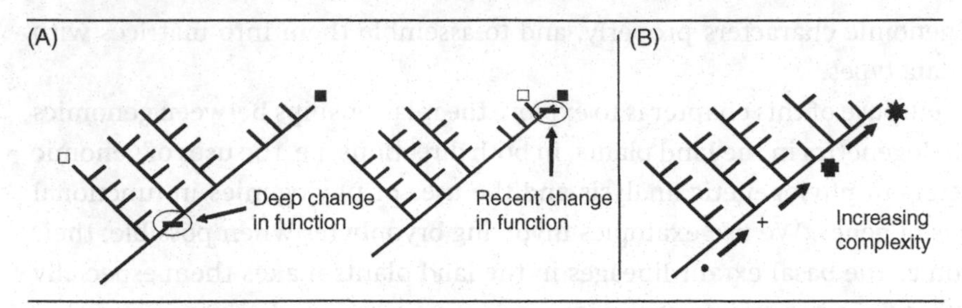

Fig. 4.1. A diagrammatic representation of two forms of phylogenetic reasoning discussed in the text. (A) Sister-group comparison. A phylogenetically distant comparison is shown on the left; there is a large background difference between these extant taxa in addition to the target functional difference, thus this is not a good natural experiment. A phylogenetically close comparison is shown on the right; we expect a low background difference except for the target functional difference, thus this a good natural experiment. (B) Ancestor–descendant comparison. Ancestral state reconstruction is used to model earlier historical stages of a process or structure. Thus one can make inferences about a complex modern process from examining simpler beginnings.

allows for an estimate of statistical significance to be developed, since each contrast is an independent replicate.

To gain a deeper understanding of the adaptive significance of a gene one can also assess its evolutionary history by using a second form of phylogenetic reasoning, *ancestor–descendant comparison* (Fig. 4.1B). In this approach, one uses modern-day species and their inferred relationships based on other data, to reconstruct ancestral states for some functional characteristics – following the algorithmic approach of Maddison (1990), and implemented in such software as MacClade (http://www.sinauer.com) and Mesquite (http://mesquiteproject.org/mesquite/mesquite.html). This allows one to infer the most likely historical sequence of events involved is assembling a modern phenotype. This can be extremely useful in dissecting a complicated endpoint into its earlier, simpler components – or conversely, in inferring how a character system has become less complex over evolutionary time, something that seems to happen often in bryophytes.

We will illustrate these approaches with two main examples of processes in which bryophytes are of special importance: desiccation-tolerance and reproduction. The bryophytes include three quite distinct lineages (which are likely not monophyletic when taken together – see below): mosses, hornworts, and liverworts. These plants, while small in stature, are very diverse and occupy most terrestrial and freshwater habitats ranging from lakes and streams to mesic forests, rain forests, arctic tundra, and desert boulders. The bryophytes have a basal phylogenetic position among the extant embryophytes, remnant lineages surviving today from the spectacular radiation of the land plants in the Devonian period, some 450 million years ago. The three main bryophyte lineages, plus a fourth extant lineage, the tracheophytes (i.e. the so-called vascular plants), make up the entirety of the monophyletic, extant, embryophytes, arguably one of the most important lineages to have arisen in Earth's history: they made possible the colonization of the land by animals, and evolved an unparalleled diversity of size, structure, chemistry, and function.

It is difficult or impossible to study many of the important physiological or genomic causes for adaptation to life on land when looking at fossils, so it is fortunate that we have the bryophyte lineages living today to study. The bryophytes are clearly a "key" to understanding how the embryophytes are related to each other and deciphering how they came to conquer the hostile land environment from their primitive home in fresh water - habitats still occupied by relatives of the land plants, the green algae (Graham 1993). Limited water availability is probably the most important environmental factor that early lands plants had to contend with. When living on at least periodically dry land, plants needed to deal with limitations inherent in their biology, due to their aquatic ancestry: the need for free water for physiological activity, and their swimming sperm.

An understanding of the complex water relationships of the tracheophytes can be gained by using ancestor-descendant comparisons as described above, by mapping traits onto the most current phylogeny of the land plants and their closest relatives in the charophycean green algae (Oliver *et al.* 2000, 2005). These authors argued that when land plants were still small and delicate, it was not possible to retain sufficient water within the plant body. These early plants had a strategy for dealing with water called *poikilohydry* (defined as the rapid equilibration of the organism's water content with its immediate external environment). They wet up quickly when free water was available in their surroundings, and dried up just as fast when it was not. Oliver *et al.* (2000, 2005) also showed that vegetative *desiccation-tolerance* (defined as the ability of cells to dry down completely to low ambient water content) was primitively present in the land plants (as seen today in nearly all bryophytes), but was then lost early in the evolution of tracheophytes. The initial evolution of vegetative desiccation-tolerance was a crucial step required for the colonization of the land, but that tolerance came at a cost: metabolic rates are lower in tolerant plants than in plants that do not maintain costly mechanisms for tolerance. Thus, a trade-off between productivity and tolerance exists. The loss of tolerance appears to have been favored as soon as an internalization of water relationships happened as the vascular plants became more complex and gained a cuticle, vascular tissue, stomata, etc. However, at least two independent evolutions (or re-evolutions) of vegetative desiccation-tolerance occurred in *Selaginella*, and a few more in the ferns. Within the angiosperms, at least eight independent cases of evolution (or re-evolution) of vegetative desiccation-tolerance occurred, often in association with spread into largely barren, very dry habitats on outcrops of rock in the tropics (Porembski & Barthlott 2000). Furthermore, a special form of irreversible desiccation-tolerance, related to reproduction, was evolved when seeds arose (Black & Pritchard 2002). The specific mechanisms were different in detail each time the general phenotype of desiccation-tolerance was re-evolved.

New phylogenomic studies suggest that most, if not all, land plants have retained the genetic potential for desiccation-tolerance, whether they can express the phenotype of desiccation-tolerance or not. Genes identified with vegetative desiccation-tolerance in mosses appear to be resident in tracheophyte species that are not vegetatively desiccation-tolerant. Oliver *et al.* (2004) produced cDNA libraries from both rehydrated and rapidly dried *Syntrichia ruralis*, and conducted extensive EST sequencing (more than 10 000 ESTs). Over 40% of the genes represented by these ESTs were classified as unknowns, but a number of the genes encode Late Embryogenesis Abundant (LEA) proteins that are normally known from drying angiosperm tissues, particularly seeds. Thus, evolutionary transitions between different levels of desiccation-tolerance may be largely controlled by changes in regulatory genes (Bartels & Salamini 2001, Bernacchia & Furini 2004).

Understanding these sorts of small-scale evolutionary changes in function may best be addressed by sister-group contrasts (as described above) between close relatives that differ in their level of desiccation-tolerance, such as within the genus *Syntrichia* where relatively close relatives to *S. ruralis* appear to be more tolerant (e.g. *S. caninervis*, a desert moss) or less tolerant (e.g. *S. norvegica*, an Arctic–Alpine moss; Oliver *et al.* 1993). Deciphering the physiological mechanisms and genes behind different desiccation-tolerant phenotypes, both vegetative and reproductive, will be an exciting endeavor in the next few years, and will clearly be aided by a comparative, phylogenetic approach (see also Chapters 5 and 7, this volume).

A similar use of comparative genomics has begun to address the other major hurdle that early land plants faced in life on land: effecting sexual reproduction. How do the free-living gametophytes bearing male and female gametangia, with free-swimming sperm, in bryophytes become the complex endosporic gametophytes encased within parent sporophytes in the more complex tracheophytes, culminating in the angiosperm flower? This is another area where ancestor-descendant comparisons can help us understand where the complicated flower comes from, with its many parts and sophisticated pollination systems moving male gametophytes close enough to the female that sperm no longer need to swim in the environment.

The Floral Genome Project (http://www.floralgenome.org/) has been working towards understanding these issues through broad phylogenetic sampling of the major lineages of angiosperms and gymnosperms. Results are promising so far within the seed plants; extensive sampling of ESTs has revealed evidence for repeated gene duplication followed by subsequent specialization (Albert *et al.* 2005). At least some members of the MADS-box gene family have been discovered in bryophytes (Henschel *et al.* 2002, Singer *et al.* 2007), indicating that there is a relationship between the simple fertilization systems of bryophytes and the much more complex ones in angiosperms. Mosses also share the KNOX gene family, which is involved in meristem patterning, with angiosperms (Champagne & Ashton 2001). Comparative genomic studies have the potential for understanding these and other morphogenetic processes (Cronk 2001); it will be exciting to extend the sampling of genes extensively through ferns, clubmosses, and bryophytes in the coming years.

4.3 The uses of comparative genomics in phylogenetic reconstruction

Some of the most intriguing questions in systematics concern the origin and relationships of diverse, ecologically dominant lineages. Such groups include metazoans, vertebrates, land plants, and flowering plants. These groups are all thought to have undergone an early rapid radiation, perhaps mediated by an ecological release or the development of a key innovation. This period of rapid diversification was followed by long periods of additional diversification, specialization, and extinction (Kenrick & Crane 1997, Qiu & Palmer 1999, Bromham 2003). The rapidity of the early diversification, compared with the length of subsequent time spent following distinct histories, makes phylogenetic reconstruction of the relative branching order of basal nodes in these groups difficult (see Chapman & Waters 2002). This pattern leaves little evidence of the order of branching due to the inferred short time periods spanned by the deepest branches, leaving few synapomorphies to define subsets of taxa. Therefore, the search has been intense for evidence to give support to phylogenies of these groups.

Several analyses spanning land plants have utilized data from different genetic markers (Hori *et al.* 1985, Mishler *et al.* 1992, 1994, Albert *et al.* 1994, Barnabas *et al.* 1995, Malek *et al.* 1996, Kallersjo *et al.* 1998, Qiu *et al.* 1998, Soltis *et al.* 1999, Duff & Nickrent 1999, Nickrent *et al.* 2000, Nishiyama *et al.* 2004). Most of these analyses show the "bryophytes" (mosses, liverworts, and hornworts; quotation marks indicate inferred paraphyly) forming a paraphyletic group at the base of the living land plants (see Goffinet 2000 for summary). However, there is disagreement about the order of branching among the major clades. Qiu *et al.* (2006) is perhaps the most inclusive phylogenetic study of this question to date; they used sequences from various chloroplast, nuclear, and mitochondrial genes and mitochondrial gene intron presence/absence characters to infer a branching order of liverworts(((mosses((hornworts(vascular plants))). The inclusion of genomic characters in the form of intron presence/absence characters is especially important, as given the apparently rapid radiation of major land plant lineages, DNA sequence data alone may not be reliable in definitively inferring phylogeny (Mishler 2000). This is particularly true of studies that use data from only one gene. In such cases, even extensive sampling cannot counteract the effects of extinction and character saturation (see discussion in Soltis *et al.* 1999). Instead, the use of genomic structural characters may contribute to minimizing the issues of extinction and character saturation in regard to phylogenetic reconstruction.

Studies using RFLPs to map the makeup of the chloroplast plastome found a large structural rearrangement that is a synapomorphy for tracheophytes (vascular plants) minus lycophytes (Raubeson & Jansen 1992). In addition, a 22 kilobase (kb) inversion in the choloroplast genome was detected in most genera of the sunflower family, Compositae, but is lacking in Barnadesioideae (Jansen & Palmer 1987), and a 50 kb inversion defines a group of Phaseoleae in the legumes, Leguminosae (Doyle 1994, 1995). Recent studies using a PCR-based strategy have shown the phylogenetic utility of chloroplast rearrangements in mosses (Sugiura *et al.* 2003, Goffinet *et al.* 2005, 2007). Likewise, structural characteristics of mitochondrial genomes such as intron presence/absence have been shown to be of phylogenetic significance (Pruchner *et al.* 2002, Knoop 2004).

Because gene rearrangements are unusual events (Raubeson & Jansen 1992) and presumably can occur anywhere in the plastome, they are viewed as being much less subject to homoplasy than sequence data (Boore *et al.* 1995, Boore & Brown 1998, Rokas & Holland 2000). Therefore, even a few potentially informative characters from the structure of the genome (herein we will call these genomic characters) can be extremely useful in reconstructing phylogenies (Dowton *et al.* 2002, Gallut & Barriel 2002). In this regard, rare genomic characters are analogous to significant morphological characters that often define major plant clades (Mishler *et al.* 1994). For example, Qiu *et al.* (1998) found that the acquisition of three mitochondrial gene introns supports liverworts as the sister group to the remainder of land plants (embryophytes). Although these represent few characters, they are quite significant given the rarity of such modifications in the organisms studied. This rarity is inferred *a priori* from the low numbers of such characters detected in the current sampling of major plant lineages, although even then homoplasy may be present (Kelch *et al.* 2004). Recent developments in genomics mean that ever-increasing amounts of DNA sequence data *and* genomic structural data will become available in the near future. Several plant nuclear genomes have been sequenced or are in the process of being sequenced. In addition, several laboratories across the U.S.A. are actively sequencing the chloroplasts and mitochondria of a large number of green plants (see table at: http://ucjeps.berkeley.edu/TreeofLife/data_table.php). Although the resultant comparative DNA sequence data no doubt will contribute much to phylogenetic reconstruction, short, deep branches could prove recalcitrant to such investigation. Thus, the use of genomic structural data may be instrumental in reconstructing such difficult phylogenies.

The land plants provide an excellent study system for this approach. There is a broad consensus, based on both morphological and molecular data, that land plants are a monophyletic group. After appearing in the Ordovician period (Gray 1993), they underwent a rapid radiation in the Silurian, with most major lineages appearing by the Early Devonian (see Kenrick & Crane 1997). Many of the groups are well defined morphologically and there is a developing consensus, based on fossil evidence and DNA sequence data, on the arrangement of many of the interior branches of the phylogeny. Most of the main clades (e.g. angiosperms, conifers, moniliforms) have a representative taxon with a complete sequence of the chloroplast genome available. The land plants as a whole are rooted somewhere among the three bryophyte lineages. Nevertheless, there is a great deal of uncertainty concerning the initial branching order at the base of the land plant phylogeny; nearly every possible branching order among the bryophyte lineages and tracheophytes has had some support (Mishler & Churchill 1984, Garbary *et al.* 1993, Garbary & Renzaglia 1998, Hedderson *et al.* 1996, Nishiyama & Kato 1999, Karol *et al.* 2001, Qiu *et al.* 2006; see an excellent summary at http://www.science.siu.edu/landplants/PhylogRelsGen.html). Therefore, land plant phylogeny, with both well-supported (e.g. vascular plants, seed plants) and ambiguous (e.g. the deepest branches in land plants) subclades, provides a suitable subject for evaluating the utility of chloroplast genomic data.

4.4 A new example of the use of characters from comparative chloroplast genomics

In order to illustrate this approach, we provide an updated analysis to Kelch *et al.* (2004). We have identified potentially informative genomic characters that may help to elucidate the branching order at the base of the land plant phylogeny. In our utilization of gene arrangement, intron, and gene presence or absence characters, we address theoretical and practical issues involved in the phylogenetic analysis of whole genome sequences.

4.4.1 *Materials and methods*

The plants examined for this study may be found in Table 4.1 and include one or more examples of charophytes, liverworts, mosses, hornworts, moniliforms, conifers, and angiosperms. Gene maps were downloaded from Genbank (Table 4.1) and manually added into a spreadsheet by using Microsoft Excel X for Mac (Microsoft Corp. 2001). We aligned the genes linearly, beginning with a section of the plastome in the large single copy region at *rpo*A that has one of the longest regions that is invariant across sampled taxa. Sequence alignment was done by hand, with particular attention to regions of putative inversions. Because inversions are uncommon across land plants, overlapping inversions are extremely rare; therefore, no special efforts were needed to minimize inversion characters. Nevertheless, large inverted sections of gene sequences were analyzed in reverse order to facilitate identification of additional gene rearrangements within the inverted region.

Visual inspection revealed that the presence, location, and order of tRNA genes were the most variable elements in the plastome. In order to prevent misalignment, tRNA genes were removed for the initial alignment and subsequently added in their appropriate positions. Characters were searched for by using basic principles of character analysis originally developed for morphological characters (Mishler & De Luna 1991, Mishler 2005). Characters comprised three types: gene rearrangement characters representing inversions of two or more genes in the plastome, gene presence/absence characters representing a gene missing from a particular position in the plastome (whether the latter missing genes are lost or transferred to other places in the genomes will be detectable only when more information is available on the genomic make-up of the studied organisms), and intron presence/absence representing the presence of a particular intron within genes in the plastome. Duplications of genes via inclusion in the IR region were included with gene rearrangement characters. Coding of inversion characters was binary and chosen to minimize the number of inversion characters. Introns were located in gene sequences and coded separately. In addition, copies of genes or pseudogenes were coded as present or absent based on synteny (their location in relation to other genes in other taxa in the analysis). In the coding of characters 0 represents absence or a putative plesiomorphy and 1 represents presence or putative apomorphy. The final comparative alignment is available as an Excel file on the Green Tree of Life website (http://ucjeps.berkeley.edu/TreeofLife).

Table 4.1 *Genbank accession numbers and sources of chloroplast gene maps for sampled taxa*

Taxon	Genbank accession no.	Reference
Charophytes		
Chara vulgaris L.	NC_008097	Turmel *et al.* 2006
Chaetosphaeridium globosum (Nordstedt) Klebahn	NC_004115	Turmel *et al.* 2002
Liverworts		
Marchantia polymorpha L.	NC_001319	Umesono *et al.* 1988
Mosses		
Physcomitrella patens (Hedwig) Bruch & W. P. Schimper	NC_005087	Sugiura *et al.* 2003
Tortula ruralis	Unpubl.	Murdock, Oliver & Mishler in prep.
Hornworts		
Anthoceros formosae Stephani	NC_004543	Kugita *et al.* 2003
Lycophytes		
Huperzia lucidula (Michx.) Trevisan	AY660566	Wolf *et al.* 2005
Isoetes flaccida Shuttlew.	Unpubl.	Karol *et al.* in prep.
Moniliphytes		
Adiantum capillis-veneris L.	NC_004766	Wolf *et al.* 2003
Angiopteris evecta	Unpubl.	Roper *et al.* 2007
Equisetum arvense L.	Unpubl.	Karol *et al.* in prep.
Psilotum nudum (L.) P.Beauv.	NC_003386	Wakasugi *et al.* unpubl.
Conifers		
Pinus koraiensis Siebold & Zucc.	NC_004677	Noh *et al.* unpubl.
Pinus thunbergiana Franco	NC_001631	Wakasugi *et al.* 1994
Angiosperms		
Acorus calamus L.	NC_007407	Goremykin *et al.* 2005
Amborella trichopoda	EMBL AJ506156	Goremykin *et al.* 2003
Arabidopsis thaliana (L.) Heynh.	NC_000932	Sato *et al.* 1999
Atropa belladonna L.	NC_004561	Schmitz-Linneweber *et al.* 2002
Epifagus virginiana L. (Bart.)	NC_001568	Wolfe *et al.* 1992
Calycanthus floridus L. var. *glaucus* (Willd.) Torrey & A. Gray (as *C. fertilis*)	NC_004993	Goremykin *et al.* 2004
Lotus japonicus (Regel) K.Larsen	NC_002694	Kato *et al.* 2000
Nicotiana tobacum L.	NC_001879	Kunnimalaiyaan & Nielsen 1997
Oenothera elata H. B.& K. ssp. *hookeri* (Torr. & A.Gray) W.Dietr. & W. L.Wagner	NC_002693	Hupfer *et al.* 2000

Table 4.1 (*cont.*)

Taxon	Genbank accession no.	Reference
Oryza sativa L.	NC_001320	Morton & Clegg 1993
Phalaenopsis aphrodite Rchb. f.	NC_007499	Chang *et al.* 2006
Spinacia oleracea L.	NC_002202	Schmitz-Linneweber *et al.* 2001
Triticum aestivum L.	NC_002762	Ogihara *et al.* 2002
Zea mays L.	NC_001666	Maier *et al.* 1995

All characters that varied among sampled taxa were included in the data matrix, including autapomorphies, as future sampling will no doubt change some autapomorphic characters into synapomorphies (for example, character 7 in ferns; see Stein *et al.* 1992). Forty-seven characters (39 potentially informative) were discovered in all (listed in Appendix 4.1), and placed in a nexus file (Table 4.2). All analyses were done with PAUP*4.0b10 (Swofford 2003). The matrix was analyzed using the branch-and-bound algorithm with the furthest addition sequence setting. The resulting trees were rooted by using the charophyte *Chara* as the outgroup taxon. A bootstrap analysis was performed using 1000 replicates of heuristic searches employing stepwise addition and TBR branch swapping. Analyses were performed (1) with all characters included and (2) with three gene copy characters excluded from analyses. The three characters excluded in some analyses comprise gene inclusion/exclusion within the inverted repeat located at the boundary of the large single copy and inverted repeat. A previous study (Kelch *et al.* 2004) indicated that these particular characters likely have been subject to homoplasy within green plants.

4.4.2 *Results*

The analysis including all characters produced 84 equally parsimonious trees (MPT; CI = 0.73, RI = 0.90). Of the 47 characters, 19 rearrangement characters, 11 of the gene presence or absence characters, and 2 of the intron characters proved parsimony-informative. Seven characters currently are autapomorphic based on current sampling and five others are synapomorphies for the two species of *Pinus* (some of these may prove to be Pinaceae or conifer synapomorphies in the future). One of these synapomorphies (Character 40: loss of *ndhJ*) represents the loss of all *ndh* genes from the plastome of *Pinus*. Although this may be the result of independent gene losses, a conservative approach was adopted here and lack of *ndh* genes was treated as a single character change.

Table 4.2 *Data matrix for 47 genomic characters (see Appendix 4.1) for 28 exemplars of land plants*

	10 20 30 40
Chara	00000100100?01000101101011000000000?0?11111101
Chaetosphaeridium	000001001000?010010011010110000000001111111010
Marchantia	000001001000?010000000010110000000001111111110
Physcomitrella	000001001000?01100000001011000000000?111111111
Tortula	000001001000?01000000001011000000000?111??????
Anthoceros	011011001000?010000000010110000001000111111111
Huperzia	000011001000?010000000010110000000001011111111
Isoetes	010011001000?0100000000101100000010001?11??????
Psilotum	1110110 1?000100000000001011000001100011111111 1
Equisetum	11000100100010000000000101100000100001111??????
Angiopteris	111011001000100000000001011000001100011111111 1
Adiantum	?11111001100100000010001011000000100011111111 1
Pinus koraiensis	????1010101010000000001011100000?1001110001111?
Pinus thunbergii	?110101010111000000000101110000 0?100111001111 11
Oenothera	1110?10000001100100000000010011011001011011111 1
Oryza	1110?1000000110000000000001111??1111010010011101
Triticum	1110?1000000110000000000001111??1111010010011101
Zea	1110?1000000110000000000001111??1111010010011101
Acorus	11101100000001000000000000111 01?1110110110111111
Phalaenopsis	11101100000001000000000000111 01?1110010110111111
Spinacia	1110110000001100000000000011001?0110010110101111
Amborella	11101100000011000000000000110011011001011 0111111
Calycanthus	11101100000011000000000000110011011001011 0111111
Arabidopsis	111011000000110000000000000100110110110110110110
Atropa	111011000000110000000000000100110110110110111111
Nicotiana	111011000000110000000000000100110110110110111111
Lotus	111011000000110000000000000001101101101101111 10
Epifagus	1110110000001100000000000010001?1110?10110111111

Notable bootstrap values included strong support for a monophyletic *Pinus* (100), monophyly of the grasses (99) and angiosperms (92), and some support for monophyly of land plants (72) and seed plants (76). Resolution and support were lacking for relationships among major lineages of vascular plants, other than the groups mentioned above (see Fig. 4.2). Five putative synapomorphies unite angiosperms; these are rearrangement and gene characters (Appendix 4.1, characters 9, 14, 30, 31, and 38). Three characters (5, 21, and 22) support the monophyly of land plants, three others (24, 37, and 41) that of seed plants, and two

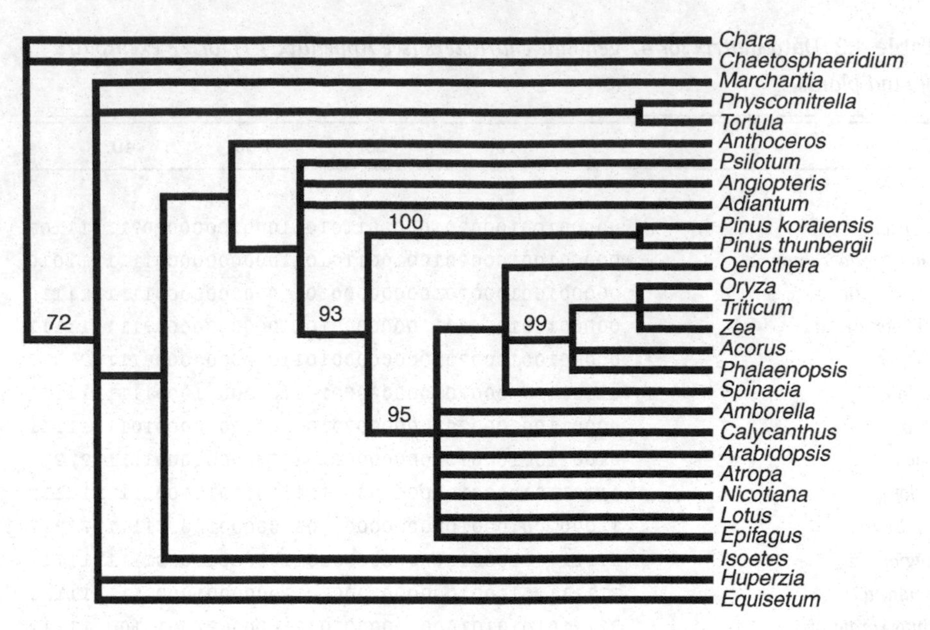

Fig. 4.2. A strict consensus tree of 84 MPT utilizing all characters. Bootstrap values above 50% appear above branches.

characters (28 and 32) support monocotyledons as monophyletic. Character 15 supports a common ancestry for euphyllophytes, 17 that of the two mosses included, and 33 the monophyly of the monilophytes. Land plants minus *Marchantia* is supported by a single intron character (46) with some homoplasy.

An analysis was carried out of the data set excluding three gene copy characters corresponding to expansion of the inverted repeat in relation to other taxa (1, 2, and 3). This resulted in six MPT. The strict consensus tree of the results of this analysis is shown in Fig. 4.3 along with the strict consensus of the 84 trees resulting from the analysis of all characters.

Discussion

The retention index for the analysis of land plant genomic data is higher than the average for published data sets (see Mishler *et al.* 1994, Sanderson & Donoghue 1989); this indicates a relatively low level of inferred homoplasy for the genomic character set. This is consistent with the expectation that genomic characters are less prone to homoplasious change than DNA sequence characters. About 11 characters (out of 47) show inferred homoplasy, higher than might be expected *a priori* for genomic characters. However, putative homoplasious characters comprise those characters involving changes in single genes. Multigene inversions are without inferred homoplasy. In some cases of inferred

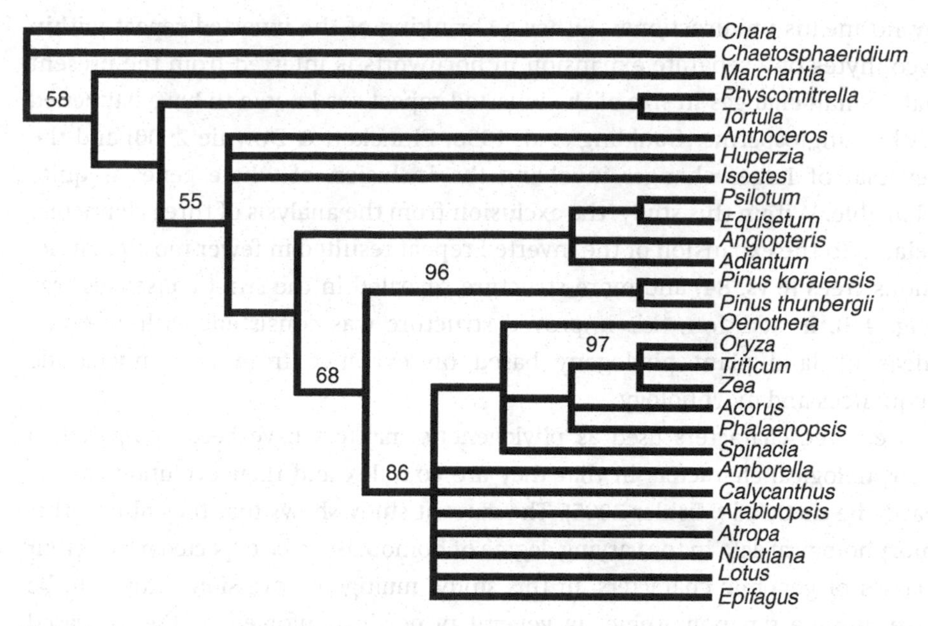

Fig. 4.3. A strict consensus tree of 6 MPT when three inverted repeat gene inclusion/exclusion characters are omitted (for explanation, see discussion in text). Bootstrap values above 50% appear above branches.

homoplasy, the loss of a particular gene in a certain location may not be true homoplasy, in that they represent independent events that result in loss of the gene or its movement to different parts of the genome. Most clades with strong bootstrap support originate within groups that have long been recognized based on morphology. These include exemplars from angiosperms, grasses, monocots, and pines. In our data set, the clade "land plants minus *Marchantia*" is supported by one synapomorphy, an intron loss (46, intron 2 missing from *ycf*3). This supports a topology in agreement with the other studies identifying liverworts as the earliest branch in land plants: studies based on morphology (Mishler & Churchill 1984, 1985) and sequence data from the chloroplast gene *rbc*L (Lewis *et al.* 1997). In addition, multiple intron presence/absence characters in the mitochondrial genome, as well as many gene sequences from chloroplasts, nuclei, and mitochondria support *Marchantia* as sister group to the rest of the land plants (Qiu *et al.* 1998, 2006).

The sister group relationship of the hornwort exemplar, *Anthoceros*, to the vascular plants, supported in a previous study by two genomic characters (Kelch *et al.* 2004), is equivocal in this study. The expansion of sampling in the free-sporing vascular plants in this study reveals that the expansion of the inverted repeat by incremental inclusion of genes from the large single copy region was by no means unidirectional. Either a shrinking of the inverted repeat within lycophytes or a separate expansion in hornworts is inferred from the present data. Small changes in size of the inverted repeat are known to have happened within angiosperms (Goulding *et al.* 1996, Plunckett & Downie 2000) and the reversal of larger changes involving the inclusion of whole genes is quite plausible. Within this study, the exclusion from the analysis of three characters related to the expansion of the inverted repeat resulted in fewer most parsimonious trees (6 vs. 84) and more structure revealed in the strict consensus tree (Fig. 4.3). In addition, this improved structure was consistent with accepted ideas of land plant phylogeny based on evidence from both nucleotide sequences and morphology.

Genomic characters used as phylogenetic markers have been compared to morphological characters in that they are complex and their evolution cannot easily be modeled (Mishler 2005). The current study shows that they also mirror morphological data in the varying degree of homoplasy to be expected by different classes of genomic characters. In this study, multigene inversions proved to be rare, unique synapomorphies (a general principle reinforced by the increased sampling in this study in regard to Kelch *et al.* 2004). Other characters, particularly those associated with the copying of genes via inclusion in the inverted repeat, show evidence of homoplasy. The presence of homoplasy in gene inclusion characters at the boundary of the IR and LSC region has been further supported by the larger sampling in this study in regard to Kelch *et al.* (2004). In particular, *Huperzia* has been shown to be more similar in IR gene inclusion to mosses than to other vascular plants included in this study. Given the comparative rarity of these genomic characters, even a small number of homoplasious characters might have significant effects on the topology of the trees resulting from phylogenetic analysis. Nevertheless, even characters subject to homoplasy can reveal phylogenetic structure within subsets of the included taxa.

4.5 Summary

The relationships of the three major lineages of bryophytes and the tracheophytes have been controversial. It is a very difficult phylogenetic problem; whatever periods of shared history there are among these lineages, they are relatively short branches a long time before the present. We need to be cautious about over-reliance on any one kind of data, including DNA sequences and genomic structural data. Clearly an extensive analysis of all appropriate nuclear and organellar DNA sequence data, plus morphology and genomic structural data, will be needed before we can confidently resolve relationships in this important region of the Tree of Life.

As we clarify the phylogeny at this deep level, the ability to use it for comparative genomics will be enhanced. In addition to all the chloroplast and mitochondrial genomes that are being sequenced, it will be essential to have some key nuclear genomes completely sequenced, ideally at least one from each of the three bryophyte lineages. Fortunately, the first of these is just completed, for the moss *Physcomitrella patens*. Its complete sequence is available on the DOE Joint Genome Institute Genome Browser at http://genome.jgi-psf.org/Physcomitrella. This species is becoming widely recognized as an experimental organism of choice not only for basic molecular, cytological, and developmental questions in plant biology, but also as a key link in understanding plant evolutionary questions, especially those related to genome evolution. It is well placed phylogenetically to provide important comparisons with the flowering plants; in terms of evolutionary distance, *Physcomitrella* is to the flowering plants what the *Drosophila* is to humans! The liverwort *Marchantia polymorpha* was proposed to the Joint Genome Institute as the next bryophyte to be completely sequenced; an announcement has recently been made that this proposal has been accepted (http://www.jgi.doe.gov/News/news_6_8_07.html). When complete, it will add a deeper anchor point for comparative genomics in land plants (as would having a genome available for hornworts). The bryophyte genomes will greatly inform bioinformatic comparisons and functional genomics in plants, just as the mouse, *Fugu*, *Drosophila*, and *Caenorhabditis* genomes have informed animal biology. As more nuclear genomes are completed, we will eventually reach the point where we can look for genomic structural characters as described above for chloroplasts. Large regions of nuclear genomes appear to be conserved in gene order and arrangement (called *synteny*). This is apparent when comparing the human genome with the mouse genome, or the maize genome with the rice genome, where a high degree of synteny is present. With better sampling, unusual and complex rearrangements should provide an exciting source of new phylogenetic characters for use in resolving deep branching events. However, the more complex nature of the nuclear genome means that automated algorithms for detecting the minimal number of rearrangements to change from one genome region to another (for example, GRAPPA, http://www.cs.unm.edu/~moret/GRAPPA/) will need to be used.

Researchers currently talk about "whole-genome phylogenetic analyses" when what they mean is an alignment of nucleotides for all the genes in a genome. A "whole-genome phylogenetic analysis" actually should look at all the genomic structural information available, in addition to nucleotide variation. These advances in truly genomic-level analyses will lead to a new era in plant phylogenetics.

Acknowledgments

The work presented here was supported in part by NSF grants DEB-9712347 (Deep Gene Research Coordination Network), DEB-0228729 (the Green Tree of Life AToL grant), and EF-0331494 (CIPRES) to BDM. We thank Bernard Goffinet and an anonymous reviewer for helpful suggestions on the manuscript.

References

Ackerly, D. (1997). Plant life histories: A meeting of phylogeny and ecology. *Trends in Ecology and Evolution*, **12**, 7–9.

Albert, V. A., Backlund, A., Bremer, K. *et al.* (1994). Functional constraints and *rbc*L evidence for land plant phylogeny. *Annals of the Missouri Botanical Garden*, **81**, 534–67.

Albert, V. A., Soltis, D. E., Carlson, J. E. *et al.* (2005). Floral gene resources from basal angiosperms for comparative genomics research. *BMC Plant Biology*, **5**, 5.

Baltimore, D. (2001). Our genome unveiled. *Nature*, **409**, 814–16.

Barnabas, S., Krishnan, S. & Barnabas, J. (1995). The branching pattern of major groups of land plants inferred from parsimony analysis of ribosomal RNA sequences. *Journal of Biosciences*, **20**, 259–72.

Bartels, D. & Salamini, F. (2001). Desiccation tolerance in the resurrection plant *Craterostigma plantagineum*: a contribution to the study of drought tolerance at the molecular level. *Plant Physiology*, **127**, 1346–53.

Bernacchia G. & Furini, A. (2004). Biochemical and molecular responses to water stress in resurrection plants. *Physiologia Plantarum*, **121**, 175–81.

Black, M. & Pritchard, H. (eds.). (2002). *Desiccation and Survival in Plants: Drying Without Dying*. Wallingford: CAB International.

Boore, J. L. & Brown, W. M. (1998). Big trees from little genomes: mitochondrial gene order as a phylogenetic tool. *Current Opinion in Genetics and Development*, **8**, 668–74.

Boore, J. L., Collins, T. M., Stanton, D., Daehler, L. L. & Brown, W. M. (1995). Deducing the pattern of arthropod phylogeny from mitochondrial DNA rearrangements. *Nature*, **376**, 163–5.

Bromham, L. (2003). What can DNA tell us about the Cambrian explosion? *Integrative and Comparative Biology*, **43**, 148–56.

Brooks, D. R. & McLennan, D. (1991). *Phylogeny, Ecology, and Behavior*. Chicago, IL: University of Chicago Press.

Champagne, C. E. M. & Ashton, N. W. (2001). Ancestry of KNOX genes revealed by bryophyte (*Physcomitrella patens*) homologs. *New Phytologist*, **150**, 23–36.

Chang, C. C., Lin, H. C., Lin, I. P. *et al.* (2006). The chloroplast genome of *Phalaenopsis aphrodite* (Orchidaceae): comparative analysis of evolutionary rate with that of grasses and its phylogenetic implications. *Molecular Biology and Evolution*, **23**, 279–91.

Chapman, R. L. & Waters, D. A. (2002). Green algae and land plants – an answer at last? *Journal of Phycology*, **38**, 237–40.

Cronk, Q. C. (2001). Plant evolution and development in a post-genomic context. *Nature Reviews Genetics*, **2**, 607–19.

Donoghue, M. J. (1989). Phylogenies and the analysis of evolutionary sequences, with examples from seed plants. *Evolution*, **43**, 1137–56.

Dowton, M., Castro, L. R. & Austin, A. D. (2002). Mitochondrial gene rearrangements as phylogenetic characters in the invertebrates: the examination of genome 'morphology'. *Invertebrate Systematics*, **16**, 345–56.

Doyle, J. J. (1994). Phylogeny of the legume family: an approach to understanding the origins of nodulation. *Annual Review of Ecology and Systematics*, **25**, 325–49.

Doyle, J. J. (1995). DNA data sets and legume phylogeny: a progress report. In *Advances in Legume Systematics*, part 7, ed. M. D. Crisp & J. J. Doyle, pp. 11–30. Kew: Royal Botanic Gardens.

Duff, R. J. & Nickrent, D. L. (1999). Phylogenetic relationships of land plants using mitochondrial small-subunit rDNA sequences. *American Journal of Botany*, **86**, 372–86.

Eisen, J. A. (1998). Phylogenomics: improving functional predictions for uncharacterized genes by evolutionary analysis. *Genome Research*, **8**, 163–7.

Funk, V. A. & Brooks, D. R. (1990). *Phylogenetic Systematics as the Basis of Comparative Biology*. Washington, D. C.: Smithsonian Institution Press.

Gallut, C. & Barriel, V. (2002). Cladistic coding of genomic maps. *Cladistics*, **18**, 526–36.

Garbary, D. J., Renzaglia, K. S. & Duckett, J. G. (1993). The phylogeny of land plants: a cladistic analysis based on male gametogenesis. *Plant Systematics and Evolution*, **188**, 237–69.

Garbary, D. J. & Renzaglia, K. S. (1998). Bryophyte phylogeny and evolution of land plants: evidence from development and ultrastructure. In *Bryology for the Twenty-First Century*, ed. J. W. Bates, N. W. Ashton & J. G. Duckett, pp. 45–63. Leeds: Maney and the British Bryological Society.

Goffinet, B. (2000). Origin and phylogenetic relationships of bryophytes. In *Bryophyte Biology*, ed. A. J. Shaw & B. Goffinet, pp. 124–49. Cambridge: Cambridge University Press.

Goffinet, B., Wickett, N. J., Werner, O. *et al.* (2007). Distribution and phylogenetic significance of a 71 kb inversion in the chloroplast genome of the Funariidae (Bryophyta). *Annals of Botany*, **99**, 747–53.

Goffinet, B., Wickett, N. J., Shaw, A. J. & Cox, C. J. (2005). Phylogenetic significance of the rpoA loss in the chloroplast genome of mosses. *Taxon*, **54**, 353–60.

Goremykin, V., Hirsch-Ernst, K. I. S., Wölfl, S. & Hellwig, F. H. (2003). Analysis of the *Amborella trichopoda* chloroplast genome sequence suggests that *Amborella* is not a basal angiosperm. *Molecular Biology and Evolution*, **20**, 1499–505.

Goremykin, V., Hirsch-Ernst, K. I. S., Wölfl, S. & Hellwig, F. H. (2004). The chloroplast genome of the "basal" angiosperm *Calycanthus fertilis* – structural and phylogenetic analyses. *Plant Systematics and Evolution*, **242**, 119–35.

Goremykin, V. V., Holland, B., Hirsch-Ernst, K. I. & Hellwig, F. H. (2005). Analysis of *Acorus calamus* chloroplast genome and its phylogenetic implications. *Molecular Biology and Evolution*, **22**, 1813–22.

Goulding, S. E., Olmstead, R. G., Morden, C. W. & Wolfe, K. H. (1996). Ebb and flow of the chloroplast inverted repeat. *Molecular and General Genetics*, **252**, 195–206.

Graham, L.-E. (1993). *The Origin of Land Plants*. New York: Wiley.

Gray, J. (1993). Major Paleozoic land plant evolutionary bio-events. *Palaeogeography, Palaeoclimatology, Palaeoecology*, **104**, 153–69.

Harvey, P. H. & Pagel, M. D. (1991). *The Comparative Method in Evolutionary Biology*. Oxford: Oxford University Press.

Hedderson, T. A., Chapman, R. L. & Rootes, W. L. (1996). Phylogenetics of bryophytes inferred from nuclear-encoded rRNA gene sequences. *Plant Systematics and Evolution*, **200**, 213–24.

Henschel, K., Kofuji, R., Hasebe, M. *et al.* (2002). Two ancient classes of MIKC-type MADS-box genes are present in the moss *Physcomitrella patens*. *Molecular Biology and Evolution*, **19**, 801–14.

Hori, H., Lim, B. L. & Osawa, S. (1985). Evolution of green plants as deduced from 5S ribosomal RNA sequences. *Proceedings of the National Academy of Sciences, U.S.A.*, **82**, 820–3.

Hupfer, H., Swiate, M., Hornung, S. *et al.* (2000). Complete nucleotide sequence of the *Oenothera elata* plastid chromosome, representing plastome I of the five distinguishable *Euoenothera* plastomes. *Molecular Genetics and Genomics*, **263**, 581–5.

Jansen, R. K. & Palmer, J. D. (1987). A chloroplast DNA inversion marks an ancient split in the sunflower family Asteraceae. *Proceedings of the National Academy of Sciences, U.S.A.*, **84**, 5818–22.

Kallersjo, M., Farris, J. S., Chase, M. W. *et al.* (1998). Simultaneous parsimony jackknife analysis of 2538 *rbc*L DNA sequences reveals support for major clades of green plants, land plants, seed plants and flowering plants. *Plant Systematics and Evolution*, **213**, 259–87.

Karol, K. G., McCourt, R. M., Cimino, M. T. & Delwiche, C. F. (2001). The closest living relatives of land plants. *Science*, **294**, 2351–3.

Kelch, D. G., Driskell, A. & Mishler B. D. (2004). Inferring phylogeny using genomic characters: a case study using land plant plastomes. *Monographs in Systematic Botany from the Missouri Botanical Garden*, **68**, 3–12.

Kenrick, P. & Crane, P. R. (1997). The origin and early evolution of plants on land. *Nature*, **389**, 33–9.

Knoop, V. (2004). The mitochondrial DNA of land plants: peculiarities in phylogenetic perspective. *Current Genetics*, **46**, 123–39.

Kugita, M., Kaneko, A., Yamamoto, Y. *et al.* (2003). The complete nucleotide sequence of the hornwort (*Anthoceros formosae*) chloroplast genome: insight into the earliest land plants. *Nucleic Acids Research*, **31**, 716–21.

Kunnimalaiyaan, M. & Nielsen, B. L. (1997). Fine mapping of replication origins (ori A and ori B) in *Nicotiana tabacum* chloroplast DNA. *Nucleic Acids Research*, **25**, 3681–6.

Lewis, L. A., Mishler, B. D. & Vilgalys, R. (1997). Phylogenetic relationships of the liverworts (Hepaticae), a basal embryophyte lineage, inferred from nucleotide sequence data of the chloroplast gene *rbc*L. *Molecular Phylogenetics and Evolution*, **7**, 377–93.

Maddison, W. P. (1990). A method for testing the correlated evolution of two binary characters: are gains or losses concentrated on certain branches of a phylogenetic tree? *Evolution*, **44**, 539–57.

Martins, E. P. (1996). *Phylogenies and the Comparative Method in Animal Behavior*. Oxford: Oxford University Press.

Maier, R. M., Neckermann, K., Igloi, G. L. & Kossel, H. (1995). Complete sequence of the maize chloroplast genome: gene content, hotspots of divergence and fine tuning of genetic information by transcript editing. *Journal of Molecular Biology*, **251**, 614–28.

Malek, O., Laettig, K., Hiesel, R., Brennicke, A. & Knoop, V. (1996). RNA editing in bryophytes and a molecular phylogeny of land plants. *EMBO (European Molecular Biology Organization) Journal*, **15**, 1403–11.

Mishler, B. D. (2000). Deep phylogenetic relationships among "plants" and their implications for classification. *Taxon*, **49**, 661–83.

Mishler, B. D. (2005). The logic of the data matrix in phylogenetic analysis. In *Parsimony, Phylogeny, and Genomics*, ed. V. A. Albert, pp. 57–70. Oxford: Oxford University Press.

Mishler, B. D. & Churchill, S. P. (1984). A cladistic approach to the phylogeny of the "bryophytes". *Brittonia*, **36**, 406–24.

Mishler, B. D. & Churchill, S. P. (1985). Transition to a land flora: phylogenetic relationships of the green algae and bryophytes. *Cladistics*, **1**, 305–28.

Mishler, B. D. & De Luna, E. (1991). The use of ontogenetic data in phylogenetic analyses of mosses. *Advances in Bryology*, **4**, 121–67.

Mishler, B. D., Thrall, P. H., Hopple, J. S. Jr., De Luna, E. & Vilgalys, R. (1992). A molecular approach to the phylogeny of bryophytes: cladistic analysis of chloroplast-encoded 16S and 23S ribosomal RNA genes. *Bryologist*, **95**, 172–80.

Mishler, B. D., Lewis, L. A., Buchheim, M. S. *et al.* (1994). Phylogenetic relationships of the "green algae" and "bryophytes." *Annals of the Missouri Botanical Garden*, **81**, 451–83.

Morton, B. R. & Clegg, M. T. (1993). A chloroplast DNA mutational hotspot and gene conversion in a noncoding region near *rbc*L in the grass family (Poaceae). *Current Genetics*, **24**, 357–65.

Nickrent, D. L., Parkinson, C. L., Palmer, J. D. & Duff, R. J. (2000). Multigene phylogeny of land plants with special reference to bryophytes and the earliest land plants. *Molecular Biology and Evolution*, **17**, 1885–95.

Nishiyama, T. & Kato, M. (1999). Molecular phylogenetic analysis among bryophytes and tracheophytes based on combined data of plastid coded genes and the 18S rRNA gene. *Molecular Biology and Evolution*, **16**, 1027–36.

Nishiyama, T., Wolf, P. G., Kugita, M. *et al.* (2004). Chloroplast phylogeny indicates that bryophytes are monophyletic. *Molecular Biology and Evolution*, **21**, 1813–19.

Ogihara, Y., Isono, K., Kojima, T. *et al.* (2002). Structural features of a wheat plastome as revealed by complete sequencing of chloroplast DNA. *Molecular Genetics and Genomics*, **266**, 740–6.

Oliver, M. J., Mishler, B. D. & Quisenberry, J. E. (1993). Comparative measures of desiccation-tolerance in the *Tortula ruralis* complex. I. Variation in damage control and repair. *American Journal of Botany*, **80**, 127–36.

Oliver, M. J., Tuba, Z. & Mishler, B. D. (2000). The evolution of desiccation tolerance in land plants. *Plant Ecology*, **151**, 85–100.

Oliver, M. J., Velten, J. & Mishler, B. D. (2005). Desiccation tolerance in bryophytes: a reflection of the primitive strategy for plant survival in dehydrating habitats? *Integrative and Comparative Biology*, **45**, 788–99.

Oliver, M. J., Dowd, S. E., Zaragoza, J., Mauget, S. A. & Payton, P. R. (2004). The rehydration transcriptome of the desiccation-tolerant bryophyte *Tortula ruralis*: transcript classification and analysis. *BMC Genomics*, **5** (89), 1–19.

Plunkett, G. M. & Downie, S. R. (2000). Expansion and contraction of the chloroplast inverted repeat in Apiaceae subfamily Apioideae. *Systematic Botany*, **25**, 648–67.

Porembski, S. & Barthlott, W. (2000). Granitic and gneissic outcrops (inselbergs) as center of diversity for desiccation-tolerant vascular plants. *Plant Ecology*, **151**, 19–28.

Pruchner, D., Beckert, S., Muhle, H. & Knoop, V. (2002). Divergent intron conservation in the mitochondrial nad2 gene: signatures for the three bryophyte classes (mosses, liverworts, and hornworts) and the lycophytes. *Journal of Molecular Evolution*, **55**, 265–71.

Qiu, Y. L. & Palmer, J. D. (1999). Phylogeny of early land plants: insights from genes and genomes. *Trends in Plant Science*, **4**, 26–30.

Qiu, Y.-L., Cho, Y., Cox, J. C. & Palmer, J. D. (1998). The gain of three mitochondrial introns identifies liverworts as the earliest land plants. *Nature*, **394**, 671–4.

Qiu, Y.-L., Li, L., Wang, B. *et al.* (2006). The deepest divergences in land plants inferred from phylogenomic evidence. *Proceedings of the National Academy of Sciences, U.S.A.*, **103**, 15511–16.

Raubeson, L. A. & Jansen, R. K. (1992). Chloroplast DNA evidence on the ancient evolutionary split in vascular land plants. *Science*, **255**, 1697–9.

Rokas, A. & Holland, P. W. H. (2000). Rare genomic changes as a tool for phylogenetics. *Trends in Ecology and Evolution*, **15**, 454–9.

Roper, J. M., Hansen, S. K., Wolf, P. G. *et al.* (2007). The complete plastid genome sequence of *Angiopteris evecta* (G. Forst.) Hoffm. (Marattiaceae). *American Fern Journal*, **97**, 95–106.

Sanderson, M. J. & Donoghue, M. J. (1989). Patterns of variation in levels of homoplasy. *Evolution*, **43**, 1781–95.

Sato, S., Nakamura, Y., Kaneko, T., Asamizu, E. & Tabata, S. (1999). Complete structure of the chloroplast genome of *Arabidopsis thaliana*. *DNA Research*, **6**, 283–90.

Schmitz-Linneweber, C., Maier, R. M., Alcaraz, J. P. *et al.* (2001). The plastid chromosome of spinach (*Spinacia oleracea*): complete nucleotide sequence and gene organization. *Plant Molecular Biology*, **45**, 307–15.

Schmitz-Linneweber, C., Regel, R., Du, T. G. *et al.* (2002). The plastid chromosome of *Atropa belladonna* and its comparison with that of *Nicotiana tabacum*: the role of RNA editing in generating divergence in the process of plant speciation. *Molecular Biology and Evolution*, **19**, 1602–12.

Singer, S. D., Krogan, N. T. & Ashton, N. W. (2007). Clues about the ancestral roles of plant MADS-box genes from a functional analysis of moss homologues. *Plant Cell Reports*, **26**, 1155–69.

Soltis, P. S., Soltis, D. E., Wolf, P. G. *et al.* (1999). The phylogeny of land plants inferred from 18S rDNA sequences: pushing the limits of rDNA signal? *Molecular Biology and Evolution*, **16**, 1774–4.

Stein, D. B., Conant, D. S., Ahearn, M. E. *et al.* (1992). Structural rearrangements of the chloroplast genome provide an important phylogenetic link in ferns. *Proceedings of the National Academy of Sciences, U.S.A.*, **89**, 1856–60.

Stuart, J. M., Segal, E., Koller, D. & Kim, S. K. (2003). A gene-coexpression network for global discovery of conserved genetic modules. *Science*, **302**, 249.

Sugiura, C., Kobayashi, Y., Aoki, S., Sugita, C. & Sugita, M. (2003). Complete chloroplast DNA sequence of the moss *Physcomitrella patens*: evidence for the loss and relocation of *rpo*A from the chloroplast to the nucleus. *Nucleic Acids Research*, **31**, 5324–31.

Swofford, D. L. (2003). *PAUP*: Phylogenetic Analysis Using Parsimony (*and Other Methods).* Sunderland, MA: Sinauer Associates.

Turmel, M., Otis, C. & Lemieux, C. (2002). The chloroplast and mitochondrial genome sequences of the charophyte *Chaetosphaeridium globosum*: insights into the timing of the events that restructured organelle DNAs within the green algal lineage that led to land plants. *Proceedings of the National Academy of Sciences, U.S.A.*, **99**, 11275–80.

Turmel, M., Otis, C. & Lemieux, C. (2006). The chloroplast genome sequence of *Chara vulgaris* sheds new light into the closest green algal relatives of land plants. *Molecular Biology and Evolution*, **23**, 1324–38.

Umesono, K., Inokuchi, H., Shiki, Y. *et al.* (1988). Structure and organization of *Marchantia polymorpha* chloroplast genome. II. Gene organization of the large single copy region from rps'12 to atpB. *Journal of Molecular Biology*, **203**, 299–331.

Wakasugi, T., Tsudzuki, J., Ito, S. *et al.* (1994). Loss of all ndh genes as determined by sequencing the entire genome of the black pine *Pinus thunbergii*. *Proceedings of the National Academy of Sciences, U.S.A.*, **91**, 9794–8.

Wanntorp, H.-E., Brooks, D. R., Nilsson, T. *et al.* (1990). Phylogenetic approaches in ecology. *Oikos*, **57**, 119–32.

Weller, S. G. & Sakai, A. K. (1999). Using phylogenetic approaches for the analysis of plant breeding system evolution. *Annual Review of Ecology and Systematics*, **30**, 167–99.

Wolf, P. G., Rowe, C. A., Sinclair, R. B. & Hasebe, M. (2003). Complete nucleotide sequence of the chloroplast genome from a leptosporangiate fern, *Adiantum capillus-veneris* L. *DNA Research*, **10**, 59–65.

Wolf, P. G., Karol, K. G., Mandoli, D. F. *et al.* (2005). The first complete chloroplast genome sequence of a lycophyte, *Huperzia lucidula* (Lycopodiaceae). *Gene*, **350**, 117–28.

Wolfe, K. H., Morden, C. W. & Palmer, J. D. (1992). Function and evolution of a minimal plastid genome from a nonphotosynthetic parasitic plant. *Proceedings of the National Academy of Sciences, U.S.A.*, **89**, 10648–52.

Appendix 4.1 Description of genomic characters

1: Inclusion of trnL from the large single copy (LSC) edge of IRa into the IR region.
2: Inclusion of *rps*7 from the LSC edge of IRa into the IR region.
3: Inclusion of *ndh*B from the LSC edge of IRa into the IR region.
4: Inversion of the gene order within the IRs.
5: Inclusion of *rps*12 from the LSC edge of IRa into the IR region.
6: Loss of IRb.
7: Inferred loss of six genes: *ndh*D, *ndh*E, *ndh*G, *ndh*I, *ndh*A, and *ndh*H.
8: Inclusion of *rpl*21, *rpl*32, trnP, and trnL.
9: Inferred loss of *chl*L and *chl*N.
10: Inversion of most genes in IR region.
11: Inferred loss of trnV, *rps*12, and *ndh*B genes in IR region.
12. Inferred loss of *rps*7 in IR region.
13: Inferred loss of *ycf*15.
14: Inclusion of *rpl*23 and *rpl*2 from IRb end of LSC into the IR region.
15: Multigene (*c.* 27 gene) inversion between trnL and *atp*F.
16: Inversion of *c.* 30 genes between trnC and *rps*11.
17: Inversion of *c.* 20 genes between *rps*16 and *pet*N.
18: Inversion of *c.* 14 genes between trnG and *ycf*3.
19: Presence/absence of *pet*A.
20: Insertion/deletion of trnD, trnY, and trnE.
21: Presence/absence of *odp*B.
22: Insertion/deletion of *c.* 18 genes from *mat*K to trnfM.
23: Insertion/deletion of 5 genes; trnS, *psb*C, *psb*D, trnT, and trnfM.
24: Presence/absence of trnS.
25: Inversion of trnfM, *rps*14, *psa*B, and *psa*A.
26: Inversion of 32 gene section from trnG to *rpo*A.
27: Presence/absence of *inf*A between *rpl*36 and *rps*8.
28: Presence/absence of *rpl*22 between *rps*3 and *rps*19.
29: Presence/absence of trnH between *rps*19 and *rpl*12.
30: Presence/absence of *ycf*2 between trnL and *ycf*15 or trnL.
31: Presence/absence of *ycf*2 in inverted repeat.
32: Presence/absence of *ycf*15 in inverted repeat.
33: Presence/absence of *ycf*15 between *ycf*2 and trnL.
34: Presence/absence of trnL between trnI or *ycf*2 and *ndh*B.
35: Presence/absence of *rps*7 between *ndh*B and *rps*12.

36: Presence/absence of *rps*15 between *ycf*1 and *ndh*H.
37: Presence/absence of *ycf*1 between trnN and *ndh*F.
38: Presence/absence of *rpl*21 between *ndh*F and *rpl*32.
39: Presence/absence of trnP between *rpl*32 and trnL.
40: Presence/absence of *ycf*1 adjacent to *rps*15.
41: Presence/absence of *ndh*J between trnF and *ndh*K.
42: Intron missing from gene (pseudogene) of *rpl*2.
43: Intron missing from gene *rps*12.
44: Intron missing from gene *atb*F.
45: Intron missing from gene *rpo*C1.
46: Second intron missing from gene *ycf*3.
47: Second intron missing from gene *clp*P2.

5

Mosses as model organisms for developmental, cellular, and molecular biology

ANDREW C. CUMING

5.1 Introduction

There is a popular genre of politically incorrect jokes on the theme of "The World's Shortest Books" (of which the least offensive example is the title "Different Ways to Spell Bob"). Until recently, it would have been fair to surmise that the title of this chapter might have qualified with ease. Certainly, that would have been the view of many *soi-disant* "mainstream" plant developmental biologists, whose Arabidocentric view of the plant kingdom had tended to ignore any organism outside the angiosperms (and most within). Thankfully, this is no longer the case. It is now appreciated that an understanding of the evolution of gene function and of the roles of genes in the programming of developmental transitions (generically known as "Evo-Devo") requires a comparative analysis of species representative of a wide range of diverse taxa. This has coincided with an explosion of molecular knowledge of at least one species of moss, *Physcomitrella patens*, the study of which is being facilitated by the complete sequencing of its genome. Consequently, we can expect to see a much greater interest in this species, and in mosses as a group of plants with their own unique features and fascination, developing within the wider plant science community. In this chapter I shall therefore concentrate on the recent discoveries made in *Physcomitrella*, and – more importantly – attempt to sketch out some of the challenges that lie ahead for researchers intending to make use of the burgeoning *Physcomitrella* resources.

The peripheralization of interest in the mosses is a comparatively recent phenomenon, for this group has long been a source of interest for botanical scholars. As is made abundantly clear in other chapters in this book, the mosses

Bryophyte Biology: Second Edition, ed. B. Goffinet & A. J. Shaw. Published by Cambridge University Press.

represent a highly species-rich group within the bryophytes – the extant representatives of the earliest group of land plants – and consequently have much to teach us about the adaptations that were necessary for the conquest of the land. Anyone who has observed mosses colonizing apparently inhospitable habitats (bare rocks, walls, roofs) will be aware of their resilience to environmental stresses, a subject discussed in Chapters 6 and 7, this volume. Through their colonization of apparently featureless substrates, and their subsequent posthumous decomposition, mosses can be fairly considered to be habitat-forming organisms that enabled the subsequent evolution of more complex land plants.

The history of research on mosses has been one of "boom and bust". Indeed, during the early years of the twentieth century, mosses represented a fertile field of discovery in genetics, cytogenetics, and developmental biology. Much of this history has been forgotten, or has been neglected owing to the relative difficulty for today's predominantly anglophone scientific community in reading and appreciating the pioneering studies, published in Latin or highly formal and archaic German, by Hedwig, Staehelin, Hofmeister, von Wettstein and others. Since the author (somewhat embarrassingly) has to number himself among the linguistically challenged majority, it is therefore a great relief to be able to recommend the illuminating account by Reski (1998) of the early history of moss research, detailing the contributions made by these pioneers.

5.2 *Physcomitrella patens*: a twenty-first century model

Early research on mosses investigated a number of species. More recently a greater focus has been concentrated on *Funaria hygrometrica*, *Physcomitrium pyriforme*, *Ceratodon purpureus*, *Physcomitrella patens*, *Tortula ruralis*, and *Sphagnum* spp. (Wood *et al.* 2004). However, it is likely that, for the foreseeable future, *Physcomitrella patens* will be at the center of attention for studies that seek to achieve a synthesis of cellular, biochemical, and molecular genetic approaches. If *Arabidopsis* can claim to be the "*Drosophila* of plant biology", then *Physcomitrella* can make a claim to be the counterpart of the nematode *Caenorhabditis*.

Physcomitrella is principally studied because of its suitability for genetic analysis. As in all mosses, the dominant phase of the life cycle – the gametophyte – is haploid. Thus mutagenesis results in the immediate revelation of mutant phenotypes. The first mutagenic studies of this species identified a number of auxotrophic and developmental mutants (Engel 1968), and subsequently Cove and his colleagues developed the use of *Physcomitrella* for the genetic analysis of such mutants, with an increasing focus on the genetic control of cell shape, morphogenesis, and polar cell growth (Ashton & Cove 1977, Ashton *et al.* 1979a,b, Grimsley *et al.* 1977, Courtice & Cove 1983, Knight *et al.* 1991, Jenkins *et al.* 1986).

These latter studies coincided with the wider development of molecular tools for the study of gene regulation, in particular the ability to undertake genetic transformation of plant cells. Schaefer *et al.* (1991) achieved the first stable transformation of *Physcomitrella*. Subsequent studies revealed that if the transforming DNA contained a sequence homologous with sequences resident within the moss genome, then the transforming DNA was preferentially integrated into the genome at the homologous site (Kammerer & Cove 1996, Schaefer & Zrÿd 1997).

The ability to undertake "gene targeting" by homologous recombination between transforming DNA and a specific locus in the host genome provides a powerful and sophisticated tool for genetic manipulation. It occurs with high frequency in bacteria and in simple eukaryotes (in yeast, gene targeting is a routine procedure for genetic analysis (Orr-Weaver *et al.* 1981)) and is used to make specific genetic alterations in a small number of vertebrate experimental systems (chicken DT40 cell lines undertake gene targeting by homologous recombination with high frequency (Sonoda *et al.* 2001), whilst mouse embryonic stem cells can not only be transformed by gene targeting, they can also be regenerated to develop into transgenic mice containing the specifically altered gene (Soriano 1995)). Gene targeting does not normally occur in flowering plants, following genetic transformation. Although transgenic plants in which gene targeting events have occurred have been isolated, the frequency with which these events occur is low (Kempin *et al.* 1997, Terada *et al.* 2002, Hanin & Paszkowski 2003). By contrast, the efficiency with which gene targeting occurs in *Physcomitrella* is high – up to 100% of transformants may exhibit gene targeting – a rate equivalent to that observed in yeast (Schaefer & Zrÿd 1997, Schaefer 2002, Kamisugi *et al.* 2005).

The ability to undertake such genetic manipulation in a plant has important consequences. First, it provides a tool for "reverse genetics": the creation of a specific mutation in any given gene permits the functional analysis of that gene through study of its mutant phenotype. Second, such reverse genetic analysis can be directly applied for the comparative analysis of gene function. What are the consequences in the moss of a mutation in a gene whose ortholog in flowering plants regulates a process specific to angiosperms (for example floral development, seed formation, etc.)? What does this tell us about the way in which any particular gene has been recruited to participate in a specific morphogenetic or developmental process in different classes of plant?

The discovery of gene targeting in *Physcomitrella* provided a spur to the more widespread adoption of *Physcomitrella* as a model species for the study of plant processes. The realization that specific genes could be manipulated with exquisite precision indicated the need for gene discovery programs that resulted in the generation of first a substantial resource of expressed sequence tags (ESTs), derived by the high-throughput sequencing of cDNA clones (Nishiyama *et al.* 2003, Rensing *et al.* 2002), followed by the establishment of a genome sequencing programme that has culminated in the release of the first draft of the complete *Physcomitrella* genome in 2007.

Thus, the early years of the twenty-first century usher in a new era of "molecular bryology", where the resources available through the genome program, allied with technical developments in transformation and high-resolution cellular analysis promise to return the study of mosses to a position of prominence as plant biologists strive to understand the processes that have shaped land plant evolution, and to manipulate these processes for applied ends.

5.3 *Physcomitrella*: life cycle and development

The life cycle of *Physcomitrella patens* is typical of mosses (Fig. 5.1). The cycle commences with the germination of a haploid spore. The spores are environmentally resilient, single-celled propagules, contained within a thick wall comprising an inner, fibrillar intine, and an outer exine composed of sporopollenin. The exine is typically covered with an outer "perine" layer produced by the developing spore capsule (by contrast with the intine and exine, which are produced by the spore during its development). Mature spores are rich in oil, the principal storage reserve, and contain several immature chloroplasts derived by the division, late during spore maturation, of the large single plastid present during the earlier stages of sporogenesis (Knoop 1984, Schulte & Renzaglia, pers. comm.)

The germ tube penetrates the spore wall, reportedly through an aperture characterized by the presence of a pectin-rich intine, to form the first protonemal filament. However, it is not unusual to observe germinating spores in which two or three filaments emerge from different parts of the surrounding wall. The protonemata consist of uniseriate filaments, which extend through elongation and successive divisions of the meristematically active apical cell. The apical cell of the filament continually divides to generate a new, mitotically active apical cell and a subapical daughter cell, thus extending the filament. Subapical cells may undergo a subsequent mitotic division to generate a side-branch initial, from which a branching filament is formed. The first filamentous cell is typically a chloronemal cell. Chloronemata are filaments that contain relatively large numbers of chloroplasts. The apical cell typically possesses a

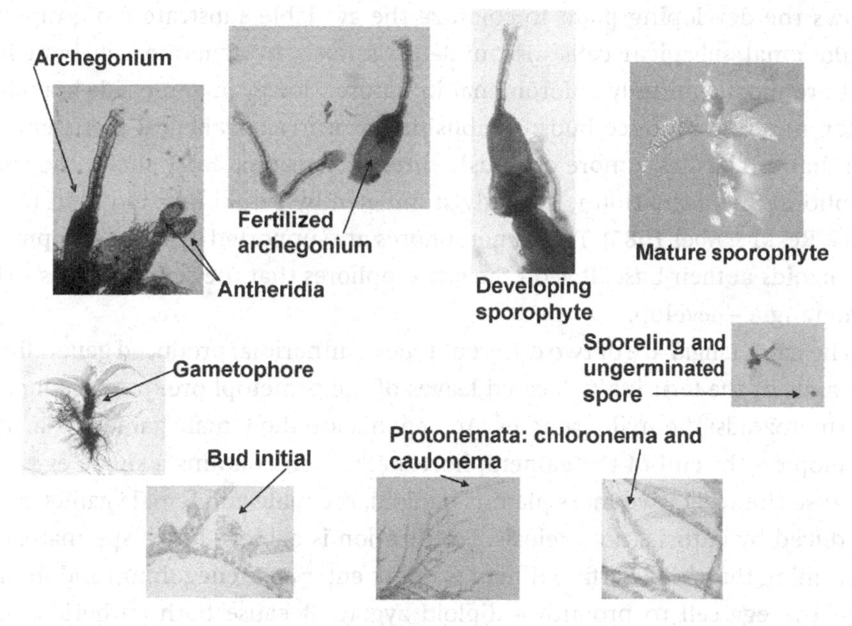

Fig. 5.1. Stages in the life cycle of *Physcomitrella patens*. Clockwise, from top left: (1) Gametangia (an archegonium, and two antheridia). (2) An archegonium with fertilized egg (arrowed) and two unfertilized archegonia. (3) Early sporophyte development: the archegonial neck is still attached to the developing sporophyte. (4) A mature sporophyte attached to the gametophore. (5) An ungerminated spore and germinated sporeling. (6, 7) Filament types: chloronemal and caulonemal filaments (arrowed). (8) A bud initial. (9) A gametophore: the dark coloration is the result of staining for GUS activity in a transgenic strain.

rounded apical dome, and the walls that separate the successive cells of the filament lie perpendicular to the long axis of the filament. The chloronemata are relatively slow-growing, and represent the first autotrophic cells of the developing plant. The apical cells elongate at a rate of 2–5 μm/h, and divide approximately every 24 h (Cove 2005).

A second filament type, caulonemata, develops by progressive differentiation of chloronemal apical cells. The induction of caulonemal differentiation is believed to be auxin-regulated (Johri & Desai 1973, Ashton *et al.* 1979a). Caulonemata grow much more rapidly than chloronemata, the caulonemal apical cell extending at a rate of 25–40 μm/h, and dividing with a reduced cell cycle time of approximately 7 h (Cove 2005). The caulonemal filaments are characterized by being relatively reduced in the numbers of chloroplasts they contain. Their apical cells have a more sharply pointed apical dome than the chloronemal apical cells, and the cross walls between the cells of the filament are oblique, rather than perpendicular to the length of the filament. The rapid growth of the caulonemata allows the developing plant to colonize the available substrate more rapidly. Caulonemal subapical cells also undergo mitosis to generate side-branches that are mostly initially chloronemal in nature. However, some side-branches differentiate to produce buds, establishing a nearly tetrahedral meristematic cell that establishes a more obviously three-dimensional leafy shoot: the gametophore. Bud formation is strongly stimulated by cytokinins and light (Cove 1984, Reski & Abel 1985). The gametophores are supported by the development of rhizoids at their base. It is on the gametophores that the sexual organs – the gametangia – develop.

The gametangia are of two different types. Antheridia, produced generally in the axils of the terminally located leaves of the gametophore, produce motile spermatozoids: the male gametes. Archegonia are the female gametangia, and develop on the end of the gametophore stalk. Each contains a single egg cell. Because the vegetative moss plant is haploid, the male and female gametes are produced by mitosis, not meiosis. Fertilization is achieved by a spermatozoid swimming through a surface film of water to enter an archegonium and so fuse with the egg cell to produce a diploid zygote. Because both antheridia and archegonia are produced in close proximity on the same gametophore, self-fertilization is a common occurrence. In nature, *Physcomitrella* is an ephemeral annual plant, appearing on the banks of ponds whose water level recedes in summer (it is sometimes known as the "reservoir moss"), and self-fertilization presumably offers an assured means of sexual reproduction. In culture, gametangial development, fertilization, and the production of sporophytes are promoted by reduced temperatures and short daylength, corresponding to the conditions prevalent during autumn, the time at which sexual reproduction in *Physcomitrella* typically occurs (Hohe *et al.* 2002). The zygote develops into the diploid sporophyte: in mosses the diploid generation is anatomically reduced and is dependent upon the dominant, vegetative, gametophyte generation. Unlike in typical mosses, such as *Funaria hygrometrica*, the sporophyte is borne on a very short seta, so that it remains closely surrounded by the terminal, or perichaetial, leaves of the gametophore during its development. It is spherical and initially green in colour, becoming orange to brown in colour as it matures following sporogenesis. Within the sporophyte, the spore mother cells enlarge and are released into the mucilaginous interior of the sporophyte, where they undergo meiosis. Each spore mother cell initially contains a single chloroplast in addition to its nucleus, and the chloroplast undergoes two cycles of division, in concert with the meiotic division of the nucleus, to deposit a single chloroplast in each cell of the meiotically derived tetrad (Schulte & Renzaglia, pers. comm.). Each of these cells matures to become a single spore, and a single spore capsule may contain approximately 4000 spores at maturity. The spore capsules may be stored dry for extended periods of time, without significant loss of spore viability, and there may be some "after-ripening" processes during dry storage, since spores germinated from stored spore capsules germinate more synchronously than do those released from freshly harvested capsules. However, to date no systematic investigation of spore dormancy has been undertaken.

5.4 The molecular biology of *Physcomitrella*: sequencing the genome

It is a truism to state that the size of an organism's genome is correlated with the degree of biological complexity it displays. However, it is also a gross oversimplification. Although this relationship holds true in general terms (e.g. *E. coli*, 4.6 Mbp; yeast, 12 Mbp; *Arabidopsis*, 150 Mbp; *Drosophila*, 165 Mbp), it is clear that among the complex multicellular eukaryotes the DNA content of the genome can vary widely (http://www.cbs.dtu.dk/databases/DOGS/index.php). Thus the largest plant genome size is an estimated 90 000 Mbp for the Easter lily, *Lilium longiflorum*, despite this plant not being markedly more complex than *Arabidopsis*. Incidentally, it should be noted that the current "world record holder" in the genome size stakes is the protist *Amoeba dubia*, with an estimated genome size of 670 000 Mbp (Winstead 2001)! Genome size, therefore, does not necessarily reflect complexity. Instead, it depends on the quantity of repetitive non-coding DNA – usually retrotransposon-derived – that the host can maintain before suffering a selective disadvantage.

Although it exhibits a structural complexity that is apparently less than that of a flowering plant, *Physcomitrella* does not have a particularly small genome. Fortunately, neither is it exceptionally large. Estimates of genome size, based on flow cytometry of propidium-iodide-stained nuclei, indicated a DNA content equivalent to 511 Mbp DNA (Schween *et al.* 2003): approximately three times greater than that of *Arabidopsis thaliana*. Cytogenetic analysis is difficult, owing to the small size of the chromosomes when mitotic figures are analyzed, but it is believed that the genome is distributed among 27 chromosomes (Reski *et al.* 1994).

The first steps towards defining the coding capacity of the *Physcomitrella* genome comprised the establishment of EST sequencing programs (Rensing *et al.* 2002, Nishiyama *et al.* 2003). Expressed sequence tags are obtained by the systematic sequencing of cDNA clones, and are collated following the single-pass sequencing of individual clones. This experimental approach enables a "snapshot" of the genes expressed in any selected cell type to be obtained. Because of the inherently error-prone nature of both cDNA synthesis by reverse transcriptase and DNA sequencing reactions using thermostable DNA polymerases, ESTs do not necessarily generate high-quality sequence information for any individual gene, unless a number of redundant sequences can be compared, but they do provide a valuable tool for gene discovery and comparative sequence analysis. Currently, nearly 200 000 ESTs have been deposited in the GenBank database, all derived from the "Gransden" laboratory strain. (This strain, the most widely used by *Physcomitrella* researchers, derives from a single spore collected near Gransden Wood, Huntingdonshire, U.K., by H. L. K. Whitehouse in 1962.) A further *c.* 100 000 ESTs have been sequenced from a second genotype (named "Villersexel", after the town in the Haute Sâone region of France near which it was collected in 2003 by Michael Lüth), which has been crossed with the Gransden genotype in order to construct a genetic linkage map: the intention is that this collection will facilitate the identification of single-nucleotide polymorphisms between the Gransden and Villersexel genotypes. In addition to these resources, a further 110 000 EST sequences have been generated as a proprietary resource in a BASF-sponsored research program (Rensing *et al.* 2002). It is expected that these will eventually be released into the public domain. The cDNA libraries from which these sequences were obtained derived from mRNA isolated from a range of different stages of moss development, including protonemata growing on defined growth medium and supplemented with different growth regulators (auxin, cytokinin, ABA), tissue subjected to drought stress, gametophores and during different stages of sporophyte development, in order to identify transcripts from the widest possible array of *Physcomitrella* genes. Bioinformatic analyses of the sequences, which entailed clustering the sequences by multiple sequence alignment to enable the assembly of individual consensus sequences, have estimated the numbers of expressed genes as approximately 25 000: a figure not very different from that estimated for the genome of *Arabidopsis*.

The formidable task of determining the whole genome sequence of *Physcomitrella* commenced in 2005, in a program undertaken for the Community Sequencing Program of the U.S. Department of Energy's Joint Genome Institute (http://www.jgi.doe.gov/sequencing/why/CSP2005/physcomitrella.html). The approach taken was a "whole-genome shotgun" approach. Essentially, nuclear DNA isolated from the Gransden strain of *Physcomitrella* was subjected to physical shearing, and a series of size-fractions were selected for cloning by blunt-end ligation into a "fosmid" vector. This generated a number of libraries of genomic DNA, with average insert sizes of 3 kb, 8 kb and 40 kb, respectively. Clones from each of these libraries were end-sequenced, and the sequences aligned in order to identify overlaps to create an assembly. By undertaking sequencing to a high level of redundancy (over 6.7 million individual sequence traces were obtained), and applying appropriate quality filters to the sequence output, it has been possible to obtain an approximately 8-fold depth of coverage of the genome sequence (i.e. all regions of the genome were sequenced, on average, 8 times: in total, nearly 5 billion bp of sequence data were obtained for an estimated genome size of *c.* 500 Mbp).

The first draft of this sequence (http://shake.jgi-psf.org/Phypa1/), released in 2007, corresponds to a length estimate for the *Physcomitrella* genome of approximately 490 Mbp, a figure quite close to the 511 Mbp estimated by flow cytometry (Schween *et al.* 2003). However, despite the average 8-fold depth of the sequencing, the sequence is incomplete. This is a common occurrence in first-draft sequences obtained by using the random shotgun approach, and the draft will undergo further revisions as more data are obtained. The first-draft version comprised over 2500 "scaffolds" – individual assemblies made up by combining overlapping sequences ("contigs") derived from clusters of clones of different lengths. Clearly, in a complete sequence, the number of scaffolds should be equal to the number of chromosomes ($n = 27$). However, at this stage in the sequence assembly, there are a number of gaps in the sequence that result from (i) lack of overlap between the individual scaffolds, (ii) uncertainty about overlapping sequences, and (iii) mis-assembly that occurs owing to the presence of highly repetitive sequences within the genome. Additionally, within the individual scaffolds, there are regions where the sequence is unknown (these are represented in the sequence as runs of "NNNNNNNN"). These unknown regions correspond to as-yet unsequenced tracts in the interior of longer clones, whose terminal sequences could otherwise be clearly aligned with those of other clones that contributed to the scaffold assembly.

Refinement of the sequence will require the incorporation of additional information. One source of this will be the inclusion of terminal sequences derived from BAC clones. These are large-insert "Bacterial Artificial Chromosome" clones (over 100 kb of DNA can be accommodated). Because of its length, a single BAC clone will contain very many smaller-insert clones within its length, and by determining the end-sequences of the BAC inserts, the long-range linkage of currently unresolved scaffolds can be achieved. Another means of linking scaffolds will be the incorporation of genetic linkage data. A genetic linkage map is currently being constructed by using molecular markers: amplified fragment length polymorphisms (AFLPs) and simple sequence repeats (SSRs) (von Stackelberg *et al.* 2006). Because these correspond, in many cases, to identifiable DNA sequences, they will provide additional long-range sequence data to assign scaffolds to individual *Physcomitrella* chromosomes.

5.5 The discovery of homologous recombination

The discovery that propelled the study of *Physcomitrella patens* from a fascinating sideshow to center stage was the discovery that transforming DNA could integrate into the *Physcomitrella* genome by homologous recombination. This discovery has led directly to the development of high-efficiency gene targeting in this organism, and indirectly provided the impetus, within the plant science community, for the establishment of the *Physcomitrella* Genome Program.

For many years prior to this discovery, *Physcomitrella* had been pioneered as a tool for the study of plant development and cellular differentiation in the laboratory of Professor David Cove, at Leeds University. These studies had been primarily genetic, involving the identification of mutant strains exhibiting altered patterns of development, and altered responses to polar growth stimuli – in particular the phototropic and gravitropic stimuli. At this time, genetic analysis in *Physcomitrella* was hampered by the low fertility exhibited by many of the existing, mutagenized laboratory strains, but complementation analysis could still be undertaken by generating somatic hybrids between mutant strains through polyethylene glycol-mediated fusion of protoplasts, to generate diploids (Grimsley *et al.* 1977).

Complementation testing is an essential and necessary first step in the analysis of the genetic basis of any given phenotype. When two genomes, containing independently isolated mutations that result in the same recognizable phenotype, are combined in the same cell, then two outcomes are possible. Either the mutant phenotype is restored ("complementation") or the mutant phenotype persists. Complementation indicates that the original mutations that gave rise to the mutant phenotype were in different genes. If the mutant phenotype persists, it indicates that the mutations are in the same gene. Such non-complementing mutations are known as a "complementation group". By assigning independently isolated mutants to complementation groups, the investigator can determine the number of different genes that might control that phenotype. However, unless genetic analysis is allied to powerful molecular tools, it is very difficult to identify the underlying genes.

In the 1990s, it was becoming clear that the genetic analysis of *Arabidopsis* mutants, coupled with the development of a high-resolution genetic linkage map provided one way in which mutant genes could be identified by "map-based cloning": through their close linkage to a genetic marker defined by a known DNA sequence that could be used as a hybridization probe to initiate a "chromosome walk" (the successive isolation of overlapping, large-insert genomic clones). However, even this technique was labor-intensive, since at this time for *Arabidopsis* the genetic linkage map was relatively sparsely marked, and its genome sequencing had only recently been initiated (Lukowitz *et al.* 2000). Consequently, alternative means were sought to identify genes regulating key developmental processes. The most successful of these are based on insertional mutagenesis either following transformation by T-DNA delivered by *Agrobacterium* - which was found to be incorporated essentially at random sites within the genome, at low copy number - or by the activation of exogenous transposons introduced into transgenic lines. Such insertional mutagenesis protocols enabled the direct cloning of the genes disrupted by these agents. Initially, the inserted DNA acted as a sequence-defined "tag", so that the insertionally mutated gene could be identified in a cloned library of genomic DNA derived from the mutated strain, by using the inserted sequence as a hybridization probe. For *Arabidopsis*, it subsequently became possible to identify the genomic sequences flanking the inserted DNA by using "inverse PCR" - a procedure illustrated in Fig. 5.2. The generation of a number of collections of publicly accessible` insertionally mutated, transgenic lines, in which the mutated sequences were deposited in searchable databases (http://signal.salk.edu/cgi-bin/tdnaexpress; http://atidb.org/; Pan *et al.* 2003) enabled the consequences of disruption of *Arabidopsis* genes to be functionally analyzed.

Because insertional mutagenesis combines the power of randomly generating mutants with the ability rapidly to isolate the disrupted sequence, it is a very attractive genetic tool, and efforts were made to develop a similar tool for the genetic analysis of *Physcomitrella* mutants. The ease with which protoplasts of *Physcomitrella* could be isolated from protonemal tissue, and their rapid and efficient regeneration to form new plants, suggested that they might also be susceptible to genetic transformation by plasmid DNA, and a program was initiated to develop a stable transformation system for this organism.

This was largely successful, and a collaboration between the Leeds group and that of Professor Jean-Pierre Zrÿd, in Lausanne, demonstrated how plasmid DNA carrying selectable marker genes could be delivered to *Physcomitrella* protoplasts in the presence of calcium ions and polyethylene glycol, and incorporated into the cells (Schaefer *et al.* 1991). Interestingly, three classes of transformed cell could be identified. Most cells were of class one: they took up the DNA and expressed the selectable marker genes, but only transiently. These protoplasts did not maintain the DNA, and soon died following the application of the selective agent (typically an aminoglycoside antibiotic, such as hygromycin or G418). The second class of transformant prospered on selective medium, but if the selective pressure was withdrawn the transgenic colonies failed to survive a subsequent exposure to selection. These so-called "unstable" transformants are thought to maintain the transforming DNA in an extrachromosomal form, such that following the relaxation of selection this DNA was lost from the cells. It was subsequently demonstrated that this transforming DNA could be maintained for very long periods (years), as extrachromosomal arrays of concatenated plasmid DNA, so long as selection

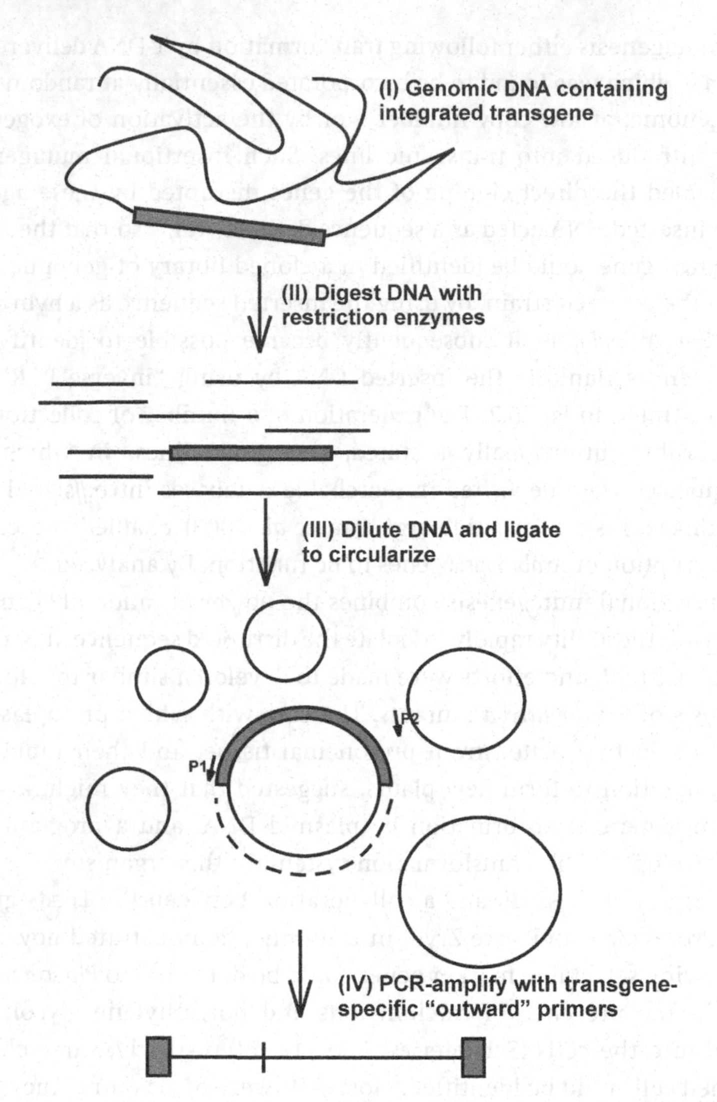

Fig. 5.2. "Inverse PCR": a method for the identification of insertionally mutagenized genes. DNA is isolated from an insertionally mutated plant (I) and is digested with a number of restriction enzymes that have defined recognition sites not found within the inserted DNA (II). The digested DNA is diluted and incubated with DNA ligase (III); ligation takes place between the ends of DNA molecules in closest proximity, and at low DNA concentration these are the two ends of the same molecule. This results in the formation of a population of circular molecules. PCR amplification with "outward-pointing" primers corresponding to the inserted DNA (IV) will amplify the flanking genomic sequences. These can then be readily identified by DNA sequence analysis.

was continuously maintained (Ashton *et al.* 2000). The third class of transformant (and the least numerous) continued to exhibit the transgenic selection marker following several cycles of alternating subculture on selective and non-selective medium, and were demonstrated by Southern blot analysis to have incorporated the transforming DNA covalently into the moss genome. Typically, the incorporated DNA corresponded to concatenated repeats of the plasmid DNA inserted at one or a few loci. Stable integration of transforming DNA is favored by transformation with linear fragments of DNA, whereas transformation with circular plasmids results in a preponderance of transformants of the "unstable" class.

The development of a reliable transformation procedure stimulated research into the development of an insertional mutagenesis technique that could be used to create tagged mutants, thereby enabling the cloning of the genes underlying developmental transitions. At this time, the characterization of a number of transposons active in maize suggested that the *Ac/Ds* transposition system might be particularly effective, since this system could be transgenically imported into *Arabidopsis* for gene tagging (Long *et al.* 1993). In maize, these transposable elements are actually two different forms of the same transposon (McClintock 1948, Coupland *et al.* 1988). The *Ac* ("Activator") element is an autonomous transposon that encodes a transposase responsible for recognition of the terminal repeats that delimit the transposon, excising the element from its genomic locus, and subsequently causing its reinsertion at another (usually linked) genomic locus. The *Ds* ("Dissociator") element is an internally deleted variant of *Ac*. It retains the terminal repeats that are the target for the transposase activity, but does not encode an active transposase element. The *Ds* element is not autonomously active, and is unable to cut itself out and reinsert elsewhere. Thus plants that carry a *Ds* element inserted within their genome are genetically stable (Coupland *et al.* 1988). This genetic system has been exploited within the *Arabidopsis* community by the construction of a number of independent transgenic lines carrying single copies of the *Ds* element at different transgenic loci scattered around the genome. These genetically stable lines can be mutagenized by introducing a second transgenic construct, carrying an active transposase gene derived from the *Ac* element. Expression of the transposase results in mobilization of the resident *Ds* elements and the consequent generation of a new series of insertion mutants (Muskett *et al.* 2003).

It seemed not unreasonable that this system might also function in *Physcomitrella*, as a means of creating tagged insertional mutants, with the added advantage that if the *Ac* transposase were to be introduced on an unstably maintained plasmid to a stably transformed line carrying the *Ds* element, then the *Ds* element could be mobilized to a new genetic locus, and then stabilized there as a consequence of the subsequent loss of the *Ac* transposase gene upon the relaxation of selection.

The generation of stably transformed lines containing the *Ds* element did not prove problematic, it being delivered on a transgenic plasmid that also conferred resistance to the antibiotic kanamycin (or G418), and integrating stably into the genome. Introduction of the *Ac* transposase on a second plasmid, conferring resistance to hygromycin, did indeed mobilize the *Ds* element, but although it appeared to be excised from the genome it was not re-inserted elsewhere (D. J. Cove, pers. comm.). Thus the maize transposition system did not function in *Physcomitrella* in the same way as it did in higher plants. However, it appeared that the frequency with which stable hygromycin-resistant transformants were generated following retransformation of the G418-resistant *Ds* lines was significantly higher than expected. Certainly it was higher than the rate at which the hygromycin-resistance-carrying plasmid generated stable transformants in previously untransformed lines.

This led to the hypothesis that the second plasmid, which shared substantial lengths of sequence homology with the first transforming plasmid (both constructs used the same basic cloning vector), might be becoming preferentially integrated in the genome by homologous recombination with the first transgenic locus, a hypothesis that was subsequently strengthened by genetic analysis of double transformants. Both the hygromycin-resistance and G418 resistance markers were found to co-segregate in independent sexual crosses, indicating their close linkage in the genome (Kammerer & Cove 1996).

Molecular confirmation of the occurrence of homologous recombination was provided by Schaefer & Zrÿd (1997) who analyzed the insertion of a number of recombinant plasmids carrying fragments of cloned *Physcomitrella* DNA. Southern blot analysis of a number of transgenic lines conclusively demonstrated that these constructs were preferentially targeted to the homologous loci with very high efficiency: in some transgenic lines, 100% of the stable transformants resulted from integration of the transforming DNA into the targeted locus. Additionally, there was an apparent association of targeting efficiency with the length of homology between the targeting construct and the targeted locus. Moreover, targeting was precise, and both genomic DNA sequences and cDNA sequences could be used to build targeting constructs. Thus, when a member of a highly homologous multigene family (a gene encoding a light-harvesting chlorophyll *a*/*b* protein) was used in a targeting experiment, it was found to target exclusively the cognate member of the gene family, and not the other family members, despite their very high nucleotide sequence similarity (Hoffman *et al.* 1999).

Gene targeting does not occur efficiently in higher plants. The rates that have been detected in *Arabidopsis* are of the order of 10^{-3} or lower (Kempin *et al.* 1997, Hanin *et al.* 2001, Hanin & Paszkowski 2003). The rates of gene targeting seen in *Physcomitrella* are more reminiscent of those that occur in the yeast *Saccharomyces cerevisiae*, in which gene targeting is routinely used as a way of generating novel mutant alleles, either by the disruption of specific genes or by the replacement of wild-type alleles with variants containing defined point mutations (Orr-Weaver *et al.* 1981). This provides a very powerful and sophisticated means of genetic manipulation, and the significance of this was not lost on the *Physcomitrella* community. The first mutant phenotype generated by gene targeting in moss was the disruption of the *ftsZ* gene: a nuclear-encoded chloroplast tubulin, the homologous recombination-mediated knockout of which resulted in the failure of chloroplast division, and the presence in each cell of a single large chloroplast (Strepp *et al.* 1998). Since that first demonstration of the utility of gene targeting in *Physcomitrella*, there has been a burgeoning of interest and activity, with the construction of large "knock-out" mutant collections generated by high-throughput transformation using randomly disrupted cDNA and genomic fragments (Egener *et al.* 2002, Schween *et al.* 2005) and the development of specific "knock-in" lines in which reporter genes are fused to specific gene sequences for targeting to the corresponding loci (Nishiyama *et al.* 2000).

The availability of the complete genome sequence will further enable the functional analysis of specific genes through the "reverse genetic" route: the creation of defined mutants in specific genes to determine the details of their regulation and function. A series of overriding questions remain. What is it about *Physcomitrella* that causes it to preferentially incorporate DNA by homologous recombination at specific sites, rather than randomly as occurs in flowering plants? What is the mechanism by which homologous recombination occurs? Can we identify the components that undertake homologous recombination in *Physcomitrella*, and use this knowledge to inform attempts to develop a high-frequency gene targeting technology for crop species?

5.6 Homologous recombination and DNA repair

The incorporation of exogenously supplied DNA into the genome is not a normal plant function. It is generally agreed that when transforming DNA is incorporated into a genome, the mechanisms responsible for its integration are those that are more commonly used for the repair of DNA damage (Schaefer 2001). The maintenance of the integrity of the genome is essential for the survival of all organisms, and a plethora of DNA damage repair systems are known to exist. The most catastrophic form of DNA damage that a cell can suffer is a double-strand break of the DNA. If both strands of the DNA are broken at the same site in a chromosome, and these are not repaired, then the loss of the telomeric fragment will occur. To prevent this, eukaryotic cells have developed two pathways for the repair of double-strand breaks that occur both as a consequence of environmental agencies (radiation or genotoxic chemicals) and routinely during DNA replication. These pathways are highly conserved among eukaryotes (Schuerman *et al.* 2005) but have been most exhaustively characterized in yeast (Fig. 5.3). They involve either non-homologous end joining between fragments (NHEJ: the ligation of two broken ends (Weterings & Van Gent 2004) or homologous recombination-mediated repair, in which the broken ends are repaired using a homologous chromosome as a template for repair (Aylon & Kupiec 2004). The frequency with which ds-DNA breaks occur during DNA replication, and the requirement to use the homologous chromosome as a template for HR-mediated repair, means that DNA damage repair by homologous recombination is usually tightly correlated with the cell cycle: DNA damage typically imposes a cell-cycle arrest at the G_2/M boundary, during which ds-DNA breaks can be repaired before the cell is allowed to divide (Lisby *et al.* 2004, Lisby & Rothstein 2004). Consequently, it is significant that in *Physcomitrella* protonemal cultures the majority of the cells have been shown to be arrested in G_2, and to contain the 2C complement of DNA (Schween *et al.* 2003). Most of these cells will be postmitotic subapical cells, and it has been proposed that if these contribute a major proportion of the protoplasts used for a transformation experiment, then the machinery may be already in place for the integration of transforming DNA by homologous recombination in the majority of protoplasts. Arrest at this stage of the cell cycle can also be seen to be a good strategy for survival for cells of a haploid organism, since the two copies of the genome will provide templates for each other's mutual repair of ds-breaks should they occur as a consequence of the organism's experiencing genotoxic stress.

The most efficient way to generate a gene targeting event is to transform protoplasts with linear fragments of DNA, rather than with circular plasmids. Although many of the first *Physcomitrella* transformation studies utilized supercoiled plasmids, it became clear that such molecules typically generated a very high ratio of unstable to stable transformants, whereas stable transformation was favored by the use of linear fragments. It is probable that when linear DNA fragments enter a cell their termini are recognized by the cellular DNA repair machinery as double-strand breaks. The very large number of such fragments that a cell will take up in the course of a transformation experiment would be recognized as a catastrophic number of ds-breaks, and is likely to elicit a massive DNA damage-repair response by the cell.

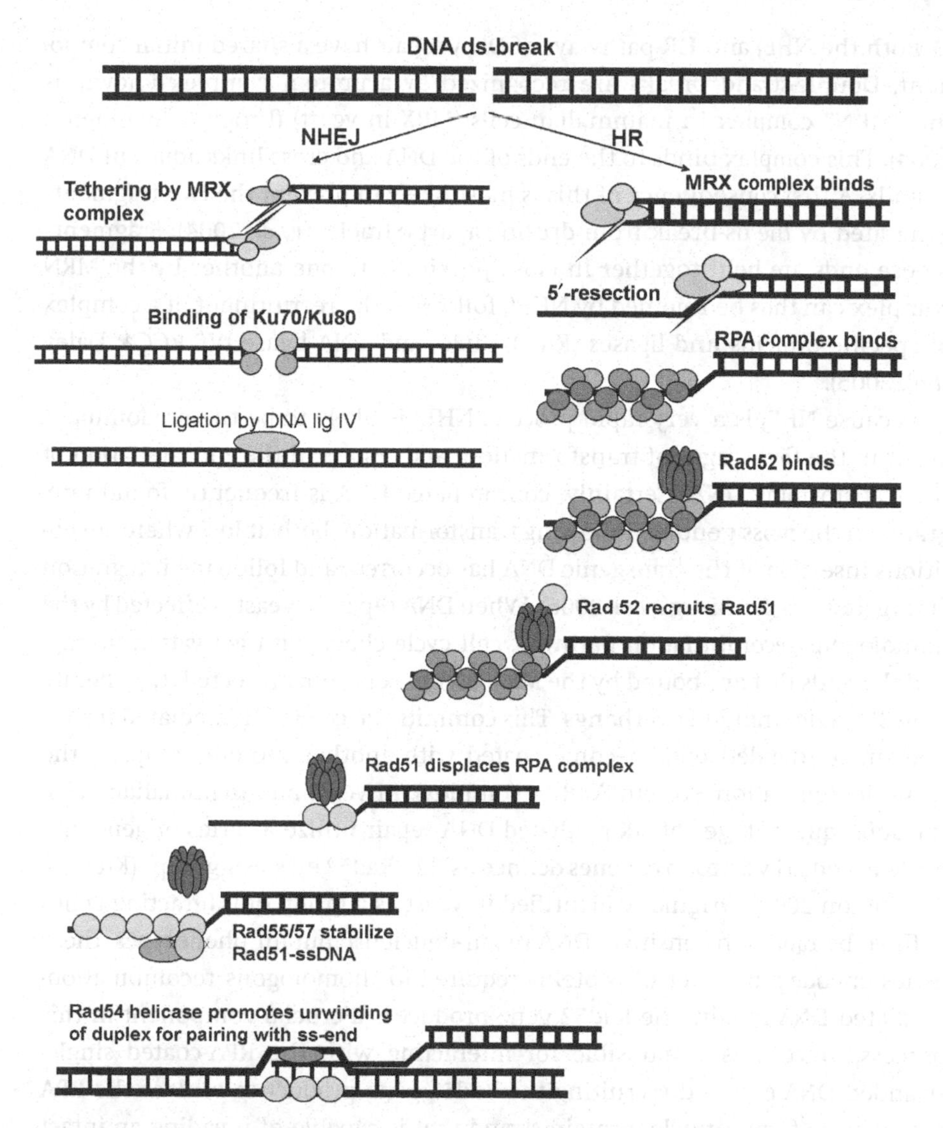

Fig. 5.3. Pathways of DNA repair by NHEJ and HR. The following model is based on our knowledge of DNA damage repair in yeast and mammalian cells, and for NHEJ, in *Arabidopsis*. A double-strand break in DNA can be repaired either by NHEJ or by HR. The first step in both pathways is the binding of the broken termini by the Mre11–Rad50–XRS1 ("MRX") complex. For the NHEJ pathway, this complex tethers the broken ends. NHEJ proceeds through the interaction of the broken ends of the DNA with the Ku70 and Ku80 proteins, then the broken ends are "polished" and rejoined by DNA ligase IV. This can result in either small sequence rearrangements at the point of ligation, or the random joining of DNA sequences of different genomic origin. HR occurs by resection of the DNA to generate a 3′-ss overhang, followed by recruitment of the Replication Protein A complex to the 3′-ss DNA, before the Rad52 protein acts to catalyze its replacement by the Rad51 protein. This forms a

Both the NHEJ and HR pathways of DNA repair have a shared initial component. Double-strand breaks are recognized by a protein complex known as the "MRN" complex in mammalian cells (MRX in yeast) (Krogh & Symington 2004). This complex binds to the ends of the DNA and cross-links adjacent DNA strands. A first consequence of this is most likely to prevent the two fragments generated by the ds-break from drifting apart (Stracker *et al.* 2004). Fragments whose ends are held together in close proximity to one another by the MRN complex can thus be rejoined by NHEJ, following the recruitment of a complex of specific proteins and ligases (Ku70, Ku80, and DNA ligase IV/XRCC4; Daley *et al.* 2005).

Because NHEJ is a very rapid process, NHEJ is likely to be the predominant event in the first stages of transformation, and results in the concatenation of the transforming DNA. Certainly, concatenated DNA is frequently found integrated in the moss genome following transformation, both at loci where adventitious insertion of the transgenic DNA has occurred, and following integration at targeted loci (Kamisugi *et al.* 2006). When DNA repair in yeast is effected by the homologous recombination pathway, cell cycle checkpoint activation occurs, and the ends that are bound by the MRN complex become resected, to generate long 3′-single-stranded overhangs. This commits the cell to HR-mediated repair. The single-stranded ends become coated with another protein complex: the trimeric Replication Protein A (RPA complex). In yeast and mammalian cells, the subsequent stages of HR-mediated DNA repair utilize a series of gene products encoded by a group of genes defined as the "Rad52 epistasis group" (Krogh & Symington 2004). Originally identified in yeast as a group of interacting genes defined by radiation-sensitive (DNA repair-deficient) mutant phenotypes, these genes encode a number of proteins required for homologous recombination-mediated DNA repair. The Rad52 gene product is a crucial component in this process, since it is responsible for interacting with the RPA-coated single-stranded DNA end and recruiting the Rad51 gene product to replace the RPA complex, to form a nucleoprotein strand that is capable of invading an intact DNA double helix and annealing with its complementary sequence (Benson *et al.*

Caption for Fig. 5.3. (cont.)
nucleoprotein filament, stabilized by other products of the Rad52 epistasis group genes (Rad55 and Rad57), that is able to invade a homologous duplex DNA. Unwinding of the duplex DNA is facilitated by the Rad54 DNA unwinding activity. The invading strand can then be extended by copying the invaded template sequence. The details of the HR pathway in plants are not known, but must differ in detail, since plants lack a recognizable *RAD52* gene.

1998, New *et al.* 1998, Shinohara & Ogawa 1998). Rad 51 is highly conserved in evolution, being the eukaryotic equivalent of the bacterial RecA protein. Typically there are several paralogous Rad51 genes in eukaryotes, all of which have counterparts in *Physcomitrella*. Interestingly, in *Physcomitrella*, there are two copies of the *RAD51* gene encoding the principal ss-DNA-interacting protein (Markmann-Mulisch *et al.* 2002) which show relatively higher levels of expression in the apical cells of protonemal filaments. (This is not surprising, since the highest incidences of DNA double-strand breakage are expected to occur in mitotically active cells.) The genes are redundant in function, knockout mutants of each having no phenotypic effect, whereas the double-mutant shows a highly radiation-sensitive phenotype, and is additionally meiotically defective (B. Reiss, pers. comm.). Most interesting is that an examination of higher plant genome sequences reveals no homolog of the *RAD52* gene. Since in both yeast and mammalian cells this gene is essential for the formation of an invasive DNA strand, it suggests that plants use an alternative mechanism for homologous recombination-based DNA repair. The *Physcomitrella* genome also lacks a *RAD52* homolog, so it remains an open question as to how homologous recombination occurs in plants. The readiness with which *Physcomitrella* incorporates transforming DNA by HR implies that homologous recombination is the default pathway for DNA repair, recommending it as a model for studies of this essential process.

Our understanding of eukaryotic HR processes stems largely from mutational analyses in yeast. In particular, the characterization of the DNA damage-deficient *Rad* mutants revealed the identity of all the components of the HR-mediated DNA repair pathway. The homologs of these genes can be identified in many organisms, and in recent years the availability of T-DNA insertion mutants of most *Arabidopsis* genes has spawned a number of "reverse genetic" analyses of DNA repair processes in mutants of the known DNA repair-related genes. However, these have failed to shed any significant light on the process of HR, probably for two reasons. First, because *Arabidopsis*, like all higher plants, is largely incompetent to undertake HR, and second because there are some key differences in the regulation of the process between yeast and plants, such as the absence of an obvious homolog of Rad52. If we are to identify novel, plant-specific components, the best way forward is to apply the power of mutational analysis in a "forward genetic" screen. This must necessarily be undertaken in an organism in which the HR-mediated DNA repair pathway predominates over the NHEJ-mediated pathway. Only *Physcomitrella* offers this opportunity. Thus, this is one fundamental plant process in which mosses may be more useful than the otherwise ubiquitous *Arabidopsis* as an experimental platform.

5.7 Homologous recombination for reverse genetics

"Reverse genetics" is the term given to the deduction of gene function, starting with knowledge of the sequence of a gene. Certainly, gaining an understanding of the mechanism of homologous recombination has intrinsic fascination, as well as ultimate strategic relevance. However, the process can also directly be used for the manipulation of the moss genome without recourse to a detailed knowledge of the precise mechanism. Gene knockouts and "knock-ins" provide tools by which gene function can be studied.

By using homologous recombination, one can construct reporter gene fusions that enable the activity of promoters to be tested within their original genomic context (unlike promoter analyses in which fusion constructs are tested in transient expression experiments, or following the stable, but ectopic insertion of transgenes). This enables observation of the cellular dynamics of individual gene products by *in vivo* imaging of proteins fused to fluorescent proteins (such as the jellyfish Green Fluorescent Protein (GFP) and its numerous spectral variants CFP (cyan) ,YFP (yellow), RFP (red), etc.) in conjunction with high resolution laser-scanning confocal microscopy. One may also examine interactions between gene products *in vivo* by using proteins fused to (i) different fluorescent reporters in techniques such as FRET (fluorescence resonance energy transfer) or (ii) to individually non-functional but combinatorially active fluorescent reporters (bimolecular fluorescence complementation: BiFC), and to isolate native multiprotein complexes from cells by "pull-down" experiments following the construction of fusion proteins by "knock-in" of affinity tags or specific epitopes.

5.8 Requirements for efficient gene targeting

All of these procedures require a minimal knowledge of the requirements for efficient gene targeting that boil down to a small number of requirements:

- What is the optimal length and design for a gene targeting construct?
- What is the best method of DNA delivery and selection of transformants?
- How can mutant lines containing targeted genes most conveniently be identified?
- What analyses are necessary to confirm that a mutation is responsible for an observed phenotype?
- What is the best way of obtaining multiply mutated lines? (For example, in cases where the existence of a family of similar genes suggests that there may be extensive redundancy.)
- Can gene targeting methodologies be applied to other moss species, or is it only possible in *Physcomitrella*?

5.8.1 *Length and design*

For a knockout construct, with the intention to inactivate a *Physcomitrella* gene, the DNA used for transformation should be a linear fragment, corresponding to the 5′- and 3′-terminal sequences of the gene, interrupted by a selection cassette: typically an antibiotic resistance gene driven by a constitutively active promoter. The most frequently used cassettes comprise a bacterial *nptII* gene conferring resistance to kanamycin or G418, under the control of the CaMV 35S promoter, and either a nopaline synthetase or CaMV-derived transcription terminator. Such cassettes are approximately 2 kb in length. Since linear fragments for transformation are most easily generated in highly pure form by PCR amplification, the overall length of the fragment should not exceed 4 kb, since fragments larger than this are less easily amplified. In a study of the minimum length requirements for high-frequency targeting, flanking targeting sequences of *c.* 500 bp each were sufficient to achieve a frequency of allele replacement of 50% of total transformants (Kamisugi *et al.* 2005). Flanking sequences 1 kb in length can generate gene targeting events in up to 100% of the stable transformants recovered. To ensure that a knockout of gene activity is achieved, it is recommended that a significant length of the coding sequence should be replaced by the selection cassette. Either genomic DNA or cDNA sequences can be used in the construction of targeting constructs, but no rigorous comparison between the efficiencies achieved with such constructs has been made. It has been suggested that for targeting sequences of less than 300 bp a cDNA-based vector may target its cognate locus with higher efficiency than the corresponding genomic sequence of the same length, if the cDNA sequence comprises a number of exons. However, this remains to be tested.

5.8.2 *DNA delivery*

DNA can be delivered either by polyethylene glycol-mediated protoplast transformation (Schaefer *et al.* 1991) or directly to intact tissue by particle bombardment (Sawahel *et al.* 1992). Both procedures are suitable for obtaining targeted gene knockouts, although our experience indicates that gene targeting is approximately two to four times as efficient (in terms of the percentage of transformants that exhibit allele replacement) using protoplast transformation. However, protoplast transformation and regeneration is a more technically demanding procedure, requiring high skill levels in tissue culture to achieve high transformation efficiencies. Protoplasts are produced by digesting protonemal tissue with the enzyme "Driselase", a commercially available cocktail of fungal cellulases. The cell walls of protonemal filaments are readily digested by the enzyme preparation, releasing protoplasts which are prevented from undergoing osmotic lysis by maintaining them in an isotonic concentration of mannitol. Individual protoplasts retain their totipotency, being able to regenerate, initially as chloronemal filaments, and then subsequently to reiterate all the stages of normal *Physcomitrella* development to generate whole plants. Although protoplast regeneration is very efficient in the hands of a skilled tissue culture worker, it may take some time to master the technique. Consequently biolistic transformation may be preferable for investigators with less experience of plant tissue culture.

"Biolistic" transformation is the delivery of transforming DNA into intact tissues by bombardment of the tissue with microprojectiles. These are small (1 μm diameter) particles of an inert metal (gold or tungsten) that can be coated with DNA and delivered by using a commercially available instrument (for example, the BioRad PDS1000 microprojectile bombardment device). Tissue bombarded in this way can be directly submitted to selection for transformed cells by incubation on the appropriate selective medium without any need for extensive manipulation, in culture.

5.8.3 *Identification of gene-targeted mutants*

PCR amplification of the targeted locus provides the most rapid assessment of the outcome of a gene targeting experiment (Fig. 5.4). Typically, following DNA delivery to protoplasts, the cells are allowed to regenerate in the absence of selection for a period of 4–5 days, before transfer to selective medium for two weeks. This is sufficient to kill off the untransformed cells, and the small colonies that survive are then subcultured onto non-selective medium for a further two weeks, in order to allow the loss of unintegrated DNA from "unstable" transformants. Stably transformed plants are identified as those that continue to grow, two weeks following their return to selective conditions. They can then be permanently grown on non-selective medium (Knight *et al.* 2002). At this stage the colonies are still small, and a further two to four weeks' growth may be necessary until a colony has grown to a size sufficient for the isolation of DNA for PCR analysis. DNA from transformants is analyzed by PCR with a pair of gene-specific, inward-pointing primers that anneal to sequences that lie outside the sequence used in the targeting construct. When these are used with outward-pointing primers that anneal to the selection cassette, it is possible to identify amplification products that result from homologous recombination events that occur (i) only between the 5′-arm of the targeting construct and its target in the genome, (ii) only between the 3′-arm of the construct and its target in the genome, and (iii) between both the 5′- and 3′-arms of the targeting construct and their targets in the genome. If no amplification is seen, it is indicative that the targeting construct has integrated adventitiously, elsewhere in the genome. Only where amplification occurs for both ends of the targeting

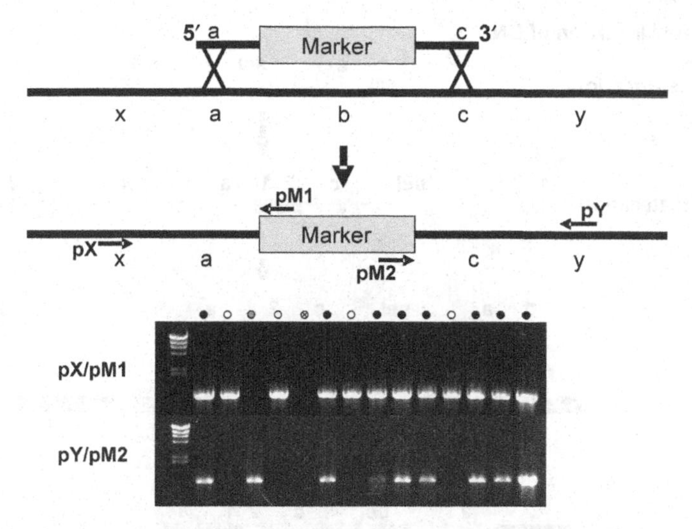

Fig. 5.4. Assaying gene targeting in *Physcomitrella*. A linear targeting construct containing two sequences ("a" and "c") corresponding to the 5′- and 3′-terminal sequences of the target gene "a-b-c". Sequence "b" is replaced in the construct by a selectable marker cassette. Homologous recombination at the 5′-end of the targeted gene is assayed by PCR by using the external gene-specific primer "pX" in combination with the selectable marker primer "pM1". Homologous recombination at the 3′-end of the targeted gene is assayed by PCR by using the external gene-specific primer "pY" in combination with the selectable marker primer "pM2". The results of such a targeting assay are shown in the agarose gel photograph. Tracks marked with open circles correspond to transgenic lines targeted at the 5′-end only. The track marked by a filled grey circle corresponds to a transgenic line targeted at the 3′-end only. Tracks marked by filled black circles indicate transgenic lines where targeted gene replacement has occurred. The track marked by a crossed circle derives from a transgenic line where integration of the transgene occurred elsewhere, in an untargeted region of the genome.

construct has an allele replacement event occurred. Such events are termed "targeted gene replacements" (TGRs) (Kamisugi *et al.* 2005, 2006). Where targeting has occurred in only one arm of the construct, this may cause a gene disruption (depending on the design of the construct), but not necessarily so. In these cases, "targeted insertion" (TI) has occurred (Fig. 5.5). This results from concatenation of the transforming DNA in the cell, prior to its integration into the targeted locus, and occurs as a consequence of homologous recombination between the targeted sequence in the genome and two repeated identical sequences in the concatemer (Kamisugi *et al.* 2006). As noted earlier, concatenation of transforming DNA is a common occurrence when large numbers of linear molecules are delivered to a protoplast, and TGR events include both single-copy replacement of the targeted locus and replacement by multiple,

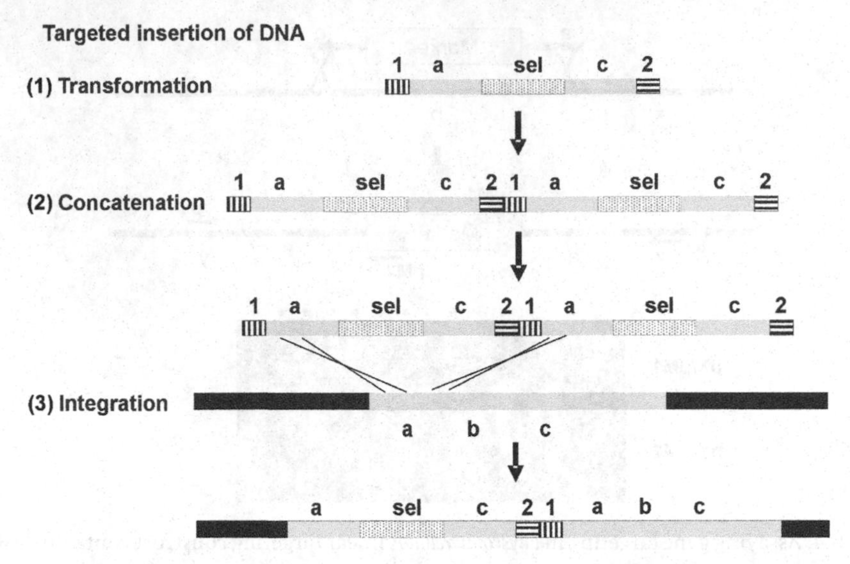

Fig. 5.5. Targeted insertion in *Physcomitrella*. Targeted insertion occurs by concatenation of (1) the transforming DNA. Here, (2) head-to-tail concatemers are integrated by (3) homologous recombination between the 5′-end of the targeted gene (sequence "a") by homologous recombination between this sequence and two repeated sequences "a" in the concatemer. This was demonstrated by sequencing the DNA integration junctions in transgenic lines resulting from transformation by using a targeting construct carrying non-homologous termini: sequences "1" and "2" in the targeting construct (Kamisugi *et al.* 2006).

concatenated copies of the targeting construct. Both will result in gene knockouts, but analysis of single-copy gene replacements is preferred. These can be identified by PCR amplification of the targeting cassette from the targeted locus by using the pair of external, gene-specific primers, and typically account for 25–50% of TGR transformants (Kamisugi *et al.* 2005).

5.8.4 *Confirmation that a targeted mutation causes a mutant phenotype*

It is clearly important to ensure that any mutant phenotype identified in the course of a gene targeting experiment results directly from the disruption of the targeted gene, and is not a consequence of some other event. The first priority is to ensure that a targeted gene replacement has not been accompanied by an additional, adventitious insertion of the targeting construct at another locus. This is most conveniently achieved by Southern blot analysis. Previous investigations of the incidence of adventitious incorporation of targeting fragments indicate that this occurs, but at a relatively low frequency: only *c.* 20% of transgenic plants exhibiting single-copy TGRs contain additional copies of the transgene inserted elsewhere in the genome (Kamisugi *et al.* 2005). Non-targeted insertion of DNA is a more serious problem where targeting fragments have been isolated from a plasmid vector by restriction enzyme digestion, and not subsequently separated from the vector backbone prior to moss transformation. Although not carrying selectable marker genes, adventitiously inserted plasmid DNA can often be detected in transgenic moss (Kamisugi *et al.* 2006) and consequently the use of PCR-amplified DNA is preferable for the generation of the transformation construct.

Even where a highly pure fragment has been delivered, and is shown to have inserted very precisely by single-copy TGR at the targeted locus alone, there is a possibility that a mutant phenotype might derive from a tissue-culture-induced artefact. Somaclonal variation is a well-attested phenomenon among plants derived from tissue culture, and likely arises through the stress-activation of retrotransposons (Soleimani *et al.* 2006). Since the *Physcomitrella* genome contains retrotransposon sequences, it would be surprising if some were not to be mobilized during protoplast regeneration, with unforeseeable consequences. Consequently, analysis of a mutant phenotype is better not to be restricted to a single individual mutant line. If many appropriately targeted independent transformants are identified in the course of a transformation experiment, then as large a number of them as is convenient should be phenotypically analyzed to ensure the association of the phenotype with the targeted gene. If this is not possible, then an alternative means of verification should be undertaken. This could include crossing the transgenic line with a wild-type strain, in order to determine whether the mutant phenotype and the transgene (the selectable marker) co-segregate. Alternatively, retransformation of the mutant with a wild-type gene should be undertaken to demonstrate that the wild-type phenotype can be restored by complementation.

5.8.5 *Analysis of a multigene family*

Many genes are members of paralogous multigene families, in which the individual gene family members may have redundant, overlapping, or partially redundant functions. In such cases, the targeted knockout of a single family member may not generate a mutant phenotype, or may have an effect whose impact is too subtle to be easily recognizable. In such cases, combinatorial mutagenesis may be required to reveal the relationships between the genes. Transformation with a construct corresponding to one member of the family appears not to target other members of the same family, even if they are only slightly divergent: such "homeologous targeting" is suppressed by the endogenous mismatch repair mechanism (Trouiller *et al.* 2006). In such cases, a series of targeting constructs, specific to each member of the gene family, will have to be prepared. These can be used either individually, to create a series of transgenic lines in which each gene family member is mutated following a single-copy TGR event, or combinatorially, in which a number of different gene-specific targeting constructs are simultaneously delivered in the same transformation (co-transformation: Hohe *et al.* 2004). In co-transformation, the incidence with which multiple transformed lines are obtained is probabilistically determined from the frequencies with which the individual genes are targeted. Both approaches have their merits and difficulties.

The first approach requires that combinations of mutants be assembled, either by crossing independent lines, or by retransformation with a second gene-specific targeting construct. In the former case, the process of establishing sexual crosses may be time-consuming (approximately 3 months are required to identify hybrid sporophytes): however, this method was used successfully to generate plants doubly mutant for the *PpRAD51* genes (B. Reiss, pers. comm.).

For retransformation, it is necessary either to use a second selectable marker, or to first remove the selectable marker used in the construction of the recipient strain. Marker removal can readily be achieved by using the "*Cre-lox*" system (Kuhn & Torres 2002). This is a technique of site-specific recombination derived from the bacteriophage P1 that enables the insertion and excision of the bacteriophage genome into and out of the genome of a bacterial host. The *Cre* gene is a site-specific recombinase (its name derives from "**c**yclization **re**combination") that recognizes a specific 34 bp sequence called *loxP* (for "**l**ocus **o**f **X**-over in **P1**"). Essentially, if a DNA sequence within a genome is flanked by two direct copies of the *loxP* sequence, then the action of the *Cre* recombinase can recombine these two sites, resulting in the deletion of any intervening sequence. This can be exploited to remove a selectable marker from within the moss genome, if the selectable marker in a targeting construct is flanked by *loxP* sites. By introducing a second plasmid carrying the *Cre* gene, under the control of a constitutively active promoter, the transient expression of the Cre recombinase (for example in an unstable transformant), will cause the deletion of the selection cassette (Trouiller *et al.* 2006).

In cotransformation experiments, a large number of transgenic lines may have to be screened in order to identify suitably targeted plants containing no adventitious transforming DNA.

A third approach to generating mutants defined by the inactivation of multiple related genes is to use an RNAi-interference (RNAi) approach. Sequence similarity between closely related genes is most strongly conserved within the mRNA coding sequences, whereas intron sequences are usually highly divergent. Multiple related mRNAs can be subjected to RNAi-mediated "knockdown", and the construction and use of such RNAi expression vectors has recently been demonstrated for *Physcomitrella* (Bezanilla *et al.* 2003, 2005). This approach has the additional advantage that it can be used to deplete cells of specific mRNAs through the expression of the RNAi construct under the control of an inducible promoter, thereby providing a means of interfering with the expression of genes whose permanent knockout might prove lethal.

5.8.6 *Is gene targeting generally applicable?*

Although proponents of *Physcomitrella* frequently refer to this organism's unique ability, among plants, to undertake homologous recombination-mediated gene targeting at high frequency, this is actually not a true statement. There is no reason not to expect that this property is shared, if not by all, then by a significant number of other moss species, and that the technique of gene targeting should therefore be applicable to a number of species that offer particular experimental advantages that *Physcomitrella* does not. Thus *Ceratodon purpureus* has been more widely used for the study of gravitropic responses, because it exhibits more vigorous growth in darkness than does *Physcomitrella*, whereas the desiccation-tolerant properties of *Tortula ruralis* commend it as a model for the study of anhydrobiosis.

Indeed, it has already been demonstrated that gene targeting can be undertaken in *C. purpureus* by the elegant targeted "knock-in" repair of a point mutation in the haem oxygenase gene, required for phytochrome synthesis and the phototropic response (Brücker *et al.* 2005). The particular disadvantage of other species, relative to *Physcomitrella*, is their comparative lack of genomic resources. With the availability of the *Physcomitrella* genome sequence, any gene can be readily amplified and mutagenic targeting constructs generated. However, it is possible to use the *Physcomitrella* genome sequence as a springboard for the isolation of the corresponding genes in other moss species. Homology-based searches of the existing sequence databases frequently demonstrate that *Physcomitrella* genes have significantly greater sequence homology with that small number of gene sequences that have been derived from other bryophytes, than with the very much greater number of angiosperm sequences. For example, BLAST searches with *Physcomitrella* sequences encoding LEA (late embryogenesis abundant) proteins implicated in desiccation tolerance frequently identify more similar sequences from the liverwort *Riccia fluitans* than from angiosperms, and comparisons between cDNAs from *Physcomitrella*, *Ceratodon* and *Tortula* exhibit nucleotide sequence identities of 80%–95% within the protein coding regions. This level of sequence identity is sufficiently high to suggest that *Physcomitrella* sequences can be used either as hybridization probes to select genes from DNA libraries of other moss species without difficulty (my laboratory has cloned a number of *Ceratodon* genes in this way), or for the design of degenerate PCR primers for the direct amplification of desired genes. For *Ceratodon purpureus*, both cDNA and genomic DNA libraries are available through the Leeds *Physcomitrella* EST Programme.

5.9 Mosses and the study of development

The beginning of the twentieth century marked the invention, in Britain, of a model construction kit called "Meccano". It gained massive popularity as a present for (almost exclusively) boys in order to inculcate them with a love for what was then the mainstay of the British economy: heavy engineering. The kit comprised a selection of strips, sheets, and brackets of perforated metal that could be fixed together with small nuts, bolts and washers. A handy 12-year-old could use a small selection of the same basic parts to construct all manner of structures: railway engines, cranes, model cars, bridges. Generations of middle-aged men (although not the author, who was distinctly unhandy) have fond recollections of childhood hours spent piecing together some miniature marvel of civil engineering.

"Meccano" provides an apt metaphor for the way in which evolution has used a limited number of molecular components in order to construct a diverse array of living structures. The diversity of life, and the different developmental strategies that are displayed by organisms from widely different taxa, are all based on their particular specialized use of a largely identical set of gene products. This is particularly true of those gene products that act as the central regulators of developmental processes: receptors, signal transduction components, and transcription factors. Understanding how such essentially similar components can be differently assembled to carry out very different functions is at the heart of comparative approaches to the study of developmental programming. "Evo-Devo" is the Meccano of biology.

This is exemplified by the evolution of transcriptional control networks crucial to plant development. Development in plants, as in all complex multicellular organisms, is under close genetic control. Thus a fertilized egg cell divides, proliferates, and undergoes cellular differentiation to generate a three-dimensional structure with an architecture that is characteristic of its particular species and that can be modified or disrupted by mutations in genes that regulate the process. However, at the same time, plant morphogenesis exhibits a high degree of plasticity, whereby the genetic programming of the formation of specific structures is responsive to external cues. Thus the timing of particular processes, for example flowering, may be determined by external stimuli such as light or temperature. Directional growth responds to light and gravity vectors, and to nutrient availability, and the ultimate size of the "adult" organs of a plant – for example the leaves – may depend on responses to external forces such as grazing. Such plasticity is a necessary adaptation for organisms with a sessile growth habit that are inescapably subject to the vagaries of the environment.

Moreover, unlike development in animals, where morphogenesis frequently entails gross changes in cellular organization brought about by the movement of cells relative to each other (gastrulation being a striking example), the architecture of plants is constrained by the specific properties of the plant cell: in particular its enclosure within a relatively rigid cell wall. Thus generation of a specific three-dimensional structure is entirely dependent on processes that control the orientation of the planes of cell division, the direction of cell expansion, and the extent of cell expansion. We can add to this complexity, different responses by plant cells to internal morphogenetic cues. Plant cells exhibit a high degree of totipotency, especially when compared with animal cells. Upon tissue culture, cells of many apparently terminally differentiated organs can become dedifferentiated and recapitulate an organogenic pathway via somatic embryogenesis in response to simple manipulation of the concentrations of morphogenetically active substances (plant hormones).

The experimental dissection of these processes can be challenging, particularly where they involve the concerted action of groups of cells that must become organized into a particular structure through intercellular communication. Thus, our current models for the formation and maintenance of complex structures such as the root and shoot meristems of flowering plants rely on the identification through mutagenesis of "master genes" that regulate developmental processes, and the identification and analysis of their "subject genes" by using promoter fusions with reporter genes such as the *E. coli* β-glucuronidase (GUS) gene, or vital reporters that permit gene expression analysis *in vivo*, in real time, such as the green fluorescent protein (GFP) and its variants. These powerful experimental tools for the study of plant developmental processes have all developed as a consequence of the intense focus on the model plant *Arabidopsis thaliana*. The availability of molecular genetic resources – in particular the ability to undertake mutagenic interrogation of the plant to identify genes responsible for specific mutant phenotypes by genetic linkage, coupled with the availability of a fully sequenced genome – has been the principal force in driving this understanding.

The deployment of such experimental approaches is now possible in mosses, using the resources that are accumulating for *Physcomitrella*. Moreover, there are certain advantages that *Physcomitrella* offers that have few parallels in *Arabidopsis*. Although the ability to conduct precise genetic manipulation by gene targeting offers one such advantage, it is arguably an even greater advantage that the architecture of *Physcomitrella* lends itself to the analysis of processes that take place at the level of the single cell in a way that is not possible in multicellular plant organs.

For much of its life, *Physcomitrella* can be regarded as an organism that is one cell thick. The protonemata comprise filaments made up of single cells joined end-to-end. These are readily observed by using high-resolution *in vivo* imaging, by laser-scanning confocal microscopy. The gametophores, although three-dimensional structures, derive from a primordium (the "bud") in which the meristematic architecture can be readily examined microscopically, and the leaves themselves are only one cell in thickness, with the possible exception of the differentiated midrib cells. This reinforces the facile comparison between *Physcomitrella* and *Caenorhabditis*, the latter having achieved prominence as a model organism in part through its particular amenability to microscopic visualization of every cell in its body.

5.10 The evolution of transcriptional networks

The "Meccano" analogy is most strikingly demonstrated in the way in which transcriptional networks have evolved in plants. Comparative analysis of different taxa – including mosses – demonstrates how different groups of transcription factors have selectively been used. One such example is discussed in Chapter 7 of this volume: the evolution of desiccation tolerance. If we examine the phenomenon of anhydrobiosis, we find that it is widely dispersed in nature, occurring in microorganisms (bacteria and yeasts) and animals (typically invertebrates: nematodes, rotifers, and tardigrades) as well as in plants. Within the plant kingdom we can recognize anhydrobiosis to be an ancient trait, widespread among the bryophytes, and doubtless an essential adaptive feature required for the conquest of the land by previously aquatic organisms (Oliver *et al.* 2000). In the tracheophytes, the property of desiccation tolerance has been lost in vegetative tissues, but retained – or rather, partitioned by an evolutionary process – in reproductive propagules (seeds and spores).

We are now able to identify the collection of genes whose expression is required for desiccation tolerance. This is a complex collection that includes a substantial number that encode the so-called "Late Embryogenesis Abundant" (LEA) proteins (Cuming 1999, Wise & Tunnacliffe 2004). These genes are present in both mosses and seed plants (and algae and animals), demonstrating their early evolutionary origin (Browne *et al.* 2002, Goyal *et al.* 2005). In plants, these genes are expressed by transcriptional activation, often mediated by abscisic acid, using a small number of transcription factors: the basic-domain leucine-zipper ("bZip") and Apetala 2 (Ap2) drought-responsive element binding (DREB) families that interact with specific *cis*-acting motifs within the promoters of the *LEA* genes, and under the control of a transcriptional activator encoded (in *Arabidopsis*) by the *ABI3* (ABA-insensitive 3) gene (Himmelbach *et al.* 2003). *LEA* genes in moss and flowering plants alike utilize the same transcriptional activation mechanisms (Knight *et al.* 1995, Kamisugi & Cuming 2005), and the transcription factors responsible for these processes in moss and flowering plants are largely interchangeable (Marella *et al.* 2006). However, in the angiosperms the transcriptional network dependent upon the *ABI3* transcriptional activator has become developmentally sequestered to later stages of seed development through the restriction of *ABI3* gene expression to this phase of development in the course of tracheophyte divergence.

This represents one method of evolutionary "capture" of a gene expression network. Other forces also act to recruit different sets of genes to the control of specific developmental activators. Thus, a transcription factor can acquire or lose "subject" genes in two ways: it can undergo modification of its DNA-binding domain, so that it recognizes a novel *cis*-acting promoter sequence, thereby potentially recruiting an entirely new collection of genes, or it can gradually acquire or lose individual subject genes through the occurrence of mutations in the respective *cis*-acting sequences of the subjects that cause them to change their "transcriptional allegiance".

Since modifications in the DNA-binding specificity of a transcription factor are likely to prove deleterious, more often than not, such modifications are most commonly found associated with gene duplication and the subsequent subfunctionalization of the duplicated genes. "Subfunctionalization" describes the evolution of gene families through gene duplication, followed by the accumulation of mutations resulting in the two copies sharing aspects of the original gene's function. Striking examples of this occur in the very well characterized MADS-box transcription factor family (Causier *et al.* 2005). The origins of this family are ancient, pre-dating the Cambrian explosion (Nam *et al.* 2003; De Bodt *et al.* 2003). In *Arabidopsis* there are over 100 MADS-box transcription factor genes (Parenicova *et al.* 2003) of which the best characterized are the so-called "MIKC" class that have been identified as regulating most aspects of floral morphogenesis, and whose complex interactions contribute to the great diversity of floral structures found in the angiosperms, and whose evolution must necessarily have underpinned the explosive speciation within this group. This subfamily of transcriptional regulators is exclusive to plants, but is still ancient in origin, representatives being found in mosses (Krogan & Ashton 2000, Henschel *et al.* 2002), ferns (Hasebe *et al.* 1998), and gymnosperms (Munster *et al.* 1997, Mouradov *et al.* 1998a, 1999, Winter *et al.* 1999, Becker *et al.* 2000). However, the amplification of this gene family, and its recruitment to the regulation of reproductive development, occurred relatively late in land plant evolution. The identification of MADS-box gene transcripts in the developing reproductive organs of *Pinus radiata* (Mouradov *et al.* 1998b) suggests that this capture occurred prior to the divergence of the gymnosperms, but after that of the ferns in the land plant lineage, since in this latter group MADS-box genes have been found to be expressed ubiquitously in both the gametophytic and sporophytic tisses (Münster *et al.* 1997). Expression of the floral homoeotic genes in angiosperms occurs only following the respecification of the vegetative meristem to an inflorescence meristem. This is mediated through the transcription factor encoded by the *FLO/LFY* gene – a homoeotic gene whose expression is essential for the transition from a vegetative shoot apical meristem to an inflorescence meristem. Whilst the *Pinus* homologue of this gene exhibits meristem-specific gene expression, and is capable of complementing *Arabidopsis lfy* mutants (Mouradov *et al.* 1998b), the *FLO/LFY* homologues of ferns are less closely associated with reproductive development. Expression does occur predominantly in the reproductive meristem, indicating that the developmental transition mediated by *FLO/LFY* had evolved at this relatively early stage in land plant evolution (Himi *et al.* 2001) but the expression of the fern MADS box genes is not closely correlated with that of *FLO/LFY* (Hasebe *et al.* 1998, Himi *et al.* 2001), implying that these genes had not yet been subordinated to *FLO/LFY* regulation. In *Physcomitrella*, there are two *FLO/LFY* paralogues, exhibiting a high degree of sequence identity. Analysis of the expression of each gene by reporter "knock-in", and by the generation of knock-out mutants, showed that the two genes have highly overlapping, largely redundant functions. This implies that the two copies result from a relatively recent gene duplication event. The *PpLFY-1* and *PpLFY-2* genes are expressed during sporophyte development, being required for the first division of the zygote (Tanahashi *et al.* 2005). This presumably represents an ancestral function from which a gene duplication in the ancestors of a subsequent lineage enabled the acquisition of a new role in the specification of reproductive development in the seed plants.

The use of powerful bioinformatic tools to undertake comparative genomic analyses of genes implicated in the regulation of development of *Arabidopsis*, coupled with the ease with which the functional analysis of their counterparts in *Physcomitrella* can be conducted, highlights the value of "molecular bryology" in gaining a fuller understanding of the evolution of plant developmental strategies.

References

Ashton, N. W. & Cove, D. J. (1977). The isolation and preliminary characterisation of auxotrophic and analogue resistant mutants of the moss *Physcomitrella patens*. *Molecular and General Genetics*, **154**, 87–95.

Ashton, N. W., Cove, D. J. & Featherstone, D. R. (1979a). The isolation and physiological analysis of mutants of the moss *Physcomitrella patens*, which over-produce gametophores. *Planta*, **144**, 437–42.

Ashton N. W., Grimsley, N. H. & Cove, D. J. (1979b). Analysis of gametophytic development in the moss *Physcomitrella patens* using auxin and cytokinin resistant mutants. *Planta*, **144**, 427–35.

Ashton, N. W., Champagne, C. E. M., Weiler, T. & Verkoczy, L. K. (2000). The bryophyte *Physcomitrella patens* replicates extrachromosomal transgenic elements. *New Phytologist*, **146**, 391–402.

Aylon, Y. & Kupiec, M. (2004). DSB repair: the yeast paradigm. *DNA Repair*, **3**, 797–815.

Becker, A., Winter, K. U., Meyer, B., Saedler, H. & Theissen, G. (2000). MADS-box gene diversity in seed plants 300 million years ago. *Molecular Biology and Evolution*, **17**, 1425–34.

Benson, F. E., Baumann, P. & West, S. C. (1998). Synergistic actions of Rad51 and Rad52 in recombination and DNA repair. *Nature*, **391**, 401–4.

Bezanilla, M., Pan, A. & Quatrano, R. S. (2003). RNA interference in the moss *Physcomitrella patens*. *Plant Physiology*, **133**, 470–4.

Bezanilla, M., Perroud, P.-F., Pan, A., Kluew, P. & Quatrano, R. S. (2005). An RNAi system in *Physcomitrella patens* with an internal marker for silencing allows rapid identification of loss of function phenotypes. *Plant Biology*, **7**, 251–7.

Browne, J., Tunnacliffe, A. & Burnell, A. (2002). Anhydrobiosis – plant desiccation gene found in a nematode. *Nature*, **416**, 38.

Brücker, G., Mittmann, F., Hartmann, E. & Lamparter, T. (2005). Targeted site-directed mutagenesis of a heme oxygenase locus by gene replacement in the moss *Ceratodon purpureus*. *Planta*, **220**, 864–74.

Causier, B., Castillo, R., Zhou, J. L. *et al.* (2005). Evolution in action: following function in duplicated floral homeotic genes. *Current Biology*, **15**, 1508–12.

Coupland, G., Baker, B., Schell, J. & Starlinger, P. (1988). Characterization of the maize transposable element *Ac* by internal deletions. *EMBO Journal*, **7**, 3653–9.

Courtice, G. R. M. & Cove, D. J. (1983). Mutants of the moss *Physcomitrella patens* which produce leaves of altered morphology. *Journal of Bryology*, **12**, 595–609.

Cove, D. J. (1984). The role of cytokinin and auxin in protonemal development in *Physcomitrella patens* and *Physcomitrium sphaericum*. *Journal of the Hattori Botanical Garden*, **55**, 79–86.

Cove, D. J. (2005). The moss *Physcomitrella patens*. *Annual Review of Genetics*, **39**, 339–58.

Cuming, A. C. (1999). LEA proteins. In *Seed Proteins*, ed. P. R. Shewry & R. Casey, pp. 753–80. Dordrecht: Kluwer Academic Publishers.

Daley, J. M., Palmbos, P. L., Wu, D. & Wilson, T. E. (2005). Nonhomologous end joining in yeast. *Annual Review of Genetics*, **39**, 431–51.

De Bodt, S., Raes, J., Van de Peer, Y. & Theissen, G. (2003). And then there were many: MADS goes genomic. *Trends in Plant Science*, **8**, 475–83.

Egener, T., Granado, J., Guitton, M.-C. *et al.* (2002). High frequency of phenotypic deviations in *Physcomitrella patens* plants transformed with a gene-disruption library. *BMC Plant Biology*, **2**, 6.

Engel, P. P. (1968). The induction of biochemical and morphological mutants in the moss *Physcomitrella patens*. *American Journal of Botany*, **55**, 438–46.

Goyal, K., Walton, L. J., Browne, J. A., Burnell, A. M. & Tunnacliffe, A. (2005). Molecular anhydrobiology: identifying molecules implicated in invertebrate anhydrobiosis. *Integrative and Comparative Biology*, **45**, 702–9.

Grimsley, N. H., Ashton, N. W. & Cove, D. J. (1977). The production of somatic hybrids by protoplast fusion in the moss, *Physcomitrella patens*. *Molecular and General Genetics*, **154**, 97–100.

Hanin, M., & Paszkowski, J. (2003). Plant genome modification by homologous recombination. *Current Opinion in Plant Biology*, **6**, 157–62.

Hanin, M., Volrath, S., Bogucki, A. *et al.* (2001). Gene targeting in *Arabidopsis*. *Plant Journal*, **28**, 671–7.

Hasebe, M., Wen, C. K., Kato, M. & Banks, J. A. (1998). Characterization of MADS homeotic genes in the fern *Ceratopteris richardii*. *Proceedings of the National Academy of Sciences, U.S.A.*, **95**, 6222–7.

Henschel, K., Kofuji, R., Hasebe, M. *et al.* (2002). Two ancient classes of MIKC-type MADS-box genes are present in the moss *Physcomitrella patens*. *Molecular Biology and Evolution*, **19**, 801–14.

Himi, S., Sano, R., Nishiyama, T. *et al.* (2001). Evolution of MADS-box gene induction by *FLO/LFY* genes. *Journal of Molecular Evolution*, **53**, 387–93.

Himmelbach, A., Yang, Y. & Grill, E. (2003). Relay and control of abscisic acid signalling. *Current Opinion in Plant Biology*, **6**, 470–9.

Hofmann, A., Codon, A., Ivascu, C. *et al.* (1999). A specific member of the *Cab* multigene family can be efficiently targeted and disrupted in the moss, *Physcomitrella patens*. *Molecular and General Genetics*, **261**, 92–9.

Hohe, A., Egener, T., Lucht, J. M. *et al.* (2004). An improved and highly standardised transformation procedure allows efficient production of single and multiple targeted gene-knockouts in a moss, *Physcomitrella patens*. *Current Genetics*, **44**, 339–47.

Hohe, A., Rensing, S. A., Mildner, M., Lang, D. & Reski, R. (2002). Day length and temperature strongly influence sexual reproduction and expression of a novel MADS-box gene in the moss *Physcomitrella patens*. *Plant Biology*, **4**, 595–602.

Jenkins, G. I., Courtice, G. R. M. & Cove, D. J. (1986). Gravitropic responses of wild-type and mutant strains of the moss *Physcomitrella patens*. *Plant, Cell and Environment*, **9**, 637–44.

Johri, M. M. & Desai, S. (1973). Auxin regulation of caulonema formation in moss protonemata. *Nature New Biology*, **245**, 223–4.

Kamisugi, Y. & Cuming, A. C. (2005). The evolution of the abscisic acid-response in land plants: comparative analysis of Group 1 LEA gene expression in moss and cereals. *Plant Molecular Biology*, **59**, 723–37.

Kamisugi, Y., Cuming, A. C. & Cove, D. J. (2005). Parameters determining the efficiency of gene targeting in the moss *Physcomitrella patens*. *Nucleic Acids Research*, **33**, e173.

Kamisugi, Y., Schlink, K., Rensing, S. A. *et al.* (2006). The mechanism of gene targeting in *Physcomitrella patens*: homologous recombination, concatenation and multiple integration. *Nucleic Acids Research*, **34**, 6205–14.

Kammerer, W. & Cove, D. J. (1996). Genetic analysis of the effects of re-transformation of transgenic lines of the moss *Physcomitrella patens*. *Molecular and General Genetics*, **250**, 380–2.

Kempin, S. A., Liljegren, S. J., Block, L. M. *et al.* (1997). Targeted disruption in *Arabidopsis*. *Nature*, **389**, 802–3.

Knight, C. D., Futers, T. S. & Cove, D. J. (1991). Genetic analysis of a mutant class of *Physcomitrella patens* in which the polarity of gravitropism is reversed. *Molecular and General Genetics*, **230**, 12–16.

Knight, C. D., Sehgal, A., Atwal, K. *et al.* (1995). Molecular responses to abscisic acid and stress are conserved between moss and cereals. *Plant Cell*, **7**, 499–506.

Knight, C. D., Cove, D. J., Cuming, A. C. & Quatrano, R. S. (2002). Moss gene technology. In *Molecular Plant Biology*, vol. 2, ed. P. M. Gilmartin & C. Bowler, pp. 285–99. Oxford: Oxford University Press.

Knoop, B. (1984). Development of bryophytes. In *The Experimental Biology of Bryophytes*, ed. A. F. Dyer & J. G. Duckett, pp. 143–76. London: Academic Press.

Krogan, N. T. & Ashton, N. W. (2000). Ancestry of plant MADS-box genes revealed by bryophyte (*Physcomitrella patens*) homologues. *New Phytologist*, **147**, 505–17.

Krogh, B. O. & Symington, L. (2004). Recombination proteins in yeast. *Annual Review of Genetics*, **38**, 233–71.

Kuhn, R. & Torres, R. M. (2002). Cre/loxP recombination system and gene targeting *Methods in Molecular Biology*, **180**, 175–204.

Lisby, M., Barlow, J. H., Burgess, R. C. & Rothstein, R. (2004). Choreography of the DNA damage response: spatiotemporal relationships among checkpoint and repair proteins. *Cell*, **118**, 699–713.

Lisby, M. & Rothstein, R. (2004). DNA damage checkpoint and repair centres. *Current Opinion in Cell Biology*, **16**, 328–34.

Long, D., Martin, M., Sundberg, E. *et al.* (1993). The maize transposable element system Ac/Ds as a mutagen in *Arabidopsis*: identification of an albino mutation induced by Ds insertion. *Proceedings of the National Academy of Sciences, U.S.A.*, **90**, 10370–4.

Lukowitz, W., Gillmor, C. S., & Scheible, W.-R. (2000). Positional cloning in *Arabidopsis*. Why it feels good to have a genome initiative working for you. *Plant Physiology*, **123**, 795–805.

Marella, H. H., Sakata, Y. & Quatrano, R. S. (2006). Characterization and functional analysis of *ABSCISIC ACID INSENSITIVE3*-like genes from *Physcomitrella patens*. *Plant Journal*, **46**, 1032–44.

Markmann-Mulisch, U., Hadi, M. Z., Koepchen, K. *et al.* (2002). The organization of *Physcomitrella patens RAD51* genes is unique among eukaryotic organisms. *Proceedings of the National Academy of Sciences, U.S.A.*, **99**, 2959–64.

McClintock, B. (1948). Mutable loci in maize. *Carnegie Institution of Washington Year Book*, **48**, 142–54.

Mouradov, A., Glassick, T. V., Hamdorf, B. A. *et al.* (1998a). Family of MADS-box genes expressed early in male and female reproductive structures of Monterey pine. *Plant Physiology*, **117**, 55–62.

Mouradov, A., Glassick, T., Hamdorf, B. *et al.* (1998b). *NEEDLY*, a *Pinus radiata* ortholog of *FLORICAULA/LEAFY* genes, expressed in both reproductive and vegetative meristems. *Proceedings of the National Academy of Sciences, U.S.A.*, **95**, 6537–42.

Mouradov, A., Hamdorf, B., Teasdale, R. D. *et al.* (1999). A *DEF/GLO*-like MADS-box gene from a gymnosperm: *Pinus radiata* contains an ortholog of angiosperm B class floral homeotic genes. *Developmental Genetics*, **25**, 245–52.

Münster, T., Pahnke, J., DiRosa, A. *et al.* (1997). Floral homeotic genes were recruited from homologous MADS-box genes preexisting in the common ancestor of ferns and seed plants. *Proceedings of the National Academy of Sciences, U.S.A.*, **94**, 2415–20.

Muskett, P., Clissold, L., Maroocco, A. *et al.* (2003). A resource of mapped dissociation launch pads for targeted insertional mutagenesis in the *Arabidopsis* genome. *Plant Physiology*, **132**, 506–16.

New, J. H., Sugiyama, T., Zaitseva, E. & Kowlaczykowski, S. C. (1998). Rad52 protein stimulates DNA strand exchange by Rad51 and Replication protein A. *Nature*, **391**, 407–10.

Nam, J., dePamphilis, C. W., Ma, H. & Nei, M. (2003). Antiquity and evolution of the MADS-box gene family controlling flower development in plants. *Molecular Biology and Evolution*, **20**, 1435–47.

Nishiyama, T., Fujita, T., Shin-I, T. *et al.* (2003). Comparative genomics of *Physcomitrella patens* gametophytic transcriptome and *Arabidopsis thaliana*: Implication for land plant evolution. *Proceedings of the National Academy of Sciences, U.S.A.*, **100**, 8007–12.

Nishiyama, T., Hiwatashi, Y., Sakakibara, K., Kato, M. & Hasebe, M. (2000). Tagged mutagenesis and gene-trap in the moss, *Physcomitrella patens*. *DNA Research*, **7**, 9–17.

Oliver, M. J., Tuba, Z. & Mishler, B. D. (2000). Evolution of desiccation tolerance in land plants. *Plant Ecology*, **151**, 85–100.

Orr-Weaver, T. L., Szostak, J. W. & Rothstein, R. J. (1981). Yeast transformation: a model system for the study of recombination. *Proceedings of the National Academy of Sciences, U.S.A.*, **78**, 6354–8.

Pan, X., Liu, H., Clarke, J. *et al.* (2003). ATIDB: *Arabidopsis thaliana* insertion database. *Nucleic Acids Research*, **31**, 1245–51.

Parenicova, L., de Folter, S., Kieffer, M. *et al.* (2003). Molecular and phylogenetic analyses of the complete MADS-box transcription factor family in *Arabidopsis*: new openings to the MADS world. *Plant Cell*, **15**, 1538–51.

Rensing, S. A., Rombauts, S., Van de Peer, Y. & Reski, R. (2002). Moss transcriptome and beyond. *Trends in Plant Science*, **7**, 535–8.

Reski, R. (1998). Development, genetics and molecular biology of mosses. *Botanica Acta*, **111**, 1–15.

Reski, R. & Abel, W. O. (1985). Induction of budding on chloronemata and caulonemata of the moss *Physcomitrella patens*, using isopentenyladenine. *Planta*, **165**, 354–8.

Reski, R., Faust, M., Wang, X. H., Wehe, M. & Abel, W. O. (1994). Genome analysis of the moss *Physcomitrella patens* (Hedw). BSG. *Molecular and General Genetics*, **244**, 352–9.

Sawahel, W., Onde, S., Knight, C. D. & Cove, D. J. (1992). Transfer of foreign DNA into *Physcomitrella patens* protonemal tissue by using the gene gun. *Plant Molecular Biology Reporter*, **10**, 315–16.

Schaefer, D. G. (2001). Gene targeting in *Physcomitrella patens*. *Current Opinion in Plant Biology*, **4**, 143–50.

Schaefer, D. G. (2002). A new moss genetics: targeted mutagenesis in *Physcomitrella patens*. *Annual Reviews of Plant Biology*, **53**, 477–501.

Schaefer, D. G. & Zrÿd, J. P. (1997). Efficient gene targeting in the moss *Physcomitrella patens*. *Plant Journal*, **11**, 1195–206.

Schaefer, D., Zrÿd, J. P., Knight, C. D. & Cove, D. J. (1991). Stable transformation of the moss *Physcomitrella patens*. *Molecular and General Genetics*, **226**, 418–24.

Schuermann, D., Molinier, J., Fritsch, O. & Hohn, B. (2005). The dual nature of homologous recombination in plants. *Trends in Genetics*, **21**, 172–81.

Schween, G., Egener, T., Fritzowsky, D. *et al.* (2005). Large-scale analysis of 73,329 *Physcomitrella* plants transformed with different gene disruption libraries: production parameters and mutant phenotypes. *Plant Biology*, **7**, 228–37.

Schween, G., Gorr, G., Hohe, A. & Reski, R. (2003). Unique tissue-specific cell cycle in *Physcomitrella*. *Plant Biology*, **5**, 1–9.

Shinohara, A. & Ogawa, T. (1998). Stimulation by Rad52 of yeast Rad51-mediated recombination. *Nature*, **391**, 404–7.

Soleimani, V. D., Baum, B. R. & Johnson, D. A. (2006). Quantification of the retrotransposon BARE-1 reveals the dynamic nature of the barley genome. *Genome*, **49**, 389–96.

Sonoda, E., Takata, M., Yamashita, Y. M., Morrison, C. & Takeda, S. (2001). Homologous DNA recombination in vertebrate cells. *Proceedings of the National Academy of Sciences, U.S.A.*, **98**, 8388–94.

Soriano, P. (1995). Gene targeting in ES cells. *Annual Review of Neuroscience* **8**, 1–18.

Stracker, T. H., Theunissen, J. W., Morales, M. & Petrini, J. H. J. (2004). The Mre11 complex and the metabolism of chromosome breaks: the importance of communicating and holding things together. *DNA Repair*, **3**, 845–54.

Strepp, R., Scholz, S., Kruse, S., Speth, V. & Reski, R. (1998). Plant nuclear gene knockout reveals a role in plastid division for the homolog of the bacterial cell division protein FtsZ, an ancestral tubulin. *Proceedings of the National Academy of Sciences, U.S.A.*, **95**, 4368–73.

Tanahashi, T., Sumikawa, N., Kato, M. & Hasebe, M. (2005). Diversification of gene function: homologs of the floral regulator *FLO/LFY* control the first zygotic cell division in the moss *Physcomitrella patens*. *Development*, **132**, 1727–36.

Terada, R., Urawa, H., Inagaki, Y., Tsugane, K. & Iida, S. (2002). Efficient gene targeting by homologous recombination in rice. *Nature Biotechnology*, **20**, 1030–4.

Trouiller, B., Schaefer, D. G., Charlot, F. & Nogue, F. (2006). MSH2 is essential for the preservation of genome integrity and prevents homeologous recombination in the moss *Physcomitrella patens*. *Nucleic Acids Research*, **34**, 232–42.

von Stackelberg, M., Rensing, S. A. & Reski, R. (2006). Identification of genic moss SSR markers and a comparative analysis of twenty-four algal and plant gene indices reveals species-specific rather than group-specific characteristics of microsatellites. *BMC Plant Biology*, **6**, 9.

Weterings, E. & van Gent, D. C. (2004). The mechanism of non-homologous end-joining: a synopsis of synapsis. *DNA Repair*, **3**, 1425–35.

Winstead, E. R. (2001). Sizing up genomes: amoeba is king! www.genomenewsnetwork.org/articles/02_01/Sizing_genomes.shtml.

Winter, K. U., Becker, A., Munster, T., Kim, J. T., Saedler, H & Theissen, G. (1999). MADS-box genes reveal that gnetophytes are more closely related to conifers than to flowering plants. *Proceedings of the National Academy of Sciences, U.S.A.*, **96**, 7342–7.

Wise, M. & Tunnacliffe, A. (2004). POPP the question: what do LEA proteins do? *Trends in Plant Science*, **9**, 13–17.

Wood, A. J., Oliver, M. J. & Cove, D. J. (eds). (2004). *New Frontiers in Bryology*. Dordrecht: Kluwer Academic Publishers.

6

Physiological ecology

MICHAEL C. F. PROCTOR

6.1 Introduction

Bryophytes are on average some two orders of magnitude smaller than vascular plants, and this difference of scale brings in its train major differences in physiology, just as many of the differences in the structural organization and physiology of insects and vertebrates are similarly scale-driven. Surface area varies as the square, and volume and mass as the cube, of linear dimensions. Hence gravity is a major limiting factor for vertebrates or trees, but trivial for insects or bryophytes. Bryophytes in general have much larger areas for evaporation in proportion to plant mass than do vascular plants. Surface tension, which operates at linear interfaces, is of little significance at the scale of the vascular plant shoot but is a powerful force at the scale of many bryophyte structures. There are also major scale-related differences in the relation of bryophytes and vascular plants to their atmospheric environment. Vascular-plant leaves are typically deployed in the turbulent air well above the ground. The diffusion resistance of the thin laminar boundary layer is small, so the epidermis with its cuticle and stomata in effect marks the boundary between (relatively slow) diffusive mass transfer within the leaf and (much faster) turbulent mixing in the surrounding air. By contrast the small leaves of many bryophytes lie largely or wholly within the laminar boundary layer of the bryophyte carpet or cushion, or of the substratum on which it grows. For these reasons it is important to approach bryophyte physiology from cell-biological and physical first principles; preconceptions and concepts carried over from vascular-plant physiology can be grossly misleading.

Raven (1977, 1984, 1995) has emphasized the importance of supracellular transport systems in the evolution of land plants, and the physiological

Bryophyte Biology: Second Edition, ed. B. Goffinet & A. J. Shaw. Published by Cambridge University Press.

correlates that we must read alongside the anatomical structures of fossil plants. But the highly differentiated supracellular conducting systems exemplified by xylem and phloem are really only a prerequisite for *large* land plants. In adapting to the erratic subaerial supply of water, vascular land plants evolved tracheids and vessels, bringing water from the soil to meet the needs of the above-ground shoots and leaves. Bryophytes in general adopted the alternative strategy of allowing free water loss (poikilohydry) and evolving desiccation tolerance, photosynthesizing and growing during moist periods and suspending metabolism during times of drought. These two patterns of adaptation are in many ways complementary. Bryophytes may appear to be limited by their lack of roots, but their poikilohydric habit means that they can colonize hard and impermeable surfaces such as tree trunks and rock outcrops, impenetrable to roots, from which vascular plants are excluded. Bryophytes typically take up water and nutrients over the whole surface of the shoots. They efficiently intercept and absorb solutes in rainwater, cloud and mist droplets, and airborne dust. This ability underlies both their conspicuous success in many nutrient-limited habitats and the vulnerability of many species to atmospheric pollution. The vascular-plant pattern of adaptation is undoubtedly optimal for a large land plant; there is much reason to believe that the poikilohydric pattern of adaptation is optimal for a small one. The divergence of bryophytes and the various vascular-plant groups goes back to the early history of plant life on land - certainly 400 million years, and probably longer (Edwards *et al.* 1998, Goffinet 2000). Mosses, Hepaticae, and Anthocerotae may well have been evolutionarily independent for equally long. Physiologically, bryophytes are neither simple nor primitive. They should be seen not as primitive precursors of vascular plants, but as the diverse and highly evolved representatives of an alternative strategy of adaptation, prominent in the vegetation of such habitats as subpolar and alpine fell-fields and tundra, bogs and fens, and the understorey of many forests from the boreal zone to the "mossy forests" of tropical mountains. They are challenged at their own scale only by the comparably adapted lichens.

The physiological ecology of bryophytes has been the subject of a number of reviews (Longton 1981, 1988, Proctor 1981a, 1982, 1990). Poikilohydry as an adaptive strategy has been discussed by Kappen & Valladares (1999) and Proctor & Tuba (2002). Mineral nutrition and pollution responses are reviewed by Brown (1982, 1984), Brown & Bates (1990), Bates & Farmer (1992) and Bates in Chapter 8 of this volume. Bryophyte production, and its responses to major environmental factors, has been reviewed by Russell (1990), Frahm (1990), Vitt (1990) and Sveinbjörnsson & Oechel (1992). The present chapter does not cover aspects (such as temperature relations of photosynthesis) that are essentially similar in all green plants, but concentrates on some ecophysiological features more particularly characteristic of bryophytes.

6.2 Water relations

Vascular plants have internal water conduction - they are endohydric - and the surface of the leaves and young stems is typically covered by an epidermis with a more or less waterproof and water-repellent cuticle, gas exchange taking place through stomata. Most bryophytes are ectohydric, free liquid water moving predominantly in capillary spaces outside the plant. In some large mosses, exemplified by the tall, robust *Dawsonia* and *Polytrichum* species and the large Mniaceae, the stems possess a well-developed central strand of water-conducting hydroids, and a substantial proportion of water conduction is internal. However, in these more or less endohydric mosses significant conduction generally takes place externally in the capillary spaces of sheathing leaf bases or rhizoid tomentum, and they have little or no control over water loss, so like other bryophytes they are poikilohydric. In all bryophytes, as in vascular plant tissues at a comparable scale, much internal water movement must be relatively diffuse, within the cell walls, through the cells themselves, or some combination of the two. Most water movement must be of this kind in the large marchantialean liverworts, and many small acrocarpous mosses must rely on a (probably always variable) balance between external and internal conduction. Bryophytes are likely to be scarcely less complex in respect of tissue water movement than vascular plants (Proctor 1979a, Steudle & Petersen 1998).

Thus, typically in bryophytes conduction of water is predominantly external, in an interconnecting network of capillary spaces on the outside surface of the plant. These include the spaces within sheathing leaf bases, in the concavities of overlapping imbricate leaves as in *Scleropodium* or *Pilotrichella*, within the felts of rhizoids or paraphyllia that cover the stem in such genera as *Philonotis* and *Thuidium*, in the interstices between the papillae that cover the leaf surfaces in, for example, *Encalypta*, *Syntrichia* and *Anomodon*, and between tightly packed shoots or between shoots and the substratum. The external water of ectohydric bryophytes is as much a part of the plant's physiological functioning as the water in the xylem of vascular plants.

The cell water relations of bryophytes are essentially the same as those of other plant cells and are illustrated by the "Höfler diagram" of Fig. 6.1(a). In a fully turgid cell the osmotic potential Ψ_π is exactly balanced by the turgor pressure Ψ_P of the cell wall; the cell is externally in equilibrium with pure liquid water, and its water potential Ψ (or Ψ_W) is zero (by definition). If the external

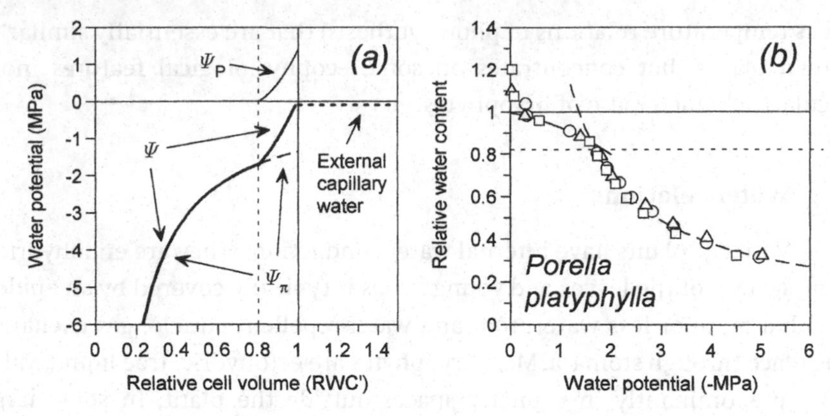

Fig. 6.1. (a) Höfler diagram for a bryophyte illustrating the relationship of cell water potential (Ψ) and its components osmotic potential (Ψ_π) and turgor pressure (Ψ_P) to relative cell volume and external capillary water. Based on the data of Fig. 6.1b. (b) The relation of relative water content to water potential for the leafy liverwort *Porella platyphylla*, from thermocouple–psychrometer measurements. Water content was originally plotted as per cent dry mass, and the full-turgor point estimated from the graph, as described by Proctor *et al.* (1998). The horizontal dotted line indicates the turgor-loss point. A rectangular hyperbola has been fitted to the data points below this, and a polynomial regression to the points between full turgor and turgor loss. This graph is in effect a Höfler diagram with water potential taken as the *x*-axis, and matches the presentation used by Proctor *et al.* (1998) and Proctor (1999). Compare Fig. 6.1a and the "pressure–volume" curve of Fig. 6.2a.

water potential becomes negative, the cell must lose water. The reduction in cell volume causes turgor pressure to fall and osmotic potential to become more negative (numerically greater). When the turgor pressure falls to zero, the water potential of the cell is equal to the osmotic potential of its contents. At any lower water content, osmotic potential and cell water potential are equal, and inversely proportional to the volume of water in the cell. The relation between osmotic potential and cell volume plots onto the Höfler diagram as a rectangular hyperbola. The relation of cell water potential to cell water content follows this hyperbola up to the turgor-loss point. It then breaks away to follow a line, generally slightly concave to the water-potential axis, to the full-turgor point, where the relative water content (RWC) = 1.0 (by definition), and $\Psi = 0$. Practical measurements are generally of tissues rather than individual cells, but if the cells all have similar properties the same principles apply. Bryophyte shoots generally carry some external water, held at small negative water potentials determined by the dimensions of the capillary spaces in which it lies. The effect of this water in a Höfler diagram is illustrated by the dotted line in Fig. 6.1a.

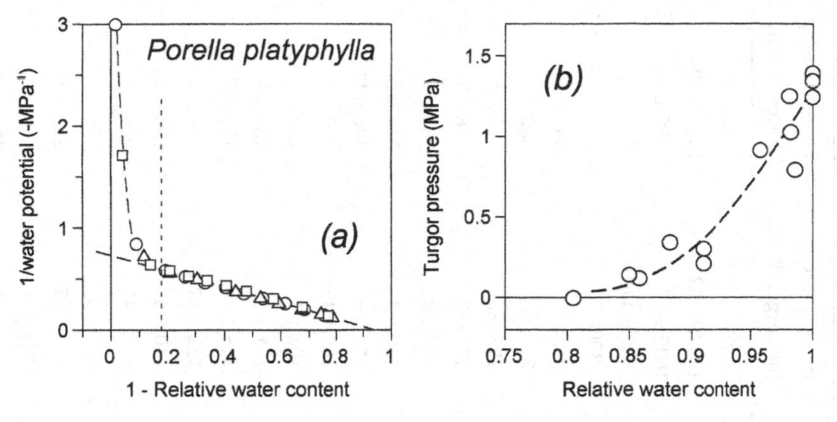

Fig. 6.2. (a) Pressure-volume graph from the same data as Fig. 6.1b. Water content is plotted as 1 – RWC and decreases from left to right; the *y*-axis is the reciprocal of water potential. Turgor loss is indicated by the vertical dotted line. A linear regression has been fitted to the points to the right of this. It intersects the *y*-axis at the reciprocal of the full-turgor osmotic potential, the turgor-loss line at the reciprocal of the osmotic potential at turgor loss, and the *x*-axis at a point which gives a measure of the effective osmotic volume of the cells. (b) The relation of turgor pressure to relative water content for *Porella platyphylla*, from thermocouple-psychrometer measurements. The curve leaves the *x*-axis at the turgor-loss point and cuts the *y*-axis at the full-turgor osmotic potential. The slope of the curve gives a measure of the bulk modulus of elasticity (ε_B) of the tissues.

If one of the axes of the graph relating water potential to water content is plotted on a reciprocal scale, the hyperbola of Fig. 6.1 becomes a straight line. The graph of $1/\Psi$ against (1 – RWC) (Fig. 6.2a) is referred to as a pressure–volume (*P–V*) curve (Jones 1992). Turgor loss is marked by the point at which the relation of $1/\Psi$ to (1 – RWC) breaks away from linearity, and the reciprocal of the osmotic potential at this point can be read from the graph. The intercept of the straight line on the $1/\Psi$ axis gives the reciprocal of the osmotic potential at full turgor. The intercept on the RWC axis is commonly taken as a measure of non-osmotic (or "apoplast") water but its exact significance is debatable (Proctor *et al.* 1998). From the data in the *P–V* curve, turgor pressure can be calculated for water contents between full turgor and the turgor-loss point (Fig. 6.2b). The steepness of slope of this curve (and the difference in water content between full turgor and turgor loss) depends on cell-wall extensibility, measured by the bulk elastic modulus, ε_B, which varies continuously between turgor loss and full turgor in a manner depending on the exact physical properties of the cell walls.

Some representative water-relations data for bryophytes are summarized in Table 6.1. Osmotic potentials at full turgor mostly lie between –1.0 and –2.0 MPa,

Table 6.1 *Water-relations parameters of bryophytes*

Figures are in general mean ± s.d. from three or four replicates. The sign * indicates a single value from the combined data of all replicates; ‡, that the values from individual replicates were not distinguishable.

Species	Osmotic potential at full turgor (−MPa)	*x*-intercept of *P*–*V* curve (RWC)	RWC at turgor loss	Bulk elastic modulus ε_B at RWC 1.0 (MPa)	Water content at full turgor (% d.m.)	Water content blotted (% d.m.)
Targionia hypophylla	0.74 ± 0.03	−0.069 ± 0.003	0.70	n.d.	1003 ± 45	940 ± 37
Conocephalum conicum	0.54 ± 0.08	−0.002 ± 0.032	0.45	2.2 ± 0.8	1400 ± 132	1277 ± 108
Marchantia polymorpha	0.38 ± 0.02	0.052 ± 0.027	0.60	1.5*	1025 ± 35	956 ± 65
Dumortiera hirsuta	0.49 ± 0.05	−0.014 ± 0.023	0.90	7.6 ± 1.2	1636 ± 118	1628 ± 109
Metzgeria furcata	1.11 ± 0.03	0.043 ± 0.017	0.75	11.3 ± 0.7	300‡	363 ± 22
Pellia epiphylla	0.72 ± 0.08	−0.031 ± 0.032	0.80	4.8*	1020‡	1046 ± 157
Bazzania trilobata	1.41 ± 0.07	0.081 ± 0.025	0.80	17.3 ± 3.5	253 ± 6	300 ± 11
Porella platyphylla	1.37 ± 0.03	0.053 ± 0.013	0.80	13.3 ± 1.2	273 ± 5	312 ± 8
Frullania tamarisci	1.78 ± 0.20	0.189 ± 0.017	0.60	7.6 ± 0.5	134 ± 3	216 ± 7
Jubula hutchinsiae	1.02 ± 0.04	0.097 ± 0.010	0.70	6.3 ± 2.8	353 ± 21	353 ± 17
Andreaea alpina	1.59 ± 0.03	0.265 ± 0.006	0.70	6.8 ± 0.4	110 ± 4	141 ± 9
Polytrichum commune	2.09 ± 0.09	0.116 ± 0.023	0.75	19.2 ± 0.4	179 ± 6	186 ± 11
Dicranum majus	1.27 ± 0.04	0.126 ± 0.025	0.80	12.2 ± 1.2	185 ± 15	193 ± 7
Tortula ruralis	1.36 ± 0.18	0.266 ± 0.093	0.75	5.8 ± 1.5	108 ± 11	n.d.
Racomitrium lanuginosum	1.29 ± 0.08	0.224 ± 0.030	0.65	5.3 ± 1.9	121 ± 4	135 ± 5
Mnium hornum	1.21 ± 0.07	0.099 ± 0.049	0.70	6.1 ± 1.7	215 ± 7	175 ± 6
Antitrichia curtipendula	1.47 ± 0.28	0.175 ± 0.033	0.65	5.9 ± 0.6	152 ± 11	174 ± 16
Neckera crispa	1.27 ± 0.09	0.271 ± 0.092	0.65	7.7 ± 1.6	140 ± 5	150 ± 13
Hookeria lucens	0.95 ± 0.03	0.021 ± 0.004	0.70	6.2 ± 1.5	571 ± 42	n.d.
Anomodon viticulosus	1.65 ± 0.07	0.230 ± 0.009	0.65	8.5 ± 2.3	133 ± 3	176 ± 10
Homalothecium lutescens	2.08 ± 0.08	0.086 ± 0.054	0.70	18.8 ± 2.9	193 ± 15	218 ± 27
Rhytidiadelphus triquetrus	1.44±0.16	0.136±0.028	0.75	9.6 ± 1.3	182‡	n.d.
Rhytidiadelphus loreus	1.34 ± 0.02	0.237 ± 0.049	0.70	5.9 ± 1.2	142 ± 10	180 ± 13

Sources: Data from Proctor *et al.* (1998) and Proctor (1999).

but are generally less negative (numerically around half these values) in thalloid liverworts. *Metzgeria furcata*, matching leafy liverworts and mosses in its unistratose thallus and tolerance of drying, is an interesting exception. The moss *Hookeria lucens* and the leafy liverwort *Jubula hutchinsiae*, both species of wet shady habitats, have notably low osmotic potentials, around −1.0. However, there is no clear indication that species of dry habitats have osmotic potentials markedly more negative than the norm; many of the more extreme older published figures based on plasmolysis are certainly wrong. The intercept of the *P–V* curve on the water-content axis correlates with cell-wall thickness relative to the cell lumen; it is high in such species as *Andreaea alpina*, *Racomitrium lanuginosum*, and *Neckera crispa*, and low in, for example, *Hookeria lucens* and the big thalloid liverworts. Water content at full turgor as a percentage of dry mass is also related to the proportion of cell-wall material, and varies widely from about 100% dry mass in small desiccation-tolerant species of sun-baked rocks to 2000% or more in thalloid liverworts of wet habitats. Both these measures change as the shoots develop and mature, and are sensitive to the inclusion of moribund older material, so they vary with the seasons and can never be very precise. Relative water content at turgor loss and ε_B are also correlated, but somewhat loosely. By vascular-plant standards, bryophyte cell walls are typically rather readily extensible (low ε_B), but some mosses (e.g. *Polytrichum commune*, *Dicranum majus*, *Homalothecium lutescens*) and leafy liverworts (e.g. *Bazzania trilobata*, *Porella platyphylla*) show ε_B values that would pass unnoticed among those of herbaceous vascular plants (Zimmerman & Steudle 1978). Cell-wall extensibility also varies with time, ε_B increasing as the shoots mature.

The division between apoplast water in the cell walls, symplast water within the cells, and external capillary water (and especially the latter two) is important for several reasons (Dilks & Proctor 1979, Beckett 1996, Proctor *et al.* 1998). First (for the physiological investigator) it is *essential* to know the full-turgor water content in order to calculate RWC values physiologically comparable with those for vascular plants. "RWC" values based on "saturated" water contents are wholly misleading, and it is much less easy to obtain an accurate estimate of the full-turgor water content of a bryophyte than of a vascular plant leaf. As Table 6.1 shows, acceptable approximate estimates of full-turgor water content can often be obtained by carefully blotting samples of saturated shoots; underestimates can arise through thumb pressure expressing symplast water from large-celled species, and overestimates through incomplete removal of external water from species with intricate external capillary spaces, or the presence of large amounts of apoplast water. When compared in terms of *true* RWC (i.e. cell water content relative to cell water content at full turgor), photosynthesis in bryophytes of widely differing adaptive types, and vascular-plant cells, responds similarly to water deficit (Fig. 6.3).

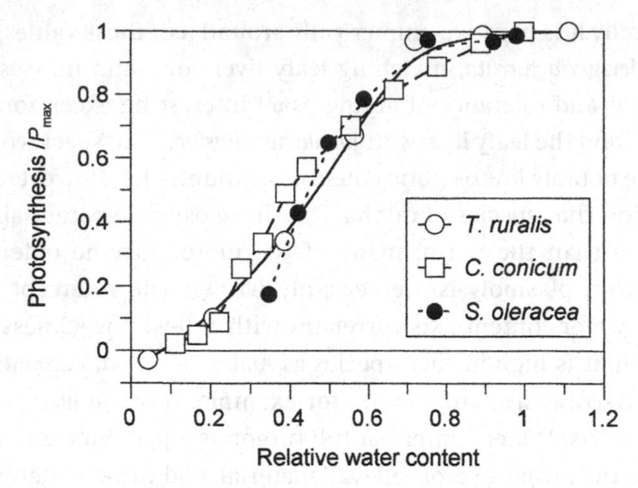

Fig. 6.3. Response of net photosynthesis to water deficit in two contrasting bryophytes, from gas-exchange measurements. The data for the desiccation-tolerant moss *Tortula (Syntrichia) ruralis* are recalculated from Tuba *et al.* (1996), taking as full turgor a value of 165% dry mass estimated from measurements at their field site in July 1998, and assuming 10% of the full-turgor water content to be apoplast water. The data for the thalloid liverwort *Conocephalum conicum* are recalculated from Slavik (1965), assuming that full-turgor water content coincides with the maximum value for net photosynthesis (900% dry mass). Measurements for spinach (*Spinacia oleracea*), a mesophytic vascular plant, are included for comparison (Kaiser 1987).

Second (for the bryophyte), the external capillary water is exceedingly important physiologically. Its significance in relation to external water movement has already been alluded to. External water is also of prime importance in relation to water storage, which in turn is a major determinant of the length of time the shoots remain turgid and able to photosynthesize and grow. It is often the largest component of water associated with the plant, and it can vary widely without affecting the water status of the cells. It is common to find that external capillary water exceeds symplast water by a factor of five or more; a not especially wet-looking sample of the pendulous African forest moss *Pilotrichella ampullacea* that I took to make measurements for a *P–V* curve turned out to have a total water content corresponding to a RWC of more than 12! Most of this water would have been held in the concavities of the overlapping "ampulla-like" leaves. The effect of external storage of large amounts of water is that for most of the time the shoots are either functioning at full turgor, or they are too dry to support metabolism, with only brief interludes at water potentials between these states. From the bryophyte's point of view, *any* habitat is "wet" during and following rain, and "dry" at other times. The primary difference is in the relative times spent wet and dry; drought stress and drought tolerance as they affect vascular plants hardly enter the picture, and the drought metabolites of vascular plants such as proline and glycine-betaine are conspicuously absent from bryophytes. We should remember that desert ephemeral vascular plants are mesophytes, which flourish following occasional periods of rain and escape drought by means of their desiccation-tolerant seeds. Bryophytes escape drought by means of their desiccation-tolerant vegetative shoots. Desiccation-tolerant bryophytes and vascular desert ephemerals may equally be seen as "drought-escaping" plants. It is a paradox that "poikilohydric" bryophytes may spend less time metabolizing at sub-optimal water content than many "homoiohydric" vascular plants! (Proctor 2000).

6.3 Bryophyte shoots as photosynthetic systems

It is easy to show by experiment in the laboratory that the rate of water loss from a vascular-plant shoot is largely determined by stomatal aperture. However, this leaves out of consideration two important factors in the field situation, one general and one particularly applicable to bryophytes. First, the latent heat of evaporation must come from the surroundings: by convective heat exchange with the air, by conduction from the substrate, or by radiative exchange with the wider environment. In a laboratory experiment with an isolated plant the amount of heat involved is small and easily left out of consideration. In the vegetation cover of a landscape it becomes a major factor in determining water loss (Jarvis & McNaughton 1986). Second, boundary-layer conditions for bryophytes are often largely determined by the extensive substrata on which they grow. Further, many bryophytes grow in the shelter of trees or smaller vascular plants which reduce the ambient windspeed to varying degrees. Thus, various environmentally determined parameters are major controls on water loss from bryophytes, and laboratory experiments on isolated bryophyte shoots or cushions that do not take this into account may have little relevance to what goes on in the field.

The small leaves of many bryophytes lie largely or wholly within the laminar atmospheric boundary layer of the bryophyte carpet or cushion, or of the substratum on which it grows. This is the layer in which the streamlines of the airflow are essentially parallel to the surface, so that transfer of heat and gases through it must take place by (slow) molecular diffusion by contrast with the much more rapid turbulent mixing in the surrounding air. The thickness of the laminar boundary layer is in the region of a few hundred micrometers at a windspeed of 1 m s^{-1}; it varies inversely as the square root of the windspeed up to the point at which the leaf or moss colony begins to generate turbulence itself. Wind-tunnel measurements (Proctor 1981b) show that at very low windspeeds a moss cushion behaves as a smooth simple object; water loss increases approximately as the square root of the windspeed, reflecting the corresponding decrease in boundary-layer thickness. Hair-points on the leaves (e.g. in *Grimmia pulvinata*) can have the effect of separating the sites of momentum and water-vapor transfer, in effect trapping an additional thickness of stagnant air between the moist leaf surfaces and the airstream, reducing the rate of water loss. (Hair-points can have other effects, increasing albedo for one.) Beyond a certain point, evaporation rises more rapidly with windspeed; the "rougher" the cushion surface (in terms of its interaction with the airstream), the lower the windspeed at which this occurs. At low windspeeds, the bryophyte colony functions, in effect, as a single "leaf", and gas exchange in the spaces between the individual leaves proceeds mainly by the comparatively slow process of molecular diffusion. Increasing evaporation at higher windspeeds reflects both the increasing tendency of the moss surface to generate turbulence in the airstream, and the fractally increasing area of the evaporating area of the cushion as measured by a boundary layer of progressively decreasing thickness. Moss or leafy-liverwort canopies operate at a scale intermediate between vascular-plant leafy canopies on the one hand, and the cells of a vascular-plant mesophyll on the other, and analogies may be sought in both directions. Bryophytes show high leaf-area index values (LAI: area of leaves divided by area occupied by the plant). A few estimates of my own gave figures of *c.* 6 in *Syntrichia intermedia*, 18 in *Mnium hornum*, and 20–25 in *Scleropodium purum* (Proctor 1979a), in the same range as the few other (unpublished) figures I have encountered. They are nearer the range of vascular-plant ratios of mesophyll area to leaf area (*c.* 14–40; Nobel 1977) than to LAIs for vascular plant canopies, which are usually less than 10 and commonly around 5. The growth forms and "life forms" of bryophytes vary greatly and in a manner certainly related to ecophysiological adaptation and microclimatic conditions in their habitats (Gimingham & Birse 1957, Mägdefrau 1982, Proctor & Smith 1995, Bates 1998, Rice *et al.* 2001).

The diffusive path for water loss is from the leaf surface to the atmosphere; that for CO_2 uptake is from the atmosphere to the chloroplasts. Therefore, CO_2 uptake encounters additional liquid-phase diffusive resistance in the cell walls and cytoplasm. As molecular diffusion is slower in water than in air by a factor of about 10^4, this additional resistance is large, even if the liquid diffusion path is only a few micrometers, and underlies the selection pressure for evolution of high LAI values in bryophytes and high mesophyll/leaf-area (A_{mes}/A) ratios in vascular-plant leaves. In addition to these diffusive resistances, the photosynthetic system of the chloroplasts may be regarded as imposing a "carboxylation resistance" to CO_2 uptake. An indication of the relative importance of these two limitations is given by the overall discrimination of photosynthesis against the heavy isotope of carbon, ^{13}C, conventionally expressed in (‰) relative to an arbitrary standard as $\delta^{13}C$ (Raven *et al.* 1987, Farquhar *et al.* 1989). The generally similar values for bryophytes (averaging around –27‰) and C3 vascular plants (Rundel *et al.* 1979, Teeri 1981) suggests that the relative magnitude of diffusion and biochemical limitations on CO_2 uptake is similar in the two groups, probably reflecting convergence on an adaptive optimum in the deployment of Rubisco relative to supporting tissues (Raven 1984, appendix 3). Substantially more negative $\delta^{13}C$ values are seen in aquatic bryophytes utilizing a proportion of respired CO_2 (e.g. *Fontinalis antipyretica* [Rundel *et al.* 1979, Raven *et al.* 1987], *Sphagnum cuspidatum* [Proctor *et al.* 1992, Price *et al.* 1997]). Less negative $\delta^{13}C$ values can be the consequence of high diffusive limitation by superincumbent water (Rice & Giles 1996, Williams & Flanagan 1996, Price *et al.* 1997, Rice 2000). Anthocerotae such as *Anthoceros* and *Phaeoceros* show consistently low discrimination against $^{13}CO_2$, giving $\delta^{13}C$ values of –15 to –20‰, because uniquely among bryophytes they have a carbon-concentrating mechanism associated with the pyrenoid (Smith & Griffiths 1996a,b, Hanson *et al.* 2002). C4 vascular plants typically have $\delta^{13}C$ values around –10 to –12‰.

Morphological adaptation in bryophytes must reconcile the potentially conflicting requirements of water conduction and storage, and free gas exchange for photosynthesis. This is achieved in various ways. Many, and probably most, bryophyte leaf surfaces carry at least a thin layer of water-repellent cuticular material, and some bear conspicuous granular or crystalline epicuticular wax (Proctor 1979b). This is most striking in some glaucous-looking species, often of moist places or shady crevices, in which water conduction must be largely internal, such as *Pohlia cruda*, *P. wahlenbergii*, *Saelania glaucescens*, many Bartramiaceae, and leafy liverworts such as *Douinia ovata* and *Gymnomitrion obtusum*. Many mosses (and some leafy liverworts) have shoot systems with closely overlapping concave leaves, the inner faces functioning for water storage, and the outer surfaces, kept free of superincumbent water by surface tension, serving for gas exchange. Striking instances of shoots of this kind are seen in, for example, *Anomobryum filiforme*, *Scleropodium* spp., *Myurium hochstetteri*, *Pleurozium schreberi*, *Pilotrichella* spp., *Weymouthia* spp., and *Nowellia curvifolia*, but there are many less extreme variations on the same theme. Densely papilla-covered or mammillate leaf surfaces are also common, and in many cases these too appear to provide a division between water conduction and gas exchange, the papilla (or mammilla) apices remaining dry while the interstices between them provide a continuous network of water-conducting channels (Buch 1945, 1947, Proctor 1979a).

A simple calculation shows that, assuming reasonable values for the liquid-phase diffusion resistance to CO_2 uptake, the rate of carbon fixation of a simple unistratose bryophyte leaf (two surfaces) would become limited by CO_2 diffusion at an irradiance of about 500 μmol m^{-2} s^{-1}, or about a quarter of full sunlight (Proctor 2005). This assumes that both leaf surfaces are completely clear for gas exchange. In reality, most bryophyte shoots or canopies consist of overlapping leaves or thallus lobes, increasing the area available for carbon fixation without greatly increasing the gas-phase diffusion path (most resistance is in the liquid phase within the leaves). On the other hand, CO_2 uptake will seldom take place over the whole leaf surface. Concave leaves holding water on the inner surface are one-sided for gas exchange, and superincumbent water will reduce gas exchange of many bryophyte shoots. In a sample of 39 mosses and 16 liverworts, chlorophyll-fluorescence estimates of 95%-saturating irradiance ranged widely, but most were <1000 μmol m^{-2} s^{-1}; the median for mosses was a little under 600 and for liverworts just over 200 μmol m^{-2} s^{-1} (Marschall & Proctor 2004). The mosses showed a bimodality between forest species peaking at 200–300 μmol m^{-2} s^{-1} and species of more open situations peaking between 500 and 700 μmol m^{-2} s^{-1}. Values higher than this came from either species of dry, very sun-exposed habitats (*Andreaea rothii, Grimmia pulvinata, Racomitrium lanuginosum, Syntrichia intermedia, S. ruralis*), or from open and sunny but constantly moist bogs and fens (*Aulacomnium palustre, Philonotis calcarea, Splachnum ampullaceum, Scorpidium scorpioides*). Photosynthesis–irradiance curves from infrared gas analysis of CO_2 uptake give 95% saturation irradiances broadly in the same range as the chlorophyll-fluorescence data (given in parentheses). For four species, Marschall & Proctor (2004) found 551 (711) μmol m^{-2} s^{-1} for *Andreaea rothii*, 583 (617) μmol m^{-2} s^{-1} for *Racomitrium aquaticum*, 832 (935) μmol m^{-2} s^{-1} for *Syntrichia ruralis*, and 228 (327) μmol m^{-2} s^{-1} for *Marchantia polymorpha*. In general, chlorophyll fluorescence would be expected to give rather higher figures than CO_2 uptake because it measures electron flow to photorespiration as well as to carbon fixation.

The leaves of Polytrichales and thalli of Marchantiales have complex ventilated photosynthetic tissues paralleling leaves of vascular plants. In the past these have tended to be seen in terms of restriction of water loss, but we should see them rather as an adaptation increasing the area for CO_2 uptake when the plant *is* adequately supplied with water. In Polytrichales, with their regular longitudinal lamellae, it is relatively easy to estimate the ratio between area for CO_2 uptake and projected leaf area, analogous to A_{mes}/A for a vascular-plant leaf. There is a very clear correlation between this value and the 95%-saturation irradiance (Fig. 6.4) (Proctor 2005). It is less easy to construct a similar graph for the Marchantiales because their photosynthetic

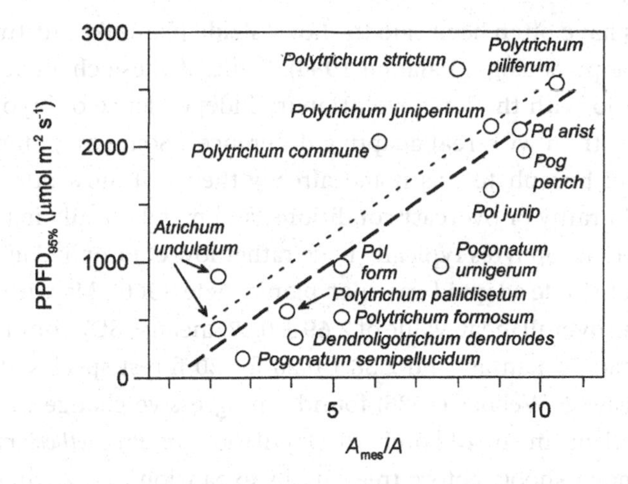

Fig. 6.4. Relation of irradiance at 95% saturation ($PPFD_{95\%}$) to the ratio of area available for CO_2 uptake to projected leaf area (A_{mes}/A) for some Polytrichaceae. The lower left-hand corner of the diagram is occupied by species of shady forest, including the common European *Atrichum undulatum* with a broad unistratose lamina and only few lamellae, *Pogonatum semipellucidum* from shady Andean cloud forest with a broader band of low lamellae in the middle of the lamina, and the New Zealand temperate-rainforest species *Dendroligotrichum dendroides* with a *Polytrichum*-like leaf but only low lamellae. The upper right-hand part of the diagram includes species of sun-exposed habitats, here four *Polytrichum* species from open moorlands in Britain, and *Pogonatum perichaetiale*, *Polytrichadelphus aristatus*, and *Polytrichum juniperinum* from the high *páramo* of the Venezuelan Andes. All these species have well-developed lamella systems. The bold pecked line is a linear regression on the data; the lighter dotted line is the theoretical limit assuming that the maximum CO_2 uptake through a single plane surface is 250 μmol m^{-2} s^{-1}. Higher values may arise from measurements on overlapping leaves, lower values from limiting factors other than CO_2 diffusion.

tissues are more irregular, but a similar relation clearly holds. Both groups include species of open, sun-exposed habitats (e.g. the moss *Polytrichum piliferum* and the liverwort *Exormotheca pustulosa*) with saturating irradiances far outside the range of most other bryophytes. Conspicuous surface wax is notably a feature of the margins of the lamellae of the leaves of Polytrichaceae (Clayton-Greene *et al.* 1985, Proctor 1992) where it serves the function of preventing the entry of water into the interlamellar spaces. The water-repellent pore margins of Marchantiales similarly prevent flooding of the ventilated photosynthetic tissues of the thallus, in the same way that the sharp water-repellent edges of the stomata prevent waterlogging of the mesophyll in vascular plants (Schönherr & Ziegler 1975).

Bryophytes have often been said to show "shade plant-like" features in their photosynthetic physiology (Valanne 1984). Some of these characteristics may have more to do with the long evolutionary independence of bryophytes and vascular plants than with real adaptive differences. Some may stem from the poikilohydry of bryophytes; as noted already they will most often be photosynthesizing in rainy or overcast conditions, and metabolically inactive in dry sunny weather. Bryophytes typically have rather low chlorophyll *a*/*b* quotients, in the range of shade-adapted vascular plants (Egle 1960). Martin & Churchill (1982) found an overall mean value of 2.69 ± 0.27 (mean $\pm$ SD) from 14 species of exposed habitats in Kansas, and 2.38 ± 0.20 for 20 forest species after canopy closure. Kershaw & Webber (1986) found a progressive change in the chlorophyll *a*/*b* quotient in an old-orchard population of *Brachythecium rutabulum* from 2.9 in young shoots before tree-canopy expansion to *c.* 2.0 in deep shade in autumn. In 39 mosses from diverse habitats mostly in southwest England the overall mean chlorophyll *a*/*b* was 2.39 ± 0.51; for 16 liverworts the corresponding figures were 1.98 ± 0.30 (Marschall & Proctor 2004). In general, sun plants tend to have lower chlorophyll/dry mass quotients, higher chlorophyll *a*/*b* quotients and higher carotenoid/chlorophyll quotients than shade plants. Correspondingly, in Marschall and Proctor's bryophyte data there were significant correlations between 95%-saturation irradiance of photosynthesis (from chlorophyll fluorescence measurements) and the quotients of chlorophyll/dry mass (negative), chlorophyll *a*/*b* (positive), and total carotenoids/total chlorophylls (positive).

Chlorophyll fluorescence (Krause & Weis 1991, Schreiber *et al.* 1995, Maxwell & Johnson 2000) is such a valuable tool in bryophyte ecophysiology that it is important to appreciate both its potentialities and its limitations. It can provide, non-invasively, much useful information about the state of the photosynthetic system in a green plant. Moreover, fluorescence measurements can be made on small amounts of material, and this makes it particularly valuable for working on small bryophytes. The energy absorbed by chlorophyll may suffer one of three fates: it may drive photochemistry, it may be re-emitted as red fluorescence, or it may be dissipated as heat. A green plant tissue will always emit a basal level of fluorescence (F_0). After a period of adaptation in the dark, the energy in a short saturating flash, too short for significant photochemistry to occur, is emitted entirely as fluorescence (F_m). At a constant level of actinic light, the steady-state fluorescence (F_s or F_t) will typically be greater than F_0 but less than F_m, and both the basal fluorescence (F_0') and the fluorescence given by a saturating flash (F_m') will be different from those measured in dark-adapted material. Measurement of these quantities is automated in modulated fluorometers, which calculate various parameters from them (Fig. 6.5). Dark-adapted

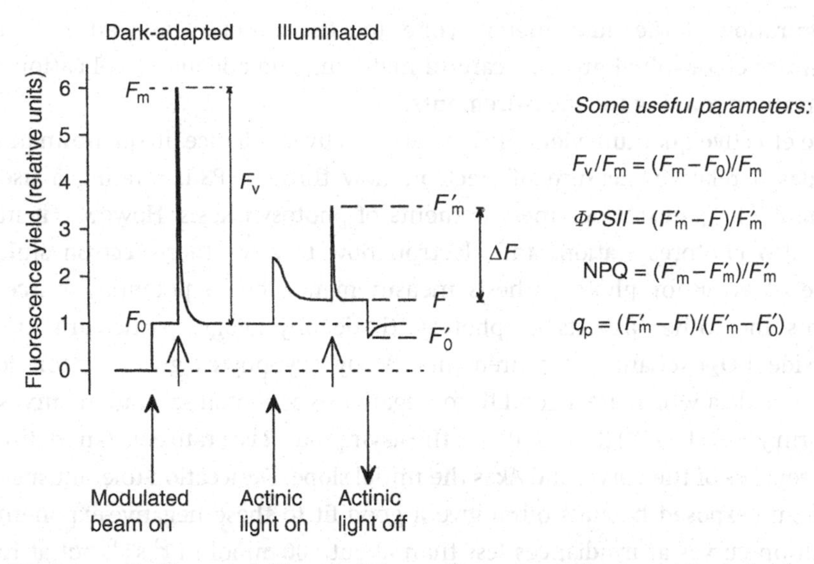

Fig. 6.5. Schematic diagram of chlorophyll-fluorescence output. Starting with a dark-adapted plant, the low-intensity modulated beam stimulates a basal level of fluorescence, F_0. A brief flash (*c.* 1 s) of saturating intensity produces a maximal peak of fluorescence (F_m), which then takes a few minutes to return to F_0. Turning on a light sufficient to drive photosynthesis (actinic light) gives an initial peak from which fluorescence settles gradually to a more or less steady level (F or F_t); a saturating flash at this point gives a peak (F'_m) lower than F_m. If the actinic light is now turned off (best in presence of far-red light) fluorescence drops to a level (F'_0) typically below the initial F_0. This gives five measurements, two initial (F_0 and F_m) and three measured in the course of the treatments of the experiment (F, F'_m and F'_0). The four most generally useful parameters calculated from them are given on the right of the diagram.

F_v/F_m is a measure of maximum quantum yield of the material (*c.* 0.75–0.85). The ratio $(F'_m - F_s)/F'_m$, often called ϕPSII, is a measure of the effective quantum yield under actinic light. The quotient $(F'_m - F_s)/(F'_m - F'_0)$, often called q_P, is a measure of the oxidation state of Q_A, the first electron acceptor of Photosystem II; it is often used in the inverse form $1 - q_P$. Non-photochemical quenching (NPQ), $(F_m - F'_m)/F'_m$, is largely a measure of the harmless dissipation of excess excitation energy as heat. Because these quantities are quotients, they are independent of the absolute value of fluorescence omitted, and hence of the quantity of material used. This is an advantage in that it allows measurements on small amounts of material, but it can also be an insidious source of error in experiments extending over a period of time, especially when they include a stress event that may lead to death of a proportion of cells in the material. It then becomes important to be sure that the absolute value of the fluorescence has remained sufficiently constant throughout the experiment. As this is sensitive to the optical configuration of the fluorometer probe and the material, as well as to the amount of chlorophyll present, careful matching and adequate replication are important in making the measurements.

The effective quantum yield ϕPSII multiplied by irradiance (in quantum units) provides a relative measure of electron flow through PSII, offering a useful alternative to gas-exchange measurements of photosynthesis. However, it measures also photorespiration, and electron flow to any other electron sink: a source of error for photosynthesis measurements, but a potential source of insights into other aspects of photosynthetic physiology, particularly taken alongside CO_2-exchange measurements. Many bryophytes give electron-flow/irradiance data which are a good fit to negative-exponential saturation curves of the form $y = A\,(1 - e^{-kx})$ (Fig. 6.6a): A is the asymptote, k is a rate constant defining the steepness of the curve, and Ak is the initial slope. Desiccation-tolerant species of dry sun-exposed habitats often give a good fit to these negative-exponential saturation curves at irradiances less than about 400 mmol $m^{-2}\ s^{-1}$, but at high irradiances fail to saturate, electron flow continuing to increase more or less linearly with irradiance (Fig. 6.6b) (Marschall & Proctor 2004). This reflects electron flow to O_2, not attributable to photorespiration but possibly due to the Mehler reaction (Asada 1999). The non-saturating electron flow at high irradiance is generally suppressed at high CO_2 concentrations, and it is not seen in Polytrichales (Proctor 2005), and rarely in Marchantiales.

Bryophytes generally give (by vascular-plant standards) high values of NPQ; the levels in desiccation-tolerant mosses of dry sun-exposed sites may reach 10–15 or more at high irradiances. These high NPQ levels relax almost completely within a few minutes in the dark (Marschall & Proctor 1999, M. C. F. Proctor, unpublished data), and are largely suppressed by the violaxanthin de-epoxidase inhibitor dithiothreitol (DTT). This suggests that they are linked to xanthophyll cycle-mediated photoprotection, dissipating harmlessly as heat excess excitation energy, which could otherwise lead to production of damaging reactive oxygen species (Björkman & Demmig-Adams 1995, Horton *et al.* 1996, Gilmore 1997, Deltoro *et al.* 1998, Smirnoff 2005, Logan 2005, Heber *et al.* 2006).

6.4 Desiccation tolerance

Desiccation tolerance is a common and characteristic but not universal feature of bryophytes (Proctor 1981a, Bewley & Krochko 1982, Oliver & Bewley 1997, Alpert & Oliver 2002, Proctor & Pence 2002, Wood 2007, Proctor *et al.* 2007b). Some species of constantly moist or shady habitats are very sensitive to drying out, and there is every gradation between these and species of sunbaked bare soil or rock surfaces, which not only survive but flourish in habitats where

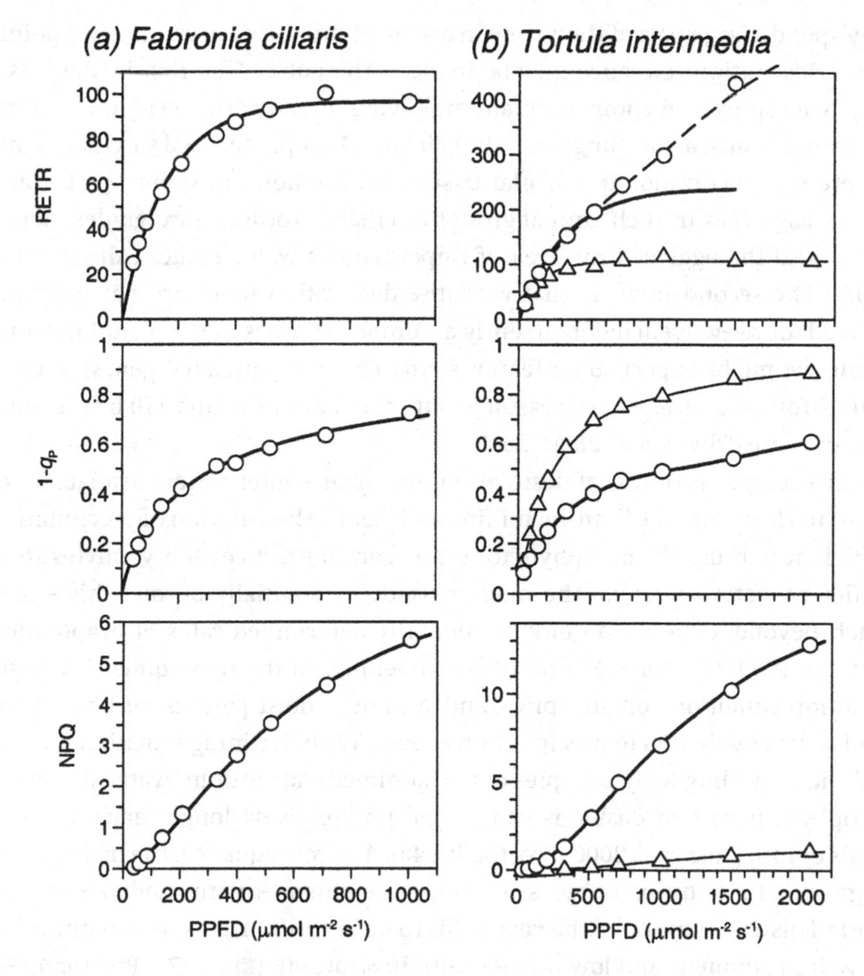

Fig. 6.6. Chlorophyll-fluorescence data from *Fabronia ciliata* (epiphyte on *Juniperus ashei*, Longhorn Cavern State Park, Texas, U.S.A.) and *Tortula* (*Syntrichia*) *intermedia* (exposed limestone rocks, Chudleigh, southwest England, U.K.). Relative electron transport rate (RETR) is often a useful surrogate for gas-exchange measurements of photosynthesis, although it includes flow to other electon sinks, e.g. photorespiration. *Fabronia pusilla* gives a good saturation curve for RETR, reaching 95% of saturation at *c.* 475 μmol m^{-2} s^{-1}(about a quarter of full sunlight). The curve of 1 – q_P does not suggest any downstream limitation on electron flow. NPQ, rising to *c.* 5.5, is high by vascular-plant standards. The curves for *Tortula intermedia* are more complex. RETR gives a good fit to a saturation curve below *c.* 400, with calculated $PPFD_{95\%} \approx 940$ μmol m^{-2} s^{-1}. However, above this point RETR continues to rise more or less linearly, implying electron flow to some as yet unknown sink. The curve for 1 – q_P suggests that *T. intermedia* can handle the electron flow generated even in full sunlight. Non-photochemical quenching (NPQ) rises to very high levels, implying a correspondingly high level of photoprotection. Most of the NPQ is suppressed by the violaxanthin de-epoxidase inhibitor dithiothreitol (DTT). Its effect is seen also in a marked raising of 1 – q_P, reduction of RETR, and complete suppression of the non-saturating electron flow at high irradiance. For further explanation see text.

they spend a large part of their time in a state of intense desiccation. Two points about desiccation tolerance may be made at the outset. The first is that it is a very widespread phenomenon among living organisms, occurring among microorganisms, algae, fungi including lichens, bryophytes, and vascular plants (where it is uncommon in vascular tissues but the norm in spores, pollen and seeds), as well as in such animal groups as ciliates, rotifers, tardigrades, nematodes, and the eggs of Crustacea of impermanent water bodies (Alpert 2000, 2005). The second point is that *vegetative* desiccation tolerance has certainly evolved (or re-evolved) independently a number of times in the plant kingdom. While we might expect some features (and desiccation-related genes) in common throughout, there is no reason to suppose that the details will be the same in every case (Oliver *et al.* 2000, 2005).

The ecological context of desiccation tolerance is intermittent availability of water to the plant. The limiting minimum is set by the duration of precipitation sufficient to bring the bryophyte to full turgor. In practice all bryophytes store sufficient water to extend the moist periods substantially beyond this - how much beyond, depends on environmentally determined rates of evaporation (Proctor 1990, Proctor & Smith 1995). Generally, in the open under the high-radiation conditions of late spring and summer, moist periods for bryophytes tend to be closely tied to precipitation events. With declining radiation income and the prevailing leafy canopies of late summer and autumn, water storage in bryophyte mats and cushions can bridge progressively longer gaps between spells of rain (Zotz *et al.* 2000, Proctor 2004a). As a consequence, the main period of growth in many bryophytes in temperate climates with moderately even rainfall distribution round the year tends to be in autumn, lack of water limiting growth in summer, and low temperature in winter (Pitkin 1975, Proctor 1990, Zotz & Rottenberger 2001). Clearly this will work out differently for different species, and in different habitats and climates. Heavy dewfall may provide sufficient water for significant early-morning photosynthesis by such species as the steppe-grassland and sand-dune moss *Tortula (Syntrichia) ruralis* (Csintalan *et al.* 1999), and the pendulous species (and other canopy epiphytes) of tropical cloud forests and temperate rainforests (e.g. species of *Meteorium*, *Pilotrichella*, *Phyllogonium*, *Weymouthia*) depend on storage of water from frequent rainfall or interception of cloudwater droplets (Proctor 2002, 2004b, León-Vargas *et al.* 2006). Even in these moist forests dry periods of a few days or longer are not uncommon, so a degree of desiccation tolerance is important for the majority of the epiphytic bryophytes. Over much of temperate northern and western Europe, or the northeastern United States, dry periods typically range from an hour or two to a few weeks. At a site in southwest England in 1989, the longest dry period recorded was 15 days (in spring); dry periods of 11 and 10 days were

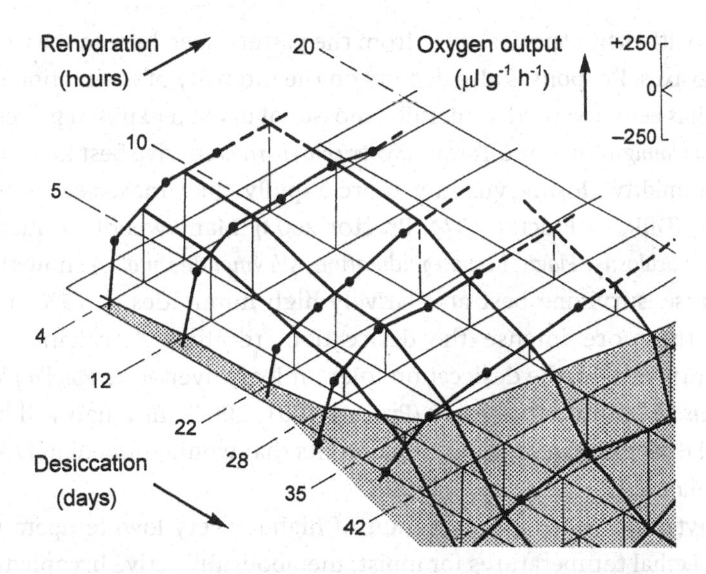

Fig. 6.7. Desiccation tolerance: the relation of net photosynthesis (measured as oxygen evolution) to desiccation time (at 50% relative humidity, *c.* 20 °C), and time after subsequent remoistening, in the moss *Anomodon viticulosus*. Shading shows the area over which net assimilation is negative. Redrawn, somewhat simplified, from the manometric data of Hinshiri & Proctor (1971). See text for further explanation.

recorded later in the same year (Proctor 2004a). Much longer dry periods occur in highly seasonal climates. In the Mojave Desert, where there is significant winter bryophyte growth, more than 100 days of continuous desiccation occurred on five occasions during 2001–2004, reaching 191 days in the drought year of 2002 (Stark 2005). In general, growth of bryophytes requires at least reasonably regular seasonal rainfall. Deserts with only irregular rainfall do not support bryophytes, even though lichen growth may survive on dewfall (Negev Desert, Israel) or fog deposition (coastal Namibia and northern Chile).

Ecophysiological responses of bryophytes to desiccation are complex. Some basic general features are illustrated in Fig. 6.7, showing the results of a series of experiments with *Anomodon viticulosus*. After short periods of desiccation (up to about 2 weeks in this case), the moss recovers rapidly and completely on remoistening and normal rates of photosynthesis are re-established within a few hours. With longer desiccation, the recovery process becomes progressively more prolonged, and final recovery less complete. In the example here, recovery is markedly slowed, but still substantially complete within 10 h after 22 days' desiccation. After 35 days' desiccation recovery on remoistening is very slow, and net photosynthesis has reached less than half its predesiccation value after 20 h remoistening. Prolonged desiccation (beyond *c.* 40 days in this instance) leads to prolonged net carbon loss, with only limited ultimate recovery.

Results with other species differ from the pattern seen here in the time scales on the two axes. Responses also depend on the intensity of desiccation to which the plant has been exposed. Generally, mosses of dry sun-exposed places such as *Racomitrium lanuginosum* and *Tortula (Syntrichia) ruralis* survive best kept at 20–50% relative humidity, losing viability more rapidly (like most seeds) at higher humidities (Dilks & Proctor 1974, Proctor 2003). Many woodland mosses (e.g. *Anomodon viticulosus*, *Plagiothecium undulatum*, *Weymouthia mollis*) show the opposite response, surviving best at relatively high humidities (*c.* 75%) and worse damaged the more intense the desiccation, recalling "recalcitrant" seeds. Perhaps surprisingly, the desiccation-tolerant leafy liverworts *Porella platyphylla* and *P. obtusata* fall into this group (Proctor 2001, 2003), although will both will withstand drying for weeks at water contents that would immediately kill most vascular plants.

Bryophytes are much more tolerant of high (or very low) temperatures dry than wet. Lethal temperatures for moist, metabolically active bryophytes are in the same range as for C3 vascular plants, about 40–50 °C (Larcher 1995, Liu *et al.* 2004). Survival of desiccation is related to temperature. The desiccation-tolerant species *Anomodon viticulosus*, *Tortula (Syntrichia) intermedia*, and *Frullania tamarisci*, gave a good straight-line "Arrhenius plot", with the logarithm of survival time linearly proportional to the reciprocal of the absolute temperature (Hearnshaw & Proctor 1982), an expected result if loss of viability follows essentially chemical degradation. In *Racomitrium lanuginosum* and *R. aquaticum*, the relation was curvilinear; it is possible that different systems are critical at different temperatures. In either case, survival times are measured in minutes at 100 °C and in months or years at 0 °C.

There clearly must be a lower limit to the time for which a bryophyte can usefully be moist. "Moist-break" experiments showed that *Hylocomium splendens* desiccated for 11 days at 32% relative humidity recovered to substantially its predesiccation photosynthetic activity after 24 h, but not after 6 h moist (Dilks & Proctor 1976). Tuba *et al.* (1996) found that *Tortula (Syntrichia) ruralis*, rehydrated after 24 h dry, attained positive net photosynthesis within 30 minutes, and recovered its predesiccation carbon balance within an hour or less. Gas-exchange measurements on *Grimmia pulvinata*, *Andreaea rothii*, and *Polytrichum formosum* have yielded similar figures (Proctor & Pence 2002, Proctor *et al.* 2007a). However, full recovery to predesiccation rates of photosynthesis may take 24 hours or more, while chloroplasts and other organelles and subcellular structures regain their normal conformation and spatial relationships (Proctor *et al.* 2007a). Studies of desiccation tolerance have most often used some aspect of photosynthesis as an index of recovery, but other systems are important too. Respiration begins immediately on rehydration. Protein synthesis is quickly re-established (Gwózdz *et al.* 1974, Oliver 1991), and is clearly a necessary part of full recovery. The photosystems become active within seconds or a minute or two of remoistening, but the enzyme systems of the "dark reactions" of photosynthesis recover more slowly. Recovery of the cell cycle seems to be much slower. Mansour & Hallet (1981) studied the effect of 24 h dehydration on the cell cycle in *Polytrichum formosum*. The first 2 h of water stress provoked completion of mitosis. On rehydration, 24 h passed before the first cell began to divide. DNA synthesis resumed in the nuclei of a few cells after 4 h, but cell-cycle activity rose steeply only after 24 h rehydration. Pressel *et al.* (2006) found a similar time scale for full re-establishment of the microtubular cytoskeleton on rehydration in food-conduction cells of *Polytrichum formosum*. There are indications that the reversible depolymerization and reassembly of the microtubular cytoskeleton may be critical in limiting the rate of drying that can be tolerated, and in full recovery of viability and cell function (Pressel 2006). Thus it is possible to envisage different levels of "recovery". Short periods at full turgor may suffice for maintenance of a positive carbon balance, yet be inadequate for cell division, or significant translocation or growth. Very little attention has been given to the desiccation tolerance of gametangia or the sporophyte generation, although sporophytes of many species must be exposed to desiccation in the early stages of their development following spring or early summer fertilization, as commonly occurs. Those that take more than a year to come to maturity (e.g. *Ulota* spp., *Pylaisia polyantha* (Lackner 1939)) face desiccation in a second summer later in their development. In *Tortula inermis* the evidence indicates either that sporophytes are more sensitive to rapid drying than the gametophytes, or that abortion of sporophytes arises from a gametophyte response to desiccation stress (Stark *et al.* 2007).

Recovery seems to be largely a matter of physical re-assembly of components conserved intact through the drying/rewetting cycle. In a number of mosses that have been studied, the recovery process itself (as measured by either chlorophyll fluorescence or gas exchange) is virtually unaffected by protein-synthesis inhibitors (Proctor & Smirnoff 2000, Proctor 2001, Proctor *et al.* 2007a). Chlorophyll-fluorescence data from *Racomitrium lanuginosum* are shown in Fig. 6.8, using F_v/F_m as an index of recovery. In the dark, there is no significant difference in the F_v/F_m values given by material recovering in water, in 3 mmol l^{-1} chloramphenicol (inhibiting chloroplast-encoded protein synthesis), in 0.3 mmol l^{-1} cycloheximide (inhibiting nuclear-encoded protein synthesis) or in 3 mmol l^{-1} dithiothreitol (which inhibits de-epoxidation of the xanthophyll violaxanthin to zeaxanthin, suppressing photoprotection). In the light, all treatments give quite good initial recovery, but after 18 h the curves have diverged widely. Cycloheximide has only a modest effect, but chloramphenicol leads to a strong

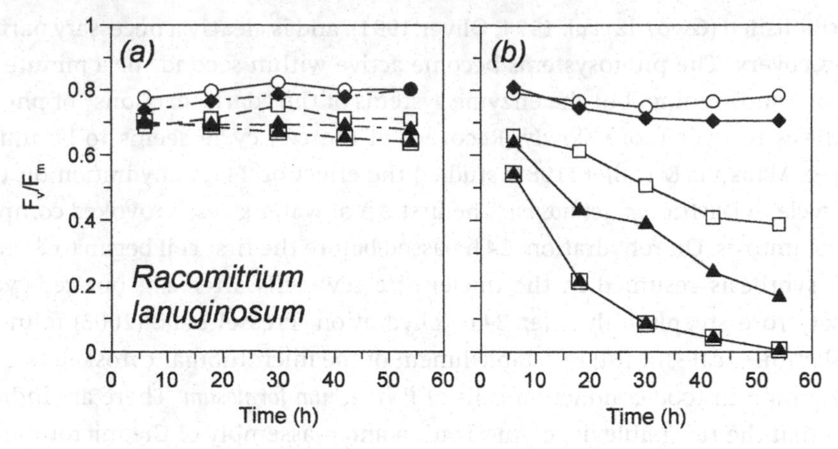

Fig. 6.8. An experiment showing the effect of metabolic inhibitors on recovery of F_v/F_m in the moss *Racomitrium lanuginosum* after *c.* 10 days' desiccation. Open circles control (water); open squares, 3 mmol l^{-1} dithiothreitol; solid triangles, 3 mmol l^{-1} chloramphenicol; solid diamonds, 0.3 mmol l^{-1} cycloheximide; superimposed squares and triangles, dithiothreitol + chloramphenicol. (a) In dark; the cycloheximide treatment is scarcely distinguishable from the control, and dithiothreitol and chloramphenicol have only modest effects alone or in combination. (b) In light (*c.* 125 μmol m^{-2} s^{-1}), cycloheximide (inhibiting nuclear-encoded protein synthesis) produces a slow but progressive fall in F_v/F_m. With dithiothreitol, F_v/F_m drops rather quickly to values around 0.4, probably reflecting suppression of the photoprotection associated with NPQ. Chloramphenicol produces a rapid and progressive drop in F_v/F_m, especially in the presence of dithiothreitol. This is likely to reflect the rapid turnover of the D1 protein of Photosystem II in the light (Anderson *et al.* 1997), with the effect of chloramphenicol inhibiting its replacement accentuated when photoprotection is suppressed by dithiothreitol.

progressive decline in dark-adapted F_v/F_m over the 54 h duration of the experiment, indicating photo-damage to a chloroplast-encoded protein, probably the D1 protein of Photosystem II, which is known to turn over rapidly in the light. Dithiothreitol has a marked effect on its own by removing xanthophyll cycle-mediated photoprotection, and it increases the effect of chloramphenicol. Other species differ in details, but the broad picture is similar in all. Recovery in the dark is complete, and very little affected by protein-synthesis inhibitors. In the light there is ongoing photo-damage, and protein synthesis is essential to repair it.

What are the essential requirements for desiccation tolerance? In very general terms, tolerance clearly requires a cell structure that can lose most of its water without disruption, and membranes that retain the essentials of their structure in the dry state or are readily and quickly reconstituted on remoistening. *All* the essential metabolic systems of the plant must remain intact or be

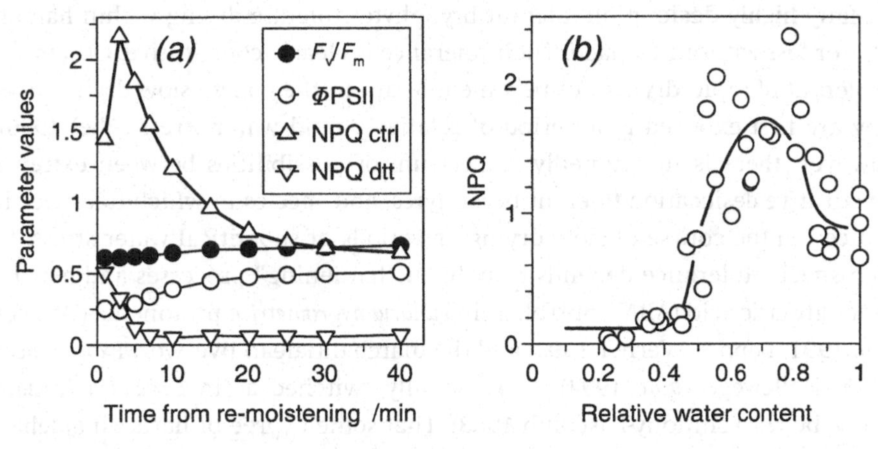

Fig. 6.9. (a) Recovery of the chlorophyll-fluorescence parameters F_v/F_m and ΦPSII (at ~50 µmol m^{-2} s^{-1} PPFD) following remoistening of the moss *Anomodon viticulosus* after 21 days' desiccation, with NPQ estimated from the mean values of F_m and F'_m of two sets of matched samples ($n = 3$). Note the rapid recovery of F_v/F_m, the slower recovery of ΦPSII, and the sharp peak of NPQ in the early minutes of recovery (redrawn from data of Csintalan *et al.* 1999). (b) Response of NPQ (at *c.* 200 µmol m^{-2} s^{-1} PPFD) to water loss in *Polytrichum formosum*; portions are shown of curves fitted to three overlapping segments of the data (redrawn from measurements of Proctor *et al.* 2007a).

readily reconstituted, and it is a reasonable *prima facie* supposition that there is unlikely to be any one critically sensitive system because selection pressure will bear heavily on any weak link. One might expect to find "molecular packaging" materials, most likely sugars (Seel *et al.* 1992, Smirnoff 1992) and/or proteins (Oliver 1996, Oliver *et al.* 2005), protecting macromolecules in the absence of water. Drying of the cell contents to a vitrified or "glassy" state is probably also important (Crowe *et al.* 1998, Buitink *et al.* 2002), both in slowing metabolic reactions and in maintaining the spatial relationships of cell components and membrane and enzyme systems - in effect, providing a reversible biological equivalent of good electron-microscopy fixation. Again both sugars and proteins are probably implicated (Buitink *et al.* 2002). Another reasonable expectation is good antioxidant and photo-protection, minimizing the production of damaging reactive oxygen species, and detoxifying those that form (Foyer *et al.* 1994, Alscher *et al.* 1997, Smirnoff 1993, 2005, Logan 2005, Heber *et al.* 2006). The high levels of NPQ seen in many desiccation-tolerant bryophytes must be important in photoprotection; NPQ also rises with desiccation stress, and peaks in the minutes following rehydration (Fig. 6.9, Csintalan *et al.* 1999, Marschall & Proctor 1999, Proctor *et al.* 2007a).

Many highly desiccation-tolerant bryophytes tolerate drying within half an hour or less without damage. Their tolerance is clearly constitutive. Others are intolerant of rapid drying but become tolerant in the course slow drying, or if they are first exposed to a period of relatively mild water stress (Abel 1956). However, there is undoubtedly a spectrum of possibilities between extreme constitutive desiccation-tolerant bryophytes, and species in which tolerance is induced in the course of slow drying or periods of subcritical water stress. In such species tolerance depends heavily on "hardening" processes and may be under abscisic acid (ABA) control, as in *Funaria hygrometrica* protonema (Werner *et al.* 1991, Bopp & Werner 1993) and the marchantialean liverwort *Exormotheca holstii* (Hellewege *et al.* 1994), or seasonally switched as in *Lunularia cruciata* (Schwabe & Nachmony-Bascomb 1963). That some degree of hardening/dehardening can occur in even the most tolerant species is shown by the results of Dilks & Proctor (1976) and Schonbeck & Bewley (1981). Much more investigation of desiccation tolerance, drought-hardening and possible effects of ABA in common forest and grassland bryophytes is greatly needed.

6.5 Overview

How can the special ecophysiological characteristics of bryophytes be summarized in a few words? First, most bryophytes carry substantial amounts of external water, which is physiologically important and can vary widely without affecting the water status of the cells, so bryophytes spend most of their time *either* fully turgid, *or* dry and metabolically inactive. It is *essential* to know (at least approximately) the true full-turgor water content for research on effects of water stress on bryophyte cells and tissues. Second, bryophytes grow and function largely within the laminar boundary layer of the atmosphere next to the ground or other substratum, and their immediate surroundings are in effect an integral part of their physiology. Third, intermittent availability of water is the norm for many bryophytes; their desiccation tolerance may be more usefully thought of as a means of *evading* drought, than as an extreme form of drought tolerance. Initial recovery from normal desiccation is rapid and seems to depend little on protein synthesis, but we still have much to learn about the fundamental basis and consequences of desiccation tolerance. Fourth, a tendency to shade-plant features in bryophytes springs mainly from their common situation in the understorey of vegetation, and their poikilohydry, photosynthesizing mainly in rainy and overcast weather. Unistratose leaves can be CO_2 diffusion-limited at high irradiance. Most bryophytes of dry sunny habitats show only moderately high light-saturation levels, but very high levels of NPQ (and photoprotection) at high irradiances. The highest saturation irradiances are found among the Polytrichales and Marchantiales with complex ventilated photosynthetic tissues. All bryophytes are C3 plants; anthocerotes are unique in having a carbon-concentrating mechanism. Finally, this last point is a reminder that bryophytes are phylogenetically diverse: the major groups – mosses, hepatics, anthocerotes – have been evolutionarily independent from one another and from vascular plants through most of the history of plant life on land.

References

Abel, W. O. (1956). Die Austrocknungsresistenz der Laubmoose. Sitzungsberichte. *Österreichische Akademie der Wissenschaften. Mathematische-Naturwissenschaftliche Klasse Abt. 1*, **165**, 619–707.

Alpert, P. (2000). The discovery, scope, and puzzle of desiccation tolerance in plants. *Plant Ecology*, **151**, 5–17.

Alpert, P. (2005). The limits and frontiers of desiccation-tolerant life. *Integrative and Comparative Biology*, **45**, 685–95.

Alpert, P. & Oliver, M. J. (2002). Drying without dying. In *Desiccation and Survival in Plants: Drying Without Dying*, ed. M. Black & H. W. Pritchard, pp. 3–43. Wallingford: CABI Publishing.

Alscher, R. G., Donahue, J. L. & Cramer, C. A. (1997). Reactive oxygen species and antioxidants: relationships in green cells. *Physiologia Plantarum* **100**, 224–33.

Anderson, J. M., Park, Y.-L. & Chow, W. S. (1997). Photoinactivation and photoprotection of photosystem II in nature. *Physiologia Plantarum*, **100**, 214–23.

Asada, K. (1999). The water-water cycle in chloroplasts: scavenging of active oxygens and dissipation of excess photons. *Annual Review of Plant Physiology and Plant Molecular Biology*, **50**, 601–39.

Bates, J. W. (1998). Is 'life-form' a useful concept in bryophyte ecology? *Oikos*, **82**, 223–37.

Bates, J. W. & Farmer, A. M. (1992). *Bryophytes and Lichens in a Changing Environment.* Oxford: Clarendon Press.

Beckett, R. P. (1996). Pressure volume analysis of a range of poikilohydric plants implies the existence of negative turgor in vegetative cells. *Annals of Botany*, **79**, 145–52.

Bewley, J. D. & Krochko, J. E. (1982). Desiccation-tolerance. In *Encyclopaedia of Plant Physiology*, New Series, vol. 12B, ed. O. L. Lange, P. S. Nobel, C. B. Osmond & H. Ziegler, pp. 325–78. Berlin: Springer-Verlag.

Björkman, O. & Demmig-Adams, B. (1995). Regulation of photosynthetic light energy capture, conversion and dissipation in leaves of higher plants. In *Ecophysiology of Photosynthesis*, ed. E.-D. Schulze & M. M. Caldwell, pp. 17–47. Berlin: Springer-Verlag.

Bopp, M. & Werner, O. (1993). Abscisic acid and desiccation tolerance in mosses. *Botanica Acta*, **106**, 103–6.

Brown, D. H. (1982). Mineral nutrition. In *Bryophyte Ecology*, ed. A. J. E. Smith, pp. 383–444. London: Chapman & Hall.

Brown, D. H. (1984). Uptake of mineral elements and their use in pollution monitoring. In *The Experimental Biology of Bryophytes*, ed. A. F. Dyer & J. G. Duckett. London: Academic Press.

Brown, D. H. & Bates, J. W. (1990). Bryophytes and nutrient cycling. *Botanical Journal of the Linnean Society*, **104**, 129–47.

Buch, H. (1945). Über die Wasser- und Mineralstoffversorgung der Moose. Part 1. *Commentationes Biologici Societas Scientiarum Fennicae*, **9**(16), 1–44.

Buch, H. (1947). Über die Wasser- und Mineralstoffversorgung der Moose. Part 2. *Commentationes Biologici Societas Scientiarum Fennicae*, **9**(20), 1–61.

Buitink, J., Hoekstra, F. A. & Leprince, O. (2002). Biochemistry and biophysics of tolerance systems. In *Desiccation and Survival in Plants: Drying Without Dying*, ed. M. Black & H. W. Pritchard, pp. 293–318. Wallingford: CABI Publishing.

Clayton-Greene, K. A., Collins, N. J., Green, T. G. A. & Proctor, M. C. F. (1985). Surface wax, structure and function in leaves of Polytrichaceae. *Journal of Bryology*, **13**, 549–62.

Crowe, J. H., Carpenter, J. F. & Crowe, L. M. (1998). The role of vitrification in anhydrobiosis. *Annual Review of Physiology*, **60**, 73–103.

Csintalan, Zs., Proctor, M. C. F. & Tuba, Z. (1999). Chlorophyll fluorescence during drying and rehydration in the mosses *Rhytidiadelphus loreus* (Hedw.) Warnst., *Anomodon viticulosus* (Hedw.) Hook. & Tayl. and *Grimmia pulvinata* (Hedw.) Sm. *Annals of Botany*, **84**, 235–44.

Deltoro, V. I., Calatayud, A., Gimeno, C., Abadia, A. & Barreno, E. (1998). Changes in chlorophyll *a* fluorescence, photosynthetic CO_2 assimilation and xanthophyll cycle interconversions during dehydration in desiccation-tolerant and intolerant liverworts. *Planta*, **207**, 224–8.

Dilks, T. J. K. & Proctor, M. C. F. (1974). The pattern of recovery of bryophytes after desiccation. *Journal of Bryology*, **8**, 97–115.

Dilks, T. J. K. & Proctor, M. C. F. (1976). Effects of intermittent desiccation on bryophytes. *Journal of Bryology*, **9**, 249–64.

Dilks, T. J. K. & Proctor, M. C. F. (1979). Photosynthesis, respiration, and water content in bryophytes. *New Phytologist*, **82**, 97–114.

Edwards, D., Wellman, C. H. & Axe, L. (1998). The fossil record of early land plants and interrelationships between primitive embryophytes: too little and too late? In *Bryology for the Twenty-First Century*, ed. J. W. Bates, N. W. Ashton & J. G. Duckett, pp. 15–43. Leeds: Maney Publishing and British Bryological Society.

Egle, K. (1960). Menge und Verhältnis der Pigmente. In *Encyclopaedia of Plant Physiology*, vol. 5, ed. W. Ruhland, pp. 444–96. Berlin: Springer-Verlag.

Farquhar, G. D., Ehleringer, J. R. & Hubick, K. T. (1989). *Annual Review of Plant Physiology and Plant Molecular Biology*, **40**, 503–37.

Foyer, C. H., Lelandais, M. & Kunert, K. J. (1994). Photooxidative stress in plants. *Physiologia Plantarum*, **92**, 696–717.

Frahm, J.-P. (1990). Bryophyte phytomass in tropical ecosystems. *Botanical Journal of the Linnean Society*, **104**, 23–33.

Gilmore, A. M. (1997). Mechanistic aspects of xanthophyll cycle-dependent photoprotection in higher plant chloroplasts and leaves. *Physiologia Plantarum*, **99**, 197–209.

Gimingham, C. H. & Birse, E. M. (1957). Ecological studies on growth-form in bryophytes. I. Correlations between growth-form and habitat. *Journal of Ecology*, **45**, 533–45.

Goffinet, B. (2000). Origin and phyletic relationships of bryophytes. In *Bryophyte Biology*, ed. A. J. Shaw & B. Goffinet, pp. 124–49. Cambridge: Cambridge University Press.

Gwóźdz, E. A., Bewley, J. D. & Tucker, E. B. (1974). Studies on protein synthesis in *Tortula ruralis*: polyribosome formation following desiccation. *Journal of Experimental Botany*, **25**, 599–608.

Hanson, D., Andrews, T. J. & Badger, M. R. (2002). Variability of the pyrenoid-based CO_2 concentrating mechanism in hornworts (Anthocerotophyta). *Functional Plant Biology*, **29**, 407–16.

Hearnshaw, G. F. & Proctor, M. C. F. (1982). The effect of temperature on the survival of dry bryophytes. *New Phytologist*, **90**, 221–8.

Heber, U., Lange, O. L. & Shuvalov, V. A. (2006). Conservation and dissipation of light energy as complementary processes: homoiohydric and poikilohydric autotrophs. *Journal of Experimental Botany*, **57**, 121–3.

Hellewege, E. M., Dietz, K. J., Volk, O. H. & Hartung, W. (1994). Abscisic acid and the induction of desiccation tolerance in the extremely xerophilic liverwort *Exormotheca holstii*. *Planta*, **194**, 525–31.

Hinshiri, H. N. & Proctor, M. C. F. (1971). The effect of desiccation on subsequent assimilation of the bryophytes *Anomodon viticulosus* and *Porella platyphylla*. *New Phytologist*, **70**, 527–38.

Horton, P., Ruban, A. V. & Walters, R. G. (1996). Regulation of light harvesting in green plants. *Annual Review of Plant Physiology and Plant Molecular Biology*, **47**, 655–84.

Jarvis, P. G. & McNaughton, K. G. (1986). Stomatal control of transpiration: scaling up from leaf to region. *Advances in Ecological Research*, **15**, 1–49.

Jones, H. G. (1992). *Plants and Microclimate*, 2nd edn. Cambridge: Cambridge University Press.

Kaiser, W. M. (1987). Effects of water deficit on photosynthetic capacity. *Physiologia Plantarum*, **71**, 142–9.

Kappen, L. & Valladares, F. (1999). Opportunistic growth and desiccation tolerance: the ecological success of poikilohydrous autotrophs. In *Handbook of Functional Plant Ecology*, ed. F. I. Pugnaire & F. Valladares, pp. 9–80. New York and Basel: Marcel Dekker.

Kershaw, K. A. & Webber, M. R. (1986). Seasonal changes in the chlorophyll content and quantum efficiency of the moss *Brachythecium rutabulum*. *Journal of Bryology*, **14**, 151–8.

Krause, G. H. & Weis, E. (1991). Chlorophyll fluorescence and photosynthesis: the basics. *Annual Review of Plant Physiology and Plant Molecular Biology*, **42**, 313–49.

Lackner, L. (1939). Über die Jahresperiodizität in der Entwicklung der Laubmoose. *Planta*, **29**, 534–616.

Larcher, W. (1995). *Physiological Plant Ecology*, 3rd edn. New York: Springer-Verlag.

León-Vargas, Y., Engwald, S. & Proctor, M. C. F. (2006). Microclimate, light adaptation and desiccation tolerance of epiphytic bryophytes in two Venezuelan cloud forests. *Journal of Biogeography*, **33**, 901–13.

Liu, Y., Li, Z., Cao, T. & Glime, J. M. (2004). The influence of high temperature on cell damage and shoot survival rates of *Plagiomnium acutum*. *Journal of Bryology*, **26**, 265–71.

Logan, B. A. (2005). Reactive oxygen species and photosynthesis. In *Antioxidants and Reactive Oxygen Species in Plants*, ed. N. Smirnoff, pp. 250–67. Oxford: Blackwell Publishing.

Longton, R. E. (1981). Physiological ecology of mosses. In *The Mosses of North America*, ed. R. J. Taylor & S. E. Leviton, pp. 77–113. Washington, D.C.: Pacific Division, American Academy of Science.

Longton, R. E. (1988). *The Biology of Polar Bryophytes and Lichens*. Cambridge: Cambridge University Press.

Mägdefrau, K. (1982). Life-forms of bryophytes. In *Bryophyte Ecology*, ed. A. J. E. Smith, pp. 45–58. London: Chapman & Hall.

Mansour, K. S. & Hallet, J. N. (1981). Effect of desiccation on DNA synthesis and the cell cycle of the moss *Polytrichum formosum*. *New Phytologist*, **87**, 315–24.

Martin, C. E. & Churchill, S. P. (1982). Chlorophyll concentrations and a/b ratios in mosses collected from exposed and shaded habitats in Kansas. *Journal of Bryology*, **12**, 297–304.

Marschall, M. & Proctor, M. C. F. (1999). Desiccation tolerance and recovery of the leafy liverwort *Porella platyphylla* (L.) Pfeiff.: chlorophyll-fluorescence measurements. *Journal of Bryology*, **21**, 257–62.

Marschall, M. & Proctor, M. C. F. (2004). Are bryophytes shade plants? Photosynthetic light responses and proportions of chlorophyll *a*, chlorophyll *b* and total carotenoids. *Annals of Botany*, **94**, 593–603.

Maxwell, K. & Johnson, G. N. (2000). Chlorophyll fluorescence – a practical guide. *Journal of Experimental Botany*, **51**, 659–68.

Nobel, P. S. (1977). Internal leaf area and cellular CO_2 resistance: photosynthetic implications of variations with growth conditions and plant species. *Physiologia Plantarum*, **40**, 137–44.

Oliver, M. J. (1991). Influence of protoplasmic water loss on the control of protein synthesis in the desiccation-tolerant moss *Tortula ruralis*. Ramifications for a repair-based mechanism of desiccation tolerance. *Plant Physiology*, **97**, 1501–11.

Oliver, M. J. (1996). Desiccation tolerance in vegetative plant cells. *Physiologia Plantarum*, **97**, 779–87.

Oliver, M. J. & Bewley, J. D. (1997). Desiccation tolerance of plant tissues: a mechanistic overview. *Horticultural Reviews*, **18**, 171–213.

Oliver, M. J., Tuba, Z. & Mishler, B. D. (2000). The evolution of vegetative desiccation tolerance in land plants. *Plant Ecology*, **151**, 85–100.

Oliver, M. J., Velten, J. & Mishler, B. D. (2005). Desiccation tolerance in bryophytes: a reflection of the primitive strategy for plant survival in dehydrating habitats. *Integrative and Comparative Biology*, **45**, 788–99.

Pitkin, P. H. (1975). Variability and seasonality of the growth of some corticolous pleurocarpous mosses. *Journal of Bryology*, **8**, 337–56.

Pressel, S. (2006). Experimental studies of bryophyte cell biology, conservation, physiology and systematics. Ph.D. Thesis, University of London.

Pressel, S., Ligrone, R. & Duckett, J. G. (2006). Effects of de- and rehydration on food-conducting cells in the moss *Polytrichum formosum*: a cytological study. *Annals of Botany*, **98**, 67–76.

Price, G. D., McKenzie, J. E., Pilcher, J. R. & Hoper, S. T. (1997). Carbon-isotope variation in *Sphagnum* from hummock-hollow complexes: implications for Holocene climate reconstruction. *The Holocene*, **7**, 229–33.

Proctor, M. C. F. (1979a). Structure and eco-physiological adaptation in bryophytes. In *Bryophyte Systematics*, ed. G. C. S. Clarke & J. G. Duckett, pp. 479-509. London: Academic Press.

Proctor, M. C. F. (1979b). Surface wax on the leaves of some mosses. *Journal of Bryology*, **10**, 531–8.

Proctor, M. C. F. (1981a). Physiological ecology of bryophytes. *Advances in Bryology*, **1**, 79–166.

Proctor, M. C. F. (1981b). Diffusion resistances in bryophytes. In *Plants and their Atmospheric Environment*, ed. J. Grace, E. D. Ford & P. G. Jarvis, pp. 219–29.

Proctor, M. C. F. (1982). Physiological ecology: water relations, light and temperature responses, carbon balance. In *Bryophyte Ecology*, ed. A. J. E. Smith, pp. 333–81. London: Chapman & Hall.

Proctor, M. C. F. (1990). The physiological basis of bryophyte production. *Botanical Journal of the Linnean Society*, **104**, 61–77.

Proctor, M. C. F. (1992). Scanning electron microscopy of lamella-margin characters and the phytogeography of the genus *Polytrichadelphus*. *Journal of Bryology*, **17**, 317–33.

Proctor, M. C. F. (1999). Water-relations parameters of some bryophytes evaluated by thermocouple psychrometry. *Journal of Bryology*, **21**, 263–70.

Proctor, M. C. F. (2000). The bryophyte paradox: tolerance of desiccation, evasion of drought. *Plant Ecology*, **151**, 41–9.

Proctor, M. C. F. (2001). Patterns of desiccation tolerance and recovery in bryophytes. *Plant Growth Regulation*, **35**, 147–56.

Proctor, M. C. F. (2002). Ecophysiological measurements on two pendulous forest mosses from Uganda, *Pilotrichella ampullacea* and *Floribundaria floribunda*. *Journal of Bryology*, **24**, 223–32.

Proctor, M. C. F. (2003). Experiments on the effect of different intensities of desiccation on bryophyte survival, using chlorophyll fluorescence as an index of recovery. *Journal of Bryology*, **25**, 215–24.

Proctor, M. C. F. (2004a). How long must a desiccation-tolerant moss tolerate desiccation? Some results of two years' data-logging on *Grimmia pulvinata*. *Physiologia Plantarum*, **122**, 21–7.

Proctor, M. C. F. (2004b). Light and desiccation responses of *Weymouthia mollis* (Hedw.) Broth. and *W. cochlearifolia* (Schwägr.) Dix., two pendulous rainforest epiphytes from Australia and New Zealand. *Journal of Bryology*, **26**, 167–73.

Proctor, M. C. F. (2005). Why do Polytrichaceae have lamellae? *Journal of Bryology*, **27**, 221–9.

Proctor, M. C. F. & Pence, V. C. (2002). Vegetative tissues: bryophytes, vascular resurrection plants and vegetative propagules. In *Desiccation and Survival in Plants: Drying Without Dying*, ed. M. Black & H. W. Pritchard, pp. 207–37. Wallingford: CABI Publishing.

Proctor, M. C. F. & Smirnoff, N. (2000). Rapid recovery of photosystems on rewetting desiccation-tolerant mosses: chlorophyll fluorescence and inhibitor experiments. *Journal of Experimental Botany*, **51**, 1695–704.

Proctor, M. C. F. & Smith, A. J. E. (1995). Ecological and systematic implications of branching patterns in bryophytes. In *Experimental and Molecular Approaches to Plant Biosystematics*, ed. P. C. Hoch & A. G. Stephenson, pp. 87–110. St. Louis, MO: Missouri Botanical Garden.

Proctor, M. C. F. & Tuba, Z. (2002). Poikilohydry and homoiohydry: antithesis or spectrum of possibilities? *New Phytologist*, **156**, 327–49.

Proctor, M. C. F., Raven, J. A. & Rice, S. K. (1992). Stable carbon isotope discrimination measurements in *Sphagnum* and other bryophytes: physiological and ecological implications. *Journal of Bryology*, **17**, 193–202.

Proctor, M. C. F., Nagy, Z., Csintalan, Zs. & Takács, Z. (1998). Water-content components in bryophytes: analysis of pressure–volume relationships. *Journal of Experimental Botany*, **49**, 1845–54.

Proctor, M. C. F., Duckett, J. G. & Ligrone, R. (2007a). Desiccation tolerance in the moss *Polytrichum formosum* Hedw.: physiological and fine-structural changes during desiccation and recovery. *Annals of Botany*, **99**, 75–93.

Proctor, M. C. F., Oliver, M. J., Wood, A. J. *et al.* (2007b). Desiccation-tolerance in bryophytes: a review. *Bryologist*, **110**, 595–621.

Raven, J. A. (1977). The evolution of land plants in relation to supracellular transport processes. *Advances in Botanical Research*, **5**, 153–219.

Raven, J. A. (1984). Physiological correlates of the morphology of early vascular plants. *Botanical Journal of the Linnean Society*, **88**, 105–26.

Raven, J. A. (1995). The early evolution of land plants: aquatic ancestors and atmospheric interactions. *Botanical Journal of Scotland*, **47**, 151–75.

Raven, J. A., Macfarlane, J. J. & Griffiths, H. (1987). The application of carbon isotope discrimination techniques. In *Plant Life in Aquatic and Amphibious Habitats*, ed. R. M. M. Crawford, pp. 129–49. Oxford: Blackwell.

Rice, S. K. (2000). Variation in carbon isotope discrimination within and among *Sphagnum* species in a temperate wetland. *Oecologia*, **123**, 1–8.

Rice, S. K., Collins, D. & Anderson, A. M. (2001). Functional significance of variation in bryophyte canopy structure. *American Journal of Botany*, **88**, 1568–76.

Rice, S. K. & Giles, L. (1996). The influence of water content and leaf anatomy on carbon isotope discrimination and photosynthesis in *Sphagnum*. *Plant, Cell and Environment*, **19**, 118–24.

Rundel, P. W., Stichler, W., Zander, R. H. & Ziegler, H. (1979). Carbon and hydrogen isotope ratios of bryophytes from arid and humid regions. *Oecologia*, **4**, 91–4.

Russell, S. (1990). Bryophyte production and decomposition in tundra ecosystems. *Botanical Journal of the Linnean Society*, **104**, 3–22.

Schonbeck, M. W. & Bewley, J. D. (1981). Responses of the moss *Tortula ruralis* to desiccation treatments. II. Variations in desiccation tolerance. *Canadian Journal of Botany*, **59**, 2707–12.

Schönherr, J. & Ziegler, H. (1975). Hydrophobic cuticular ledges prevent water entering the air pores of liverwort thalli. *Planta*, **124**, 51–60.

Schreiber, U., Bilger, W. & Neubauer, C. (1995). Chlorophyll fluorescence as a nonintrusive indicator for rapid assessment of in vivo photosynthesis. In *Ecophysiology of Photosynthesis*, ed. E.-D. Schulze & M. M. Caldwell, pp. 49–70. Berlin: Springer-Verlag.

Schwabe, W. & Nachmony-Bascomb, S. (1963). Growth and dormancy in *Lunularia cruciata* (L.) Dum. II. The response to daylength and temperature. *Journal of Experimental Botany*, **14**, 353–78.

Seel, W. E., Hendry, G. A. F. & Lee, J. A. (1992). Effects of desiccation on some activated oxygen processing enzymes and anti-oxidants in mosses. *Journal of Experimental Botany*, **43**, 1031–7.

Slavik, B. (1965). The influence of decreasing hydration level on photosynthetic rate in the thalli of the hepatic *Conocephalum conicum*. In *Water Stress in Plants. Proceedings of a Symposium held in Prague, September 30–October 4, 1963*, ed. B. Slavik, pp. 195–201. The Hague: W. Junk.

Smirnoff, N. (1992). The carbohydrates of bryophytes in relation to desiccation tolerance. *Journal of Bryology*, **17**, 185–91.

Smirnoff, N. (1993). The role of active oxygen in the response of plants to water deficit and desiccation. *New Phytologist*, **125**, 27–58.

Smirnoff, N. (ed.) (2005). *Antioxidants and Reactive Oxygen Species in Plants*. Oxford: Blackwell.

Smith, E. C. & Griffiths, H. (1996a). The occurrence of the chloroplast pyrenoid is correlated with the activity of a CO_2-concentrating mechanism and carbon isotope discrimination in lichens and bryophytes. *Planta*, **198**, 6–16.

Smith, E. C. & Griffiths, H. (1996b) A pyrenoid-based carbon-concentrating mechanism is present in terrestrial bryophytes of the class Anthocerotae. *Planta*, **200**, 203–12.

Stark, L. R. (2005). Phenology of patch hydration, patch temperature and sexual reproductive output in the desert moss *Crossidium crassinerve*. *Journal of Bryology*, **27**, 231–40.

Stark, L. R., Oliver, M. J., Mishler, B. D. & McLetchie, D. N. (2007). Generational differences in response to desiccation stress in the desert moss *Tortula inermis*. *Annals of Botany*, **99**, 53–60.

Steudle, E. & Peterson, C. A. (1998). How does water get through roots? *Journal of Experimental Botany*, **49**, 775–88.

Sveinbjörnsson, B. & Oechel, W. C. (1992). Controls on growth and productivity of bryophytes: environmental limitations under current and anticipated conditions. In *Bryophytes and Lichens in a Changing Environment*, ed. J. W. Bates & A. M. Farmer, pp. 77–102. Oxford: Clarendon Press.

Teeri, J. A. (1981). Stable carbon isotope analysis of mosses and lichens growing in xeric and moist habitats. *Bryologist*, **84**, 82–4.

Tuba, Z., Csintalan, Zs. & Proctor, M. C. F. (1996). Photosynthetic responses of a moss, *Tortula ruralis*, ssp. *ruralis*, and the lichens *Cladonia convoluta* and *C. furcata* to water deficit and short periods of desiccation, and their ecophysiological significance: a baseline study at present CO_2 concentration. *New Phytologist*, **133**, 353–61.

Valanne, N. (1984). Photosynthesis and photosynthetic products in mosses. In *The Experimental Biology of Bryophytes*, ed. A. F. Dyer & J. G. Duckett, pp. 257–73. London: Academic Press.

Vitt, D. H. (1990). Growth and production dynamics of boreal mosses over climatic, chemical and topographic gradients. *Botanical Journal of the Linnean Society*, **104**, 35–59.

Werner, O., Espin, R. M. R., Bopp, M. & Atzorn, R. (1991). Abscisic-acid induced drought tolerance in *Funaria hygrometrica* Hedw. *Planta*, **186**, 99–103.

Williams, T. G. & Flanagan, L. B. (1996). Effect of changes in water content on photosynthesis, transpiration and discrimination against $^{13}CO_2$ and $C^{18}O^{16}O$ in *Pleurozium* and *Sphagnum*. *Oecologia*, **108**, 38–46.

Wood, A. J. (2007). The nature and distribution of vegetative desiccation tolerance in hornworts, liverworts and mosses. *Bryologist*, **110**, 163–77.

Zimmermann, U. & Steudle, E. (1978). Physical aspects of water relations of plant cells. *Advances in Botanical Research*, **6**, 46–117.

Zotz, G. & Rottenberger, S. (2001). Seasonal changes in diel CO_2 exchange of three Central European moss species: a one-year field study. *Plant Biology*, **3**, 661–9.

Zotz, G., Schweikert, A., Jetz, W. & Westerman, H. 2000. Water relations and carbon gain in relation to cushion size in the moss *Grimmia pulvinata* (Hedw.) Sm. *New Phytologist*, **148**, 59–67.

7

Biochemical and molecular mechanisms of desiccation tolerance in bryophytes

MELVIN J. OLIVER

7.1 Introduction

Bryophytes, because they descend from the earliest branching events in the phylogeny of land plants, hold an important position in our investigations into the mechanisms by which plants respond to dehydration and by what paths such mechanisms have evolved. This is true regardless of what aspect of plant responses to dehydration one is interested in; whether it be mild water deficit stress as seen in most plants including those of agronomic importance, or desiccation as seen in orthodox seeds or in the leaves of desiccation-tolerant (or resurrection) plants. It is quite possible that the mechanisms by which bryophytes tolerate dehydration closely reflect the way that the first land plants coped with the rigors of a drying atmosphere as they began their colonization of the land. In a recent phylogenetic synthesis of the evolution of desiccation tolerance within the land plants (Oliver *et al.* 2000), it was postulated that vegetative desiccation tolerance was required for plants to transition from an aqueous environment to the dry land. In the initial ventures into dehydrating atmospheres, plants were of a very simple architecture and had yet to evolve the complex strategies to prevent water loss that we see in modern day plants. Once the cells of these plants were no longer surrounded by liquid water they would rapidly lose water and dry. Thus it is highly likely that primitive plants spent much of their time in the air-dried state, which generally means they would experience water deficits of −100 MPa or more (Alpert & Oliver 2002). To survive such water deficits (most modern day plants do not survive −3 to −5 MPa) these plants had to evolve vegetative desiccation tolerance. Given that, as we believe, desiccation tolerance first evolved in spores as a means of surviving the rigors of dispersal, it is quite likely that the type of

Bryophyte Biology: Second Edition, ed. B. Goffinet & A. J. Shaw. Published by Cambridge University Press.

desiccation tolerance that these early land plants exhibited was co-opted for use in vegetative cells (Oliver *et al.* 2005). Spores appear to utilize, if one can extrapolate from studies of desiccation tolerance in pollen (Hoekstra 2002), a relatively simple mechanism of cellular protection to survive drying to very low water potentials. As plants evolved to fill the multitude of different types of habitats and environmental niches that dry land offered, the ways in which plants were able to tolerate dehydration also evolved from a simple protective mechanism to ones of greater complexity and efficiency. It is clear, however, that as plants became more complex and gained the capability to transport water and nutrients through specialized conduits or vascular tissues, they lost the ability to tolerate desiccation of their vegetative tissues (Oliver *et al.* 2000). Bryophytes, whose ancestors marked the transition to land and hence preceded the development of tracheophytes in the land plant phylogeny, present a unique opportunity to look at mechanisms of dehydration tolerance, in particular vegetative desiccation tolerance, that have directly evolved from the earliest stages of land plant evolution.

7.2 Phenotypic considerations

Dehydration tolerance describes the ability of plants to survive and recover from the loss of cellular water. Desiccation tolerance, however, is the ability of plants (or plant tissues) to withstand the severest dehydration stress, namely to dry to equilibrium with moderately dry air and to fully recover when rehydrated. Vegetative desiccation tolerance is reported to be relatively common, but not universal, in the bryophytes (Proctor 1990, Proctor & Pence 2002); it is rare in vascular plants with only 330 species listed so far (Porembski & Barthlott (2000). The commonality of vegetative desiccation tolerance in bryophytes is a widely held belief but is derived from anecdotal reports and personal observations in the field. Wood (2007) attempts to bring a more scientific accounting to the question of the occurrence of desiccation tolerance in the bryophytes by only reporting those species for which there has been an experimental demonstration of tolerance. By this strict cataloging Wood identifies 158 species of mosses, 51 liverworts, and 1 species of hornwort as being vegetatively desiccation-tolerant, about 1% of all bryophytes. Wood (2007) also provides a protocol for the experimental determination of desiccation tolerance for bryophytes and as this is more widely applied the number of species that can be documented as desiccation tolerant will undoubtedly rise. Although many bryophytes seem to be able to survive desiccation, very few have been the subjects of in-depth studies into how such a phenotype is mechanistically delivered. Indeed, much of what we know about the mechanistic aspects of desiccation tolerance in bryophytes comes from the study of single species, the moss *Tortula ruralis* (*Syntrichia ruralis*).

Modern day bryophytes face much the same environment, with regards to water relations, as primitive plants did when first they occupied dry land habitats. The leaves of bryophytes are generally one cell layer thick and rely on an external supply of liquid water to remain turgid. Bryophyte cells freely lose water to the atmosphere when liquid water is no longer present on their surface and as a consequence they rapidly and directly equilibrate with the water potential of the air, which is generally dry, and thus they desiccate. The speed at which desiccation is achieved is dependent upon the relative humidity of the surrounding air: the dryer the air the more rapid the rate of dehydration. The speed of desiccation has important consequences for the cells of bryophytes and has a direct impact on the type of mechanism these plants have evolved to survive the air-dried state (as will be discussed below). The extent of water loss, which is dependent upon both the relative water content of the air and the temperature of the habitat, is also an important factor in how bryophytes tolerate desiccation. For example, equilibration at 50% relative humidity (RH) at 28 °C would generate a −100 MPa water deficit in the cells of a bryophyte. Most bryophytes can survive dehydration to −20 to −40 MPa for a short time but these are not considered desiccation-tolerant, even though such water deficits are an order of magnitude greater than what most plants can survive (Proctor & Pence 2002). Desiccation-tolerant bryophytes can survive rapid drying rates, more extensive water loss, and remain dry for longer periods than those that are simply dehydration tolerant. *Tortula caninervis* (*Syntrichia caninervis*), a desert relative of *Tortula ruralis*, can survive rapid desiccation (within 30 min) to approximately −540 MPa (equilibrated to the atmosphere above activated silica gel; 2%–4% RH) for up to six years, returning to normal metabolic activity upon rehydration (Oliver *et al.* 1993, unpublished data (pers. obs.)). Some desiccation-tolerant bryophytes also increase their level of desiccation tolerance if they experience mild dehydration events prior to desiccation, a process known as hardening (Proctor & Pence 2002). This is true for most of the *Tortula* (*Syntrichia*) species, including *Tortula ruralis* (Schonbeck & Bewley 1981, unpublished data). Rehydration, as one would expect, is almost instantaneous and creates in its own right a stressful cellular event. Thus the phenotype of desiccation tolerance is not a simple one, as speed of drying, time in the dried state, how dry the tissue becomes, previous drying history, and the rigors of rehydration all play a role in determining how desiccation tolerance is achieved.

7.3 General aspects of desiccation tolerance as they relate to bryophytes

Early observations of bryophyte tissues undergoing desiccation and rehydration led workers to postulate that it was the mechanical properties of cells that formed the underlying basis of vegetative desiccation tolerance in plants. Structural features such as small cells, flexible cell walls, small vacuoles, and a lack of plasmodesmata were all suggested as important aspects of desiccation tolerance in plants (Iljin 1957). Although this viewpoint has been largely discarded as a means of explaining desiccation tolerance in bryophytes (Oliver & Bewley 1984), the ability to tolerate the mechanical stresses of desiccation and rehydration is an important factor for all plants but is more evident in the desiccation-tolerant angiosperms (Vicré *et al.* 2004). Bewley (1979) proposed the current theory that desiccation tolerance is primarily an inherent property of the cellular contents (protoplasm), an hypothesis that has so far stood the test of experimental inquiry. Bewley (1979), based on available microscopic and biochemical evidence, further defined what the properties of a desiccation-tolerant protoplasm must encompass: it must (a) limit damage to a repairable level; (b) maintain integrity when in the dried state; and (c) upon rehydration, rapidly mobilize mechanisms that repair any damage that has been sustained as a result of both desiccation and from the rapid influx of water. Basically this means that desiccation tolerance can be achieved by protection of cellular structures and components, damage control during the dried state, and an active cellular repair activity that functions during and following rehydration (Bewley & Oliver 1992). Thompson and co-workers (Platt *et al.* 1994, Thompson & Platt 1997), based on freeze fracture and freeze substitution electron microscopy of dried and drying *Tortula ruralis* (freeze fracture only) and *Selaginella lepidophylla* (a fern relative) vegetative tissues, extended the Bewley (1979) view of desiccation tolerance to include the requirement that a critical level of physical cell order be maintained in the dry state. This conservation of cell order requires a high degree of effective packing and shape fitting of cellular constituents driven by the compaction forces of dehydration. Alpert and Oliver (2002) elaborated further by contending that desiccation tolerance must also require an orderly shutdown of metabolism during drying so as to avoid the possibility of a build up in toxic intermediates and the generation of reactive oxygen species (ROS), a major stress associated with desiccation (Smirnoff 1993).

The effectiveness of the cellular protection aspects of a particular mechanism for desiccation tolerance, and to some extent the repair processes, determines the intensity of the desiccation (the minimum water potential) that a particular species can survive. In general, desiccation-tolerant bryophytes are capable of surviving water potentials below −300 MPa, and most can survive long periods in air that is dried to <2% RH (over silica gel) that generates water potentials closer to −500 MPa. In contrast, desiccation-tolerant pteridophytes cannot survive such treatment (Gaff 1977). The effectiveness of cellular protection and the efficiency of repair processes probably combine to determine the length of time desiccation-tolerant plants can remain in the dried state. Bryophytes are particularly hardy in this regard. *Tortula caninervis* and *Tortula ruralis* have been documented to resume normal metabolic activity after being stored dry for three years (Oliver *et al.* 1993), and *Grimmia laevigata* grew when hydrated after ten years of storage in a herbarium (Keever 1957). However, *Tortula norvegica*, although capable of tolerating desiccation to a water potential of −540 MPa, was severely damaged following storage for 12 months or more (Oliver *et al.* 1993), suggesting that the level of cellular protection and efficiency of cellular repair are not as well developed in this species compared to its close relatives. Most desiccation tolerant bryophytes survive best and thus longer if kept equilibrated to atmospheres between 20% and 50% RH at 20 °C (−100 to −400 MPa), storage at lower RH leads to a quicker loss in viability (Proctor & Tuba 2002).

Levels of cellular protection vary not only between desiccation-tolerant species of bryophytes but also seasonally and between generations. The South African moss *Atrichum androgynum* is more tolerant, as measured by ion leakage assays, during the dry months of the year and less tolerant when moisture is readily available (Beckett & Hoddinott 1997). This may partially reflect a seasonal hardening/dehardening process but this in turn is directly related to the availability of cellular protectants stored within the cells; in times when water is plentiful more energy is directed towards growth than to the sequestration and biosynthesis of protectants. Stark *et al.* (2007) demonstrate that the sporophytes of *Tortula inermis*, a highly tolerant desert species, are clearly less desiccation-tolerant than the gametophytic generation that supports them. This may be generally true for most desiccation-tolerant bryophytes although evidence is generally anecdotal; gametangia of desiccation-tolerant mosses are apparently desiccation-tolerant as they persist over several months during which the mosses spend a considerable time dry (Mishler & Oliver 1991, Stark 1997).

7.4 Biochemical and molecular aspects of desiccation tolerance in bryophytes

7.4.1 *Constitutive cellular protection*

The ability of desiccation-tolerant bryophytes to withstand and recover from very rapid water loss, to water potentials below −500 MPa within 30 min, indicates that they have a constitutive cellular protection mechanism that is ready and waiting to be challenged (or activated) by desiccation. The effectiveness of this constitutive cellular protection is evidenced by the fact that the plasma-membrane and organellar membranes of *Tortula ruralis* cells remain intact and maintain their normal bilayer organization in the dried state (Platt *et al.* 1994).

The loss of water in very dry atmospheres is far too rapid to allow for the induction and *de novo* establishment of cellular protection processes as this would require not only a switch in metabolic activity to generate protective components, but also the induction of transcription and synthesis of novel proteins that are required for tolerance: both of which require a significant amount of time. This notion is substantiated by the overwhelming evidence that in *Tortula ruralis* (and other moss species) protein synthesis, in particular the translation initiation process, is extremely sensitive to the loss of water from the cytoplasm and quickly ceases, rendering the synthesis of proteins for cellular stability impossible (reviewed by Bewley 1979, Bewley & Krochko 1982). Furthermore, even if drying is slow (within 6 h) novel transcripts are not recruited into the protein synthetic machinery during dehydration and so desiccation tolerance in bryophytes, at least in the moss *Tortula ruralis*, does not require the synthesis of novel proteins during the drying phase of a desiccation event (Oliver 1991). The inference from these observations is that proteins, whether protective or involved in establishing protective components, are already present in the cells in sufficient quantities to establish desiccation tolerance. This is not the case for any other mechanism or cellular strategy for vegetative desiccation tolerance where such proteins are synthesized *de novo* during drying (Bartels & Sunkar 2005); it is, however, similar to the strategy for developmentally programmed desiccation tolerance where cells are primed for a desiccation event prior to dispersal from the mature plant. This may further emphasize the evolutionary link between desiccation tolerance in spores and mechanisms for vegetative desiccation tolerance in bryophytes.

At this point it is worth, for the sake of contrast, briefly discussing mechanisms of vegetative desiccation tolerance that are seen in the angiosperms. As mentioned above, vegetative desiccation-tolerant angiosperms can only survive desiccation if the rate of dehydration allows for the establishment of cellular protective measures (Oliver & Bewley 1997, Phillips *et al.* 2002, Proctor & Pence 2002). The early phase of drying in the desiccation-tolerant angiosperms is marked by a major switch in transcriptional activity, resulting in the generation of a large number of novel transcripts, which in turn generate proteins specific for the drying phase (Ingram & Bartels 1996, Oliver *et al.* 2000, Alpert & Oliver 2002, Phillips *et al.* 2002, Collett *et al.* 2004). The induced change in transcriptional control reveals a broad range of dehydration-regulated genes that have an equally broad range of putative functions, testament to the severity of the stress and the complexity of the interactions between cellular processes designed to deliver tolerance (Phillips & Bartels 2000). The induction of gene expression in response to dehydration in desiccation-tolerant angiosperms involves two classes of signal induction pathway, one that is controlled by the plant hormone abscisic acid

(ABA) and the other ABA-independent (Phillips *et al.* 2002, Ramanjulu & Bartels 2002). Endogenous ABA concentrations increase in vegetative tissues in response to dehydration in all angiosperms studied to date, even those that are sensitive to this stress (Bray 1997). Although some signaling pathways involved in the response of plants to dehydration are ABA-independent, application of exogenous ABA is sufficient to induce desiccation tolerance in desiccation-sensitive callous tissue derived from the leaves of the desiccation-tolerant angiosperm *Craterostigma plantagineum* (Bartels *et al.* 1990). Pretreatment of fronds of the resurrection fern *Polypodium virginianum* with exogenous ABA allows the plant to survive otherwise lethal rates of dehydration (Reynolds & Bewley 1993).

In orthodox seeds and vegetative tissues of desiccation-tolerant angiosperms two cellular components, the Late Embryogenesis Abundant (LEA) proteins and soluble sugars, accumulate in response to desiccation (for review see Phillips *et al.* 2002, Buitink *et al.* 2002, Kermode & Finch-Savage 2002). These two components are generally considered critical to the acquisition of cellular desiccation tolerance, although the actual function of the LEA proteins remains unclear (Cuming 1999). The types of gene that are induced by desiccation (and in many cases ABA treatment) code for proteins that fall into several categories: proteins involved in signal transduction pathways that regulate genes during stress responses, proteins participating in carbohydrate metabolism (generally involved in sucrose accumulation), protective proteins such as chloroplast stabilizing proteins, LEAs and heat shock proteins, aquaporins, and proteins involved in anti-oxidant biosynthesis and ROS scavenging (Phillips & Bartels 2000, Ramanjulu & Bartels 2002, Phillips *et al.* 2002, Collett *et al.* 2004, Illing *et al.* 2005). The physiological significance of these gene products will be discussed in more detail where relevant to the discussion of desiccation tolerance in bryophytes.

How is constitutive desiccation tolerance achieved in the bryophytes? This is a question for which we have lots of theories but few answers. Much of the uncertainty is a result of the fact that we do not fully understand desiccation tolerance *per se* and much of what we do know comes mainly from inference and observation of what components are present in desiccated cells. Given this, it is generally accepted that there are at least three components necessary for desiccation tolerance in bryophytes: (1) sugars, in general sucrose, (2) protective proteins, in particular proteins with homology to the LEA proteins of angiosperms, and (3) antioxidants and enzymes involved in protection from ROS.

Sugars: The accumulation of soluble sugars has long been correlated with the acquisition of desiccation tolerance in plants and other organisms (Gaff 1977, Scott 2000, Phillips *et al.* 2002, Walters *et al.* 2002). Orthodox seeds, pollen, and most plants that accumulate soluble sugars in response to desiccation utilize the disaccharide sucrose. Sucrose makes up approximately 10% of the dry mass of

Tortula ruralis gametophytes and does not change in amount during desiccation or rehydration in the dark or light (Bewley *et al.* 1978). This is apparently common for desiccation-tolerant bryophytes (Smirnoff 1992). The current hypotheses are that sugars either protect the cell via a process known as vitrification, i.e. the formation of a biological glass (a supersaturated liquid with the mechanical properties of a solid that prevents the crystallization of cellular solutes and slows chemical activity), and/or by maintaining hydrogen bonds within and between macromolecules, thus stabilizing their structure, e.g. membranes (Hoekstra *et al.* 2001). Sucrose has long been associated with glass formation and it is currently thought that LEA proteins are also required for the vitrification process (Buitink *et al.* 2002). However, biological glass formation during desiccation has not been verified for bryophytes as yet.

Proteins: The major group of proteins associated with cellular protection during drying is the LEA proteins. LEA proteins, of which there are at least five major groups, have functions that remain largely unknown, but are assumed to be important in the establishment of desiccation tolerance in seeds and, by inference, vegetative tissues (Cuming 1999, Buitink *et al.* 2002, Kermode & Finch-Savage 2002). The most compelling evidence for their role in desiccation tolerance comes from work with *Arabidopsis* mutants, a double mutant of ABA-deficient (aba – lesion in ABA biosynthesis) and ABA-insensitive (abi3 – lesion in responsiveness to ABA), that lack many of the LEA proteins and which do not tolerate desiccation (Koorneef *et al.* 1989, Meurs *et al.* 1992). Also, overexpression of a single LEA protein, HVA1 from barley, in transgenic rice plants increased their tolerance to water deficit stress (Xu *et al.* 1996). LEA proteins have been identified in the vegetative tissues of all desiccation-tolerant plants studied so far (Ingrams & Bartels 1996, Oliver & Bewley 1997, Blomstedt *et al.* 1998) and one class in particular, the Group 2 LEAs (dehydrins) have been associated with the vegetative response of non-tolerant plants to water stress (Skriver & Munday 1990, Bray 1997, Bartels & Sunkar 2005). Using antibodies specific for a conserved motif within dehydrin protein sequences, Bewley *et al.* (1993) demonstrated that dehydrins are sequestered in leaf cells of *Tortula ruralis* under hydrated conditions and are unaffected by desiccation and rehydration, except for a slight loss of the main dehydrin during slow drying. This is consistent with a constitutive cellular protection strategy and in stark contrast to the induction of synthesis and sequestration of dehydrins during dehydration in both sensitive and desiccation-tolerant angiosperms (Bartels *et al.* 1990, Cuming 1999, Bartels 2005, Illing *et al.* 2005).

Many hypotheses as to how LEA proteins protect cells from the rigors of dehydration have been proposed (Cuming 1999) but as of yet there is no definitive mechanism that can be attributed to any of the LEA proteins or groups.

Studies have provided evidence for LEA proteins acting as hydration buffers, in ion sequestration, renaturation of unfolded proteins, direct protection of membranes, and binding to DNA to stabilize chromatin (Crowe *et al.* 1992, Cuming 1999). *In vitro* studies point to a role for LEA proteins in the stabilization of enzymes under denaturating conditions, either by preventing aggregation (Goyal *et al.* 2005) or by direct protection (Grelet *et al.* 2005). Of note is the suggestion, driven from data derived from molecular modeling of Group 3 LEA proteins, that LEA proteins could form cytoskeletal filaments that could perhaps aid in the ordered drying of the cytoplasm to prevent damage or to simply stabilize membranes (Wise & Tunnacliffe 2004).

Other protective proteins that have been associated with vegetative desiccation tolerance in angiosperms, such as the low-molecular-mass heat shock proteins (Alamillo *et al.* 1995), have not been investigated in bryophytes as yet. These proteins have been shown to have chaperone-like properties that help proteins fold and maintain their active structures, and are thought to play this role *in vivo* during dehydration (Bartels & Sunkar 2005).

7.4.2 *Reactive oxygen scavenging pathways*

The generation of ROS (e.g. singlet oxygen, hydroxyl radicals, hydrogen peroxide, and superoxide anions) during dehydration occurs mainly from the inhibition of photosynthesis, by oxidation of the D1 protein and inhibition of the repair of Photosystem II reaction centers, and is potentially the major source of cellular damage for desiccation-tolerant bryophytes (and all plants) as they dry (Smirnoff 1993, Apel & Hurt 2004). ROS accumulation has many damaging effects on cellular components besides the effect on Photosystem II components, including protein denaturation via the oxidation of sulfhydryl groups, pigment loss, and lipid peroxidation and free fatty acid accumulation in membranes (McKersie 1991, Smirnoff 1993, Apel & Hurt 2004).

Tortula ruralis appears to have the capability of preventing the peroxidation of membrane fatty acids during dehydration by suppressing lipoxygenase activity, an enzyme that if released from sequestering peroxisomes or the vacuole catalyzes the peroxidation of fatty acids within the cells. Part of the protection may be afforded by the observed maintenance of oil droplets within the cytoplasm of *Tortula ruralis* gametophyte cells, which may act as a sort of peroxidation buffer (Stewart & Bewley 1982). This is not true of all mosses, as dehydration of the moderately desiccation-tolerant moss *Atrichum androgynum* results in significant peroxidation of membrane phospholipids (Guschina *et al.* 2002). Oxidative denaturation of sulfhydryl-containing enzymes during dehydration has been reported for several desiccation-tolerant mosses (Stewart & Lee 1972). These workers also reported that they could alleviate this damage by incubating the mosses in reduced glutathione (GSH) solutions; reduced glutathione is a natural antioxidant present in all plant cells. The pool of GSH in *Tortula ruralis* is reduced to 30% of its hydrated cell level during slow drying, indicating a reduced ability of the moss to buffer ROS and therefore withstand oxidative injury when in the dried state (Dhindsa 1987). *Tortula ruraliformis*, on the other hand, does not exhibit a reduction of its GSH pool during desiccation (Seel *et al.* 1992b) but it does suffer a decline in another cellular antioxidant, ascorbate. Thus in *Tortula ruralis* maintaining a significant pool of GSH in the dried state may not be an important strategy for protection whereas it is for *Tortula ruraliformis*. These studies are somewhat difficult to reconcile because the ascorbate and glutathione systems are closely linked in plants (Foyer *et al.* 1994, Alscher *et al.* 1997, Smirnoff 2005). Obviously the antioxidant pathways and their synthetic and catabolic activities during desiccation in desiccation-tolerant (and sensitive) bryophytes are a ripe target for further research.

When plant tissues are desiccated in the light, ROS levels increase dramatically and oxidative damage increases (Smirnoff 1993). The increased generation of ROS by light in drying or dried plant tissues results from the inability of the light harvesting complexes of the photosynthetic machinery to pass on the energy absorbed from light into the normal photochemical pathways which are inactive under dehydrated conditions. The unchanneled energy, if not dissipated in some manner, can directly result in the formation of reactive oxygen species from water. Seel *et al.* (1992a) demonstrated that *Tortula ruraliformis* was capable of preventing lipid peroxidation and pigment loss under high light conditions whereas a more desiccation-sensitive moss, *Dicranella palustris*, suffered both increased oxidative damage and an inability to recover from dehydration under the same conditions. It thus appears that for desiccation-tolerant mosses, photoprotective mechanisms that limit ROS production or detoxify them are important if survival, especially long-term survival in high-light environments such as deserts, is to be achieved. One obvious means of limiting light-induced ROS production is to dissipate the excess excitation energy generated by the photosynthetic machinery as heat, and in plants this is achieved via the xanthophyll cycle (Demmig-Adams & Adams 1992). Several studies point to this as a mechanism for photoprotection in both liverworts (Deltoro *et al.* 1998, Marschall & Proctor 1999) and mosses (Heber *et al.* 2001, 2006).

It is obvious from this discourse that we have a limited knowledge of how bryophytes develop and maintain cellular protection to limit desiccation-induced damage and thus achieve desiccation tolerance. The process of vitrification in desiccating vegetative tissues is largely underexplored and bryophytes offer an ideal model for such studies. How biological glasses form and what is required (e.g. sugar content, types of sugars, protein–sugar interactions, level of dehydration, etc.) for their formation would greatly increase our understanding of the constitutive protection system in desiccation-tolerant bryophytes. In addition, much more work is needed before we fully understand how bryophytes protect themselves from the oxidative damage associated with desiccation and rehydration.

7.5 Induced desiccation tolerance in bryophytes

As mentioned earlier, bryophytes can exhibit an increase in desiccation tolerance if they experience mild dehydration events prior to desiccation (hardening). This result implies, at least in part, that desiccation tolerance, and by inference cellular protection strategies, are inducible under some circumstances. Beckett (1999) demonstrated that desiccation tolerance of the mesic moss *Atrichum androgynum* could be increased by a previous drying treatment and that addition of abscisic acid (ABA) to the hydrated moss could produce the same increase in tolerance. ABA treatment of *Funaria hygrometrica* not only increases tolerance to desiccation, allowing for survival of rapid desiccation that is normally lethal, but also results in the induction of the synthesis of a number of proteins that accumulate during drying (Werner *et al.* 1991, Bopp & Werner 1993). Werner *et al.* (1991) also determined that endogenous levels of ABA increased 5–6-fold during slow drying, indicating that an induced protection system operates in this moss to develop desiccation tolerance. However, in these experiments the protonemal cultures were dried for only short periods of time in air of unknown water content and it is unclear whether equilibrium with the water potential of the air was achieved. ABA has similar effects on the tolerance of some liverworts to desiccation (Hellewege *et al.* 1994) and its precursor, lunularic acid, controls the switch from a sensitive to a tolerant stage of *Lunularia cruciata* (Schwabe & Nachmony-Bascomb 1963). These reports are of interest here because they indicate the possibility that in some bryophytes, those that normally do not face extreme dehydrating conditions for any length of time, vegetative desiccation tolerance can be induced via a mechanism that involves ABA and is, at least on the surface, similar to that seen in the vegetative desiccation-tolerant angiosperms. It is clear that bryophytes have the ability and the molecular machinery to respond to ABA, a point that was elegantly demonstrated by Knight *et al.* (1995) who introduced a β-glucuronidase reporter gene driven by the Em ABA-responsive promoter from wheat into *Physcomitrella patens*, a desiccation-sensitive moss. Not only was the reporter gene responsive to ABA, but the researchers were also able to demonstrate the presence of DNA binding proteins that were specific to the ABA Responsive Elements (ABREs) present in the Em promoter. The reporter gene was also able to respond to osmotic stress caused by exposing the transgenic moss to mannitol solutions, albeit to a lesser degree than when exposed to exogenous ABA.

Parenthetically, ABA was undetectable in *Tortula ruralis* gametophytes in either the hydrated or the drying stages of the wet–dry–wet cycle and was not detected in dried tissues (Bewley *et al.* 1993). Furthermore, ABA does not induce dehydrin accumulation in hydrated gametophytes (Bewley *et al.* 1993, M. J. Oliver, unpublished observations). However, in contrast, treatment of *Tortula ruralis* gametophytes with ABA does result in the induction of two early light-inducible protein (ELIP) genes such that the transcripts for these proteins accumulate (Zeng *et al.* 2002). This clearly indicates that *Tortula ruralis* can also respond to exogenous ABA. Oliver *et al.* (2005) interpret these observations as illustrating the complexity of desiccation tolerance and posed two interesting evolutionary questions: (1) Did mesic bryophytes forego the constitutive cellular protection aspect of desiccation tolerance in favor of an inducible system that allows them to better compete in a mesic habitat? Or (2) did the constitutive cellular protection system evolve from a primitive developmental system in spores that allowed some mosses to move into progressively more extreme xeric habitats? Only a phylogenetic approach to desiccation tolerance can address the hypotheses these questions generate.

7.6 Cellular recovery

An equally important aspect to desiccation tolerance in bryophytes, in conjunction with constitutive cellular protection, is the ability to quickly recover cellular and metabolic activity when rehydration occurs. It would be easy to assume that the cellular protection capabilities of bryophytes are sufficiently effective that cells could simply pick up where they left off when water returns. Indeed, some aspects of cellular activity appear to do just that, in particular the Photosystem II activity of photosynthesis (an important site to protect, as discussed above), which recovers normal function within minutes following rehydration, as measured by chlorophyll fluorescence measurement (F_v/F_m) (Proctor & Smirnoff 2000; see Chapter 6, this volume). Nevertheless, considerable evidence indicates that desiccation-tolerant bryophytes do suffer cellular damage that becomes evident when they are rehydrated. Whether the damage occurs directly as a result of the process of desiccation or is caused by the rapid influx of water into bryophyte cells when free water becomes available is still a question for debate.

We do know that changes in desiccated cells occur. We have already discussed physical and structural changes in membranes associated with the generation of ROS production associated with desiccation. We also know that the composition of membranes is altered in desiccation-tolerant tissues during desiccation; for example, Buitink *et al.* (2000) demonstrated that tolerant tissues differ from sensitive tissues in the partitioning of amphiphilic substances into membranes. As cells dry, amphiphilic substances partition from the soluble cytoplasmic compartment in to the membranes, resulting in a deleterious change in the integrity of the membrane. Buitink *et al.* (2000) found that transfer of amphiphilic molecules occurred at higher water contents in desiccation-sensitive tissues than in those that are desiccation-tolerant. They suggest that the amphiphilic molecules that enter membranes at the higher water content lead to a disruption of membrane integrity, whereas those compounds that enter at low water content increase membrane stability, perhaps acting as anti-oxidants. Regardless, at least for *Tortula ruralis*, visible and extensive membrane damage is not evident (Platt *et al.* 1994).

The rehydration of dried cells is also known to result in significant injury; all desiccation tolerance mechanisms have to include a process that attempts to prevent or limit such damage (Osborne *et al.* 2002). Orthodox seeds manage to lessen the effect of rehydration by slowing it down and allowing water to re-enter cells in a controlled and somewhat ordered fashion. The structure of bryophytes precludes this possibility and they rehydrate almost instantaneously when water is added. In light of this one would hypothesize that bryophytes, in order to combat rehydration (or desiccation) induced damage, would have to rely on a process that was induced as water enters the cells to protect them from damage or rapidly repair any damage that was inflicted by the inrush of water. The phenological evidence would suggest that this hypothesis is a valid one. The first sign that damage has occurred is the leakage of solutes from the protoplasm, possibly as a result of an alteration in membrane structure even though it is not visible. In desiccation-tolerant tissues this leakage is transient, however, indicating that damage has been limited to a level at which mechanisms activated by rehydration can effectively repair it. The extent of leakage is dependent upon the rate at which the prior desiccation event occurred. The faster the drying rate, the more solutes are leaked during rehydration (Bewley & Krochko 1982, Oliver & Bewley 1984, Oliver *et al.* 1993). The cessation of solute leakage occurs relatively quickly, within minutes, suggesting that the repair components are present at all times and are activated by rehydration. It is generally accepted that membrane phase transitions are the cause of rehydration leakage in vegetative tolerant tissues (for review see Crowe *et al.* 1992).

When water enters the dried cells of *Tortula ruralis* the condensed cytoplasm rapidly expands to fill the cavity that formed as a result of plasmolysis (Tucker *et al.* 1975). Over the next five minutes, as seen by using Nomarsky optics (without fixation), the chloroplasts swell and take on a globular appearance and thylakoids appear disrupted, as confirmed by electron microscopy after fixation of cells five minutes after rehydration (Tucker *et al.* 1975). The extent of thylakoid disruption is directly related to the speed at which desiccation occurred prior to the rehydration event: the faster the desiccation, the greater the disruption (Oliver & Bewley 1984). Mitochondria also swell and exhibit disruption of their inner membranes but the level of damage to these organelles appears to be unrelated to the rate of prior drying. Such events seem common for desiccation-tolerant mosses in the few minutes following rehydration (Oliver & Bewley 1984) and in all cases normal cellular structure is achieved within 24 h. It has been suggested that such cellular observations are artefacts caused by the fixation process in preparation for microscopy (Wesley-Smith 2001). Such explanation seems, however, in this case very unlikely as fixation occurred five minutes after rehydration when cells were fully hydrated and hydrated controls show no such abnormalities. In addition, chloroplast swelling was clearly evident by Nomarsky optics and dried cells of desiccation-sensitive species exhibit structural abnormalities upon rehydration identical to those seen in desiccation-tolerant mosses, but in the sensitive species the cells do not recover, but die (Bewley & Pacey 1978, Krochko *et al.* 1978).

Recently, Proctor *et al.* (2007) elegantly demonstrated the effectiveness of the constitutive protection mechanism for desiccation tolerance in the moss *Polytrichum formosum*. Using moss that had been slowly dried to equilibrium with ambient air at 40%–50% RH, they were able to demonstrate that the fine structure of the moss was unaffected by the desiccation event and that physiologically both photosynthesis and respiration recover rapidly and reach predesiccation levels 24 h post-rehydration. Furthermore, this study provides evidence that under these conditions the initial recovery is independent of protein synthesis. Earlier work by Pressel *et al.* (2006) also indicated little or no damage in the leptoids and specialized parenchymal cells, which are involved in nutrient transport, of *Polytrichum formosum* under the same drying conditions. These authors also demonstrate a recovery to the predesiccation state within 12–24 h following rehydration and, interestingly, provide strong evidence for a key role of the microtubular cytsokeleton in the rapid cellular recovery process. These studies confirm those of Platt *et al.* (1994) for *Tortula ruralis* and *Selaginella lepidophylla* in that desiccation *per se* does not appear to result in cellular injury, perhaps as a result of an ordered collapse of the cells during drying, for which the activities of the cytoskeletal components are key as suggested by Pressel *et al.* (2006) and supported by Proctor *et al.* (2007). It is clear from these studies that the constitutive protection mechanism present in this species is sufficiently efficient to prevent any major or observable cellular damage under relatively mild drying conditions.

These studies taken in isolation tend to make it difficult to invoke the need for a cellular repair aspect to the desiccation tolerance mechanism in bryophytes as proposed by Bewley (1979) and expanded upon by Bewley & Oliver (1992), a point that is argued by Proctor *et al.* (2007). However, the conditions under which these plants were dried are ideal and invoke a drying rate that has proven to be the least stressful for bryophytes (Bewley 1979, Bewley & Oliver 1992, Oliver & Wood 1997) and does not reflect reliable conditions in the field, especially with regards to the more tolerant desert species. As discussed previously, there is a great deal of evidence for cellular damage during desiccation, and the level of that damage depends upon the rate of dehydration, the extent of dehydration, and the length of time that the plant remains dry (as evidenced by a loss of viability over time). Hence it is clear that cellular repair is a necessary part of the mechanism of tolerance in these plants. The need to invoke this particular aspect of the desiccation tolerance mechanism and the extent to which it is operative may vary with the particular circumstance, however. As we move forward to better understand the roles and activities of the gene products that are induced during rehydration in bryophytes, as I discuss in the following sections, the relative importance and depth of the repair process in desiccation tolerance mechanisms in bryophytes will become evident.

7.7 Biochemical and molecular aspects of recovery

The complexity of the recovery of bryophytes from desiccation is only just becoming clear and it is equally clear that understanding this complexity will be a lengthy and intricate endeavor. Over the years, *Tortula ruralis* has been the model for studies directed at understanding the recovery from desiccation in bryophytes and much has been achieved.

All desiccation-tolerant bryophytes rapidly recover synthetic metabolism following rehydration although, as was the case for structural components, the speed of the recovery of synthetic metabolism is affected by the rate at which desiccation occurred (Oliver & Bewley 1997). Protein synthesis, one of the more sensitive processes to desiccation, recovers to normal levels within the first two hours following the addition of water to dried gametophytes of *Tortula ruralis* (Gwózdz *et al.* 1974), closer to three to four hours if the moss is rapidly dried (M. J. Oliver, unpublished data). The recovery of protein synthesis is fast because many of the components necessary for active protein synthesis are conserved in the dried moss, the only exception being an unidentified translation initiation factor(s) (Gwózdz & Bewley 1975, Dhindsa & Bewley 1976) which presumably has to be synthesized during the initial recovery phase (probably from sequestered transcripts; see below).

Although protein synthesis recovers relatively rapidly, the moss does not simply start where it left off when desiccation occurred. Within the first two hours following rehydration the pattern of protein synthesis is dramatically altered from that seen in hydrated moss prior to drying (Oliver 1991). This marks a major switch in the control of gene expression such that the synthesis of a number of proteins is reduced or terminated (termed hydrins) and the synthesis of other proteins is initiated or substantially increased (termed rehydrins). Hydrin and rehydrin are functional terms and do not refer to any common sequence motif or structural property nor imply a common enzymatic function, and so a protein is designated as one or the other by virtue of its biosynthetic response to desiccation and rehydration. Oliver (1991) was able to detect 25 hydrins and 74 rehydrins by using radioactive labeling of proteins and 2-D gel analysis, and was also able to demonstrate that the controls of the changes in the synthesis of the two groups of protein are not mechanistically linked. It takes a certain amount of prior water loss to fully activate the synthesis of rehydrins upon rehydration, whereas the synthesis of hydrins responds almost immediately upon the initiation of drying. The observation that there is a critical point in the loss of water from the gametophytic cells that will trigger the synthesis of rehydrins when water becomes available again suggests that the moss has evolved a strategy that allows it to respond only when it is likely that the plant will desiccate rather than simply experience a short period of dehydration. This would make sense as a means of conserving energy in an uncertain environment.

The change in gene expression, as observed by the change in the pattern of protein synthesis upon rehydration in *Tortula ruralis*, occurs within a background of a qualitatively unaltered mRNA population (Oliver 1991), suggesting that mosses respond to desiccation and rehydration primarily by an alteration in the control of translation rather than transcription as is the case with the stress and desiccation responses of angiosperms (Ingram & Bartels 1996, Phillips *et al.* 2002). In other words, the genes that code for proteins involved in the response to rehydration are already transcribed into mRNA. But what prevents their immediate translation? The main role of translational controls in regulating the gene expression response to rehydration was confirmed when rehydrin cDNAs were constructed and used to follow transcript accumulations by using northern blot analysis (Scott & Oliver 1994) and has since been further validated in genomic level studies (see below, Wood *et al.* 1999, Oliver *et al.* 2004, 2005) including expression profile analysis (M. J. Oliver, unpublished data). The manner in which the translational response to dehydration is controlled, as with most other processes, appears to be dependent upon the rate at which the previous desiccation event was achieved. If desiccation is rapid, rehydrin transcripts accumulate during the first hour following rehydration, apparently to replenish the pool of transcripts that has been reduced by rapid drying (Scott & Oliver 1994, Oliver & Wood 1997, Velten & Oliver 2001). One could argue that for rehydration following rapid desiccation the change in gene expression is transcription driven, however no novel transcripts are made, i.e. there appear to be no genes whose transcription is initiated during rehydration as seen in the angiosperm response, nor is transcription rapid enough to drive the change in protein synthesis one observes in these samples. Clearly translational controls still override the response. If desiccation occurs slowly, in contrast, rehydrin transcripts have accumulated to peak levels in the dried moss prior to the rehydration event such that they are readily available for translation as water returns to the cells (Scott & Oliver 1994, Oliver & Wood 1997, Wood & Oliver 1999). The accumulation of transcripts occurs during the slow drying phase, not by a process of increased transcription of selected genes, but by the sequestration of rehydrin mRNAs in a stable form, packaged with proteins that must be pre-existent in the moss (Wood & Oliver 1999). The sequestered rehydrin transcripts are maintained in an untranslatable form, presumably as a result of protein–RNA interactions during packaging, as messenger ribonucleoprotein particles (mRNPs) in the slow-dried gametophytic cells. Upon rehydration, it is postulated that the inrush of water results in the rapid release of the rehydrin transcripts from their associated proteins and their rapid recruitment into the protein synthetic complex for the biosynthesis of the various rehydrins. This is supported by their rapid inclusion in the polysomal fraction of rehydrated gametophytes (Scott & Oliver 1994, Wood & Oliver 1999). The sequestration of rehydrin transcripts during drying in preparation for the rigors of rehydration support the notion that the repair mechanism in rehydrated slow-dried *Tortula ruralis*, perhaps reflecting the natural response, is activated rather than induced (which implies a *de novo* assembly of components). The prevention of the storage of rehydrin transcripts by rapid desiccation may, at least in part, explain the slower recovery rate of rapid dried moss, as these gametophytes will have to rely upon what little was stored prior to dehydration until new components can be synthesized and assembled.

Some physiological studies into the recovery of photosynthesis following rehydration of desiccation-tolerant mosses have questioned the need for protein synthesis and repair with respect to the recovery of chloroplasts and chloroplast functions (reviewed by Proctor & Pence 2002, Proctor & Tuba 2002). Essentially, recovery of chloroplastic function, specifically Photosystem II, is extremely rapid (10–20 min) and is unaffected by protein synthesis inhibitors in the dark for rehydrated dried gametophytes (dried to −70 MPa) of several desiccation-tolerant mosses (Proctor & Smirnoff 2000, Proctor 2001). Protein synthesis is, however, required if rehydration takes place in the light, presumably to repair photo-oxidative damage. CO_2 uptake and assimilation do not recover instantaneously and require protein synthesis. Measurements of the rapid recovery of photosynthesis upon rehydration for dried *T. ruralis* also suggest that protein synthesis may not be required for the recovery of chloroplast structure and function (Tuba *et al.* 1996, Csintalan *et al.* 1999). These studies suffer, however, from some caveats. The effectiveness of protein synthesis inhibitors is difficult to assess as they can also affect the rate of uptake of the radiolabeled amino acids used to assess protein synthetic rates (Jacobyj & Sutcliffe 1962, Parthier *et al.* 1964). Protein synthesis inhibitors also rarely prevent protein synthesis completely, even when used in combination. Proctor & Smirnoff (2000) report an effective inhibition of 90% after 20 min in the presence of two inhibitors (to inhibit both cytoplasmic and organellar synthesis); however, no uptake measurements or amino acid pool size measurements were reported. The level of desiccation obtained in the mosses used in these investigations is relatively moderate, approximately −70 MPa, and as discussed above the rate and depth of desiccation has a major effect on the amount of cellular damage seen in bryophytes. If damage is limited, 10% of normal protein synthetic rates may be sufficient to effect repair of the chloroplasts in these studies. This remains to be tested, however. Even with these considerations it does appear that parts of the photosynthetic machinery are well protected and may require little repair. It is interesting to speculate that, because of the need to rapidly utilize the time that water is available and the need for energy to repair other cellular damage, part of the bryophyte mechanism of tolerance lies in a focused and effective protection of the chloroplast, especially those processes involved in the generation of ATP. This may explain why transcripts for ELIPs, proteins that are involved in photo-protection of the chloroplast (Lindahl *et al.* 1997, Montane & Kloppstech 2000), are stable to rapid desiccation (Zeng *et al.* 2002).

7.8 Genomic approach to desiccation tolerance

In more recent times a genomic approach has been taken in order to fully understand the complexity of the desiccation-rehydration response of desiccation-tolerant mosses, again using *Tortula ruralis* as the bryophyte model (Wood & Oliver 2004). The genomic level approach best suited to this endeavor, at least at this stage, is to build an expressed sequence tag (EST) collection and use this as a resource for an expression profiling strategy using cDNA-based microarrays. Scott and Oliver (1994) isolated the first 18 rehydrin cDNAs; Wood *et al.* (1999) subsequently recovered 152 ESTs from a cDNA expression library constructed from mRNAs extracted from the mRNP fraction of slow-dried gametophytes. Of these cDNAs only 29% exhibited any sequence similarity to previously identified nucleotide and/or peptide sequences deposited in public databases. Of those ESTs that could be annotated with a putative identity or function, several were identified as encoding ribosomal proteins, indicating the importance of protein synthesis in the response, and others encoded protective proteins such as ELIPs and LEA proteins. The inability to annotate a significant portion of the isolated ESTs may be due perhaps to the dearth of bryophyte sequences in the public databases (a situation that is rapidly improving with the sequencing of the *Physcomitrella patens* genome: www.jgi.doe.gov/sequencing/why/CSP2005/physcomitrella.html), but may also indicate, as is seen with many of the desiccation-tolerant tracheophyte genomic resources (Illing *et al.* 2005, Iturriaga *et al.* 2006), that some genes associated with desiccation tolerance are truly novel and represent processes for which little is known. The large percentage of unknowns in the small *Tortula ruralis* EST collection was underscored by the more recent addition of 10 368 ESTs from a non-normalized rehydration specific library to the collection (Oliver *et al.* 2004). The 10 368 ESTs represent 5563 clusters (contig groups representing individual genes), 40.3% of which could not be assigned an identity by comparison to annotated sequences in the public databases (Oliver *et al.* 2004). The larger EST collection was derived from a non-normalized rehydration-specific library to allow for a qualitative look at transcript abundance during the recovery phase. Genome ontology (GO) mapping of the *Tortula* clusters gave a broad look at what cellular activities appear to be emphasized in the rehydrated gametophytes and confirmed that the protein synthetic machinery, membrane structure and metabolism, and plastid integrity are central to the response. The GO analysis also revealed previously unresearched areas of cellular recovery such as membrane transport processes, phosphorylation, and signal transduction. Signal transduction is especially intriguing given that translational controls appear more important in the alteration of gene expression than are transcriptional activities that respond at the culmination of a signal transduction. However, the rehydration cDNA library was derived from rehydrated rapid-dried gametophytes and these rely, at least in part, on the replenishment of rehydrin transcripts via transcription.

The qualitative aspects of the large *Tortula ruralis* EST collection allowed for a general look at the most abundant transcripts present in the early phases of recovery following rehydration. Surprisingly, seven of the 30 most abundant transcripts present in the rehydrated moss encode LEA or LEA-like proteins (Oliver *et al.* 2004, 2005), including one that we consider to be a "primitive" dehydrin like LEA, Tr288 (Velten & Oliver 2001). This observation led to the suggestion that in desiccation-tolerant mosses some LEA proteins may play a dual role by protecting cells during both dehydration and rehydration. Alternatively, a more provocative hypothesis would be that LEA proteins in desiccation-tolerant plants and seeds are not sequestered during dehydration to protect cells from the effects of water loss but are accumulated to deal with the rigors of rehydration/imbibition. The significance of the relatively crude bioinformatics measure of transcript abundance was given a strong boost by an expression profile study using a cDNA microarray constructed from the 5563 individual clusters derived from the large EST collection (Oliver *et al.* 2005). Twenty-four of the clusters (representing individual genetic elements) that exhibited at least a two-fold increase in transcripts accumulation levels within gametophytes that had been rehydrated for between 1 and 2 h have sequence similarity to known LEA protein sequences. These transcripts are also elevated greater than two-fold in the polysomal RNA fraction indicating their recruitment into the translational mRNA pool. One of these transcripts represents Tr288, a previously identified rehydrin that has been extensively characterized as belonging to the LEA protein group (Velten & Oliver 2001). Since each cluster represents a unique nucleotide sequence it would suggest that following rehydration *Tortula ruralis* has a wide range of LEA-like proteins available for use in the recovery process, again suggesting that these proteins play an important role in cellular protection from rehydration-induced damage or directly in its repair. Koag *et al.* (2003) demonstrated that a maize dehydrin (DHN1) was capable of binding to lipid vesicles with some degree of specificity, preferring lipid bilayers that contained a significant proportion of acidic phospholipids. Oliver *et al.* (2005) report some preliminary work with the protein encoded by Tr288 that suggests that it too is capable of selective binding to lipid vesicles, also preferring acidic phospholipid mixtures. This leads us to the hypothesis that in rehydrating *Tortula ruralis* gametophytes LEA proteins may serve to stabilize membranes, or perhaps serve a role in lipid transport for reconstitution of damaged membranes, during the cellular upheaval that results from the inrush of water to dried tissues.

7.9 Final comments

The genomic approach to understanding the very complex trait of vegetative desiccation tolerance offers much in the way of cataloging those genes whose products play some role in the response of bryophytes to desiccation and rehydration. Coupled with bioinformatics and expression profiling, the genomic approach can also, by inference, provide clues as to which cellular processes and activities should be targeted for further studies. It may even point towards important regulatory proteins or processes that offer the promise of controlling the response and ultimately, perhaps, by a biotechnological approach, be useful in establishing or improving dehydration tolerance (not desiccation tolerance) in an important agronomic species. However, there is reason for caution when considering these hopes and assertions. The genomic approach is only going to bear fruit if the hard cellular, metabolic, and biochemical studies accompany the identification of possible key targets. We must also understand which genes and processes are truly adaptive and are central to the phenomenon of cellular desiccation tolerance. Bray (1997) suggests that the change in expression of specific genes resulting from a dehydration stress can be misleading. It is possible that changes in gene expression result from cellular injury, a scenario not unlikely during desiccation and rehydration. Injury may trigger the upregulation of specific genes or gene products that are not directly involved in promoting adaptation to dehydration but rather simply responding to the injury and therefore of secondary importance. One way to approach this problem is to look at desiccation tolerance in a phylogenetic context as discussed by Oliver *et al.* (2000, 2005). Ongoing studies using a comparative genomic approach to alterations in gene expression as a result of desiccation and rehydration may form the foundation for such studies. Using a combination of bioinformatics and expression profiling to compare the response to desiccation and rehydration in a number of species that occupy significant and key positions in the phylogeny of land plants, we can start to sort out the details of the evolution of desiccation tolerance and gain an insight into what genes are truly adaptive and central to this phenotype. A recent report by Illing *et al.* 2005 is a small but important first step in this endeavor, revealing the evolutionary link between vegetative desiccation tolerances in angiosperms with the mechanism of desiccation tolerance exhibited in orthodox seeds. Comparisons using bryophyte models will be invaluable in this process and will underpin the evolutionary conclusions that can be drawn from such studies, which are now underway.

The overall picture of desiccation tolerance in bryophytes is rapidly growing and details are being flushed out in all areas of plant biology, from ecology to physiology, from physiology to cellular biology, from cellular biology to biochemistry, and from biochemistry to genetics (in all its forms). At the moment it is still in its infancy, as can be seen in Fig. 7.1. What we know is not overwhelming but we have many hypotheses to test and we have established a solid structure on which to build: exciting times lie ahead. The most exciting bryological breakthrough that offers much in these efforts is the sequencing of the full genome of the moss *Physcomitrella patens* (see above). Although this is not a desiccation-tolerant bryophyte the full genomic sequence of a moss, and the expanding genomic and genetic tools that accompany it, will revolutionize how

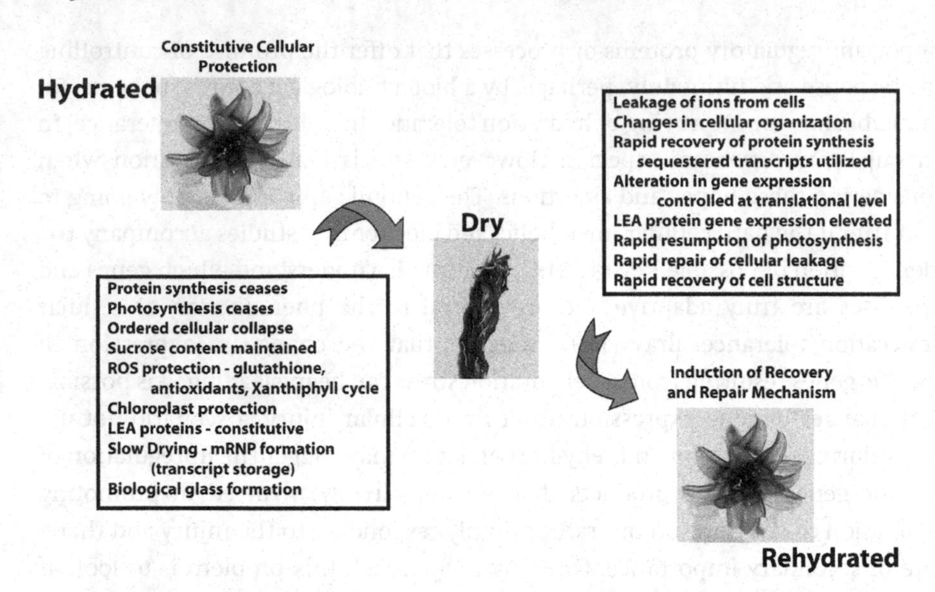

Fig. 7.1. Summary depiction of the current view of desiccation tolerance in bryophytes from a biochemical and molecular perspective.

we study important traits, not only in bryophytes but also in all plants. The ability to introduce, knock out, and replace genes via homologous recombination truly make this bryophyte model not only unique in plant biology at present but also very powerful for the analysis of gene function and importance in many plant specific processes (Frank *et al.* 2005). The full genome sequence increases the power of this plant model exponentially.

References

Alamillo, J., Almogura, C., Bartels, D. & Jordano, J. (1995). Constitutive expression of small heat shock proteins in vegetative tissues of the resurrection plant *Craterostigma plantagenium. Plant Molecular Biology*, **29**, 1093–9.

Alpert, P. & Oliver, M. J. (2002). Drying without dying. In *Desiccation and Survival in Plants: Drying Without Dying*, ed. M. Black & H. W. Pritchard, pp. 3–43. Wallingford: CABI Publishing.

Alscher, R. G., Donahue, J. L. & Cramer, C. L. (1997). Reactive oxygen species and antioxidants: relationship in green cells. *Physiologia Plantarum*, **100**, 224–33.

Apel, K. & Hurt, H. (2004). Reactive oxygen species: Metabolism, oxidative stress, and signal transduction. *Annual Review of Plant Biology*, **55**, 373–99.

Bartels, D. (2005). Desiccation tolerance studied in the resurrection plant *Craterostigma plantagineum. Integrative and Comparative Biology*, **45**, 696–701.

Bartels, D. & Sunkar, R. (2005). Drought and salt tolerance in plants. *Critical Reviews in Plant Sciences*, **24**, 23–58.

Bartels, D., Schneider, K., Terstappen, G., Piatkowski, D. & Salamini, F. (1990). Molecular cloning of abscisic acid-modulated genes which are induced during desiccation of the resurrection plant *Craterostigma plantagineum*. *Planta*, **181**, 27–34.

Beckett, R. P. (1999). Partial dehydration and ABA induce tolerance to desiccation-induced ion leakage in the moss *Atrichum androgynum*. *South African Journal of Botany*, **65**, 1–6.

Beckett, R. P. & Hoddinott, N. (1997). Seasonal variations in tolerance to ion leakage following desiccation in the moss *Atrichum androgynum* from a KwaZulu-Natal afromontane forest. *South African Journal of Botany*, **63**, 276–9.

Bewley, J. D. (1979). Physiological aspects of desiccation tolerance. *Annual of Review Plant Physiology*, **30**, 195–238.

Bewley, J. D. & Krochko, J. E. (1982). Desiccation tolerance. In *Encyclopedia of Plant Physiology*, vol. 12B, *Physiological Ecology II*, ed. O. L. Lange, P. S. Nobel, C. B. Osmond & H. Ziegler, pp. 325–78. Berlin: Springer-Verlag.

Bewley, J. D. & Oliver, M. J. (1992). Desiccation-tolerance in vegetative plant tissues and seeds: protein synthesis in relation to desiccation and a potential role for protection and repair mechanisms. In *Water and Life: A Comparative Analysis of Water Relationships at the Organismic, Cellular and Molecular Levels*, ed. C. B. Osmond & G. Somero, pp. 141–60. Berlin: Springer-Verlag.

Bewley, J. D. & Pacey, J. (1978). Desiccation-induced ultrastructural changes in drought-sensitive and drought-tolerant plants. In *Dry Biological Systems*, ed. J. H. Crowe & J. S. Clegg, pp. 53–73. London: Academic Press.

Bewley, J. D., Halmer, P., Krochko, J. E. & Winner, W. E. (1978). Metabolism of a drought-tolerant and a drought-sensitive moss: respiration, ATP synthesis and carbohydrate status. In *Dry Biological Systems*, ed. J. H. Crowe & J. S. Clegg, pp. 185–203. London: Academic Press.

Bewley, J. D., Reynolds, T. L. & Oliver, M. J. (1993). Evolving strategies in the adaptation to desiccation. In *Plant Responses to Cellular Dehydration During Environmental Stress. Current Topics in Plant Physiology: American Association of Plant Physiologists Series*, vol. 10, ed. T. J. Close & E. A. Bray, pp. 193–201. Rockville: American Association of Plant Physiologists.

Blomstedt, C. K., Neale, A. D., Gianello, R. D., Hamill, J. D. & Gaff, D. F. (1998). Isolation and characterization of cDNAs associated with the onset of desiccation tolerance in the resurrection grass, *Sporobolus stapfianus*. *Plant Growth Regulation*, **24**, 219–28.

Bopp, M. & Werner, O. (1993). Abscisic acid and desiccation tolerance in mosses. *Botanica Acta*, **106**, 103–6.

Bray, E. A. (1997). Plant responses to water deficit. *Trends in Plant Science*, **25**, 48–54.

Buitink, J., Hemmings, M. A. & Hoekstra, F. A. (2000). Is there a role for oligosaccharides in seed longevity? An assessment of intracellular glass stability. *Plant Physiology*, **122**, 1217–24.

Buitink, J., Hoekstra, F. A. & Leprince, O. (2002). Biochemistry and biophysics of tolerance systems. In *Desiccation and Survival in Plants: Drying Without Dying*, ed. M. Black & H. W. Pritchard, pp. 293–318. Wallingford: CABI Publishing.

Collett, H., Shen, A., Gardner, M. *et al.* (2004). Towards transcript profiling of desiccation tolerance in *Xerophyta humilis*: Construction of a normalized 11 k X. *humilis* cDNA set and microarray expression analysis of 424 cDNAs in response to dehydration. *Physiologia Plantarum*, **122**, 39–53.

Crowe, J. H., Hoekstra, F. A. & Crowe, L. M. (1992). Anhydrobiosis. *Annual Review of Physiology*, **54**, 579–99.

Csintalan, Z., Proctor, M. C. F. & Tuba, Z. (1999). Chlorophyll fluorescence during drying and rehydration in the mosses *Rhytidiadelphus loreus* (Hedw.) Warnst., *Anomodon viticulosus* (Hedw.) Hook. & Tayl. and *Grimmia pulvinata* (Hedw.) Sm. *Annals of Botany*, **84**, 235–44.

Cuming, A. C. (1999). LEA proteins. In *Seed Proteins*, ed. P. R. Shewry & R. Casey, pp. 753–80. Dordrecht: Kluwer Academic Publishers.

Deltoro, V. I., Calatayud, A., Gimeno, C., Abadia, A. & Barreno, E. (1998). Changes in chlorophyll *a* fluorescence, photosynthetic CO_2 assimilation and xanthophylls cycle interconversions during dehydration in desiccation-tolerant and intolerant liverworts. *Planta*, **207**, 224–8.

Demmig-Adams, B. & Adams, W. W. (1992). Photoprotection and other responses of plants to high light stress. *Annual Review of Plant Physiology and Plant Molecular Biology*, **43**, 599–626.

Dhindsa, R. (1987). Glutathione status and protein synthesis during drought and subsequent rehydration of *Tortula ruralis*. *Plant Physiology*, **83**, 816–19.

Dhindsa, R. & Bewley, J. D. (1976). Plant desiccation: Polysome loss not due to ribonuclease. *Science*, **191**, 181–2.

Frank, W., Decker, E. L. & Reski, R. (2005). Molecular tools to study *Physcomitrella patens*. *Plant Biology*, **7**, 220–7.

Foyer, C. H., Lelandais, M. & Kunert, K. J. (1994). Photooxidative stress in plants. *Physiologia Plantarum*, **92**, 696–717.

Gaff, D. F. (1977). Desiccation-tolerant vascular plants of Southern Africa. *Oecologia*, **31**, 95–109.

Goyal, K., Walton, L. J. & Tunnacliffe, A. (2005). LEA proteins prevent protein aggregation due to water stress. *Biochemical Journal*, **388**, 151–7.

Grelet, J., Benamar, A., Teyssier, E. *et al.* (2005). Identification in pea seed mitochondria of a late-embryogenesis abundant protein able to protect enzymes from drying. *Plant Physiology*, **137**, 157–67.

Guschina, I. A., Harwood, J. L., Smith, M. & Beckett, R. P. (2002). Abscisic acid modifies the changes in lipids brought about by water stress in the moss *Atrichum androgynum*. *New Phytologist*, **156**, 255–64.

Gwózdz, E. A. & Bewley, J. D. (1975). Plant desiccation and protein synthesis: II. On the relationship between endogenous adenosine triphosphate levels and protein synthesizing capacity. *Plant Physiology*, **55**, 1110–14.

Gwózdz E. A., Bewley, J. D. & Tucker, E. B. (1974). Studies on protein synthesis in *Tortula ruralis*: Polyribosome reformation following desiccation. *Journal of Experimental Botany*, **25**, 599–608.

Heber, U., Bukhov, N. G., Shuvalov, V. A., Koyabishi, Y. & Lange, O. L. (2001). Protection of the photosynthetic apparatus against damage by excessive illumination in homoihydric leaves and poikilohydric mosses and lichens. *Journal of Experimental Botany*, **52**, 1999–2006.

Heber, U., Lange, O. L. & Shuvalov, V. A. (2006). Conservation and dissipation of light energy as complementary processes: homoihydric and poikilohydric autotrophs. *Journal of Experimental Botany*, **57**, 1211–23.

Hellewege, E. M., Dietz, K. J., Volk, O. H. & Hartung, W. (1994). Abscisic acid and the induction of desiccation tolerance in the extremely xerophytic liverwort *Exormotheca holstii*. *Planta*, **194**, 525–31.

Hoekstra, F. A. (2002). Pollen and spores: desiccation tolerance in pollen and the spores of lower plants and fungi. In *Desiccation and Survival in Plants: Drying Without Dying*, ed. M. Black & H. W. Pritchard, pp. 185–205. Wallingford: CABI Publishing.

Hoekstra, F. A., Golvina, E. A. & Buitink, J. (2001). Mechanisms of plant desiccation tolerance. *Trends in Plant Science*, **6**, 431–8.

Iljin, W. S. (1957). Drought resistance in plants and physiological processes. *Annual Review of Plant Physiology*, **8**, 257–74.

Illing, N. A., Denby, K. J., Collett, H., Shen, A. & Farrant, J. M. (2005). The signature of seeds in resurrection plants: A molecular and physiological comparison of desiccation tolerance in seeds and vegetative tissues. *Integrative and Comparative Biology*, **45**, 771–87.

Ingram, J. & Bartels, D. (1996). The molecular basis of dehydration tolerance in plants. *Annual Review of Plant Physiology and Plant Molecular Biology*, **47**, 377–403.

Iturriaga, G., Cushman, M. A. F. & Cushman, J. C. (2006). An EST catalogue from the resurrection plant *Selaginella lepidophylla* reveals abiotic stress-adaptive genes. *Plant Science*, **170**, 1173–84.

Jacobyj, B. F. & Sutcliffe, J. F. (1962). Effects of chloramphenicol on the uptake and incorporation of amino-acids by carrot root tissue. *Journal of Experimental Botany*, **13**, 335–47.

Keever, C. (1957). Establishment of *Grimmia laevigata* on bare granite. *Ecology*, **38**, 422–9.

Kermode, A. R. & Finch-Savage, W. E. (2002). Desiccation sensitivity in orthodox and recalcitrant seeds in relation to development. In *Desiccation and Survival in Plants: Drying Without Dying*, ed. M. Black & H. W. Pritchard, pp. 149–84. Wallingford: CABI Publishing.

Knight, C. D., Sehgal, A., Atwal, K. *et al.* (1995). Molecular responses to abscisic acid and stress are conserved between moss and cereals. *Plant Cell*, **7**, 499–506.

Koag, M.-C., Fenton, R. D., Wilkins, S. & Close, T. J. (2003). The binding of maize DHN1 to lipid vesicles. Gain of structure and lipid specificity. *Plant Physiology*, **131**, 309–16.

Koorneef, M., Hanhart, C. J., Hilhorst, H. W. M. & Karssen, C. M. (1989). *In vivo* inhibition of seed development and reserve protein accumulation in recombinants of abscisic acid biosynthesis and responsiveness mutants in *Arabidopsis thaliana*. *Plant Physiology*, **90**, 463–9.

Krochko, J. E., Bewley, J. D. & Pacey, J. (1978). The effects of rapid and very slow speeds of drying on the ultrastructure and metabolism of the desiccation-sensitive moss *Cratoneuron filicinum*. *Journal of Experimental Botany*, **29**, 905–17.

Lindahl, M., Funk, C., Webster, J. *et al.* (1997). Expression of ELIPs and PSIIs protein in spinach during acclimative reduction of the photosystem II antenna in response to increased light intensities. *Photosynthesis Research*, **54**, 227–36.

Marschall, M. & Proctor, M. C. F. (1999). Desiccation tolerance and recovery of the leafy liverwort *Porella platyphylla* (L.) Pfeiff.: chlorophyll-fluorescence measurements. *Journal of Bryology*, **21**, 261–7.

McKersie, B. (1991). The role of oxygen free radicals in mediating freezing and desiccation stress in plants. In *Active Oxygen and Oxidative Stress and Plant Metabolism, Current Topics in Plant Physiology: American Association of Plant Physiologists Series*, vol. 10, ed. E. Pell & K. Staffen, pp. 107–18. Rockville: American Association of Plant Physiologists.

Meurs, C., Basra, A. S., Karssen, C. M. & van Loon, L. C. (1992). Role of abscisic acid in the induction of desiccation tolerance in developing seeds of *Arabidopsis thaliana*. *Plant Physiology*, **98**, 1484–93.

Mishler, B. D. & Oliver, M. J. (1991). Gametophytic phenology of *Tortula ruralis*, a desiccation-tolerant moss, in the Organ Mountains of Southern New Mexico. *Bryologist*, **94**, 143–53.

Montane, M. H. & Kloppstech, K. (2000). The family of light-harvesting-related proteins (LHCs, ELIPs, HLIPs): was the harvesting of light their primary function? *Gene*, **258**, 1–8.

Oliver, M. J. (1991). Influence of protoplasmic water loss on the control of protein synthesis in the desiccation-tolerant moss *Tortula ruralis*: ramifications for a repair-based mechanism of desiccation-tolerance. *Plant Physiology*, **97**, 1501–11.

Oliver, M. J. & Bewley, J. D. (1984). Desiccation and ultrastructure in bryophytes. *Advances in Bryology*, **2**, 91–131.

Oliver, M. J. & Bewley, J. D. (1997). Desiccation-tolerance of plant tissues: A mechanistic overview. *Horticultural Reviews*, **18**, 171–214.

Oliver, M. J. & Wood, A. J. (1997). Desiccation tolerance in mosses. In *Stress Induced Processes in Higher Eukaryotic Cells*, ed. T. M. Koval, pp. 1–26. New York: Plenum.

Oliver, M. J., Mishler, B. D. & Quisenberry, J. E. (1993). Comparative measures of desiccation-tolerance in the *Tortula ruralis* complex. I. Variation in damage control and repair. *American Journal of Botany*, **80**, 127–36.

Oliver, M. J., Tuba, Z. & Mishler, B. D. (2000). Evolution of desiccation tolerance in land plants. *Plant Ecology*, **151**, 85–100.

Oliver, M. J., Dowd, S. E., Zaragoza, J., Mauget, S. A. & Payton, P. R. (2004). The rehydration transcriptome of the desiccation-tolerant bryophyte *Tortula ruralis*: Transcript classification and analysis. *BMC Genomics*, **5.89**, 1–19.

Oliver, M. J., Velten, J. & Mishler, B. D. (2005). Desiccation tolerance in bryophytes: a reflection of the primitive strategy for plant survival in dehydrating habitats? *Integrative and Comparative Biology*, **45**, 788–99.

Osborne, D. J., Boubriak, I. & Leprince, O. (2002). Rehydration of dried systems: Membranes and the nuclear genome. In *Desiccation and Survival in Plants: Drying Without Dying*, ed. M. Black & H. W. Pritchard, pp. 343–64. Wallingford: CABI Publishing.

Parthier, B., Malaviya, B. & Mothes, K. (1964). Effects of chloramphenicol and kinetin on uptake and incorporation of amino acids by tobacco leaf disks. *Plant and Cell Physiology*, **5**, 401–11.

Phillips, J. R. & Bartels, D. (2000). Gene expression during dehydration in the resurrection plant *Craterostigma plantagineum*. In *Plant Tolerance to Abiotic Stresses in Agriculture: Role of Genetic Engineering*, ed. J. H. Cherry, pp. 195–9. Dordrecht: Kluwer Academic Publishers.

Phillips, J. R., Oliver, M. J. & Bartels, D. (2002). Molecular genetics of desiccation-tolerant systems. In *Desiccation and Survival in Plants: Drying Without Dying*, ed. M. Black & H. W. Pritchard, pp. 319–41. Wallingford: CABI Publishing.

Platt, K. A., Oliver, M. J. & Thomson, W. W. (1994). Membranes and organelles of dehydrated *Selaginella* and *Tortula* retain their normal configuration and structural integrity: freeze fracture evidence. *Protoplasma*, **178**, 57–65.

Porembski, S. & Barthlott, W. (2000). Granitic and gneissic outcrops (inselbergs) as center of diversity for desiccation-tolerant vascular plants. *Plant Ecology*, **151**, 19–28.

Pressel, S., Ligrone, R. & Duckett, J. G. (2006). Effects of de- and rehydration on food-conducting cells in the moss *Polytrichum formosum*: a cytological study. *Annals of Botany*, **98**, 67–76.

Proctor, M. C. F. (1990). The physiological basis of bryophyte production. *Botanical Journal of the Linnean Society*, **104**, 61–77.

Proctor, M. C. F. (2001). Patterns of desiccation tolerance and recovery in bryophytes. *Plant Growth Regulation*, **35**, 147–56.

Proctor, M. C. F. & Pence, V. C. (2002). Vegetative tissues: bryophytes, vascular resurrection plants and vegetative propagules. In *Desiccation and Survival in Plants: Drying Without Dying*, ed. M. Black & H. W. Pritchard, pp. 207–37. Wallingford: CABI Publishing.

Proctor, M. C. F. & Smirnoff, N. (2000). Rapid recovery of photosystems on rewetting desiccation-tolerant mosses: chlorophyll fluorescence and inhibitor experiments. *Journal of Experimental Botany*, **51**, 1695–704.

Proctor, M. C. F. & Tuba, Z. (2002). Poikilohydry and homoihydry: antithesis or spectrum of possibilities? *New Phytologist*, **156**, 327–49.

Proctor, M. C. F., Ligrone, L. & Duckett, J. G. (2007). Desiccation tolerance in the moss *Polytrichum formosum*: physiological and fine-structural changes during desiccation and recovery. *Annals of Botany*, **99**, 75–93.

Ramanjulu, S. & Bartels, D. (2002). Drought- and desiccation-induced modulation of gene expression in plants. *Plant, Cell and Environment*, **25**, 141–51.

Reynolds, T. L. & Bewley, J. D. (1993). Characterization of protein synthetic changes in a desiccation-tolerant fern, *Polypodium virginianum*. Comparison of the effects of drying, rehydration and abscisic acid. *Journal of Experimental Botany*, **44**, 921–8.

Schwabe, W. & Nachmony-Bascomb, S. (1963). Growth and dormancy in *Lunularia cruciata* (L.) Dum. II. The response to daylength and temperature. *Journal of Experimental Botany*, **14**, 353–78.

Scott, P. (2000). Resurrection plants and the secrets of eternal life. *Annals of Botany*, **85**, 159–66.

Scott, H. B. II, & Oliver, M. J. (1994). Accumulation and polysomal recruitment of transcripts in response to desiccation and rehydration of the moss *Tortula ruralis*. *Journal of Experimental Botany*, **45**, 577–83.

Seel, W. E., Hendry, G. A. F. & Lee, J. E. (1992a). Effects of desiccation on some activated oxygen processing enzymes and anti-oxidants in mosses. *Journal of Experimental Botany*, **43**, 1031–7.

Seel, W. E., Hendry, G. A. F. & Lee, J. E. (1992b). The combined effects of desiccation and irradiance on mosses from xeric and hydric habitats. *Journal of Experimental Botany*, **43**, 1023–30.

Schonbeck, M. W. & Bewley, J. D. (1981). Responses of the moss *Tortula ruralis* to desiccation treatments. II. Variations in desiccation tolerance. *Canadian Journal of Botany*, **59**, 2707–12.

Skriver, K. & Mundy, J. (1990). Gene expression in response to abscisic acid and osmotic stress. *Plant Cell*, **2**, 503–12.

Smirnoff, N. (1992). The carbohydrates of bryophytes in relation to desiccation-tolerance. *Journal of Bryology*, **17**, 185–91.

Smirnoff, N. (1993). Role of active oxygen in the response of plants to water deficit and desiccation. *New Phytologist*, **125**, 27–58.

Smirnoff, N. (ed.) (2005). *Antioxidants and Reactive Oxygen Species in Plants*. Oxford: Blackwell Publishing.

Stark, L. R. (1997). Phenology and reproductive biology of *Synthrichia inermis* (*Bryopsida, Pottiaceae*) in the Mojave Desert. *Bryologist*, **100**, 13–27.

Stark, L. R., Oliver, M. J., Mishler, B. D. & McLetchie, D. N. (2007). Generational differences in response to desiccation stress in the desert moss *Tortula inermis*. *Annals of Botany*, **99**, 53–60.

Stewart, G. R. & Lee, J. A. (1972). Desiccation-injury in mosses. II. The effect of moisture stress on enzyme levels. *New Phytologist*, **71**, 461–6.

Stewart, R. R. C. & Bewley, J. D. (1982). Stability and synthesis of phospholipids during desiccation and rehydration of a desiccation-tolerant and a desiccation-intolerant moss. *Plant Physiology*, **69**, 724–7.

Thompson, W. W. & Platt, K. A. (1997). Conservation of cell order in desiccated mesophyll of *Selaginella lepidophylla* ([Hook & Grev.] Spring). *Annals of Botany*, **79**, 439–47.

Tuba, Z., Csintalan, Zs. & Proctor, M. C. F. (1996). Photosynthetic responses of a moss, *Tortula ruralis* ssp. *ruralis*, and the lichens *Cladonia convoluta* and *C. furcata* to water deficit and short periods of desiccation, and their ecophysiological significance: a baseline study at present CO_2 concentration. *New Phytologist*, **133**, 353–61.

Tucker, E. B., Costerton, J. W. & Bewley, J. D. (1975). The ultrastructure of the moss *Tortula ruralis* on recovery from desiccation. *Canadian Journal of Botany*, **53**, 94–101.

Velten, J. & Oliver, M. J. (2001). Tr288: A rehydrin with a dehydrin twist. *Plant Molecular Biology*, **45**, 713–22.

Vicré, M., Lerouxel, O., Farrant, J. M., Lerouge, P. & Driouich, A. (2004). Composition and desiccation-induced alternations of the cell wall in the resurrection plant *Craterostigma wilmsii*. *Physiologia Plantarum*, **120**, 229–39.

Walters, C., Farrant, J. M., Pammenter, N. W. & Berjak, P. (2002). Desiccation stress and damage. In *Desiccation and Plant Survival*, ed. M. Black & H. W. Pritchard, pp. 263–91. Wallingford: CABI Publishing.

Werner, O., Espin, R. M. R., Bopp, M. & Atzorn, R. (1991). Abscisic acid-induced drought tolerance in *Funaria hygrometrica* Hedw. *Planta*, **186**, 99–103.

Wesley-Smith, J. (2001). Freeze-substitution of dehydrated plant tissues: artefacts of aqueous fixation revisited. *Protoplasma*, **218**, 154–67.

Wise, M. J. & Tunnacliffe. A. (2004). POPP the question: what do LEA proteins do? *Trends in Plant Science*, **9**, 13–17.

Wood, A. J. (2007). Frontiers in bryological and lichenological research. The nature and distribution of vegetative desiccation tolerance in hornworts, liverworts and mosses. *Bryologist*, **110**, 163–7.

Wood, A. J. & Oliver, M. J. (1999). Translational control in plant stress: the formation of messenger ribonucleoprotein particles (mRNPs) in response to desiccation of *Tortula ruralis* gametophytes. *Plant Journal*, **18**, 359–70.

Wood, A. J., Duff, J. R. & Oliver, M. J. (1999). Expressed sequence tags (ESTs) from desiccated *Tortula ruralis* identify a large number of novel plant genes. *Plant, Cell and Physiology*, **40**, 361–8.

Wood, A. J. & Oliver, M. J. (2004). Molecular biology and genomics of the desiccation-tolerant moss *Tortula ruralis*. In *New Frontiers in Bryology: Physiology, Molecular Biology and Functional Genomics*, ed. A. J. Wood, M. J. Oliver & D. J. Cove, pp. 71–90. Dordrecht: Kluwer Academic Publishers.

Xu, D., Duan, X., Wang, B. *et al.* (1996). Expression of a late embryogenesis abundant protein gene, HVA1, from barley confers tolerance to water deficit and salt stress in transgenic rice. *Plant Physiology*, **110**, 249–57.

Zeng, Q., Chen, X. & Wood, A. J. (2002). Two early light-inducible protein (ELIP) cDNAs from the resurrection plant *Tortula ruralis* are differentially expressed in response to desiccation, rehydration, salinity, and high light. *Journal of Experimental Botany*, **53**, 1197–205.

8

Mineral nutrition and substratum ecology

JEFF W. BATES

8.1 Introduction

Bryophytes do not appear to differ fundamentally from higher plants and green algae in their basic requirements for mineral macronutrients and trace elements. However, bryophytes differ significantly from vascular plants in pathways for nutrient acquisition and these may sometimes have far-reaching consequences for the ecosystems in which they grow. Owing to their specific modes of nutrient capture, bryophytes frequently accumulate chemicals to concentrations far exceeding those in the ambient environment. This property has led to the development of moss biomonitoring methods, which have taken hold firmly in the wider scientific community since the first edition of this book appeared.

As in the earlier edition, this chapter describes the special problems that bryophytes encounter in obtaining essential mineral nutrients, and in dealing with non-essential elements and compounds. Far more is known now than in the earlier edition about nitrogen deposition and utilization by bryophytes, and hence the chapter will focus on these aspects of mineral nutrition and substrate ecology.

The substratum on which a bryophyte grows can be a source of nutrients and other chemicals that may cause stresses. I have retained the useful distinction between "substrate", used for the substance on which an enzyme or biochemical process works (as in Section 8.3.1), and "substratum", used for the surface supporting a plant or lichen, although the etymological grounds for this are slight. Substratum specificity and chemical specialisms are considered in some detail but aspects involving competition and population dynamics are now largely covered in Chapter 10 by Rydin.

Bryophyte Biology: Second Edition, ed. B. Goffinet & A. J. Shaw. Published by Cambridge University Press.

8.2 Mineral nutrition

Knowledge about the elemental requirements of bryophytes has accrued slowly from cultivation experiments employing defined nutrient solutions, from chemical analyses of tissues and by studies of ion uptake and cell electrophysiology. Over the past ten years or so, modern developments for manipulating genes and identifying and visualizing their protein products have further revolutionized our understanding of the apparatus by which ions and molecules pass through plant cell membranes (e.g. Roberts 2006). Much of this work has involved *Arabidopsis thaliana*, now routinely employed as the "model" vascular plant. Fortunately, some comparable work has been undertaken in cultures (usually protonematal) of the "model" bryophyte *Physcomitrella (Aphanorrhegma) patens*; however, we are still largely ignorant about the diversity of molecular transport mechanisms present among bryophytes.

8.2.1 *Cell transport processes*

Transporter proteins

The thoughtful reviews of J. A. Raven (e.g. Raven 1977, 2003, Raven *et al.* 1998) about solute transport and molecular transport systems in bryophytes in comparison with other plant groups form useful starting points for anyone interested in this topic. In all living cells the lipid bilayer comprising the plasmalemma, the tonoplast, and other membranous organelles is penetrated by an array of proteins with specific functions. Among those with a solute transporter function are ion channels, envisaged as charge-lined tubes admitting hydrated ions or solutes of a specific size and charge, and carrier proteins that may change shape to pass the substrate ion or molecule through the membrane providing that it correctly fits the active site(s). In either case the direction of this "facilitated diffusion" depends on the existing gradients of concentration and electrical charge across the membrane. Control is provided by "gating" and "ungating" of channels in response to signals provided by chemical messengers, light, charge, etc. Passage of the substrate is only possible when the channels are ungated. Only a few simple molecules, including ammonia, carbon dioxide, and oxygen, are believed to be capable of diffusing through lipid membranes without such aids.

Aquaporins

Water is now known to diffuse through cell membranes predominantly via special channel proteins termed aquaporins (Chaumont *et al.* 2005). At least 12 different aquaporin proteins are believed to occur in *Physcomitrella patens* (Borstlap 2002). These include representatives of all four main subfamilies of aquaporins found in vascular plants, indicating that they had evolved prior to the diversification of tracheophytes and bryophytes. As Borstlap (2002) notes, the surprising diversity of aquaporin proteins in land plants is probably related to the central importance of hydraulics in their everyday functioning.

Active transport of ions and solutes

Active transport is required when the direction of movement is against a transmembrane gradient in concentration and/or electrical charge. Bryophyte cells closely resemble vascular plant and green algal cells of Characeae in maintaining an electronegative cell interior, in part due to active efflux of H^+, probably catalyzed by a "P" type ATPase, which expends one mole of ATP per mole of H^+ pumped out. This proton pump also regulates cytoplasmic pH in bryophytes at around 7.3–7.6 and provides the driving force for a number of specific symporter proteins, allowing passage through the plasmalemma and against a concentration/charge gradient for NH_4^+, sugars and amino acids. Related mechanisms probably operate for active entry of K^+, NO_3^-, SO_4^{2-} and $H_2PO_4^-$ and for efflux of Ca^{2+} and Na^+ (via antiporter proteins) but not all have yet been unequivocally demonstrated in bryophytes. Much less is known about the specific details of tonoplast transport in bryophytes.

Transfer cells

Of relevance to the uptake and cell to cell fluxes of nutrients is the occurrence in bryophytes of specialized cells known as transfer cells. These are characterized by extensive and often complex ingrowths of the secondary cell wall into the cell lumen. This is necessarily accompanied by an increased area of the lining plasma membrane, which also becomes richly furnished with transporter proteins, and thus enhances rates of nutrient transfer with neighboring cells. Offler *et al.* (2003) note that these cells are known in all major plant taxonomic groups including bryophytes, algae, and fungi, and occur at sites where there are uptake bottlenecks, often involving apoplastic–symplastic transfers. In bryophytes, transfer cells have so far been detected only at the relatively protected gametophyte–sporophyte junction (Ligrone & Gambardella 1988). In vascular plant roots they also occur under certain types of nutrient deficiency and might be expected to occur in bryophytes under similar conditions. However, it can be imagined that cell plasmolysis during desiccation of poikilohydric plants would damage the convoluted membrane, which may explain their scarcity in bryophytes.

Significant intracellular (symplastic) conduction of photosynthate and other nutrients has only recently been claimed to be widespread in bryophytes (Ligrone & Duckett 1994, 1996, Ligrone *et al.* 2000), although largely on the basis of ultrastructural evidence (Raven 2003). Apart from the long-known and relatively well differentiated leptoids of some ectohydric mosses implicated in transport of photosynthate, many less obviously differentiated conducting parenchyma cells are believed to fulfil a similar function in bryophytes (Ligrone *et al.* 2000).

Action potentials

Action potentials are losses (depolarization) of the normal transmembrane potential difference that last in plant cells typically for a few seconds. They have been observed in the liverwort *Conocephalum conicum* and the hornwort *Anthoceros*. The depolarization in *C. conicum* is caused by Ca^{2+} influx and Cl^- efflux. The following repolarization is connected with entry of K^+ and efflux of H^+ (Trebacz *et al.* 1994). In aquatic green algae of the Characeae the ion fluxes (measured in large nodal cells) are believed to permit osmotic regulation of cell turgor (e.g. Shepherd *et al.* 2002). Virtually nothing is known about the occurrence of action potentials in the major groups of bryophytes or whether they have any involvement in membrane repair as implicated in higher plants. Action potentials could conceivably have an important role in osmotic adjustment as poikilohydric bryophyte cells undergo cycles of dehydration and rehydration.

8.2.2 *Mineral nutrient requirements*

Growth on defined media

An example is provided by Hoffman's (1966) study of the nutrient relations of the cosmopolitan moss *Funaria hygrometrica*. The effects of nutrient deficiencies were investigated in protonemata grown on agar containing Hoagland's solution with individual elements lacking. The results showed that *F. hygrometrica* has macronutrient and micronutrient requirements closely similar to those of vascular plants. Hoffman also performed "nutrient triangle" experiments to define the optimal ratios of the major anions (N:P:S) and cations (K:Ca:Mg) for growth. These overly complex factorial experiments are difficult to analyze effectively. Nevertheless, an important conclusion was that the growth of protonemata and the development of numerous gametophores were favored by quite different cation combinations. By contrast, in the anion experiments, the absence of any one of the elements resulted in poor protonemal growth and no gametophores. Hoffman and many subsequent investigators have used mature shoots (and protonemata) or thalli as the experimental material. Compared with experiments starting with spores, these have the potential problem that elements initially present in the plants may be in sufficient quantity to mask or confound the effects of the applied nutrient treatments.

Bryophytes need relatively dilute nutrient media, in contrast to those required for optimal growth of crop plants. Working with the thalloid liverwort

Marchantia polymorpha, Voth (1943) varied the dilution of a basic nutrient solution to alter osmotic pressure but not the ratios of the elements. When a concentrated solution was used (solution 1), many of the thallus tips and wings were killed, the thallus dry mass and area were small, and production of gemmae cups was low. Over the intermediate concentration range (solutions 3–5), the plants increased in size, were darker, had more ascending tips, and developed more rhizoids in response to greater dilution of the nutrients. At the lowest concentrations (solutions 6–10) a greater intensity of red-purple coloration developed in the rhizoids, scales, and lower epidermis, and rhizoids were especially numerous whereas gemmae cups became fewer. Cell walls were extremely thin in the strongest solutions, with many collapsed cells seen, but a maximum thickness of cell walls was seen in the most dilute solutions, with most cells appearing healthy. Although survival of *M. polymorpha* is possible over a wide range of extreme concentrations, the species clearly grows best in dilute media. Only a handful of bryophytes has been subjected to scrutiny in solution culture experiments (Brown 1982) and further careful work is desirable.

Chemical analysis of tissues

Chemical analyses of bryophytes can provide useful clues about mineral requirements and about tolerance of non-essential elements, and offer a means of biomonitoring the deposition of elements such as heavy metals. Numerous studies have investigated total element concentrations of bryophyte tissues, usually employing dry-ashing or wet-ashing techniques to solubilize the minerals for analysis by spectrophotometric methods. Many authors have discussed the protocols for preparing materials for analysis and particularly the need to remove surface contamination by soil particles and rock fragments (e.g. Shacklette 1965, Woollon 1975, Brown 1982). However, washing of previously dried material is not recommended because of the risk of leakage of cell solutes during rehydration (Brown & Buck 1979). Washing with tapwater, a potentially mineral rich solution, is also likely to alter element levels through cation exchange (below). Therefore, before embarking upon a program of chemical analyses, it is important to consider the possible cellular locations of the elements in question and to design the sampling and extraction method to provide the maximum information for the effort involved.

8.2.3 *Mineral uptake by whole bryophytes*

Kinetics of ion uptake

The few studies of the kinetics of absorption of ions by bryophytes have concentrated upon heavy metal pollutants and radionuclides (Brown 1984). For metals, zinc absorption by the aquatic moss *Fontinalis antipyretica*

(Pickering & Puia 1969) is typical in showing a phase of rapid (30 minutes) uptake of 50% of the absorbed Zn. The remainder is absorbed slowly over several days; the absorption is sensitive to light, temperature and metabolic inhibitors. The rapid uptake represents passive sorption of zinc ions onto the extracellular cation exchange sites of the moss tissue (see below), whereas the slower phase is believed to represent true uptake into the cells. Wells & Richardson (1985) reported that a range of physiological anions and non-essential analogs displayed similar saturation kinetics in *Hylocomium splendens*. Cadmium uptake by *Rhytidiadelphus squarrosus* exhibited Michaelis–Menten kinetic constants (K_m, V_{max}) that differed quite markedly between field populations exhibiting slightly different morphologies (Brown & Beckett 1985). This is particularly relevant to the use of bryophytes in monitoring heavy metal deposition.

Cation exchange

Clymo (1963) emphasized the importance of *cation exchange* in accumulation of cations by *Sphagnum* in mires. In fact, the cell walls of most plants possess a net negative charge owing to the ionization of weak acid moieties built into their fibrillar structure. The *cation exchange capacity* (CEC) can be determined by saturating these sites with a suitable cation (A^{n+}) and then displacing this with another cation (B^{n+}) and measuring the quantity of A^{n+} released into solution. The plasmalemma is probably not exposed directly to the ionic composition of the exterior solution as the negative charges tend to repel anions and alter the ratio of cations entering the wall environment. Detailed study (Richter & Dainty 1989a) of the cation-exchanger in *Sphagnum russowii* suggests that polymeric uronic acids account for over half the CEC, phenolic compounds are responsible for about 25%, and amino acids, silicates, and sulfate esters deposited in the wall all make lesser contributions. Dependent on the pH of the external solution, all or a fraction of the acid moieties may ionize, and thereby release a proton; e.g., for carboxyls in uronic acids and amino acids,

$$R.COOH = R.COO^- + H^+.$$

Under strongly acid conditions the reaction is driven to the left and CEC falls as ionization is suppressed, but progressively through less acidic, neutral, and alkaline conditions the net negative charge increases. The extent of ionization of a given weak acid group is indicated by its pK, i.e. the pH at which 50% has ionized and 50% remains un-ionized. By varying pH stepwise in the presence of metals with contrasted valencies (Na^+, Ca^{2+}, La^{3+}), Richter & Dainty (1989a) showed that *S. russowii* possesses two classes of cation binding site. One, with a low pK (2–4), appears to be principally due to uronic acids and amino acids; the other, with a high pK (>5), is almost certainly due mainly to weak phenolic acids.

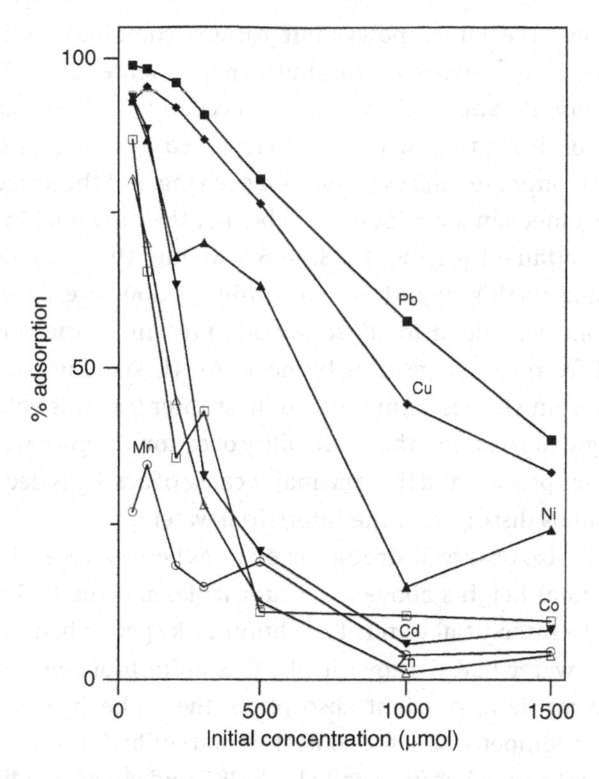

Fig. 8.1. Percentage absorption of metal ions onto the cation-exchanger of *Hylocomium splendens* from solutions containing equimolar concentrations of seven metal ions. The solutions were supplied (5 g of air-dried moss to 500 ml) at a range of initial concentrations and their final metal contents were determined after incubation for 2 h. Redrawn from Rühling & Tyler (1970).

Metals and other cations permeating the cell wall easily displace the protons from the ionized weak acids and may become relatively firmly held by the negative charges. At the same time the external medium receives the displaced protons and, in some circumstances, this may lead to its acidification. Exchangeably bound cations are readily displaced by other cations in the external medium, particularly if the latter: (a) are present at higher concentration, (b) have larger hydrated atomic radii, or (c) possess a higher valency. The data presented in Fig. 8.1 were obtained by incubating the moss *Hylocomium splendens* in a mixture of cations (Rühling & Tyler 1970). They reveal an order of binding affinity for several heavy metal cations (Cu, Pb > Ni > Co > Zn, Mn) that appears to be widespread. The heavy ions Cu and Pb were adsorbed preferentially onto the exchange sites even when supplied in the presence of much higher concentrations of the lighter cations Ca, K, Mg and Na. The behavior of the exchange sites varies with the species of cation employed to determine CEC. This is probably because the larger polyvalent cations combine strongly with and "condense" the fixed anions to varying extents (Sentenac & Grignon 1981, Richter & Dainty 1989b). Wide variations occur in CEC between bryophyte taxa and some of this variation appears to have ecological significance.

In *Sphagnum*-dominated mires the cation-exchanger of the *Sphagnum* plants is believed to be a mechanism, albeit probably not the only one, by which acidic conditions are attained (Clymo 1963, 1967, Brehm 1971, Clymo & Hayward 1982). According to this hypothesis, incoming cations are adsorbed and the released protons are added to those already present in the mire water, the production of fresh exchange sites by new growth keeping pace with cation inputs. More recent studies, summarized in Chapter 9 of this volume by Vitt & Wieder, strongly suggest that the availability of cations in mire waters is insufficient to drive this process and that the major cause of acidity is decomposition of humic compounds dissolved in the interstitial water.

Clymo (1963) also observed strong correlations between the CEC of *Sphagnum* spp., their optimal heights above the water table, and the hydrogen ion concentration of the interstitial water. Thus hummock species had the highest CEC and hummock water had the lowest pH. It is quite probable that the cation-exchanger has a role in nutrient absorption, the higher values of hummock species perhaps compensating for shorter periods of hydration in this position. Among plants *Sphagnum* has unusually high CEC under acid conditions, a factor that coincidentally favors heavy metal accumulation; however, an elevated CEC is also a characteristic of calcicole bryophytes (see Section 8.3.3).

Element location within the tissues

Much of the natural variability in total cation contents of bryophytes appears to reflect extracellular accumulations by the cation-exchanger rather than wide variations in the living cells. In many situations a clearer picture can be obtained if the intracellular and cation-exchanger compartments are analyzed separately. This can be achieved by employing a sequential elution technique as described by Brown & Wells (1988) and Bates (1992a).

Clear patterns emerge for the major cations when bryophyte taxa from different habitats are compared. Those with a clear metabolic function are accumulated within the cells at consistently high concentrations: a relatively high K concentration is believed to be essential for the normal folding of cytoplasmic enzymes; Mg is present in chlorophyll and is an activator of several enzymes; Ca is believed to act primarily as a "messenger" in plant cells and is largely absent from the cytoplasm but it is often the predominant cation externally on the cation-exchanger, reflecting its abundance in many natural situations as well as its importance as a stabilizer of cell membranes and cell

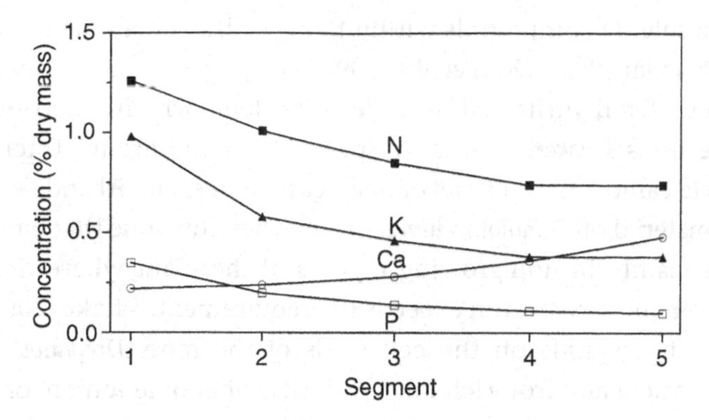

Fig. 8.2. Concentrations of some major nutrient elements in the annual stem segments of *Hylocomium splendens* on 7 August 1948 in boreal forest at Grenholmen, Uppland, Sweden. Segment 1 was initiated in 1948, segment 2 in 1947, and so on. After Tamm (1953).

walls (e.g. Hirschi 2004). In response to environmental stresses (low temperature, osmotic stress, abscisic acid), a signal is issued in the form of a transient release of Ca^{2+} that binds to the protein calmodulin. In *Physcomitrella patens* the calmodulin then binds to transporter-like proteins that catalyze ion fluxes that in turn may help alleviate the imposed stresses (Takezawa & Minami 2004). Roughly half of the total Mg in bryophyte tissues may also be exchangeable. Many other metals and some other cations, including the ammonium ion and cationic pesticides may also enter the cation-exchanger.

Element concentrations alter as tissues age. Tamm (1953) neatly demonstrated this in *Hylocomium splendens*. The shoots or "fronds" of *H. splendens* consist of chains of annual "segments". Each segment is normally clearly demarcated from its forbears and offspring owing to a predominantly sympodial pattern of growth that makes dating of the tissues comparatively simple. N, P, and K reached their highest concentrations in the young shoot apices and declined in older segments (Fig. 8.2). Ca, however, increased in the older segments on a dry mass basis. According to Bates (1979) this is partly an artefact arising through an increase in the cell wall : protoplasm ratio owing to slow degradation of the cell walls. Eckstein & Karlsson (1999) have provided a more detailed analysis of nitrogen dynamics in the segment chains of *H. splendens*.

Element concentrations in bryophytes also exhibit seasonal fluctuations that may be related to changes in the supply rates from the various sources and also biological factors such as growth dilutions in the plants themselves (e.g. Lewis Smith 1978, Bates 1987, Markert & Weckert 1989, Martínez-Abaigar *et al.* 2002a).

There are a few recent reports of the occurrence of biomineralization in bryophytes. This is the process whereby soluble elements combine to form crystals of insoluble compounds within living cells, a phenomenon that is well known in vascular plants. Ron *et al.* (1999) described a range of minerals including bohemite, ferrihydrite, gibbsite, jarosite, lepidocrocite, and pirolusite in cells of the moss *Hookeria lucens* in Spain. The same group (Estébanez *et al.* 2002) reported amorphous crusts of opal, carbonates, and Al and Fe hydroxides in healthy material of *Homalothecium sericeum*. They conclude that biomineralization occurs mainly in non-growing regions of the plant where the supply of elements from the substratum exceeds the requirement. Satake (2000) reported iron containing crystals on the cell walls of the moss *Drepanocladus fluitans* growing in an acid and iron-rich lake. A similar phenomenon, involving deposits containing Fe and Mn on the leaves of *Fontinalis antipyretica*, was reported from a stream polluted with mine effluent (Sérgio *et al.* 2000).

Mineral supply to the sporophyte

Mineral nutrients appear to reach the developing sporophyte from the gametophyte via conducting cells in the central strand of the seta. When Chevalier *et al.* (1997) supplied radioactively labeled orthophosphate to gametophores of *Funaria hygrometrica*, a proportion of the ^{32}P was eventually detected in the capsule and its spores. The proportion translocated was highest (18% of total absorbed) when the capsule was green without recognizable spores, but fell to zero in plants with mature brown capsules. Uptake of ^{32}P also occurred when the solution was applied directly to the capsule, indicating that absorption of nutrients from wet deposition by young sporophytes may occur in nature. Brown & Buck (1978) used an analytical approach to infer a similar pattern of nutrient cation movements from the leafy gametophores of *F. hygrometrica* to the developing sporophyte. By contrast, Basile *et al.* (2001), employing X-ray microanalysis, demonstrated that conduction of the heavy metals Pb and Zn is largely blocked in its passage from gametophyte to sporophyte by the transfer cells of the placenta. Zinc, an essential micronutrient, was able to pass this barrier more effectively than the inessential Pb. Rydin (1997) suggested that the production of sporophytes may be an important sink for nutrient resources in bryophyte populations but this needs verification for mineral nutrients.

8.2.4 *Nutrient inputs in nature*

Figure 8.3 shows the three most likely sources of nutrients for terrestrial bryophyte gametophores in nature: (1) the substratum; (2) wet deposition, i.e. precipitation including leachates from any plant or other surfaces over which it flows; (3) dry deposition, i.e. dust and gases (e.g. NH_3, SO_2, NO_2). Bryophytes may utilize several sources for the different essential elements. Techniques that have been used to study nutrient supply include: (i) analysis

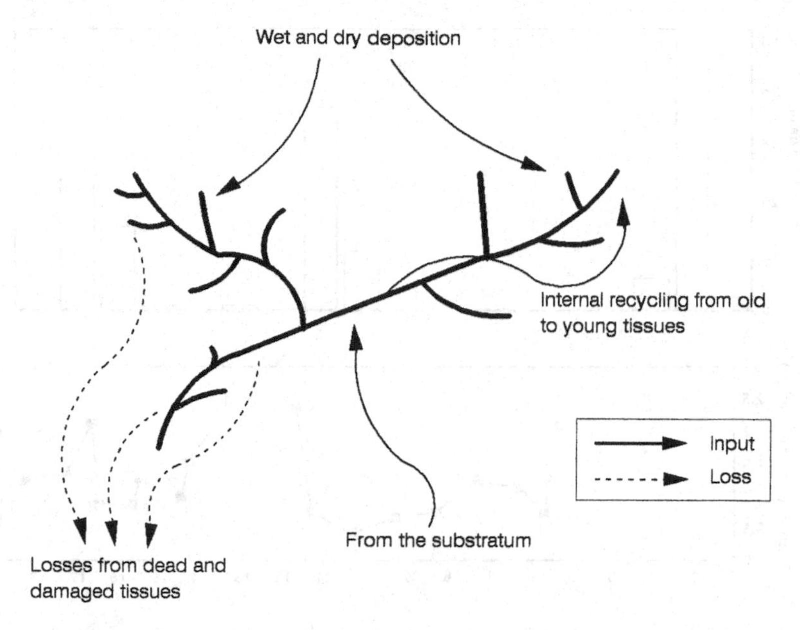

Fig. 8.3. A dynamic model of the potential inputs and losses of nutrients and non-essential elements to a bryophyte. Reproduced from Bates (1992a) with permission of the British Bryological Society.

of tissues and of precipitation (including canopy throughfall and tree stemflow) before and after passing through a bryophyte layer; (ii) nutrient application experiments.

Analytical studies

Tamm's (1953) study of growth and nutrition of the boreal forest moss *Hylocomium splendens* is widely regarded as a "classic" in bryology, having provided the foundation for many later investigations. Uptake of water from the soil by the ectohydric "fronds" of *H. splendens* is poor and he considered it unlikely that mineral nutrients were input by this pathway. Growth rates of *H. splendens* were higher under the forest canopy than in clearings, and particularly rapid in the zone under the boundaries of tree canopies. This was also the region where Tamm demonstrated the greatest nutrient enrichment of throughfall by leachates from the tree canopy. Thus, a major conclusion was that *H. splendens* received mineral nutrients predominantly as wet deposition. Canopy leachates appeared to be important as a source of P, which is present at very low concentrations in precipitation. Tamm also deduced that, despite a strong dependency of growth on moisture supply, nutrient limitation was the most important obstacle to the productivity of *H. splendens* in Norwegian forests.

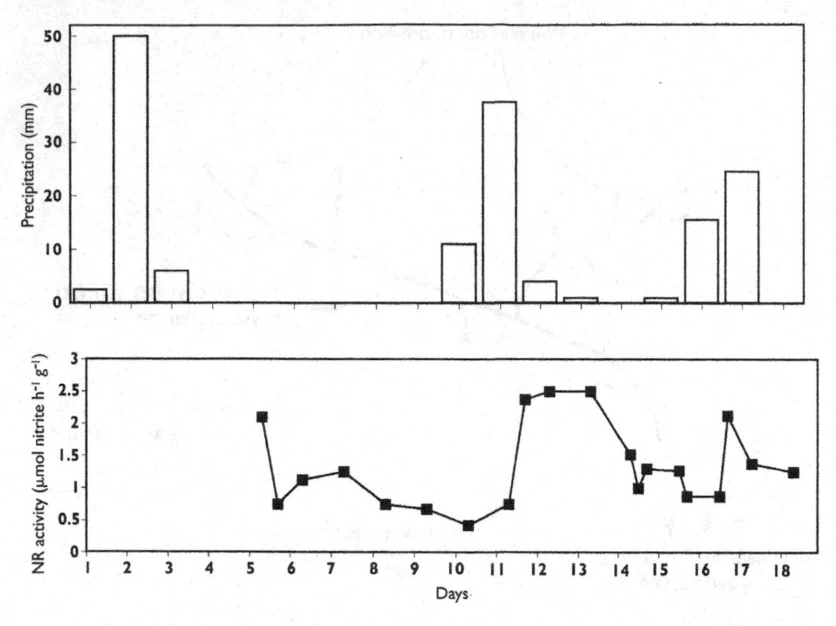

Fig. 8.4. Nitrate reductase (NR) activity in *Sphagnum fuscum* in relation to natural precipitation (upper graph) on an ombrotrophic, subarctic mire at Abisko, Sweden. Day 1 corresponds to 18 July 1983. NR activities are means of four replicates. Redrawn from Woodin *et al.* (1985).

The importance of wet deposition in supplying mineral elements to *Sphagnum* species in Scandinavian ombrotrophic ("rain-fed") mires was also inferred by Malmer (1988). Total concentrations of several elements and especially N and S showed, in some cases, correlations with known wet depositions of the elements at the mires. With regard to N utilization by ombrotrophic peat-mosses, Woodin *et al.* (1985) demonstrated a remarkably close coupling between the atmospheric supply of nitrate ions in wet deposition and the assimilation of this N source by the nitrate reductase enzyme. During dry periods the nitrate reductase activity in *S. fuscum* at an unpolluted site in Northern Sweden remained low, but activity rapidly increased during natural precipitation containing dilute NO_3^- (Fig. 8.4), or during experimental treatments of the moss carpet with 1 mM NO_3^-. Woodin *et al.* (1985) noted that by efficient capture of the NO_3^- in rainwater the *Sphagnum* plants deprive higher plants rooted in the peat of this nutrient supply.

An experimental approach to the problem of determining the sources of mineral nutrients to *Calliergonella cuspidata* in Dutch chalk grassland was adopted by van Tooren *et al.* (1990). They determined nutrient concentrations in rainfall, in the water dripping from *C. cuspidata*, and in the shoots of the moss.

In this instance, NH_4^+ was the only ion that appeared to be absorbed by the moss in significant amounts from the natural wet and dry deposition received. Interestingly, N and P concentrations were significantly higher at the end of the experiment in plants on soil compared to those on acid-washed sand, and uptake of N, P and K were all significantly higher in shoots on soil, and these also had a higher growth rate than the sand plants, when they were maintained in a humid garden frame. These results, although achieved under artificial conditions, show that nutrient uptake from soil cannot be ignored in ectohydric bryophytes.

Nutrient application experiments

The application of nutrients to natural swards of bryophytes has been increasingly used to examine the effectiveness of different supply pathways and to assess the likely impacts of anthropogenic increases in nutrient (mainly N) supply (see Section 8.2.9).

Elements present at elevated concentrations in the substratum are often found in high concentrations in bryophytes, indicating direct uptake. Hébrard *et al.* (1974) provided a unique demonstration of this by fashioning artificial boulders from a concrete mixture into which a solution of the radionuclide ^{90}Sr had been mixed. The isotope readily entered shoots of *Grimmia orbicularis* and *Leucodon sciuroides* later implanted into cracks in the boulders. Maximal concentrations in the shoots were attained during prolonged wet periods, these providing the most suitable conditions for solubilization and uptake of the ^{90}Sr. Although pleurocarpous mosses in forests generally have a poorer contact with their underlying soil, Bates & Farmer (1990) eventually found elevated Ca concentrations in the young apices of *Pleurozium schreberi* plants growing over a layer of calcium carbonate powder. It was concluded that Ca^{2+} ions had moved to the apices through the cell wall (apoplast) system under the influence of an evaporative moisture flow. Similar conclusions were reached by Brūmelis *et al.* (2000) following a transplant study with turves of *Hylocomium splendens*. Nutrient flow from underlying litter is also implied in a study of grassland bryophytes by Rincón (1988).

The importance of wet deposition in supplying macronutrients to ectohydric mosses was investigated by Bates (1987, 1989a,b) in *Pseudoscleropodium (Scleropodium) purum* in nutrient application experiments performed under field conditions. This pleurocarpous species, like *Hylocomium splendens* and *Pleurozium schreberi*, forms monospecific carpets that are separated from the underlying soil by a layer of accumulated litter in grassland, scrub and open forest habitats. Addition of K and Ca caused immediate increases of these metals in the cation-exchanger, moreover the addition of Ca displaced natural exchangeable Mg, but

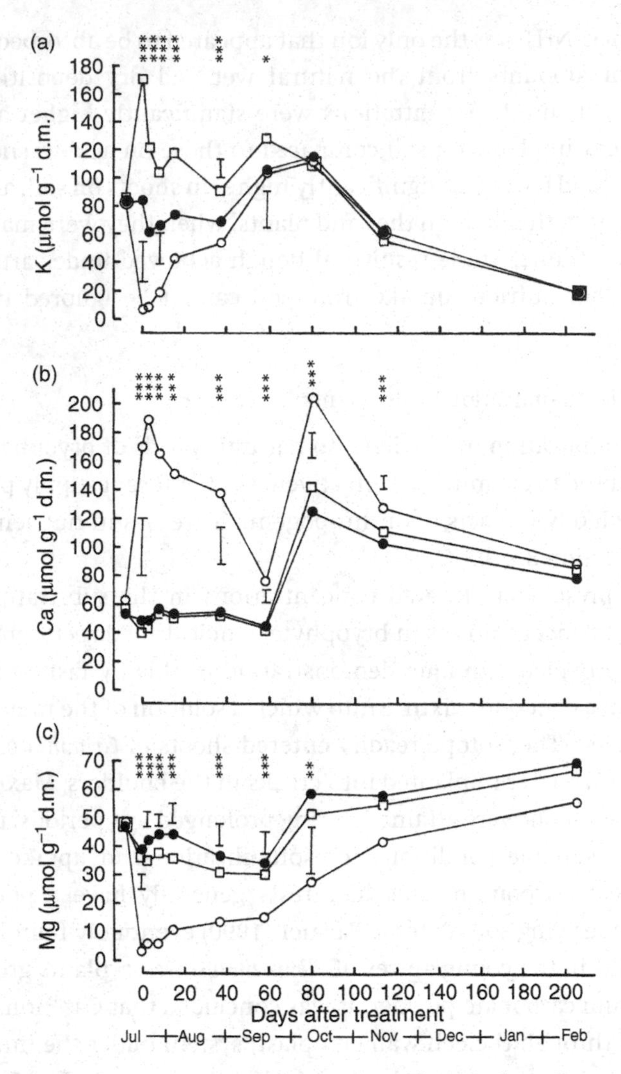

Fig. 8.5. Metal concentrations on the cation exchanger of *Pseudoscleropodium purum* immediately before nutrient application, and at intervals afterwards, in Windsor Forest, Berkshire, England. Filled circles, untreated; squares, KH_2PO_4-treated; open circles, $CaCl_2$-treated. (a) Potassium, (b) calcium, (c) magnesium. Significance of treatment effect at each harvest: ***, $p<0.001$; ** $p<0.01$; * $p<0.05$. Vertical bars represent $LSD_{0.05}$. Reproduced from Bates (1989a) with permission of the British Bryological Society.

levels of all three cations gradually equilibrated with their ambient availabilities as revealed by the control (Fig. 8.5). Interestingly, the Ca-treated shoots, in which exchangeable Mg had been displaced, experienced a period of significantly lowered Mg concentration in the intracellular fraction (Fig. 8.6). This result indicates

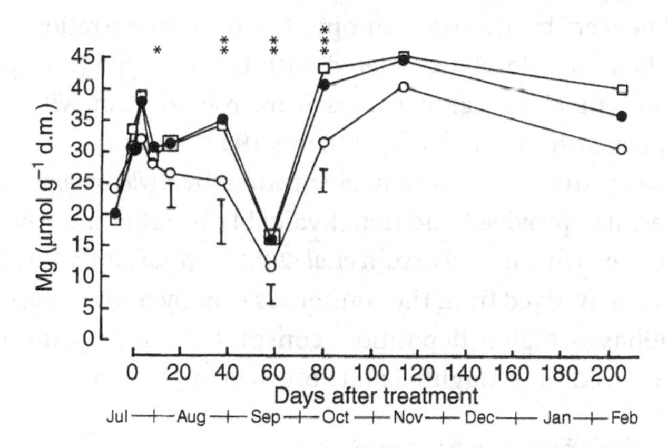

Fig. 8.6. Intracellular Mg concentration of *Pseudoscleropodium purum* before nutrient application, and at intervals afterwards. See Fig. 8.5 for details. Reproduced from Bates (1989a) with permission of the British Bryological Society.

that sequestration onto the cation exchanger is probably an important first stage in the absorption of Mg into the living protoplasts. In these experiments, a marked and protracted rise in the level of intracellular P was observed, but cellular absorption of K occurred only under conditions of prolonged moisture availability.

The ecological importance of the nutrient-retaining capacity of *Pseudoscleropodium purum* has been emphasized in a comparative study with *Brachythecium* (Bates 1994). In field conditions *B. rutabulum* normally maintains a higher productivity than *P. purum* (Rincón 1988, Rincón & Grime 1989) by exploiting nutrients in plant litter (see "litter species", below). However, when nutrients were supplied in an initial short "pulse" and the plants cultivated in a nutrient-free environment, the relative growth rate of *P. purum* was higher than that of *B. rutabulum*. It was concluded that *P. purum* utilizes the unpredictable nutrient supply in wet deposition in an efficient and opportunistic manner, whereas *B. rutabulum* relies upon a more or less continuous input of nutrients from its litter substratum.

Hoffman (1972) made a detailed study of ^{137}Cs transfers in a *Liriodendron* forest stand at Oak Ridge, Tennessee, that is extremely revealing about sources of elements to bryophytes. He introduced the radionuclide, which behaves as an analog of K in plant cells, into the tree stems through vertical slits. Eventually radioactivity was recovered in epiphytes and woodland floor bryophytes, the main pathway being via leachates from the tree canopy in the throughfall. On a dry mass basis the levels were ultimately higher in the bryophytes than in the tree foliage. A study of element concentrations in *Rhytidiadelphus triquetrus* transplants also suggested that the chemistry of this terricolous moss was strongly influenced by the tree canopy. Cation concentrations in the moss appeared to be more closely correlated with the soil type (calcareous or non-calcareous) in which the canopy trees were rooted than with the soil type immediately beneath the moss carpet (Bates 1993).

An eight-year study of nutrient absorption by *Hylocomium splendens* in a Swedish forest has provided additional valuable insights into the sources and utilization of N by this moss (Forsum *et al.* 2006). Importantly it was discovered that amino acids, derived from the coniferous canopy of the forest, particularly under conditions of high N deposition, constituted a significant proportion of the total N received in throughfall and absorbed by the moss.

8.2.5 *Desiccation effects on nutrient retention*

Bryophytes leak cell solutes during rehydration following a period of desiccation (e.g. Gupta 1977, Brown & Buck 1979). Until recently this process had been investigated only under laboratory conditions, but Coxson (1991) reported major losses of electrolytes from tropical forest epiphytes during rehydration. Wilson & Coxson (1999) described pulse releases of solutes from *Hylocomium splendens* in coniferous forest. The extent to which these losses may be disadvantageous or benefit other organisms merits further investigation.

The effects of desiccation–rehydration cycles on ion uptake and nutrient utilization have been little studied, but results from several field studies (Bates 1987, 1989a,b, van Tooren *et al.* 1990, Bakken 1994) suggest that desiccation may often impair the capacities of bryophytes to benefit from favorable nutrient regimes. Bates (1997) compared the growth and nutrient accumulation from regular applications of NPK in the mosses *Brachythecium rutabulum* and *Pseudoscleropodium purum* growing continuously hydrated or subject to intermittent drying. *Pseudoscleropodium purum* proved to be significantly more tolerant of desiccation than *B. rutabulum* and was able to absorb nutrients (e.g. P) almost as well under intermittent desiccation as under continuous hydration. Growth and nutrient uptake were both strongly suppressed by intermittent desiccation in *B. rutabulum*, whose high productivity depends on long periods of continuous hydration. These observations have been extended by Bates & Bakken (1998) and Badacsonyi *et al.* (2000).

Brown & Buck (1979) were among the first investigators to demonstrate convincingly that the cell wall cation exchanger of mosses probably functions to sequester and aid, via an apoplast–symplast route, the reabsorption of cations, such as K^+ and Mg^{2+}, leaked from the protoplasts during rehydration episodes. Supportive data are also presented by Bates (1997), Bates & Bakken (1998) and Badacsonyi *et al.* (2000).

In the case of nitrogen uptake as NO_3^-, assimilation by the enzyme nitrate reductase (NR) is a key process and one that is often sensitive to drought in vascular plants. In a comparison of NR activity following desiccation and rehydration in the highly desiccation-tolerant *Tortula ruralis* and *Porella platyphylla*, Marschall (1998) found some major differences. In the liverwort, NR activity remained unchanged in the light, but rose progressively within the first hour after rehydration in the dark. By contrast, NR activity declined sharply over the same period in the moss *T. ruralis*. These differences may result from fundamental differences in the pools of available sugars and reductants in mosses and liverworts upon rehydration, however, further work on NR in bryophytes is needed to clarify the situation.

8.2.6 *Evidence for internal recycling of nutrients*

Several writers (e.g. Malmer 1988, Brown & Bates 1990, Bates 1992a) have speculated that internal recycling of essential elements (i.e. from old to young tissues) may occur in bryophytes, thus removing the need for continued ion absorption. What evidence is there for such recycling?

Older work (e.g. Collins & Oechel 1974, Callaghan *et al.* 1978) had suggested that translocation of resources such as photosynthates did not occur in bryophytes, except in taxa like *Polytrichum* with obvious conducting tissues. However, Alpert (1989) demonstrated movement of photoassimilate (but not mineral nutrients) from leaves to stem bases and underground stems of the ectohydric moss *Grimmia laevigata*. Rydin & Clymo (1989) also obtained evidence of movement, from old to young tissues of *Sphagnum recurvum*, of both carbon and phosphorus compounds, and demonstrated the presence of numerous plasmodesmata linking stem cells and thus affording a possible symplast pathway. Neither of these species possesses recognizable conducting tissues in the sense of Hébant (1977). Ligrone & Duckett (1994, 1996) have described "food conducting" cells in many mosses that lack conventional conducting tissues but it is currently unknown whether these are involved in translocation of inorganic nutrients.

Wells & Brown (1996) devised an ingenious method for testing the recycling hypothesis employing shoots of *Rhytidiadelphus squarrosus* cut to different initial sizes (4 and 8 cm) and cultivated in nutrient-free conditions. Nutrient contents were determined in the new growth (N) and in the existing 2 cm segment. New growth was being supported entirely by the nutrient content of the existing growth, with elements moving from the latter for this purpose. When the shorter (4 cm) segments were used, the withdrawal of nutrients from the parent segments was proportionately greater in response to the smaller overall pool size. Brümelis & Brown (1997) working with segment chains of *Hylocomium splendens* also adjusted the internal nutrient pool available to the developing juvenile segment by removing branches from the parent segment. Branch removal led to a lowered K, Mg, Ca and Zn content of the juveniles. A similar approach was employed by Bates & Bakken (1998) except that nutrient pools were altered by killing (steaming) sections of stem and the two ecologically contrasted mosses *Brachythecium rutabulum* and *Pseudoscleropodium purum* were compared. Internal relocation of nutrients was important in the "low productivity" *P. purum* but not in the "high productivity" *B. rutabulum*.

An important contribution has been made by Eckstein & Karlsson (1999) who compared recycling of ^{15}N among segments of *H. splendens* with that occurring in ramets of *Polytrichum commune* in arctic Sweden. Young growth of both species was an important "sink" for N. In late summer all older segments in *P. commune* showed a net loss as the element was moved to subterranean stems. In *H. splendens* the dynamics can be summarized as follows: current-year's segments are totally dependent upon older segments for nitrogen and received a disproportionate supply of ^{15}N (i.e. the most recently absorbed N); one-year-old segments still import ^{15}N from older segments but also absorb external N and so act largely as conduits for N-supply to the juveniles; two-year-old segments act as storage sites for resources; three-year-old segments are degenerating and act only as sources of N. In a related study on *H. splendens* growing in subarctic birch woodland, Eckstein (2000) demonstrated that mean retention times for nitrogen in this moss varied from 3 to 10 years owing to effective acropetal transport and relocation of the element in younger tissues. Aldous (2002) has also demonstrated substantial translocation of nitrogen from older to younger tissues in *Sphagnum capillifolium* by using the tracer $^{15}NH_4{}^{15}NO_3$ added to field plots in four North American bogs. She applied an analysis of ^{15}N dynamics in the capitulum (1 cm) and top 2 cm of the stem and concluded that translocation mainly offered a second opportunity for the young tissues to benefit from recently deposited N and possibly also from N mineralized within the peat. This investigation did not provide conclusive information about whether the N translocation in *S. capillifolium* occurs in the symplast or by another pathway.

Collectively, these studies indicate that the relocation of elements within growing bryophytes is probably a widespread and important facet of their mineral nutrition.

8.2.7 *Role of bryophytes in ecosystem nutrient dynamics*

Bryophytes assume long-term dominance only in peatlands and some tundra environments where competition from higher plants is absent (Bates 1998). Nevertheless, they can also form conspicuous components of ecosystems dominated by higher plants, notably in moist forests, or become dominant for short periods in successional or ephemeral communities. In these situations they may have an importance in the overall nutrient economy of the ecosystem

Table 8.1 *Mineral inputs and accumulation in two forests with luxuriant bryophyte ground covers*

(a) Coed Cymerau oakwood, Wales (after Rieley et al. *1979)*

	kg ha^{-1} yr^{-1}			
	Ca	Mg	K	Na
Throughfall	10.0	13.9	19.0	103.8
Litterfall	21.0	4.2	10.2	3.1
Total input (T + L)	31.0	18.1	29.2	106.9
Bryophyte accumulation	4.1	3.9	14.3	1.6

(b) Washington Creek black spruce forest, Alaska (after Oechel & Van Cleve 1986)

	meq m^{-2} per season				
	N	P	Ca	Mg	K
Combined throughfall and litterfall	24.0	0.6	29.0	5.0	4.0
Bryophyte accumulation	92.0	5.0	14.0	12.0	16.0

that is disproportionate in relation to their often modest biomass (see reviews by Longton 1988, 1992, Slack 1988, Brown & Bates 1990, Bates 1992a).

In a successional community on glacial sands dominated by the mosses *Polytrichum juniperinum* and *Polytrichum piliferum* in New Hampshire, U.S.A., Bowden (1991) concluded that measurements of N in bulk precipitation accounted for 58%, and nitrogen fixation and coarse organic N for 7% of the total N input. The remaining 35% of N was input as wet deposited organic N, dry deposition, and dew. Nitrogen was retained, with only small losses, by both the mosses and the accumulating organic matter in the soil. This moss-ecosystem was extremely efficient at removing N from precipitation; when the moss was removed experimentally, N losses from the ecosystem temporarily exceeded inputs.

Bryophytes may be important in more complex ecosystems by absorbing nutrients in precipitation, dust, and litter before they can be taken up by the roots of higher plants (Oechel & Van Cleve 1986). This has already been mentioned with respect to utilization of wet-deposited NO_3^- by *Sphagnum* in ombrotrophic mires (Section 8.2.4). Detailed estimates of nutrient inputs by throughfall and litterfall to the bryophyte layer in two forests are summarized in Table 8.1. Nutrient accumulation by bryophytes, which formed about 90% of the ground flora in the Welsh oakwood, was comfortably exceeded by the inputs, but there is little excess K for tree and other higher plant growth. In the nutrient-poor Alaskan black spruce forest (Table 8.1b) the bryophyte layer appears to have accumulated more of every element except Ca than was input as throughfall and litter. Moreover, the moss layer retained considerable further potential for nutrient sequestering on its cation exchange complex. Additional nutrients may have been obtained by the bryophytes from the underlying soil (Oechel & Van Cleve 1986). Despite this efficiency in nutrient capture, two species (*Hylocomium splendens*, *Sphagnum nemoreum*) responded with higher photosynthesis when fertilized with a complete nutrient (Hoagland's) solution, suggesting that they had been nutrient limited. Bryophytes may also have significance in the nutrient economies of other communities where they are less conspicuous than in forest. In Dutch chalk grassland they grow and absorb nutrients during autumn and winter when higher plants are inactive, and they release nutrients by decomposition in spring and summer, which are utilized by the higher plants (van Tooren *et al.* 1988).

Bryophytes may also influence ecosystems by retaining nutrients for long periods in their undecomposed dead matter (Longton 1988, van Tooren 1988, Brown & Bates 1990). Bryophyte tissues decompose at much slower rates than those of higher plants, major factors being low temperatures, waterlogging, acidity, high cation exchange capacity, presence of high contents of lignin-like compounds, accumulation of lipids, and high carbon:nitrogen ratios. Insufficient data on bryophyte decomposition rates exist for many key habitats (Brown & Bates 1990).

Whether fungi are involved in mineral nutrient cycling between bryophytes and other ecosystem components remains uncertain. Chapin *et al.* (1987) presented data suggesting that mycorrhizal fungi of the dominant tree *Picea mariana* (black spruce) in Alaskan forest stimulated the release of phosphorus from the overlying bryophyte carpet to the tree roots. Quite a different picture was obtained by Wells & Boddy (1995) who observed translocation of ^{32}P from pieces of inoculated wood (buried in the leaf litter) to the living apices of the moss *Hypnum cupressiforme*. This occurred via the saprotrophic basidiomycete *Phanerochaete velutina*, which was observed to connect to the older parts of *H. cupressiforme*. Although true mycorrhizas involving mosses are unknown, this type of association might account for the uptake of scarce elements such as P from the underlying soil. Liverwort–fungus symbioses, in contrast, are relatively common (e.g. Duckett *et al.* 1991, Duckett & Read 1995) but little is known about their importance in mineral nutrition. Further important discoveries can be anticipated in this field.

8.2.8 *Effects of nutrient scarcity and nutrient excess*

Among vascular plants the often low availabilities of the two macronutrients, nitrogen and phosphorus, in natural habitats have provided a major stimulus to the evolution of a range of nutritional specialisms which often involve symbioses with micro-organisms. These include N_2-fixing root nodules, the carnivorous habit, mycorrhizas, and the "cluster roots" of Proteaceae and some other plant families. It is pertinent to ask whether any similarities exist among bryophytes. Ironically, the release of large quantities of anthropogenic nitrogen and phosphorus compounds into the atmosphere and waterways has come to characterize the modern age. There has been a recent surge of interest among plant scientists about the effects of excess N inputs on bryophytes and bryophyte-containing plant communities that had originally established by coping with shortages of these elements. Although pollution is outside the scope of this chapter, some mention of this work is required as it is informative about nutrient assimilation pathways and physiological effects of macronutrients.

Nitrogen

No carnivorous bryophytes are known, neither do bryophytes appear to enter into symbioses with soil bacteria such as *Rhizobium*, presumably because they lack roots and a sophisticated vascular system. Nevertheless, nitrogen fixation, by micro-organisms (sometimes unidentified) containing the nitrogenase enzyme, in "biological soil crusts" represents the main input of this element into a number of bryophyte-rich ecosystems, and particularly in polar tundra in both the Arctic and Antarctic (Belnap 2001). Studies in the High Arctic have revealed a high species diversity of nitrogen-fixing cyanobacteria growing epiphytically on the leaves and stems of mosses and probably benefiting from the moisture held amongst these structures (Zielke *et al.* 2002, 2005). N_2-fixation in other ecosystems may be responsible for smaller but still significant inputs (e.g. Brasell *et al.* 1986, Matzek & Vitousek 2003). Whether any of these associations represents specific mutualisms is currently unknown.

Associations of nitrogen-fixing cyanobacteria with hornworts and liverworts have long been known, but it is only recently that a glimpse of the biochemical sophistication of the symbiotic relationship has been obtained (Meeks 1998, Adams 2000). Based on studies of the *Anthoceros–Nostoc* and *Blasia–Nostoc* associations it is clear that the bryophyte, when starved of nitrogen compounds, releases one or more chemical signals that induce the formation of short, infective filaments called hormogonia by the *Nostoc* and may also have an attractant function towards these motile filaments (Meeks *et al.* 1999, Adams 2002). The hormogonia have smaller cells than parent *Nostoc* filaments, lack the characteristic N_2-fixing heterocysts, and exhibit gliding motion, which is vital for infection. They eventually inhabit mucilage-filled cavities on the ventral surfaces of the hornworts and the ventral auricles of *Blasia*. Upon infection, developmental changes lead to the generation of *Nostoc* filaments with greater densities of heterocysts than in the free-living parent filaments and the cyanobacteria begin to leak about 80% of their fixed nitrogen to the host plant in the form of NH_3. As in cyanobacterial lichens, where a similar leakage of reduced N occurs to the fungal symbiont, this release is connected with a reduced activity of glutamine synthetase in the symbiotic *Nostoc*, which partly blocks its entry into normal nitrogen metabolism within the cyanobacterium. Considerable progress has been made in identifying the genes in *Nostoc* (the *hrmUA* operon) switching the changeover from hormogonia to heterocyst-containing filaments in response to a hormogonium-repressing factor (HRF) discovered in aqueous extracts of *Anthoceros* (Meeks *et al.* 1999, Adams 2002).

Studies on the effects of increased wet deposited nitrogen to bryophytes have concentrated upon ombrotrophic mires, but have also involved minerotrophic (including calcareous) fens (Bergamini & Pauli 2001, Paulissen *et al.* 2005), acid grasslands (Morecroft *et al.* 1994, Carroll *et al.* 2000), boreal forest (Bakken 1994, Skrindo & Økland 2002, Forsum *et al.* 2006), arctic heath (Gordon *et al.* 2001), montane species (Woolgrove & Woodin 1996a,b, Pearce *et al.* 2003) and epiphytes (Mitchell *et al.* 2004, 2005). Mires are among the most nutrient deficient habitats and also locations where the peaty substratum contains a large body of fixed carbon that might be released into the atmosphere as greenhouse gases should conditions alter to favor its decomposition. Major shifts in vegetation composition, especially the ousting of bryophytes by higher plants, invariably accompany the application of macronutrients to natural bryophyte-rich ground communities (Mickiewicz 1976, Jäppinen & Hotanen 1990, Kellner & Mårshagen 1991, Virtanen *et al.* 2000, Aude & Ejrnæs 2005). However, much depends on the nature of the initial community, its degree of isolation from sources of potential invaders, and the type and intensity of fertilization.

One of the first effects of the quickly dissolving solid fertilizers used to improve timber yield in northern European forests is to cause "burning" of the bryophyte tissues contacted (Jäppinen & Hotanen 1990). Most of the common species (*Pleurozium schreberi*, *Hylocomium splendens*, *Dicranum* spp., *Sphagnum* spp.) decreased markedly in these studies, but *Polytrichum commune* appeared more resistant. The decline of *Rhytidiadelphus squarrosus* in acidic and calcareous grassland plots observed by Morecroft *et al.* (1994) and Carroll *et al.* (2000) in response to ammonium nitrate or ammonium sulfate additions was not accompanied by a detectable increase in higher plant cover and appears to have resulted from direct disturbance of the moss's nitrogen metabolism. This may also have been the cause of a loss of bryophyte biomass and diversity in response to N and NPK additions (as solids) in the calcareous fens studied in Switzerland by Bergamini & Pauli (2001).

In originally pristine mires, applications of combined-N may cause a stimulation of the elongation growth of *Sphagnum* species in the first year (e.g. Aerts *et al.*

1992, Gunnarsson & Rydin 2000) but have generally reduced growth of bog-mosses in the longer term (however, see Vitt *et al.* 2003). At sites already experiencing high N deposition, additions of the nutrient generally do not cause further growth increases of *Sphagnum*, instead other factors such as P availability or low temperature appear to limit growth (Aerts *et al.* 1992, Limpens *et al.* 2003a, Gunnarsson *et al.* 2004). No evidence was found that the fen moss *Calliergonella cuspidata* was directly damaged by high N deposition (Bergamini & Peintinger 2002). Much of the experimental work has used ammonium nitrate as the N source. Where ammonium and nitrate ions have been supplied separately there is evidence that the former is more detrimental to bryophytes (Pearce *et al.* 2003, Paulissen *et al.* 2005). Although there is strong evidence that *Sphagnum* and other bryophytes are competitively displaced by faster-growing vascular plants under high N deposition (e.g. Limpens *et al.* 2003b), there may also be detrimental effects owing to stimulation of the fungal parasite *Lyophyllum palustre* and of epiphyllic algae (Limpens *et al.* 2003c).

Baxter *et al.* (1992) showed that when *Sphagnum* species accumulate excess N, they do so in the form of increased free cytosolic amino acids. Production of these "useless" amino acids requires carbon "skeletons" and may eventually become harmful by depriving metabolism of fixed carbon. Nordin & Gunnarsson (2000) showed that where amino acid-N accumulation exceeded $2\,mg\,g^{-1}$ dry mass in the capitulum, growth began to be retarded. Paulissen *et al.* (2005) demonstrated large differences in the abilities of some fen bryophytes to form amino acids as a detoxification mechanism when challenged with excess NH_4^+: *Calliergonella cuspidata*, the most sensitive species in terms of growth reduction, did not accumulate N and produced only moderate amounts of arginine; *Sphagnum squarrosum* and *Polytrichum commune*, the most tolerant taxa, showed strong accumulation of total N and several amino acids. For further information about this complex and rapidly expanding subject also see the review by Turetsky (2003).

Phosphorus

P is often scarce in natural ecosystems or else present in unavailable forms within organic matter. Press & Lee (1983) demonstrated significant acid phosphatase activity in 11 species of *Sphagnum* surveyed in Britain and Sweden. Acid phosphatase activity was negatively correlated with the total P concentration of the plants and, in experiments, increased under conditions of phosphate starvation. The study was presented against the background of increased P supply due to atmospheric pollution, but the results imply that *Sphagnum* may often utilize simple organic forms of P in the peatland environment, the supply of inorganic P in precipitation being poor.

Christmas & Whitton (1998) have described appreciable surface phosphatase activity in the aquatic mosses *Fontinalis antipyretica* and *Rhynchostegium riparioides*, implying that these mosses can also utilize simple organic forms of P. Activities of phosphomonoesterase (PMEase) were highest in the nutrient-poor headwaters of a stream where tissue P concentrations were low. Downstream the activity of PMEase declined progressively as concentrations of P in water (and moss) increased. The authors suggest that assays of PMEase in aquatic mosses could provide a simple and reliable indication of nutrient status in streams where nutrient concentrations may fluctuate widely in the short-term.

Extending this study to a range of upland and mainly terrestrial mosses in northern England, Turner *et al.* (2003) demonstrated wide seasonal variations in phosphatase activities (usually highest in winter and lowest in summer) whereas tissue P concentrations remained relatively constant through the year. Noting a negative correlation between enzyme activity and nutrient availability, these authors suggest that information about past nutrient (P) conditions could be obtained by determining phosphatase activities of herbarium specimens.

Although P is one of the key nutrients responsible for anthropogenic eutrophication of waterways, its effects on aquatic bryophytes have received comparatively little attention (e.g. Bowden *et al.* 1994, Steinman 1994). In a laboratory study, Martínez-Abaigar *et al.* (2002b) demonstrated "luxury" accumulation of P by the aquatic liverwort *Jungermannia exsertifolia* subsp. *cordifolia* from culture solutions enriched with KH_2PO_4 and significant negative effects of the absorbed nutrient on net photosynthesis and pigment concentrations.

8.2.9 *Biomonitoring of mineral deposition*

As a result of their efficient mineral-absorbing capabilities, bryophytes have become popular organisms for biomonitoring levels and identifying sources of elements, especially pollutants, by analyzing their tissues. Indeed, based on the sheer numbers of surveys and published studies appearing in all parts of the world, a strong case can be made that biomonitoring is currently the primary justification for scientists spending valuable public resources studying bryophytes!

Bryophytes commonly sequester mineral elements, including those of major physiological importance, those that are required only in trace quantities (e.g. Cu, Fe, Mn, Zn) and those that are non-essential (e.g. Cd, Cr, Hg, Ni, Pb, Se, Sr, Ti, V). Some of these elements may be picked up from the substratum and others from wind-blown particles or in wet deposition. The absorption appears to involve three separate processes. First, there is passive adsorption onto the bryophyte's cation-exchanger. Generally, heavy metals are more effectively adsorbed than physiological cations like K^+ and Ca^{2+} and they may "condense" the exchange sites so that they are effectively immobilized. Second, some mineral elements are capable of entering the cells via transporter proteins (e.g. Brown & Beckett 1985, Brown & Sidhu 1992, Basile *et al.* 1994). Third, the numerous small leaves and intricate surfaces of bryophytes offer many possibilities for entrapment of metal-containing soil and ash particles. Collectively, these processes allow bryophytes to *bioaccumulate* heavy elements to concentrations far in excess of ambient concentrations or of concentrations found in most vascular plants.

Two main approaches have been used: (1) surveys involving analysis of indigenous bryophytes; (2) surveys with transplanted mosses and "moss bags". Both have the advantage that the basic materials are cheap and widely available, but there are also problems of sample reproducibility that must be carefully addressed if the results are to be meaningful. Besides revealing patterns in metal deposition, any survey of concentrations in living plant material is likely to be distorted to some extent by variations in moss growth, element uptake, and losses imposed by microhabitat variations (Damman 1978, Gerdol *et al.* 2002, Zechmeister *et al.* 2003, Leblond *et al.* 2004). In surveys with indigenous bryophytes, widespread pleurocarpous mosses have become the preferred subjects from a practical (and conservation) viewpoint. Careful protocols are required to standardize the samples used for analysis (see reviews of Brown 1984, Burton 1986, 1990, Tyler 1990, Berg & Steinnes 1997, Onianwa 2001).

Monitoring heavy metal deposition

Countrywide surveys of metals in mosses (usually *Hylocomium splendens* or *Pleurozium schreberi*) have been repeated over a long period in Norway and Sweden (Rühling & Tyler 2004). Figure 8.7 shows some results from repeated surveys in Norway. The tightly clustered Ni-isopleths in the north and south are explained by emissions from copper–nickel smelters on the Kola Peninsula and by long-range transport from industrial Europe, respectively. Lead reaches Norway principally through long-range atmospheric transport; the data suggest that deposition has decreased by 30%–40% since the first survey in 1977. Further reductions have been reported in the latest survey (Rühling & Tyler 2004). Similar surveys have now been carried out at five-yearly intervals in a coordinated way in the majority of European countries under the "Heavy Metals in European Mosses" monitoring program of the United Nations Economic Commission for Europe since 1990 (e.g. Markert *et al.* 1996, Herpin *et al.* 1996, Gombert *et al.* 2004). Multivariate analyses of these multielement surveys enable the "signatures" from different sources to be distinguished.

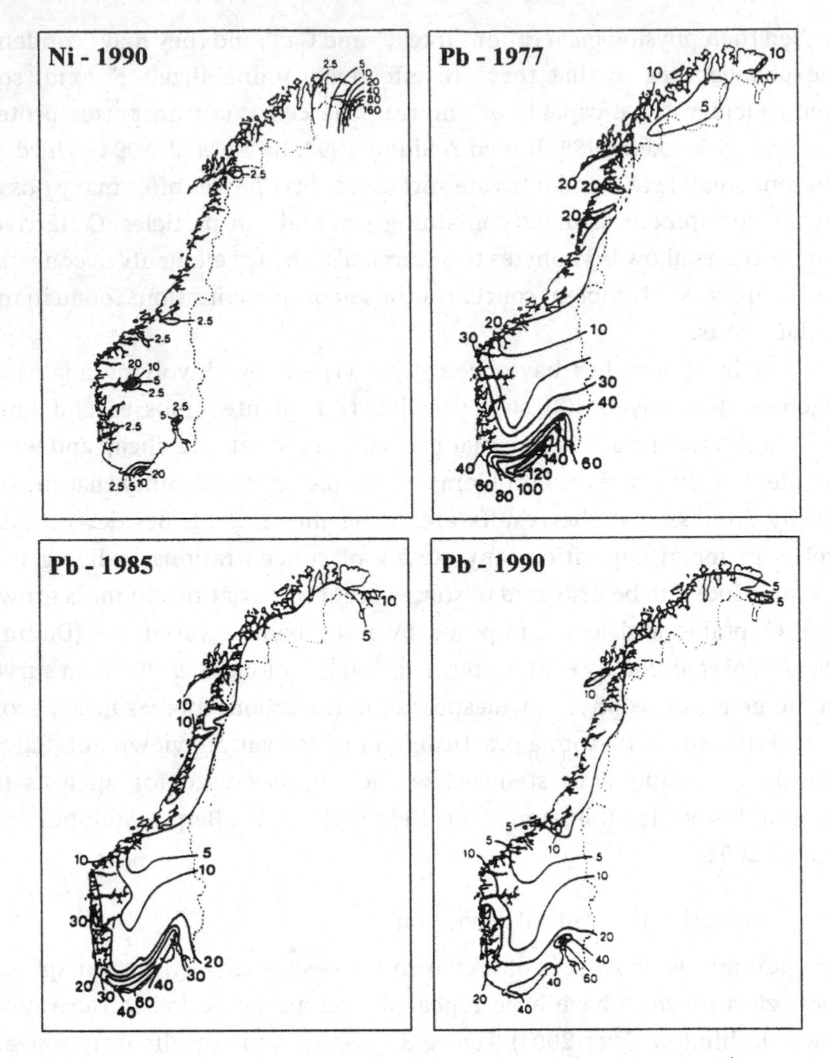

Fig. 8.7. Contour maps showing average nickel and lead concentrations ($\mu g\,g^{-1}$) of *Hylocomium splendens* in Norway based on samples collected at 495 sites (redrawn from Berg et al. 1995).

Berg *et al.* (1995) subjected the 1990 Norwegian data to principal components analysis and obtained the following major "axes": (1) with highest values in the south, representing long-range transport from other European countries of many elements (Bi, Pb, Sb, Mo, Cd, V, As, Zn, Tl, Hg, Ga); (2) elements associated with mineral particles in soil dust (e.g. Y, La, Al, Fe, V, Cr); (3) related to copper-nickel smelters producing Ni, Cu, Co and As; (4) marine influence (Mg, B, Na, Sr, Ca); (5) explained by a zinc smelter in the southwest (Zn, Cd, Hg); (6) related to iron mining in the far north (Fe, Cr, Al); (7) believed to reflect leachates from vascular plants, Cs and Rb are absorbed by roots and later transfer to mosses. Similar conclusions were reached by Kuik & Wolterbeek (1995) from a comparable analysis of heavy metal data from *Pleurozium schreberi* samples in the Netherlands.

Measurements of metal concentrations and their isotopes in herbarium specimens provide an alternative perspective on temporal changes in deposition in regions where there has been a long history of bryological exploration (Johnsen & Rasmussen 1977, Farmer *et al.* 2002).

In areas where the natural bryophyte vegetation is poor owing to atmospheric pollution or other stresses, metal deposition can be surveyed by transplanting bryophytes or employing moss bags. The latter consist of small samples of moss enclosed in an inert mesh bag that can be tied to tree branches or otherwise exposed for standard periods. The moss in these bags will usually die quickly from drought stress, if it has not already been killed by acid washing to remove contamination. Therefore very long exposure times will lead to disintegration and need to be avoided. Metal accumulation depends mainly on particulate trapping and cation exchange capacity, and although several species have been used, *Sphagnum* spp. have proved most popular (Brown 1984, Burton 1986). Other types of exposure may be necessary where longer exposure periods are required. Tuba & Csintalan (1993) described the exposure of living cushions of *Tortula ruralis* within wooden boxes for three months to determine metal deposition in and around an industrial town in Hungary.

Monitoring with aquatics

Submerged aquatics accumulate metals and other substances from water to a much greater extent than vascular plants, partly because their nutrient uptake is less seasonal, and partly because they can absorb over their entire surface. Drainage from disused metal mines is a common cause of water contamination but metals may also be in solution and accumulated owing to natural geological features (Samecka-Cymerman *et al.* 2000). In mid-Wales (U.K.) the liverwort *Scapania undulata* has proved to be one of the most metal-tolerant taxa among the common species of acid upland streams. Metal concentrations in its shoots reflect those in the water (McLean & Jones 1975). In many situations levels of metals in bryophytes appear to be at equilibrium with those in the water. Kelly & Whitton (1989) established the nature of these relationships for three mosses and a liverwort based on measurements of Zn, Cd, and Pb accumulation in many European streams. Each species and element exhibited a different pattern; moreover, the bryophytes absorbed greater quantities of metals than three algae with which they were compared. García-Álvaro *et al.* (2000) established relationships between water concentrations and those in

Rhynchostegium riparioides for five major nutrient elements, Fe, and Na. Uptake of metals is rapid, probably largely representing adsorption onto the cation-exchanger, and levels in bryophytes may also decrease (depuration) following a concentration spike (Mouvet *et al.* 1993, Claveri *et al.* 1994, Martins & Boaventura 2002). However, uptake of mercury by *Jungermannia vulcanicola* and *Scapania undulata* from an acid stream in Japan involved formation of crystals of HgS in the cell walls (Satake *et al.* 1990). Both indigenous aquatics and transplanted samples have been widely used to monitor heavy metal pollution in rivers. Mouvet (1985) and Mouvet *et al.* (1986) give examples where pollutant releases from industrial premises have been precisely identified by monitoring metal levels in mosses at intervals along watercourses. Chlorinated hydrocarbons and pharmacological compounds derived from faeces and urine may also enter water courses and be accumulated by aquatic bryophytes (Mouvet *et al.* 1993, Delépée *et al.* 2003).

Monitoring nitrogen deposition

Several recent studies indicate that bryophytes may also make effective biomonitors for assessing deposition of atmospheric N. Herbarium specimens were analyzed by Baddeley *et al.* (1994) to demonstrate increases in N content of the upland moss *Racomitrium lanuginosum* in Britain. Solga *et al.* (2005) and Solga & Frahm (2006) found significant correlations of N content and negative correlations of biomass with N deposition in some common pleurocarpous mosses in Germany. δ^{15}N ratios in the moss tissue were correlated with the ratios of ammonium-N to nitrate-N deposition among the sites. Pearson *et al.* (2000) also reported that δ^{15}N values calculated from isotopic assays of mosses growing on roofs and walls discriminated between the reduced and oxidized forms of atmospheric nitrogen in urban and rural environments.

8.3 Substratum ecology

8.3.1 *Range of substrata occupied*

Bryophytes grow on a wide range of natural substrata: soil, rock, bark, rotting wood, dung, animal carcases and leaf cuticles (Smith 1982a). From an ecological viewpoint (During 1979, 1992, Bates 1998) the main properties that determine whether a substratum can be colonized by a particular bryophyte species are: (1) the lifespan of the surface; (2) its chemical properties; (3) its water-holding capacity. It will be appreciated that if each of these properties offered just a few distinct habitat classes, collectively they would yield a range of contrasted ecological niches. In fact many bryophytes are faithful indicators for particular sets of substratum-related conditions.

8.3.2 *Longevity of substrata*

During (1979) emphasized the importance of the lifespan of the substratum (or its surface) in determining the kinds of bryophytes that might colonize it and successfully reproduce. It had already been established in higher plants that certain integrated sets of morphological and physiological characteristics favored particular *life-strategies* (e.g. Grime 1974). During (1979, 1992) argued that the most important habitat properties shaping the evolution of bryophytes are (1) longevity of the substratum, determining the effort to be put into rapid reproduction, and (2) the need for long-range dispersal to colonize new substratum patches, influencing the number and size of spores produced (many, small spores give the greatest chance of successful long-range dispersal). Where there is little need for long-range dispersal, a third factor, the evolutionary option to avoid any unfavorable season as a dormant spore rather than a more tender gametophyte, presents itself and favors large spore size.

The main life-strategies recognized by During (1992) are shown in Table 8.2. The first column shows types that, before the end of their lives, must colonize new substratum patches at some distance from the existing patch. Thus they produce many light spores to increase the chance of success. *Funaria hygrometrica* is the best known example of a *fugitive*, a mobile species that colonizes briefly available habitat patches (e.g. gaps in turf), then spreads to other, often distant sites. *Colonists* occupy similar unpredictably appearing habitats that persist for longer. Local population multiplication may be brought about by gemmae and rhizoid tubers (e.g. *Bryum bicolor*). Bryophytes that form long-lasting carpets on relatively stable forest floors (e.g. *Hylocomium splendens*) are *perennial stayers*. Their annual spore output is small but occurs over many successive years. The right-hand column includes species with larger spores where there is less need for long-range dispersal. *Annual shuttle* species are ephemeral plants that

Table 8.2 *Bryophyte life-strategies based on the revised system of During (1992)*

	Spores		
Potential life span	Numerous, small (<20 μm)	Few, large (>20 μm)	Reproductive effort
<1 year	*Fugitives*	*Annual shuttle*	High
A few years	*Colonists*[a]	*Medium shuttle*[c]	
Many years	*Perennial stayers*[b]	*Dominants*	Low

[a] Consisting of *Ephemeral colonists*, *Colonists s.s.* and *Pioneers*.

[b] Consisting of *Competitive perennials* and *Stress-tolerant perennials*.

[c] Consisting of *Short-lived shuttle* and *Long-lived shuttle*.

recolonize almost the same place (suitable "microsites") year after year from large immobile spores left by the previous generation(s). Their wider dispersal is often positively hindered by production of cleistocarpic capsules (e.g. *Ephemerum*, *Riccia*). Some of these species possess long-lived spores that remain dormant and enter a *diaspore bank* in the soil from which they may germinate in any of several successive years (During 1997). *Short-lived shuttle* and *long-lived shuttle* species occupy longer-lived microsites. Examples are provided by *Splachnum* spp. on dung patches and many epiphytes on twigs and branches. The longer-lived types also commonly have asexual propagules. Lastly, *dominants* refers to large-spored bryophytes that dominate certain ecosystems. *Sphagnum* species in peatlands are the only clear example (During 1992).

During's classification rests upon a subjective assessment of the correlations between bryophyte habitats and their morphological and physiological attributes. Hedderson & Longton (1995) employed multivariate analysis to study these plant and habitat concordances more objectively in three large orders of mosses (see also Longton 1997). They concluded that many of the life strategies did exist in fact, but that these should be regarded as "noda" in a continuous network of life history variation.

8.3.3 *Substratum and chemical specialists*

It is a well-established paradigm that many bryophytes are faithful indicators of particular microhabitats (Birks *et al.* 1998). One suspects that these associations often have a chemical basis as far as the substratum is concerned, although in other cases there appears to be some other ecophysiological explanation. Unraveling the causes of these relations is at the very heart of bryophyte ecology, yet in surprisingly few instances do we have a clear understanding of the reasons for substratum specificity (Cleavitt 2001, Pharo & Beattie 2002). Wiklund & Rydin (2004) remind us that the critical constraints on niche occupation may operate during establishment from spores rather than in mature gametophores. Using the examples of *Buxbaumia viridis* and *Neckera pennata*, rare mosses in Sweden, they show how environmental factors (in this case pH and moisture availability) interact to provide favorable but time-limited windows for successful colonization of the substratum. The following sections describe some of the more important specializations found among bryophytes.

Epiphytes

Plants that grow upon the stems of other plants without deriving sustenance from their living tissues are called epiphytes. The "host" is termed a *phorophyte*. The bark of trees in many parts of the world supports a diverse flora of epiphytic bryophytes, although they become scarce in very deep forest shade, on very acid surfaces (e.g. some conifers), under atmospheric pollution, and where the bark is abraded by winter ice or rubbed by livestock. The epiphytic flora reaches greatest luxuriance under continuously moist conditions, notably in high-altitude cloud forest (Pócs 1982). Numerous earlier studies of epiphytic communities have been reviewed by Smith (1982b). He separated epiphytes into "obligate" and "facultative" kinds, the latter also occurring in other habitats. Trees at maturity frequently support distinct vertical zones of communities (e.g. Trynoski & Glime 1982, Cornelissen & ter Steege 1989). Young twigs often support open communities with desiccation-tolerant taxa of Orthotrichaceae, *Frullania*, and small Lejeuneaceae, whereas the lower trunk may become completely swathed in a carpet of more desiccation-sensitive Brachytheciaceae and Hypnaceae. Phorophyte axes probably support a successional progression of communities as they age, but most studies have inferred this from measurements made at only one time (Tewari *et al.* 1985, Stone 1989, Lara & Mazimpaka 1998). Direct studies of epiphyte successional dynamics are badly needed. Obligate epiphytes mostly appear to be early successional species, whereas the luxuriant climax communities of the trunk base are usually dominated by facultative epiphytes (Smith 1982b, Bates *et al.* 1997).

Although a degree of "host specificity" is encountered among epiphyte communities, it is now clear that individual epiphytes respond to the nature of the environment rather than "recognize" a particular phorophyte species (Palmer 1986, Schmitt & Slack 1990). Much ecological work has been directed at discovering the main environmental factors affecting epiphytic communities. Numerous earlier data are brought together in the monumental *Phytosociology and Ecology of Cryptogamic Epiphytes* (Barkman 1958), which may still be consulted with profit. Sampling problems are also considered by Bates (1982a) and John & Dale (1995). Major substratum factors influencing community composition include longevity of the tree, rate of renewal of the bark surface, water-holding capacity of the bark, and its acidity and nutrient content. Lifespan of the tree becomes important if one likens it to an island that is progressively acquiring a flora. This concept has mostly been discussed in the context of conservation of "old forest" taxa by sympathetic forest management (Rose 1992). Trees with rapidly flaking or peeling bark like many conifers, *Eucalyptus* and *Betula* will clearly only be able to support rapidly establishing bryophytes with colonist and short-lived shuttle life-strategies. In dry climates a high water-holding capacity of the bark (e.g. as in *Sambucus nigra*) may be critical in allowing some species to survive as epiphytes, but this aspect has been little studied. Among trees with similar physical properties, bark acidity assumes major importance in determining community composition (Studlar 1982, Bates 1992b). Bark pH ranges from neutrality (e.g. *Ulmus*) to markedly acid (pH 3.5 or less in many conifers), but there is much intraspecific

Fig. 8.8. Leaves bearing epiphyllous liverworts and lichens from tropical-montane forest, near Ruhija, Bwindi Impenetrable Forest National Park, Uganda.

variation. Acidity and nutrient content of bark appear to be at least partially influenced by soil conditions (Bates 1992b, Gustafsson & Eriksson 1995) as well as by acid atmospheric pollutants, which can easily overwhelm the limited buffering capacity (Farmer *et al.* 1991). Nutrients in precipitation, canopy leachate, and dust are likely to be the main sources for epiphytes, together with any from decomposing bark (Bengstrom & Tweedie 1998, Hietz *et al.* 2002). Trees and their epiphytic coverings may also be highly effective in scavenging aerosol droplets from mists, especially in cloudy upland situations (Romero *et al.* 2006). Luxuriant epiphytes must frequently have a significant effect on forest nutrient dynamics (Rieley *et al.* 1979, Nadkarni 1984, Clark *et al.* 1998).

Epiphylls

In certain constantly humid forests bryophytes can colonize almost any relatively stable surface. In these conditions some species (called epiphylls) are capable of growing on the leaves of higher plants (Fig. 8.8). The phenomenon is most noticeable in the tropics and subtropics where large-leaved evergreens are prominent (e.g. Sjögren 1975, Pócs 1982). The most frequent epiphylls are tiny leafy liverworts of the Lejeuneaceae, but larger taxa of *Frullania, Plagiochila, Radula* and mosses frequently occur facultatively. Most of the specialist epiphylls are short-lived shuttle species (Table 8.2). Surprisingly, the most long-lived leaves support the lightest epiphyllous coverings and possibly these have adaptations that inhibit epiphyll growth (Coley *et al.* 1993). Following a long debate, it is generally agreed that epiphylls do not significantly reduce the photosynthetic output of host leaves. Coverings of epiphylls may actually deter herbivores but colonization by fungal pathogens is probably increased. Little is known about their nutrient relationships; however, Berrie & Eze (1975) showed movement of water and phosphate from host leaves to the epiphyllous liverwort *Radula flaccida*. Some of the rhizoids of *R. flaccida* were observed to penetrate the host's cuticle and contact the walls of epidermal and mesophyll cells. This does not appear to be an instance of outright parasitism, however, as no transfer of photosynthate was detected (Eze & Berrie 1977). The epiphyll–host relationship evidently merits fuller investigation.

Epiliths

Bryophytes inhabiting rocks have received less attention than epiphytes (Smith 1982b). Once again it is convenient to recognize "obligate" and "facultative" types, the genera *Gymnomitrion*, *Marsupella*, *Andreaea*, *Grimmia*, and *Racomitrium* containing many obligate epiliths. Such species possibly require a considerably more permanent and less water-retentive substratum than is provided by bark, but the reasons for their substratum selection are not well understood. Their competitive exclusion from more benign habitats on other substrata by faster-growing species is probably a factor. Bates *et al.* (1997) speculated that reduced competition, following atmospheric pollution, accounted for occurrences as epiphytes of normally epilithic bryophytes in parts of southern Britain. However, the balance between precipitation amounts and water-holding capacity of bark may also be important, with bark in drier areas being best able to support epiliths (Bates *et al.* 2004). Aho & Weaver (2006) present some useful ideas on methods for studying water relations and acidity on rock surfaces.

Most work on epiliths has aimed at delimiting the niches of species. Alpert (1985, 1988) studied the ability of *Grimmia laevigata* to colonize xeric microsites on rock surfaces. These were not colonized naturally, but adult plants transplanted to xeric sites survived without impairment, suggesting that a greater desiccation-sensitivity of the establishment phase limits a wider distribution. Jonsgard & Birks (1993) and Heegaard (1997) investigated the physical and chemical niches of *Racomitrium* and *Andreaea* species, respectively, in western Norway, using multivariate analyses and generalized linear modeling methods.

Litter species

Litter, meaning undecomposed dead plants, constitutes substrata ranging from the ephemeral (leaves) to the reasonably long-lasting (fallen tree trunks). All are potentially rich in plant nutrients although this may often be unavailable to bryophytes.

It is now evident that several bryophytes originally thought of as normal inhabitants of the soil surface are, at least seasonally, exploiters of litter deposited by dominant vascular plants. Rincón (1988) investigated the effects of a range of plant litter types on growth of some common grassland bryophytes. Nutrient-rich litter of the stinging nettle (*Urtica dioica*) stimulated the growth of all species and notably that of *Brachythecium rutabulum*, a moss that is frequently associated with the dense stands of *U. dioica* and other tall herbs. Rhizoidal attachments to the litter are important in nutrient exploitation (Rincón 1990).

A number of bryophytes, sometimes called *epixylic* species, occur more often on rotting logs than on other types of substratum. Most studies have been concerned with description of the succession of their bryophyte communities as fallen logs and cut stumps decay (Muhle & LeBlanc 1975). These go through a number of physical changes, such as loss of bark, softening of the wood, and break-up, and they may present a moister environment in later stages than initially. In a Swedish spruce forest Söderström (1988) recognized four stages in the succession: (1) facultative epiphytes that had fallen with the log (mostly lichens and *Ptilidium pulcherrimum*); (2) early epixylics (*Anastrophyllum hellerianum*, *Lophozia* spp., *Drepanocladus unciniatus*, *Cladonia* lichens) colonize soon after the log falls; (3) late epixylics (e.g. *Lepidozia reptans*, *Brachythecium starkei*, *Dicranum scoparium*, *Plagiothecium denticulatum*) do not colonize until decay is advanced; (4) ground flora species (e.g. *Hylocomium splendens*, *Pleurozium schreberi*, *Ptilidium crista-castrensis*) colonize as the log becomes indistinguishable. No evidence appears to have been obtained of mineral nutrient transfers from rotting logs to bryophytes. As some of the commonest species (e.g. *Brachythecium rutabulum*, *Eurhynchium praelongum*) also exploit leaf litter, however, this seems highly likely. Saprotrophic fungi could be involved in cryptic nutrient transfers (cf. Wells & Boddy 1995).

Decaying logs provide a classic example of a patchily distributed habitat of limited duration and their specialist bryophytes present opportunities to test several hypotheses in population biology (e.g. Herben & Söderström 1992). Kimmerer (1994) compared the dissemination of *Dicranum flagellare*, an asexually reproducing species, with *Tetraphis pellucida*, which produces both spores and gemmae. *Tetraphis pellucida* was highly successful at rapidly colonizing new logs and stumps, whereas *D. flagellare* persisted mainly by rapidly colonizing local gaps appearing through disturbance rather than by finding wholly new surfaces. Slugs appeared to be a major dispersal vector for the detachable branches of *D. flagellare* (Kimmerer & Young 1995).

Fire mosses

Fire is a natural event in many forest and grassland ecosystems but today humans are also responsible for a very large number of accidental and deliberately started fires. These range from small bonfires, through rejuvenating "burns" (e.g. of moor and scrub) to conflagrations that engulf wildlife over vast areas of countryside. Bryophytes are usually readily destroyed in fires, but they are often a conspicuous element in the early succession on burned land. Southorn (1976) pointed out that *Funaria hygrometrica* is associated with old fire sites throughout the world. At seven experimental bonfire sites in Surrey (England) she first observed bryophyte protonemata nine weeks after burning in spring-burnt sites but only after 25 weeks when burning was carried out in winter. Eventually a community composed of *F. hygrometrica, Ceratodon purpureus, Bryum argenteum*, and tuberous *Bryum* spp. became established in the first year after burning. This pioneer community was progressively deposed by recolonizing angiosperms in the second year, and bryophytes had vanished by the third year. In burnt *Picea mariana* forests in Labrador, Foster (1985) also recorded *F. hygrometrica* and *Ceratodon purpureus* as colonists of charred unstable surfaces, together with *Polytrichum juniperinum*. These were gradually replaced by normal forest mosses like *Pleurozium schreberi*, *Ptilium crista-castrensis*, and *Hylocomium splendens* when the tree cover had re-established, enabling light intensities to decrease and atmospheric humidity to increase. Bell & Newmaster (2002) and Newmaster *et al.* (2003), working in western Canadian forests, observed wide variations in the temperatures of wildfires and the degree to which the original bryophyte vegetation survived. Southorn (1976) demonstrated that under hot bonfires and slow forest fires the soil is greatly changed. Organic matter in the soils is burnt off and much soluble inorganic matter is deposited as ash, including the plant nutrients K, Mg, Ca, and P. The bases cause a dramatic rise in pH (up to 10.1 units in the Surrey bonfires). From culture experiments Southorn (1977) concluded that a requirement of *F. hygrometrica* for relatively high concentrations of nitrate-N and P was a major reason for its success on bonfire sites. Brown (1982), however, deduced that the raised pH may also be critical. The lag in colonization mentioned above was probably due to the presence in the ash of large quantities of ammonium-N, which was found in culture work (Southorn 1977, Dietert 1979) to be detrimental to growth. Detoxification of this by leaching and the action of nitrifying micro-organisms appears to be a prerequisite for colonization by *F. hygrometrica*. Southorn (1977) speculated that soluble organic toxins in ash might be responsible for delaying the colonization by vascular plants that enables *F. hygrometrica* to flourish in the first year. Brasell *et al.* (1986) reported high rates of nitrogen fixation from bryophyte/soil cores taken from *Eucalyptus* forest fire sites in Tasmania that are presumably a result of microbial activity.

Fast litter fires of grassland and heathland cause much less damage to the original vegetation and alter soil conditions comparatively little, and regrowth may occur from surviving underground parts. Following heathland fires in Scotland, Hobbs & Gimingham (1984) found that much variation in the pattern of recovery reflected the varying diversities of the preburn communities. The latter was determined partly by compositional differences during the heather (*Calluna vulgaris*) growth cycle. A bryophyte-dominated early recovery phase (*Campylopus paradoxus, Ceratodon purpureus, Polytrichum juniperinum, P. piliferum*) was obtained when heather stands in the "pioneer" and "building" stages were burnt. However, the pleurocarpous mosses *Hylocomium splendens, Hypnum jutlandicum* and *Pleurozium schreberi* also regrew quickly from partly combusted mats in older stands in some cases.

Under conditions of acute drought heath fires may ignite the underlying peat with much more serious consequences for the preburn vegetation (Clément & Touffet 1990, Maltby *et al.* 1990, Gloaguen 1990). Here a bonfire type of succession is initiated; however, the poor conditions and lack of propagules can retard higher plant recolonization for ten years or more. On the North York Moors (England) this was mainly due to droughting of the *Calluna* seedlings (Legg *et al.* 1992), but after some fires colonization by the invasive southern hemisphere moss *Campylopus introflexus* prevented *Calluna* regeneration (Equihua & Usher 1993).

Dung and cadaver mosses

Dung and the decomposing corpses of animals are sometimes colonized by a distinctive group of coprophilous bryophytes. Obligate coprophiles are restricted to the moss family Splachnaceae. Coprophiles are most frequent in otherwise nutrient-poor environments such as ombrotrophic mires, moors, and alpine and polar tundras. Among the commoner genera *Splachnum* is favoured by wetter ground conditions than *Tetraplodon*. In north Wales *Tetraplodon mnioides* is most abundant beneath dangerous mountain crags where there is a steady supply of the carcases of unfortunate sheep (Hill 1988). Under ideal conditions these mosses are highly successful so that in central Alberta uncolonized droppings are rare (Marino 1997). Useful comments on substratum preferences of individual Splachnaceae are given in Smith (1982a).

It is likely that these mosses are exploiting a rich nutrient source but surprisingly little appears to have been published on the specific nutrient requirements of the Splachnaceae. Some *Tayloria* species like *T. lingulata* are not coprophilous but grow in basic flushes. This may indicate a general high base requirement in members of the family. Those that grow on the pellets of birds of prey and carcases may endure when only the bones remain, probably indicating a high phosphorus requirement. However, in a laboratory experiment no difference was found in the abilities of dung- and bone-inhabiting species to grow on moose (herbivore) and wolf (carnivore) dung (Marino 1991a). Marino

Fig. 8.9. The curious capsules of *Splachnum luteum* Hedw., one of the so-called "dung mosses" that are habitat specialists on decaying animal remains.

(1997) suggested that moose droppings may lose nutrients quicker than wolf dung under field conditions by leaching. The latter author also marshaled evidence suggesting that high pH is not always characteristic of dung and carcases, thus weakening an earlier hypothesis of Cameron & Wyatt (1989).

One of the essential traits of coprophilous mosses is their entomophilous behavior: a dependence on insects for spore dispersal (Koponen & Koponen 1978). The sporophyte of coprophiles is highly adapted with some flower-like properties that appear to lure dung-flies (Koponen 1978, Cameron & Troili 1982). The seta is long and often unusually thick, and the capsule is usually brightly colored: red in *Splachnum rubrum*; yellow in *S. luteum*; purplish in *S. ampullaceum* and *Tetraplodon mnioides*. Its apophysis region is swollen and in some cases (e.g., *S. luteum*) it is drawn out radially into a disk-like structure (Fig. 8.9). Dung-flies are also believed to be attracted by the emission of volatile attractants by the capsule. A range of volatile octane derivatives, organic acids, aldehydes, ketones, and alcohols has been identified but it is not known which are active (Pyysalo *et al.* 1983). The spores are sticky and readily dispersed by the flies to fresh dung or corpses.

The substrata colonized by Splachnaceae represent spectacularly small and short-lived targets for spore dispersal, but they possess a highly focused dispersal mechanism. Commonly, several Splachnaceae will colonize a single dung patch and a competitive struggle may ensue (Marino 1991a). Differences in spore maturation dates of species and the types of fly vector, however, probably enable several coprophiles to coexist in an area (Marino 1991b). A successional sequence may be observed with members of the Splachnaceae colonizing first, followed by less specialized colonists like *Ceratodon purpureus, Bryum* spp., and *Pohlia* spp., and finishing with common pleurocarpous mosses of the surrounding community (e.g. Webster & Sharp 1973, Lloret 1991). Marino (1997) has summarized what is currently known about the competitive hierarchies and population dynamics of these highly specialized and fascinating mosses.

In some countries Splachnaceae are noticeably much rarer today than in the first half of the twentieth century. Discussing reasons why *Splachnum ampullaceum* had become rare in lowland Britain, Crundwell (1994) concluded that widespread land drainage was to blame, and this may indeed be a contributory factor. Ironically, however, the main reason is likely to be an insidious, and equally well-targeted, destruction of the dung-flies that disperse Splachnaceae spores by modern pesticides, especially ivermectins (D. A. Holyoak, pers. comm.), although this requires confirmation in specific instances. These chemicals are widely used to treat horses, cattle, and sheep against a range of internal and external parasites. They are excreted in the dung and destroy flies and other invertebrates that feed upon it (Wall & Strong 1987, Strong & Wall 1988, 1994, Strong *et al.* 1996). Here is a classic instance of a conflict between the interests of bryophytes (and bryologists) on the one hand, and progress in animal welfare and agriculture on the other, for conservationists to resolve!

Calcicoles and calcifuges

The distinction between calcicole ("calcium-loving") and calcifuge ("calcium-hating") species can be the principal dichotomy in a regional bryophyte flora (e.g. Bates 1995). Calcicoles are restricted to rocks and soils containing calcium carbonate, or inhabit waters that have flowed over or percolated through these substrata. Some calcicoles also occur on the least acid types of tree bark. Calcifuges live on substrata with an acid reaction, or in soft waters. Many bryophytes are apparently indifferent to the acidity of their substratum (e.g. the common grassland mosses *Pseudoscleropodium purum* and *Rhytidiadelphus squarrosus*) whereas others may need near neutral conditions ("neutrocline" taxa). Much less is known about the specific adaptations of calcicole and calcifuge bryophytes than their vascular plant equivalents. We may suspect that aluminum and iron, if present in the substratum, will be relatively mobile under acid conditions, but become extremely immobile under mildly alkaline conditions. This is partly supported by analyses of bryophytes growing on

Table 8.3. *The cation exchange capacity in a range of epilithic calcicole and calcifuge mosses from western England and south Wales*

Values were determined by using unbuffered $CaCl_2$ solution (25 mM) to saturate the exchange sites and $SrCl_2$ solution (25 mM) to elute the adsorbed Ca^{2+} ions. Values are means of three replicates and the estimated 95% confidence interval.

	Rock	Ca adsorbed (μg g^{-1} dry wt)
Calcicoles		
Ctenidium molluscum	Carboniferous limestone	15510 ± 3497
Homalothecium sericeum	Carboniferous limestone	12460 ± 319
Orthotrichum cupulatum	Carboniferous limestone	12250 ± 1382
Schistidium apocarpum	Carboniferous limestone	12940 ± 955
Tortella tortuosa	Carboniferous limestone	15160 ± 679
Tortula ruralis	Carboniferous limestone	10160 ± 684
	Mean	**13080**
Calcifuges		
Andreaea rothii	Granite	2660 ± 124
Dicranoweisia cirrata	Old Red Sandstone	3200 ± 287
Grimmia donniana	Vitrified lead slag	2610 ± 114
Ptychomitrium polyphyllum	Old Red Sandstone	6690 ± 160
Racomitrium fasciculare	Old Red Sandstone	3330 ± 287
Racomitrium lanuginosum	Old Red Sandstone	2330 ± 287
	Mean	**3470**

From Bates 1982b.

contrasted rock types in Scotland. A selection of calcifuges, and notably *Andreaea rothii*, contained large concentrations of Fe, whereas the calcicoles contained up to 17 times more Ca (Bates 1978, 1982b). The highest Al concentrations were also found in *A. rothii*, but concentrations of this element were not consistently higher in calcifuges than in calcicoles.

The differences in total metal concentrations correlate with marked differences in CEC of calcicole and calcifuge bryophytes (Bates 1982b). Among epiliths, CEC is 3–4 times higher in calcicoles than calcifuges (Table 8.3). Following experiments using EDTA to remove Ca^{2+} from the tissues, Bates (1982b) hypothesized that calcicoles had inherently leakier cell membranes than calcifuges. The elevated CEC of calcicoles was hypothesized to be necessary to ensure adequate Ca^{2+} adsorption for permeability control. Working with bryophytes of woodland soils, Büscher *et al.* (1990) found similar differentiation of CEC between calcifuge and calcicole species but reached a different conclusion. In laboratory experiments they investigated the selectivity of ion adsorption onto the cation-exchanger from mixtures of cations resembling soil solutions. They concluded that lower CEC in calcifuge mosses was an adaptation allowing avoidance of high Al^{3+} uptake, as plants with low CEC absorbed relatively less Al^{3+} from a mixed Al–Fe–Mn–Ca solution than those with high CEC. Although excessive Fe has been shown to be toxic to the calcicole bryophyte *Fissidens cristatus* (Woollon 1975), we still know very little about the susceptibilities of bryophytes to elevated Al^{3+}. Indeed, the ombrotrophic peatland environments favored by many *Sphagnum* spp. may often be acid but contain relatively little available Al^{3+}, which possibly explains the high CECs found in this genus.

By analogy with vascular plants, we may expect bryophytes to encounter Fe-deficiency in calcareous habitats. Structures analogous to the rhizodermal transfer cells of dicotyledonous calcicoles (e.g. Marschner 1986) have not yet been demonstrated in calcicole bryophytes but physiological adaptations to improve Fe-mobility from the substratum are likely to be present.

Importantly, calcium status, when isolated from variables like pH, appears to be relatively unimportant to bryophytes. *Calliergonella cuspidata*, a common moss of chalk downland and calcareous fen, shows a strong preference for pH values around neutrality and will not grow below pH 6 even if the calcium concentration is increased (Streeter 1970). Clymo (1973) demonstrated that high calcium concentrations were relatively harmless to most *Sphagnum* spp., as were high pH values, but the combination of high pH and high Ca proved to be lethal to all taxa except those characteristic of base-rich flushes (e.g. *S. squarrosum*). Hummock species like *S. capillifolium* are the most susceptible to these conditions. Similar conclusions were drawn by Vanderpoorten & Klein (1999) based on field measurements of water chemistry and distribution patterns of some common aquatic bryophytes in waterfalls of the River Rhine. The putative calcifuges *Marsupella emarginata* and *Scapania undulata* only grew in waters with low solute content, but were relatively indifferent to pH. Conversely, the "calcicoles" *Chiloscyphus polyanthos, Cratoneuron filicinum, Rhynchostegium riparioides*, and *Thamnobryum alopecurum* usually occurred in base-rich water but they were also found when Ca^{2+} concentration was low. Although relatively indifferent to calcium status, these species are all strongly intolerant of low pH.

Building on recent studies of organic acid metabolism, Lee (1999) has proposed a general model for vascular plants in which ungating of channel proteins, permitting release of malate and citrate anions, is provided by Al^{3+} in calcifuges and by Ca^{2+} in calcicoles. In the soil solution these anions either precipitate Al^{3+} and thus ameliorate its toxicity (acid conditions), or they chelate Fe^{3+} and promote its uptake (alkaline conditions). The main tenets of Lee's model (see also Roberts 2006) are consistent with many of the foregoing observations; however, much further work is needed to validate it generally for bryophytes.

Halophytes

No bryophytes live permanently submerged in the oceans although the aquatic moss *Fontinalis dalecarlica* is able to grow in the northern Baltic Sea owing to its low salinity (Söderlund *et al.* 1988). On land very few species are true halophytes. Even in coastal dunes where deposition of saltwater spray is likely, Boerner & Forman (1975) found that none of the beach mosses survived a spray treatment with natural seawater. They concluded that survival in nature depended on dilution of the incoming salt before it contacted the mosses. In British saltmarshes, Adam (1976) recorded 66 bryophyte taxa living in situations where at least occasional tidal immersion would be experienced. Most of these species are widely distributed in non-saline habitats and it was suggested that some may represent halophytic ecotypes, but many probably experience relatively low salinities, which are less harmful than full oceanic strength seawater. A few bryophytes occur in association with inland salt deposits (Zechmeister 2005). One of the best-known coastal halophytes is *Schistidium maritimum* (syn. *Grimmia maritima*) which grows in the splash zone on non-calcareous seashore rocks (and occasionally on stones in saltmarshes) on shores around the Northern Atlantic (Fig. 8.10). On very sheltered shores it may be immersed by the highest spring tides. It is commonly accompanied by an epilithic form of *Ulota phyllantha* and *Tortella flavovirens*, which also appear to be highly salt-tolerant (Bates 1975).

Fig. 8.10. Dark cushions of the halophytic moss *Schistidium maritimum* growing with lichens of the "splash zone" of a rocky seashore at Montrose, northeast Scotland (photo M. C. F. Proctor).

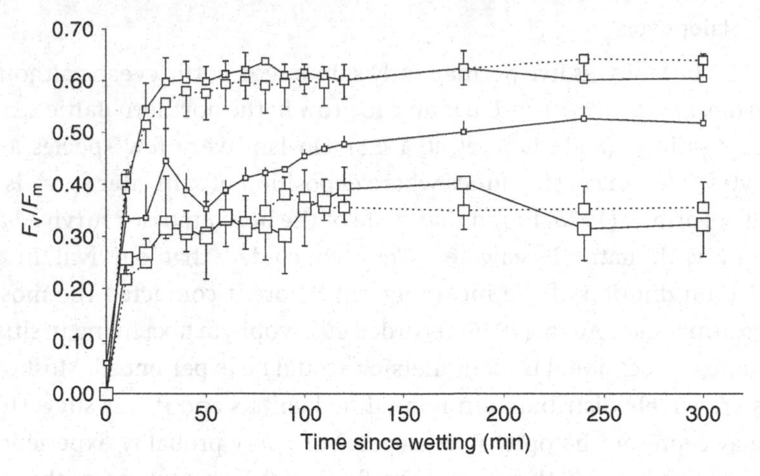

Fig. 8.11. Recovery of quantum efficiency (F_v/F_m) in *Schistidium maritimum* following rehydration in different dilutions of artificial seawater. Sizes of squares represent dilution (0, 25, 50, 75 or 100% seawater): smallest = 0% (distilled water), largest = 100% artificial seawater. Data are means and standard errors of three replicate determinations. Material: exposed granite seashore rocks, Cap de Flamanville, Manche, France. Treatment: material was air dried at room temperature for two weeks and then stored over saturated NaOH (6% R.H.) for five days prior to rehydration in the stated seawater dilution. Measurements of quantum efficiency were made with a Hansatech FMS1 fluorometer. See Bates & Brown (1974) for artificial seawater recipe. (X.-Y. Phoon & J. W. Bates, unpublished data.)

Salt tolerance in *S. maritimum* appears to depend principally on the exclusion of salt by means of a markedly impermeable cell membrane. There is also some evidence for the existence of a metabolically active Na^+ efflux pump (Bates 1976). In contrast, glycophytic mosses like *Grimmia pulvinata* suffer a major loss of cell K^+ and influx of Na^+ if confronted with seawater (Bates & Brown 1974). This increased permeability in *G. pulvinata* is probably caused by Na^+ ions in the seawater competing with and displacing Ca^{2+} ions that normally stabilize the polar heads of the membrane lipids. *Schistidium maritimum* withstands the normal ratio of Ca : Na in seawater but if this is experimentally lowered it also suffers some K-loss and Na-entry.

The simple "salt exclusion" mechanism of salt tolerance operating in *S. maritimum* should make it highly susceptible to osmotic water loss ("physiological drought") to the salty external solution. Indeed, most cushions of this moss can be demonstrated to contain appreciable quantities of sea salt. Except during periods of heavy rainfall, one imagines that full turgor of the cells is constantly challenged by outward osmosis. Laboratory measurements of chlorophyll fluorescence of this species in a range of seawater dilutions (Fig. 8.11) show that the higher external salinities are indeed inhibitory to quantum efficiency (F_v/F_m), whereas lower salinities (25%, 50% seawater) are stimulatory. *Schistidium maritimum* is also highly desiccation-tolerant and it probably frequently avoids the consequences of low external solute potentials by entering a metabolically inactive dehydrated state, something that vascular halophytes cannot do. These observations may explain why *S. maritimum* does not extend very far southwards on the Atlantic coast of Europe and also why it appears to luxuriate in rainier climates and on shores where there is some freshwater seepage.

Metallophytes

Some bryophytes are demonstrably "metal-tolerant", being able to withstand levels of heavy metals that are toxic to other species. One famous group of species is known as the copper mosses. These (e.g. *Grimmia atrata, Mielichhoferia* spp., *Scopelophila cataractae*) generally occur on rocks rich in copper sulfide and may be of some use in prospecting for copper ore. It is doubtful that copper mosses have a definite nutritional requirement for Cu, but quite likely that they are extremely poor competitors with an unusually high tolerance of Cu and/or its associated sulfide-generated acidity (Brown 1982). Metal-tolerant populations have also been recognized in some wide-ranging bryophytes (e.g. *Marchantia polymorpha, Solenostoma crenulata, Ceratodon purpureus, Funaria hygrometrica;* see Jules & Shaw 1994) but the underlying physiological mechanisms remain obscure and require further investigation (Shaw 1994).

References

Adam, P. (1976). The occurrence of bryophytes on British saltmarshes. *Journal of Bryology*, **9**, 265–74.

Adams, D. G. (2000). Symbiotic interactions. In *Ecology of Cyanobacteria: their Diversity in Time and Space*, ed. B. Whitton & M. Potts, pp. 523–61. Dordrecht: Kluwer.

Adams, D. G. (2002). The liverwort-cyanobacterial symbiosis. *Proceedings of the Royal Irish Academy*, **102B**, 27–9.

Aerts, R., Wallén, B. & Malmer, N. (1992). Growth-limiting nutrients in *Sphagnum*-dominated bogs subject to low and high atmospheric nitrogen supply. *Journal of Ecology*, **80**, 131–40.

Aho, K. & Weaver, T. (2006). Measuring water relations and pH of cryptogam rock-surface environments. *Bryologist*, **109**, 348–57.

Aldous, A. R. (2002). Nitrogen translocation in *Sphagnum* mosses: effects of atmospheric deposition. *New Phytologist*, **156**, 241–54.

Alpert, P. (1985). Distribution quantified by microtopography in an assemblage of saxicolous mosses. *Vegetatio*, **64**, 131–9.

Alpert, P. (1988). Survival of a desiccation-tolerant moss, *Grimmia laevigata*, beyond its observed microdistributional limits. *Journal of Bryology*, **15**, 219–27.

Alpert, P. (1989). Translocation in the nonpolytrichaceous moss *Grimmia laevigata*. *American Journal of Botany*, **76**, 1524–9.

Aude, E. & Ejrnæs, R. (2005). Bryophyte colonisation in experimental microcosms: the role of nutrients, defoliation and vascular vegetation. *Oikos*, **109**, 323–30.

Badacsonyi, A., Bates, J. W. & Tuba, Z. (2000). Effects of desiccation on phosphorus and potassium acquisition by a desiccation-tolerant moss and lichen. *Annals of Botany*, **86**, 621–7.

Baddeley, J. A., Thompson, D. B. A. & Lee, J. A. (1994). Regional and historical variation in the nitrogen content of *Racomitrium lanuginosum* in Britain in relation to atmospheric nitrogen deposition. *Environmental Pollution*, **84**, 189–96.

Bakken, S. (1994). Growth and nitrogen dynamics of *Dicranum majus* under two contrasting nitrogen deposition regimes. *Lindbergia*, **19**, 63–72.

Barkman, J. J. (1958). *Phytosociology and Ecology of Cryptogamic Epiphytes*. Assen: Van Gorcum.

Basile A., Cogoni, A.E., Bassi, P. *et al.* (2001). Accumulation of Pb and Zn in gametophytes and sporophytes of the moss *Funaria hygrometrica* (Funariales). *Annals of Botany*, **87**, 537–43.

Basile, A., Giordano, S., Cafiero, G., Spagnuolo, V. & Castaldo-Cobianchi, R. (1994). Tissue and cell localization of experimentally-supplied lead in *Funaria hygrometrica* Hedw. using X-ray SEM and TEM analysis. *Journal of Bryology*, **18**, 69–81.

Bates, J. W. (1975). A quantitative investigation of the saxicolous bryophyte and lichen vegetation of Cape Clear Island, County Cork. *Journal of Ecology*, **63**, 143–62.

Bates, J. W. (1976). Cell permeability and regulation of intracellular sodium concentration in a halophytic and a glycophytic moss. *New Phytologist*, **77**, 15–23.

Bates, J. W. (1978). The influence of metal availability on the bryophyte and macro-lichen vegetation of four types on Skye and Rhum. *Journal of Ecology*, **66**, 457–82.

Bates, J. W. (1979). The relationship between physiological vitality and age in shoot segments of *Pleurozium schreberi* (Brid.) Mitt. *Journal of Bryology*, **10**, 339–51.

Bates, J. W. (1982a). Quantitative approaches in bryophyte ecology. In *Bryophyte Ecology*, ed. A. J. E. Smith, pp. 1–44. London: Chapman & Hall.

Bates, J. W. (1982b). The role of exchangeable calcium in saxicolous calcicole and calcifuge mosses. *New Phytologist*, **90**, 239–52.

Bates, J. W. (1987). Nutrient retention by *Pseudoscleropodium purum* and its relation to growth. *Journal of Bryology*, **14**, 565–80.

Bates, J. W. (1989a). Retention of added K, Ca and P by *Pseudoscleropodium purum* growing under an oak canopy. *Journal of Bryology*, **15**, 589–605.

Bates, J. W. (1989b). Interception of nutrients in wet deposition by *Pseudoscleropodium purum*: an experimental study of uptake and retention of potassium and phosphorus. *Lindbergia*, **15**, 93–8.

Bates, J. W. (1992a). Mineral nutrient acquisition and retention by bryophytes. *Journal of Bryology*, **17**, 223–40.

Bates, J. W. (1992b). Influence of chemical and physical factors on *Quercus* and *Fraxinus* epiphytes at Loch Sunart, western Scotland: a multivariate analysis. *Journal of Ecology*, **80**, 163–79.

Bates, J. W. (1993). Regional calcicoly in the moss *Rhytidiadelphus triquetrus*: survival and chemistry of transplants at a formerly SO_2-polluted site with acid soil. *Annals of Botany*, **72**, 449–55.

Bates, J. W. (1994). Responses of the mosses *Brachythecium rutabulum* and *Pseudoscleropodium purum* to a mineral nutrient pulse. *Functional Ecology*, **8**, 686–92.

Bates, J. W. (1995). Numerical analysis of bryophyte-environment relationships in a lowland English flora. *Fragmenta Floristica et Geobotanica*, **40**, 471–90.

Bates, J. W. (1997). Effects of intermittent desiccation on nutrient economy and growth of two ecologically contrasted mosses. *Annals of Botany*, **79**, 299–309.

Bates, J. W. (1998). Is 'life-form' a useful concept in bryophyte ecology? *Oikos*, **82**, 223–37.

Bates, J. W. & Bakken, S. (1998). Nutrient retention, desiccation and redistribution in mosses. In *Bryology for the Twenty-first Century*, ed. J. W. Bates, N. W. Ashton & J. G. Duckett, pp. 293–304. Leeds: Maney and British Bryological Society.

Bates, J. W. & Brown, D. H. (1974). The control of cation levels in seashore and inland mosses. *New Phytologist*, **73**, 483–95.

Bates, J. W. & Farmer, A. M. (1990). An experimental study of calcium acquisition and its effects on the calcifuge moss *Pleurozium schreberi*. *Annals of Botany*, **65**, 87–96.

Bates, J. W., Proctor, M. C. F., Preston, C. D., Hodgetts, N. G. & Perry, A. R. (1997). Occurrence of epiphytic bryophytes in a 'tetrad' transect across southern Britain. 1. Geographical trends in abundance and evidence of recent change. *Journal of Bryology*, **19**, 685–714.

Bates, J. W., Roy, D. B. & Preston, C. D. (2004). Occurrence of epiphytic bryophytes in a 'tetrad' transect across southern Britain. 2. Analysis and modelling of epiphyte-environment relationships. *Journal of Bryology*, **26**, 181–97.

Baxter, R., Emes, M. J. & Lee, J. A. (1992). Effects of an experimentally applied increase in ammonium on growth and amino-acid metabolism of *Sphagnum cuspidatum* Erhr. ex Hoffm. from differently polluted areas. *New Phytologist*, **120**, 265–74.

Bell, F. W. & Newmaster, S. G. (2002). The effects of silvicultural disturbances on the diversity of seed-producing plants in the boreal mixedwood forest. *Canadian Journal of Forestry Research*, **32**, 1180–91.

Belnap, J. (2001). Factors influencing nitrogen fixation and nitrogen release in biological soil crusts. In *Biological Soil Crusts: Structure, Function and Management*, ed. J. Belnap & O. L. Lange, pp. 241–61. Berlin & Heidelberg: Springer-Verlag.

Bengstrom, D. M. & Tweedie, C. E. (1998). A conceptual model for integration studies of epiphytes: nitrogen utilisation, a case study. *Australian Journal of Botany*, **46**, 273–80.

Berg, T., Røyset, O., Steinnes, E. & Vadset, M. (1995). Atmospheric trace element deposition: principal component analysis of ICP-MS data from moss samples. *Environmental Pollution*, **88**, 67–77.

Berg, T. & Steinnes, E. (1997). Use of mosses (*Hylocomium splendens* and *Pleurozium schreberi*) as biomonitors of heavy metal deposition: from relative to absolute deposition values. *Environmental Pollution*, **98**, 61–71.

Bergamini, A. & Pauli, D. (2001). Effects of increased nutrient supply on bryophytes in montane calcareous fens. *Journal of Bryology*, **23**, 331–9.

Bergamini, A. & Peintinger, M. (2002). Effects of light and nitrogen on morphological plasticity of the moss *Calliergonella cuspidata*. *Oikos*, **96**, 355–63.

Berrie, G. K. & Eze, J. M. O. (1975). The relationship between an epiphyllous liverwort and host leaves. *Annals of Botany*, **39**, 955–63.

Birks, H. J. B., Heegaard, E., Birks. H. H. & Jonsgard, B. (1998). Quantifying bryophyte-environment relationships. In *Bryology for the Twenty-first Century*, ed. J. W. Bates, N. W. Ashton & J. G. Duckett, pp. 305–19. Leeds: Maney and British Bryological Society.

Boerner, R. E. & Forman, R. T. T. (1975). Salt spray and coastal dune mosses. *Bryologist*, **78**, 57–63.

Borstlap, A. C. (2002). Early diversification of plant aquaporins. *Trends in Plant Science*, **7**, 529–30.

Bowden, R. D. (1991). Input, outputs, and accumulation of nitrogen in an early successional moss (*Polytrichum*) ecosystem. *Ecological Monographs*, **61**, 207–23.

Bowden, W. B., Finlay, J. C. & Maloney, P. E. (1994). Long-term effects of PO_4 fertilization on the distribution of bryophytes in an arctic river. *Freshwater Biology*, **32**, 445–54.

Brasell, H. M., Davies, S. K. & Mattay, J. P. (1986). Nitrogen fixation associated with bryophytes colonizing burnt sites in Southern Tasmania, Australia. *Journal of Bryology*, **14**, 139–49.

Brehm, V. K. (1971). Ein *Sphagnum*-Bult als Beispiel einer natürlichen Ionenaustauschersäule. *Beiträge zur Biologie der Pflanzen*, **47**, 287–312.

Brown, D. H. (1982). Mineral nutrition. In *Bryophyte Ecology*, ed. A. J. E. Smith, pp. 383–444. London: Chapman & Hall.

Brown, D. H. (1984). Uptake of mineral elements and their use in pollution monitoring. In *The Experimental Biology of Bryophytes*, ed. A. F. Dyer & J. G. Duckett, pp. 229–55. London: Academic Press.

Brown, D. H. & Bates, J. W. (1990). Bryophytes and nutrient cycling. *Botanical Journal of the Linnean Society*, **104**, 129–47.

Brown, D. H. & Beckett, R. P. (1985). Intracellular and extracellular uptake of cadmium by the moss *Rhytidiadelphus squarrosus*. *Annals of Botany*, **55**, 179–88.

Brown, D. H. & Buck, G. W. (1978). Distribution of potassium, calcium and magnesium in the gametophyte and sporophyte generations of *Funaria hygrometrica* Hedw. *Annals of Botany*, **42**, 923–9.

Brown, D. H. & Buck, G. W. (1979). Desiccation effects and cation distribution in bryophytes. *New Phytologist*, **82**, 115–25.

Brown, D. H. & Sidhu, M. (1992). Heavy metal uptake, cellular location, and inhibition of moss growth. *Cryptogamic Botany*, **3**, 82–5.

Brown, D. H. & Wells, J. M. (1988). Sequential elution technique for determining the cellular location of cations. In *Methods in Bryology*, ed. J. M. Glime, pp. 227–33. Nichinan: Hattori Botanical Laboratory.

Brümelis, G. & Brown, D. H. (1997). Movement of metals to new growing tissue in the moss *Hylocomium splendens* (Hedw.) BSG. *Annals of Botany*, **79**, 679–86.

Brümelis, G., Lapiņa, L. & Tabors, G. (2000). Uptake of Ca, Mg and K during growth of annual segments of the moss *Hylocomium splendens* in the field. *Journal of Bryology*, **22**, 163–74.

Burton, M. A. S. (1986). *Biological Monitoring*. MARC Report Number 32. London: Monitoring and Assessment Research Centre, King's College London.

Burton, M. A. S. (1990). Terrestrial and aquatic bryophytes as monitors of environmental contaminants in urban and industrial habitats. *Botanical Journal of the Linnean Society*, **104**, 267–80.

Büscher, P., Koedam, N. & van Spreybroeck, D. (1990). Cation-exchange properties and adaptation to soil acidity in bryophytes. *New Phytologist*, **115**, 177–86.

Callaghan, T. V., Collins, N. J. & Callaghan, C. H. (1978). Photosynthesis, growth and reproduction of *Hylocomium splendens* and *Polytrichum commune* in Swedish Lapland. *Oikos*, **31**, 73–88.

Cameron, R. G. & Troili, D. (1982). Fly-mediated spore dispersal in *Splachnum ampullaceum* (Musci). *Michigan Botanist*, **21**, 59–65.

Cameron, R. G. & Wyatt, R. (1989). Substrate restriction in entomophilous Splachnaceae. II. Effects of hydrogen ion concentration on establishment of gametophytes. *Bryologist*, **92**, 397–404.

Carroll, J. A., Johnson, D., Morecroft, M. *et al.* (2000). The effect of long-term nitrogen additions on the bryophyte cover of upland acidic grasslands. *Journal of Bryology*, **22**, 83–9.

Chapin, F. S. III, Oechel, W. C., Van Cleve, K. & Lawrence, W. (1987). The role of mosses in the phosphorus cycling of an Alaskan black spruce forest. *Oecologia*, **74**, 310–15.

Chaumont, F., Moshelion, M. & Daniels, M. J. (2005). Regulation of plant aquaporin activity. *Biology of the Cell*, **97**, 749–64.

Chevalier, D., Nurit, F. & Pesey, H. (1977). Orthophosphate absorption by the sporophyte of *Funaria hygrometrica* during maturation. *Annals of Botany*, **41**, 527–31.

Christmas, M. & Whitton, B. A. (1998). Phosphorus and aquatic bryophytes in the Swale-Ouse river system, north England. 1. Relationship between ambient phosphate, internal N:P ratio and surface phosphatase activity. *Science of the Total Environment*, **210**, 389–99.

Clark, K. L., Nadkarni, N. M. & Gholz, H. L. (1998). Growth, net production, litter decomposition, and net nitrogen accumulation by epiphytic bryophytes in a tropical montane forest. *Biotropica*, **30**, 12–23.

Claveri, B., Morhain, E. & Mouvet, C. (1994). A methodology for the assessment of accidental copper pollution using the aquatic moss *Rhynchostegium riparioides*. *Chemosphere*, **28**, 2001–10.

Cleavitt, N. (2001). Disentangling moss species limitations: the role of physiologically base substrate specificity for six species occurring on substrates with varying pH and percent organic matter. *Bryologist*, **104**, 59–68.

Clément, B. & Touffet, J. (1990). Plant strategies and secondary succession on Brittany heathlands after severe fire. *Journal of Vegetation Science*, **1**, 195–202.

Clymo, R. S. (1963). Ion exchange in *Sphagnum* and its relation to bog ecology. *Annals of Botany*, **27**, 309–24.

Clymo, R. S. (1967). Control of cation concentrations, and in particular of pH, in *Sphagnum* dominated communities. In *Chemical Environment in the Aquatic Habitat*, ed. H. L. Golterman & R. S. Clymo, pp. 273–84. Amsterdam: North Holland.

Clymo, R. S. (1973). The growth of *Sphagnum*: some effects of environment. *Journal of Ecology*, **61**, 849–69.

Clymo, R. S. & Hayward, P. M. (1982). The ecology of *Sphagnum*. In *Bryophyte Ecology*, ed. A. J. E. Smith, pp. 229–89. London: Chapman & Hall.

Coley, P. D., Kursar, T. A. & Machado, J.-L. (1993). Colonization of tropical rain forest leaves by epiphylls: effects of site and host plant leaf lifetime. *Ecology*, **74**, 619–23.

Collins, N. J. & Oechel, W. C. (1974). The pattern of growth and translocation of photosynthate in a tundra moss, *Polytrichum alpinum*. *Canadian Journal of Botany*, **52**, 355–63.

Cornelissen, J. H. C. & ter Steege, H. (1989). Distribution and ecology of epiphytic bryophytes and lichens in dry evergreen forest of Guyana. *Journal of Tropical Ecology*, **5**, 131–50.

Coxson, D. S. (1991). Nutrient release from epiphytic bryophytes in tropical montane rain forest (Guadeloupe). *Canadian Journal of Botany*, **69**, 2122–9.

Crundwell, A. C. (1994). *Splachnum ampullaceum* Hedw. In *Atlas of the Bryophytes of Britain and Ireland*, vol. 3, Mosses (Diplolepideae), ed. M. O. Hill, C. D. Preston & A. J. E. Smith, p. 48. Colchester: Harley.

Damman, A. W. H. (1978). Distribution and movement of elements in ombrotrophic peat bogs. *Oikos*, **30**, 480–95.

Delépée, R., Pouliquen, H. & Le Bris, H. (2003). The bryophyte *Fontinalis antipyretica* Hedw. bioaccumulates oxytetracycline, flumequine and oxolinic acid in the freshwater environment. *Science of the Total Environment*, **322**, 243–53.

Dietert, M. F. (1979). Studies on the gametophyte nutrition of the cosmopolitan species *Funaria hygrometrica* and *Weissia controversa*. *Bryologist*, **82**, 417–31.

Duckett, J. G. & Read, D. J. (1995). Ericoid mycorrhizas and rhizoid-ascomycete associations in liverworts share the same mycobiont: isolation of the partners and resynthesis of the associations *in vitro*. *New Phytologist*, **129**, 439–47.

Duckett, J. G., Renzaglia, K. S. & Pell, K. (1991). A light and electron microscope study of rhizoid-ascomycete associations and flagelliform axes in British hepatics. *New Phytologist*, **118**, 233–57.

During, H. J. (1979). Life strategies of bryophytes; a preliminary review. *Lindbergia*, **53**, 2–18.

During, H. J. (1992). Ecological classifications of bryophytes and lichens. In *Bryophytes and Lichens in a Changing Environment*, ed. J. W. Bates & A. M. Farmer, pp. 1–31. Oxford: Clarendon Press.

During, H. J. (1997). Bryophyte diaspore banks. *Advances in Bryology*, **6**, 103–34.

Eckstein, R. L. (2000). Nitrogen retention by *Hylocomium splendens* in a subarctic birch woodland. *Journal of Ecology*, **88**, 506–15.

Eckstein, R. L. & Karlsson, P. S. (1999). Recycling of nitrogen among segments of *Hylocomium splendens* as compared with *Polytrichum commune* – implications for clonal integration in an ectohydric bryophyte. *Oikos*, **86**, 87–96.

Equihua, M. & Usher, M. B. (1993). Impact of carpets of the invasive moss *Campylopus introflexus* on *Calluna vulgaris* regeneration. *Journal of Ecology*, **81**, 359–65.

Estébanez, B., Alfayate, C., Ballesteros, T. *et al.* (2002). Amorphous mineral incrustations in the moss *Homalothecium sericeum*. *Journal of Bryology*, **24**, 25–32.

Eze, J. M. O. & Berrie, G. K. (1977). Further investigations into the physiological relationships between an epiphyllous liverwort and its host leaves. *Annals of Botany*, **41**, 351–8.

Farmer, A. M., Bates, J. W. & Bell, J. N. B. (1991). Seasonal variations in acidic pollutant inputs and their effects on the chemistry of stemflow, bark and epiphyte tissues in three oak woodlands in N.W. Britain. *New Phytologist*, **118**, 441–51.

Farmer, J. G., Eades, L. J., Atkins, H. & Chamberlain, D. F. (2002). Historical trends in the lead isotopic composition of archival *Sphagnum* mosses from Scotland (1838–2000). *Environmental Science and Technology*, **36**, 152–7.

Forsum, A., Dahlman, L., Näsholm, T. & Nordin, A. (2006). Nitrogen utilization by *Hylocomium splendens* in a boreal forest fertilization experiment. *Functional Ecology*, **20**, 421–6.

Foster, D. R. (1985). Vegetation development following fire in *Picea mariana* (Black Spruce) – *Pleurozium* forests of south-eastern Labrador, Canada. *Journal of Ecology*, **73**, 517–34.

García-Álvaro, M. A., Martínez-Abaigar, J., Núñez-Olivera, E. & Beaucourt, N. (2000). Element concentrations and enrichment ratios in the aquatic moss *Rhynchostegium riparioides* along the River Iregua (La Rioja, Northern Spain). *Bryologist*, **103**, 518–33.

Gerdol, R., Bragazza, L. & Marchesini, R. (2002). Element concentrations in the forest moss *Hylocomium splendens*: variation associated with altitude, net primary production and soil chemistry. *Environmental Pollution*, **116**, 129–35.

Gloaguen, J. C. (1990). Post-burn succession on Brittany heathlands. *Journal of Vegetation Science*, **1**, 147–52.

Gombert, S., Traubenberg, C., Losno, R., Leblond, S., Collin, J. & Cossa, D. (2004). Biomonitoring of element deposition using mosses in the 2000 French survey: identifying sources and spatial trends. *Journal of Atmospheric Chemistry*, **49**, 479–502.

Gordon, C., Wynn, J. M. & Woodin, S. J. (2001). Impacts of increased nitrogen supply on High Arctic heath: the importance of bryophytes and phosphorus availability. *New Phytologist*, **149**, 461–71.

Grime, J. P. (1974). Vegetation classification by reference to strategies. *Nature*, **250**, 26–31.

Gunnarsson, U., Granberg, G. & Nilsson, M. (2004). Growth, production and interspecific competition in *Sphagnum*: effects of temperature, nitrogen and sulphur treatments on a boreal mire. *New Phytologist*, **163**, 349–59.

Gunnarsson, U. & Rydin, H. (2000). Nitrogen fertilization reduces *Sphagnum* production in bog communities. *New Phytologist*, **147**, 527–38.

Gupta, R. K. (1977). A study of photosynthesis and leakage of solutes in relation to the desiccation effects in bryophytes. *Canadian Journal of Botany*, **55**, 1186–94.

Gustafsson, L. & Eriksson, I. (1995). Factors of importance for the epiphytic vegetation of aspen *Populus tremula* with special emphasis on bark chemistry and soil chemistry. *Journal of Applied Ecology*, **32**, 412–24.

Hébant, C. (1977). *The Conducting Tissues of Bryophytes*. Vaduz: J. Cramer.

Hébrard, J.-P., Foulquier, L. & Grauby, A. (1974). Approche expérimentale sur les possibilités de transfert du ^{90}Sr d'un substrat solide à une mousse terrestre: *Grimmia orbicularis* Bruch. *Bulletin de la Société Botanique de France*, **121**, 235–50.

Hedderson, T. A. & Longton, R. E. (1995). Patterns of life history variation in the Funariales, Polytrichales and Pottiales. *Journal of Bryology*, **18**, 639–75.

Heegard, E. (1997). Ecology of *Andreaea* in western Norway. *Journal of Bryology*, **19**, 527–636.

Herben, T. & Söderström, L. (1992). Which habitat parameters are most important for the persistence of a bryophyte species on patchy, temporary substrates? *Biological Conservation*, **59**, 121–6.

Herpin, U., Berlekamp, J., Markert, B. *et al.* (1996). The distribution of heavy metals in a transect of the three states the Netherlands, Germany and Poland, determined with the aid of moss monitoring. *The Science of the Total Environment*, **187**, 185–98.

Hietz, P., Wanek, W., Wania, R. & Nadkarni, N. (2002). Nitrogen-15 natural abundance in a montane cloud forest canopy as an indicator of nitrogen cycling and epiphyte abundance. *Oecologia*, **131**, 350–5.

Hill, M. O. (1988). A bryophyte flora of north Wales. *Journal of Bryology*, **15**, 377–491.

Hirschi, K. D. (2004). The calcium conundrum: both versatile nutrient and specific signal. *Plant Physiology*, **136**, 2438–42.

Hobbs, R. J. & Gimingham, C. H. (1984). Studies of fire in Scottish heathland communities. II. Post-fire vegetation development. *Journal of Ecology*, **72**, 585–610.

Hoffman, G. R. (1966). Observations on the mineral nutrition of *Funaria hygrometrica* Hedw. *Bryologist*, **69**, 182–92.

Hoffman, G. R. (1972). The accumulation of cesium-137 by cryptogams in a *Liriodendron tulipifera* forest. *Botanical Gazette*, **133**, 107–19.

Jäppinen, J.-P. & Hotanen, J.-P. (1990). Effect of fertilization on the abundance of bryophytes in two drained peatland forests in eastern Finland. *Annales Botanici Fennici*, **27**, 93–108.

John, E. & Dale, M. R. T. (1995). Neighbor relations within a community of epiphytic lichens and bryophytes. *Bryologist*, **98**, 29–37.

Johnsen, I. & Rasmussen, L. (1977). Retrospective study (1944–1976) of heavy metals in the epiphyte *Pterogonium gracile* collected from one phorophyte. *Bryologist*, **80**, 625–9.

Johnsgard, B. & Birks, H. J. B. (1993). Quantitative studies on saxicolous bryophyte–environment relationships in western Norway. *Journal of Bryology*, **17**, 579–611.

Jules, E. S. & Shaw, A. J. (1994). Adaptation to metal-contaminated soils in populations of the moss *Ceratodon purpureus* – vegetative growth and reproductive expression. *American Journal of Botany*, **81**, 791–7.

Kellner, O. & Mårshagen, M. (1991). Effects of irrigation and fertilization on the ground vegetation in a 130-year-old stand of Scots pine. *Canadian Journal of Forestry Research*, **21**, 733–8.

Kelly, M. G. & Whitton, B. A. (1989). Interspecific differences in Zn, Cd and Pb accumulation by freshwater algae and bryophytes. *Hydrobiologia*, **175**, 1–11.

Kimmerer, R. W. (1994). Ecological consequences of sexual versus asexual reproduction in *Dicranum flagellare* and *Tetraphis pellucida*. *Bryologist*, **97**, 20–5.

Kimmerer, R. W. & Young, C. C. (1995). The role of slugs in dispersal of the asexual propagules of *Dicranum flagellare*. *Bryologist*, **98**, 149–53.

Koponen, A. M. (1978). The peristome and spores in Splachnaceae and their evolutionary and systematic significance. *Bryophytorum Bibliotheca*, **13**, 535–67.

Koponen, A. M. & Koponen, T. (1978). Evidence of entomophily in Splachnaceae (Bryophyta). *Bryophytorum Bibliotheca*, **13**, 569–77.

Kuik, P. & Wolterbeek, H. T. (1995). Factor analysis of atmospheric trace-element deposition data in the Netherlands obtained by moss monitoring. *Water, Air and Soil Pollution*, **84**, 323–46.

Lara, F. & Mazimpaka, V. (1998). Succession of epiphytic bryophytes in a *Quercus pyrenaica* forest from the Spanish Central Range (Iberian Peninsula). *Nova Hedwigia*, **67**, 125–38.

Leblond, S., Gombert, S., Colin, J., Losno, R. & Traubenberg, C. (2004). Biological and temporal variations of trace element concentrations in the moss species *Scleropodium purum* (Hedw.) Limpr. *Journal of Atmospheric Chemistry*, **49**, 95–110.

Lee, J. A. (1999). The calcicole-calcifuge problem revisited. *Advances in Botanical Research*, **29**, 2–30.

Legg, C. J., Maltby, E. & Proctor, M. C. F. (1992). The ecology of severe moorland fire on the North York Moors: seed distribution and seedling establishment of *Calluna vulgaris*. *Journal of Ecology*, **80**, 737–52.

Lewis Smith, R. I. (1978). Summer and winter concentrations of sodium, potassium and calcium in some maritime antarctic cryptogams. *Journal of Ecology*, **66**, 891–909.

Ligrone, R. & Duckett, J. G. (1994). Cytoplasmic polarity and endoplasmic microtubules associated with the nucleus and organelles are ubiquitous features of food-conducting cells in bryoid mosses (Bryophyta). *New Phytologist*, **127**, 601–14.

Ligrone, R. & Duckett, J. G. (1996). Polarity and endoplasmic microtubules in food-conducting cells of mosses: an experimental study. *New Phytologist*, **134**, 503–16.

Ligrone, R. & Gambardella, R. (1988). The sporophyte-gametophyte junction in bryophytes. *Advances in Bryology*, **3**, 225–74.

Ligrone, R., Duckett, J. G. & Renzaglia, K. R. (2000). Conducting tissues and phyletic relationships of bryophytes. *Philosophical Transactions of the Royal Society of London*, B**355**, 795–813.

Limpens, J., Tomassen, H. B. M. & Berendse, F. (2003a). Expansion of *Sphagnum fallax* in bogs: striking the balance between N and P availability. *Journal of Bryology*, **25**, 83–90.

Limpens, J., Berendse, F. & Klees, H. (2003b). N deposition affects N availability in interstitial water, growth of *Sphagnum* and invasion of vascular plants in bog vegetation. *New Phytologist*, **157**, 339–47.

Limpens, J., Raymakers, J. T. A. G., Baar, J., Berendse, F. & Zijlstra, J. D. (2003c). The interaction between epiphytic algae, a parasitic fungus and *Sphagnum* as affected by N and P. *Oikos*, **103**, 59–68.

Lloret, F. (1991). Population-dynamics of the coprophilous moss *Tayloria tenuis* in a Pyrenean forest. *Holarctic Ecology*, **14**, 1–8.

Longton, R. E. (1988). *Biology of Polar Bryophytes and Lichens*. Cambridge: Cambridge University Press.

Longton, R. E. (1992). The role of bryophytes and lichens in terrestrial ecosystems. In *Bryophytes and Lichens in a Changing Environment*, ed. J. W. Bates & A. M. Farmer, pp. 32–76. Oxford: Clarendon Press.

Longton, R. E. (1997). Reproductive biology and life-history strategies. *Advances in Bryology*, **6**, 65–101.

Malmer, N. (1988). Patterns in the growth and the accumulation of inorganic constituents in the *Sphagnum* cover on ombrotrophic bogs in Scandinavia. *Oikos*, **53**, 105–20.

Maltby, E., Legg, C. J. & Proctor, M. C. F. (1990). The ecology of severe moorland fire on the North York Moors: effects of the 1976 fires, and subsequent surface and vegetation development. *Journal of Ecology*, **78**, 490–518.

Marino, P. C. (1991a). Competition between mosses (Splachnaceae) in patchy habitats. *Journal of Ecology*, **79**, 1031–46.

Marino, P. C. (1991b). Dispersal and coexistence of mosses (Splachnaceae) in patchy habitats. *Journal of Ecology*, **79**, 1047–60.

Marino, P. C. (1997). Competition, dispersal and coexistence of Splachnaceae in patchy habitats. *Advances in Bryology*, **6**, 241–63.

Markert, B., Herpin, U., Siewers, U., Berlkamp, J. & Lieth, H. (1996). The German heavy metal survey by means of mosses. *The Science of the Total Environment*, **182**, 159–68.

Markert, B. & Weckert, V. (1989). Fluctuations of element concentrations during the growing season of *Polytrichum formosum* Hedw. *Water, Air, and Soil Pollution*, **43**, 177–89.

Marschall, M. (1998). Nitrate reductase activity during desiccation and rehydration of the desiccation tolerant moss *Tortula ruralis* and the leafy liverwort *Porella platyphylla*. *Journal of Bryology*, **20**, 273–85.

Marschner, H. (1986). *Mineral Nutrition of Higher Plants*. London: Academic Press.

Martins, R. J. E. & Boaventura, R. A. R. (2002). Uptake and release of zinc by aquatic bryophytes (*Fontinalis antipyretica* L. ex Hedw.). *Water Research*, **36**, 5005–12.

Martínez-Abaigar, J., García-Álvaro, M. A., Beaucourt, N. & Núnez-Olivera, E. (2002a). Combined seasonal and longitudinal variations of element concentrations in two aquatic mosses (*Fontinalis antipyretica* and *F. squamosa*). *Nova Hedwigia*, **74**, 349–64.

Martínez-Abaigar, J., Núnez-Olivera, E. & Beaucourt, N. (2002b). Short-term physiological responses of the aquatic liverwort *Jungermannia exsertifolia* subsp. *cordifolia* to KH_2PO_4 and anoxia. *Bryologist*, **105**, 86–95.

Matzek, V. & Vitousek, P. (2003). Nitrogen fixation in bryophytes, lichens and decaying wood along a soil-age gradient in Hawaiian montane rain forest. *Biotropica*, **35**, 12–19.

McLean, R. O. & Jones, A. K. (1975). Studies of tolerance to heavy metals in the flora of the rivers Ystwyth and Clarach, Wales. *Freshwater Biology*, **5**, 431–44.

Meeks, J. C. (1998). Symbiosis between nitrogen-fixing cyanobacteria and plants. *BioScience*, **48**, 266–76.

Meeks, J. C., Campbell, E. L., Hagen, K. *et al.* (1999). Developmental alternatives of symbiotic *Nostoc punctiforme* in response to its plant partner *Anthoceros punctatus*. In *The Phototropic Prokaryotes*, ed. G. A. Peschek, W. Löffelhardt & G. Schmetterer, pp. 665–78. New York: Kluwer/Plenum.

Mickiewicz, J. (1976). Influence of mineral fertilization on the biomass of moss. *Polish Ecological Studies*, **2**, 57–62.

Mitchell, R. J., Sutton, M. A., Truscott, A. M. *et al.* (2004). Growth and tissue nitrogen of epiphytic Atlantic bryophytes: effects of increased and decreased atmospheric N deposition. *Functional Ecology*, **18**, 322–9.

Mitchell, R. J., Truscot, A. M., Leith, I. D. *et al.* (2005). A study of the epiphytic communities of Atlantic oakwoods along an atmospheric nitrogen deposition gradient. *Journal of Ecology*, **93**, 482–92.

Morecroft, M. D., Sellers, E. K. & Lee, J. A. (1994). An experimental investigation into the effects of atmospheric nitrogen deposition on two semi-natural grasslands. *Journal of Ecology*, **82**, 475–83.

Mouvet, C. (1985). The use of aquatic bryophytes to monitor heavy metals pollution of freshwaters as illustrated by case studies. *Verhein Internationale Verein Limnologie*, **22**, 2420–5.

Mouvet, C., Morhain, E., Sutter, C. & Couturieux, N. (1993). Aquatic mosses for the detection and follow-up of accidental discharges in surface waters. *Water, Air and Soil Pollution*, **66**, 333–48.

Mouvet, C., Pattée, E. & Cordebar, P. (1986). Utilisation des mousses aquatiques pour l'identification et la localisation précise de sources de pollution métallique multiforme. *Acta Oecologia*, **7**, 77–91.

Muhle, H. & LeBlanc, F. (1975). Bryophyte and lichen succession on decaying logs. I. Analysis along an evaporational gradient in eastern Canada. *Journal of the Hattori Botanical Laboratory*, **39**, 1–33.

Nadkarni, N. (1984). Epiphyte biomass and nutrient capital of a Neotropical elfin forest. *Biotropica*, **16**, 249–56.

Newmaster, S. G., Belland, R. J., Arsenault, A. & Vitt, D. H. (2003). Patterns of bryophyte diversity in humid coastal and inland cedar-hemlock forests of British Columbia. *Environmental Review*, **11**, S159–85.

Nordin, A. & Gunnarsson, U. (2000). Amino acid accumulation and growth of *Sphagnum* under different levels of N deposition. *Ecoscience*, **7**, 474–80.

Oechel, W. C. & Van Cleve, K. (1986). The role of bryophytes in nutrient cycling in the taiga. In *Forest Ecosystems in the Alaskan Taiga*, ed. K. Van Cleve, F. S. Chapin III, P. W. Flanagan, L. A. Viereck & C. T. Dyrness, pp. 121–37. New York: Springer-Verlag.

Offler, C. E., McCurdy, D. W., Patrick, J. W. & Talbot, M. J. (2003). Transfer cells: cells specialized for a special purpose. *Annual Review of Plant Biology*, **54**, 431–54.

Onianwa, P. C. (2001). Monitoring atmospheric metal pollution: a review of the use of mosses as indicators. *Environmental Monitoring and Assessment*, **71**, 13–50.

Palmer, M. W. (1986). Pattern in corticolous bryophyte communities of the North Carolina Piedmont: do mosses see the forest or the trees? *Bryologist*, **89**, 59–65.

Paulissen, M. P. C. P., Besalú, L. E., de Bruijn, H., van der Ven, P. J. M. & Bobbink, R. (2005). Contrasting effects of ammonium enrichment on fen bryophytes. *Journal of Bryology*, **27**, 109–17.

Pearce, I. S. K., Woodin, S. J. & van der Wal, R. (2003). Physiological and growth responses of the montane bryophyte *Racomitrium lanuginosum* to atmospheric nitrogen deposition. *New Phytologist*, **160**, 145–55.

Pearson, J., Wells, D. M., Seller, K. J. *et al.* (2000). Traffic exposure increases natural ^{15}N and heavy metal concentrations in mosses. *New Phytologist*, **147**, 317–26.

Pharo, E. J. & Beattie, A. J. (2002). The association between substrate variability and bryophyte and lichen diversity in eastern Australian forests. *Bryologist*, **105**, 11–26.

Pickering, D. C. & Puia, I. L. (1969). Mechanism for the uptake of zinc by *Fontinalis antipyretica*. *Physiologia Plantarum*, **22**, 653–61.

Pócs, T. (1982). Tropical forest bryophytes. In *Bryophyte Ecology*, ed. A. J. E. Smith, pp. 59–104. London: Chapman & Hall.

Press, M. C. & Lee, J. A. (1983). Acid phosphatase activity in *Sphagnum* species in relation to phosphate nutrition. *New Phytologist*, **93**, 567–73.

Pyysalo, H., Koponen, A. & Koponen, T. (1983). Studies on entomophily in Splachnaceae (Musci). II. Volatile compounds in the hypophysis. *Annales Botanici Fennici*, **21**, 335–8.

Raven, J. A. (1977). The evolution of land plants in relation to supracellular transport processes. *Advances in Botanical Research*, **5**, 314–19.

Raven, J. A. (2003). Long-distance transport in non-vascular plants. *Plant, Cell and Environment*, **26**, 73–85.

Raven, J. A., Griffiths, H., Smith, E. C. & Vaughn, K. C. (1998). New perspectives in the biophysics and physiology of bryophytes. In *Bryology for the Twenty-First Century*, ed. J. W. Bates, N. W. Ashton & J. G. Duckett, pp. 261–75. Leeds: Maney and British Bryological Society.

Richter, C. & Dainty, J. (1989a). Ion behavior in plant cell walls. I. Characterization of the *Sphagnum russowii* cell wall ion exchanger. *Canadian Journal of Botany*, **67**, 451–9.

Richter, C. & Dainty, J. (1989b). Ion behavior in plant cell walls. II. Measurement of the Donnan free space, anion-exclusion space, anion-exchange capacity, and cation-exchange capacity in delignified *Sphagnum russowii* cell walls. *Canadian Journal of Botany*, **67**, 460–5.

Rieley, J. O., Richards, P. W. & Bebbington, A. D. L. (1979). The ecological role of bryophytes in a north Wales woodland. *Journal of Ecology*, **67**, 497–527.

Rincón, E. (1988). The effect of herbaceous litter on bryophyte growth. *Journal of Bryology*, **15**, 209–17.

Rincón, E. (1990). Growth responses of *Brachythecium rutabulum* to different litter arrangements. *Journal of Bryology*, **16**, 120–2.

Rincón, E. & Grime, J. P. (1989). Plasticity and light interception by six bryophytes of contrasted ecology. *Journal of Ecology*, **77**, 439–46.

Roberts, S. K. (2006). Plasma membrane anion channels in higher plants and their putative functions in roots. *New Phytologist*, **169**, 647–66.

Romero, C., Putz, F. E. & Kitajima, K. (2006). Ecophysiology in relation to exposure of pendant epiphytic bryophytes in the canopy of a tropical montane oak forest. *Biotropica*, **38**, 35–41.

Ron, E., Estébanez, B., Alfayate, C., Marfil, R. & Corttella, A. (1999). Mineral deposits in cells of *Hookeria lucens*. *Journal of Bryology*, **21**, 281–8.

Rose, F. (1992). Temperate forest management: its effects on bryophyte and lichen floras and habitats. In *Bryophytes and Lichens in a Changing Environment*, ed. J. W. Bates & A. M. Farmer, pp. 211–33. Oxford: Clarendon Press.

Rühling, Å. & Tyler, G. (1970). Sorption and retention of heavy metals in the woodland moss *Hylocomium splendens* (Hedw.) Br. et Sch. *Oikos*, **21**, 92–7.

Rühling, Å. & Tyler, G. (2004). Changes in the atmospheric deposition of minor and rare elements between 1975 and 2000 in south Sweden, as measured by moss analysis. *Environmental Pollution*, **131**, 417–23.

Rydin, H. (1997). Competition among bryophytes. *Advances in Bryology*, **6**, 135–68.

Rydin, H. & Clymo, R. S. (1989). Transport of carbon and phosphorus compounds about *Sphagnum*. *Proceedings of the Royal Society, London*, B**237**, 63–84.

Samecka-Cymerman, A., Kempers, A. J. & Kolon, K. (2000). Concentrations of heavy metals in aquatic bryophytes used for biomonitoring in rhyolite and trachybasalt areas: a case study with *Platyhypnidium rusciforme* from the Sudety Mountains. *Annales Botanici Fennici*, **37**, 95–104.

Satake, K. (2000). Iron accumulation on the cell wall of the aquatic moss *Drepanocladus fluitans* in an acid lake at pH 3.4–3.8. *Hydrobiologia*, **433**, 25–30.

Satake, K., Shibata, K. & Bando, Y. (1990). Mercury sulphide (HgS) crystals in the cell walls of the aquatic bryophytes, *Jungermannia vulcanicola* Steph. and *Scapania undulata* (L.) Dum. *Aquatic Botany*, **36**, 325–41.

Schmitt, C. K. & Slack, N. G. (1990). Host specificity of epiphytic lichens and bryophytes: a comparison of the Adirondack Mountains (New York) and the Southern Blue Ridge Mountains (North Carolina). *Bryologist*, **93**, 257–74.

Sentenac, H. & Grignon, C. (1981). A model for predicting ionic equilibrium concentrations in cell walls. *Plant Physiology*, **68**, 415–19.

Sérgio, C., Figueira, R. & Viegas Crespo, A. M. (2000). Observations of heavy metal accumulation in the cell walls of *Fontinalis antipyretica*, in a Portuguese stream affected by mine effluent. *Journal of Bryology*, **22**, 251–5.

Shacklette, H. T. (1965). Element content of bryophytes. *Geological Survey Bulletin*, **1198-D**, D1–D21.

Shaw, A. J. (1994). Adaptation to metals in widespread and endemic plants. *Environmental Health Perspectives*, **102**, 105–8.

Shepherd, V. A., Beilby, M. J. & Shimmen, T. (2002). Mechanosensory ion channels in charophyte cells: the response to touch and salinity stress. *European Biophysics Journal*, **31**, 341–55.

Sjögren, E. (1975). Epiphyllous bryophytes of Madeira. *Svensk Botaniska Tidskrift*, **69**, 217–88.

Skrindo, A. & Økland, R. H. (2002). Effects of fertilization on understorey vegetation in a Norwegian *Pinus sylvestris* forest. *Applied Vegetation Science*, **5**, 167–72.

Slack, N. G. (1988). The ecological importance of lichens and bryophytes. *Bibliotheca Lichenologica*, **30**, 23–53.

Smith, A. J. E. (ed.) (1982a). *Bryophyte Ecology*. London: Chapman & Hall.

Smith, A. J. E. (1982b). Epiphytes and epiliths. In *Bryophyte Ecology*, ed. A. J. E. Smith, pp. 191–227. London: Chapman & Hall.

Söderlund, S., Forsberg, A. & Pedersén, M. (1988). Concentrations of cadmium and other metals in *Fucus vesiculosus* L. and *Fontinalis dalecarlica* Br. Eur. from the northern Baltic Sea and the southern Bothnian Sea. *Environmental Pollution*, **51**, 197–212.

Söderström, L. (1988). Sequence of bryophytes and lichens in relation to substrate variables of decaying coniferous wood in northern Sweden. *Nordic Journal of Botany*, **8**, 89–97.

Solga, A., Burkhardt, J., Zechmeister, H. G. & Frahm, J.-P. (2005). Nitrogen content, 15N natural abundance and biomass of the two pleurocarpous mosses *Pleurozium schreberi* (Brid.) Mitt. and *Scleropodium purum* (Hedw.) Limpr. in relation to atmospheric nitrogen deposition. *Environmental Pollution*, **134**, 465–73.

Solga, A. & Frahm, J.-P. (2006). Nitrogen accumulation by six pleurocarpous moss species and their suitability for monitoring nitrogen deposition. *Journal of Bryology*, **28**, 46–52.

Southorn, A. L. D. (1976). Bryophyte recolonization of burnt ground with particular reference to *Funaria hygrometrica*. I. Factors affecting the pattern of recolonization. *Journal of Bryology*, **9**, 63–80.

Southorn, A. L. D. (1977). Bryophyte recolonization of burnt ground with particular reference to *Funaria hygrometrica*. II. The nutrient requirements of *Funaria hygrometrica*. *Journal of Bryology*, **9**, 361–73.

Steinman, A. D. (1994). The influence of phosphorus enrichment on lotic bryophytes. *Freshwater Biology*, **31**, 53–63.

Stone, D. F. (1989). Epiphyte succession on *Quercus ganyana* branches in the Willamette Valley of Western Oregon. *Bryologist*, **92**, 81–94.

Streeter, D. T. (1970). Bryophyte ecology. *Science Progress*, **58**, 419–34.

Strong, L. & Wall, R. (1988). Invermectin in cattle: non-specific effects on pastureland ecology. *Aspects of Applied Biology*, **17**, 231–8.

Strong, L. & Wall, R. (1994). Effects of ivermectin and moxidectin on the insects of cattle dung. *Bulletin of Entomological Research*, **84**, 403–9.

Strong, L., Wall, R., Woodford, A. & Djeddou, D. (1996). The effect of faecally excreted ivermectin and fenbendazole on the insect colonization of cattle dung following the oral administration of sustain-release boluses. *Veterinary Parasitology*, **62**, 253–66.

Studlar, S. M. (1982). Succession of epiphytic bryophytes near Mountain Lake, Virginia. *Bryologist*, **85**, 51–63.

Takezawa, D. & Minami, A. (2004). Calmodulin-binding proteins in bryophytes: identification of abscisic acid-, cold-, and osmotic stress-induced genes encoding novel membrane-bound transporter-like proteins. *Biochemistry and Biophysics Research Communications*, **317**, 428–36.

Tamm, C. O. (1953). Growth, yield and nutrition in carpets of a forest moss (*Hylocomium splendens*). *Meddelanden från Statens Skogsforskningsinstitut*, **43**, 1–140.

Tewari, M., Upreti, N., Pandey, P. & Singh, S. P. (1985). Epiphytic succession on tree trunks in a mixed oak-cedar forest, Kumaun Himalaya. *Vegetatio*, **63**, 105–12.

Trebacz, K., Simonis, W. & Schönknecht, G. (1994). Cytoplasmic Ca^{2+}, K^{+}, Cl^{-} and NO_3^{-} activities in the liverwort *Conocephalum conicum* L. at rest and during action potentials. *Plant Physiology*, **106**, 1073–84.

Trynoski, S. E. & Glime, J. M. (1982). Direction and height of bryophytes on four species of northern trees. *Bryologist*, **85**, 281–300.

Tyler, G. (1990). Bryophytes and heavy-metals: a literature review. *Botanical Journal of the Linnean Society*, **104**, 231–53.

Tuba, Z. & Csintalan, Z. (1993). The use of moss transplantation technique for bioindication of heavy metal pollution. In *Plants as Biomonitors. Indicators for Heavy Metals in the Terrestrial Environment*, ed. B. Markert, pp. 253–9. Weinheim: VCH.

Turner, B. L., Baxter, R., Ellwood, N. T. W. & Whitton, B. A. (2003). Seasonal phosphatase activities of mosses from Upper Teesdale, northern England. *Journal of Bryology*, **25**, 189–200.

Turetsky, M. R. (2003). The role of bryophytes in carbon and nitrogen cycling. *Bryologist*, **106**, 395–409.

van Tooren, B. F. (1988). Decomposition of bryophyte material in two Dutch chalk grasslands. *Journal of Bryology*, **15**, 343–52.

van Tooren, B. F., den Hertog, J. & Verhaar, J. (1988). Cover, biomass and nutrient content of bryophytes in Dutch chalk grasslands. *Lindbergia*, **14**, 47–58.

van Tooren, B. F., van Dam, D. & During, H. J. (1990). The relative importance of precipitation and soil as sources of nutrients for *Calliergonella cuspidata* in a chalk grassland. *Functional Ecology*, **4**, 101–7.

Vanderpoorten, A. & Klein, J.-P. (1999). Variations of aquatic bryophyte assemblages in the Rhine Rift related to water quality. 2. The waterfalls of the Vosges and the Black Forest. *Journal of Bryology*, **21**, 109–15.

Virtanen, R., Johnston, A. E., Crawley, M. J. & Edwards, G. R. (2000). Bryophyte biomass and species richness on the Park Grass Experiment, Rothamsted, UK. *Plant Ecology*, **151**, 129–41.

Vitt, D. H., Wieder, K., Halsey, L. A. & Turetsky, M. (2003). Response of *Sphagnum fuscum* to nitrogen deposition: a case study of ombrogenous peatlands in Alberta, Canada. *Bryologist*, **106**, 235–45.

Voth, P. D. (1943). Effects of nutrient-solution concentration on the growth of *Marchantia polymorpha*. *Botanical Gazette*, **104**, 591–601.

Wall, R. & Strong, L. (1987). Environmental consequences of treating cattle with the antiparasitic drug ivermectin. *Nature*, **327**, 418–21.

Webster, H. J. & Sharp, A. J. (1973). Bryophytic succession on caribou dung in Arctic Alaska. *American Biological Society Bulletin*, **20**, 90.

Wells, J. M. & Boddy, L. (1995). Phosphorus translocation by saprotrophic basidiomycete mycelial cord systems on the floor of a mixed deciduous woodland. *Mycological Research*, **99**, 977–80.

Wells, J. M. & Brown, D. H. (1996). Mineral nutrient recycling within shoots of the moss *Rhytidiadelphus squarrosus* in relation to growth. *Journal of Bryology*, **19**, 1–17.

Wells, J. M. & Richardson, D. H. S. (1985). Anion accumulation by the moss *Hylocomium splendens*: uptake and competition studies involving arsenate, selenate, selenite, phosphate, sulphate and sulphite. *New Phytologist*, **101**, 571–83.

Wiklund, K. & Rydin, H. (2004). Ecophysiological constraints on spore establishment in bryophytes. *Functional Ecology*, **18**, 907–13.

Wilson, J. A. & Coxon, D. S. (1999). Carbon flux in a subalpine spruce-fir forest: pulse release from *Hylocomium splendens* feather-moss mat. *Canadian Journal of Botany*, **77**, 564–9.

Woodin, S., Press, M. C. & Lee, J. A. (1985). Nitrate reductase activity in *Sphagnum fuscum* in relation to wet deposition of nitrate from the atmosphere. *New Phytologist*, **99**, 381–8.

Woolgrove, C. E. & Woodin, S. J. (1996a). Effects of pollutants in snowmelt on *Kiaeria starkei*, a characteristic species of late snowbed bryophyte dominated vegetation. *New Phytologist*, **133**, 519–29.

Woolgrove, C. E. & Woodin, S. J. (1996b). Current and historical relationships between the tissue nitrogen content of a snowbed bryophyte and nitrogenous air pollution. *Environmental Pollution*, **91**, 283–8.

Woollon, F. B. M. (1975). Mineral relationships and ecological distribution of *Fissidens cristatus* Wils. *Journal of Bryology*, **8**, 455–64.

Zechmeister, H. G. (2005). Bryophytes of continental salt meadows in Austria. *Journal of Bryology*, **27**, 297–302.

Zechmeister, H. G., Hohenwallner, D., Ross, A. & Hanus-Illner, A. (2003). Variations in heavy metal concentrations in the moss species *Abietinella abietina* (Hedw.) Fleisch. according to sampling time, within site variability and increase in biomass. *The Science of the Total Environment*, **301**, 55–65.

Zielke, M., Ekker, A. S., Olsen, R. A., Spjelkavik, S. & Solheim, B. (2002). The influence of abiotic factors on biological nitrogen fixation in different types of vegetation in the High Arctic. *Arctic, Antarctic, and Alpine Research*, **34**, 293–9.

Zielke, M., Solheim, B., Spjelkavik, S. & Olsen, R. A. (2005). Nitrogen fixation in the High Arctic: role of vegetation and environmental conditions. *Arctic, Antarctic, and Alpine Research*, **37**, 372–8.

9

The structure and function of bryophyte-dominated peatlands

DALE H. VITT AND R. KELMAN WIEDER

9.1 Introduction

Peatlands are unbalanced ecosystems where plant production exceeds decomposition of organic material. As a result, considerable quantities of organic material, or peat, accumulate over long periods of time: millennia. This organic material is composed primarily of plant fragments remaining after partial decomposition of the plants that at one time lived on the surface of the peatland. Decomposition occurs through the action of micro-organisms that have the ability to utilize dead plant components as sources of carbon for respiration (Thormann & Bayley 1997) in both the upper, aerobic peat column (the acrotelm) and the lower, anaerobic peat (the catotelm) (Ingram 1978, Clymo 1984, Wieder *et al.* 1990, Kuhry & Vitt 1996). Labile cell contents, cellulose, and hemicellulose are more readily available sources of carbon than recalcitrant fractions that contain lignin-like compounds, with these latter compounds being concentrated in peat by decomposition (Williams *et al.* 1998, Turetsky *et al.* 2000). The vascular plant-dominated, tree, shrub, and herb layers produce less biomass (Campbell *et al.* 2000) and decompose more readily than the bryophyte-dominated ground layer (Moore 1989). Surfaces of northern peatlands are almost always completely covered by a continuous mat of moss (National Wetlands Working Group 1988, Vitt 1990), and the large amount of biomass contained in this layer is composed of cell wall material that decomposes slowly. This slow decomposition, coupled with water-saturated, anaerobic conditions in the peat, cool climate, and a cool moist growing season conducive to bryophyte growth, allows organic matter to accumulate over large areas. In northern peatlands, the peat that accumulates is generally composed of a high percentage of material derived from bryophytes.

Bryophyte Biology: Second Edition, ed. B. Goffinet & A. J. Shaw. Published by Cambridge University Press.

9.2 Structure and peatland types

Wetlands are, in general, ecosystems that have accumulated some organic matter and have an abundance of hydrophytic vegetation. They can be divided into five basic types, three of which are non-peat forming systems that are often defined as areas with less than 40 cm of peat (Zoltai & Vitt 1995). These three non-peat-forming wetland types may have a well-developed tree or shrub layer (swamps), be dominated by sedges, grasses, and rushes without trees and shrubs (marshes), or contain emergent vegetation in less than a meter of water (shallow open water) (National Wetlands Working Group 1988, Zoltai & Vitt 1995). All of these wetland types have seasonally fluctuating water tables that are strongly influenced by surrounding surface and ground waters (Zoltai & Vitt 1995, Vitt 2006). Nitrogen mineralization rates are high (Bridgham *et al.* 1998) and surface inflows may be nutrient-enriched, so that these wetlands are often eutrophic (Mitch & Gosselink 1993) (Fig. 9.1). The lack of a well-developed bryophyte-dominated ground layer, coupled with abundant vascular plant litter, allows relatively rapid decomposition and results in little peat accumulation (Thormann *et al.* 1999). Peat-forming wetlands (often termed "mires" in Europe and "peatlands" in North America) are ecosystems that accumulate organic matter. Although only two basic types of peatlands have generally been recognized (bogs and fens), peatlands can be more precisely defined by using a combination of hydrologic, chemical, and floristic criteria (Zoltai & Vitt 1995).

9.2.1 *Hydrology*

Peatlands that derive their water and nutrient supplies from precipitation and from water that has been in contact with upland soils are termed fens. Water flows through fens via one to several inflows and outflows (Fig. 9.2). The chemistry of surrounding upland soil water, and/or groundwater, influences the water chemistry of fens (Siegel & Glaser 1987) and often causes them to be relatively rich in base cations. However, if the surrounding uplands are relatively acidic, the fens may be poor in base cations (Halsey *et al.* 1997a,b).

Peatlands that derive their water and nutrient supplies solely from precipitation are termed bogs. These peatlands are somewhat elevated above the surrounding area, such that water flows from the raised bog surface on to the surrounding wetland or upland (Fig. 9.2). Therefore, bogs have relatively stagnant waters that do not reflect the surrounding soil conditions. For this reason, the chemistry of the precipitation has the most important influence on the chemistry of the bog waters (Malmer 1962, Vitt *et al.* 1990, Malmer *et al.* 1992). If hydrology is considered the most significant criterion for peatland

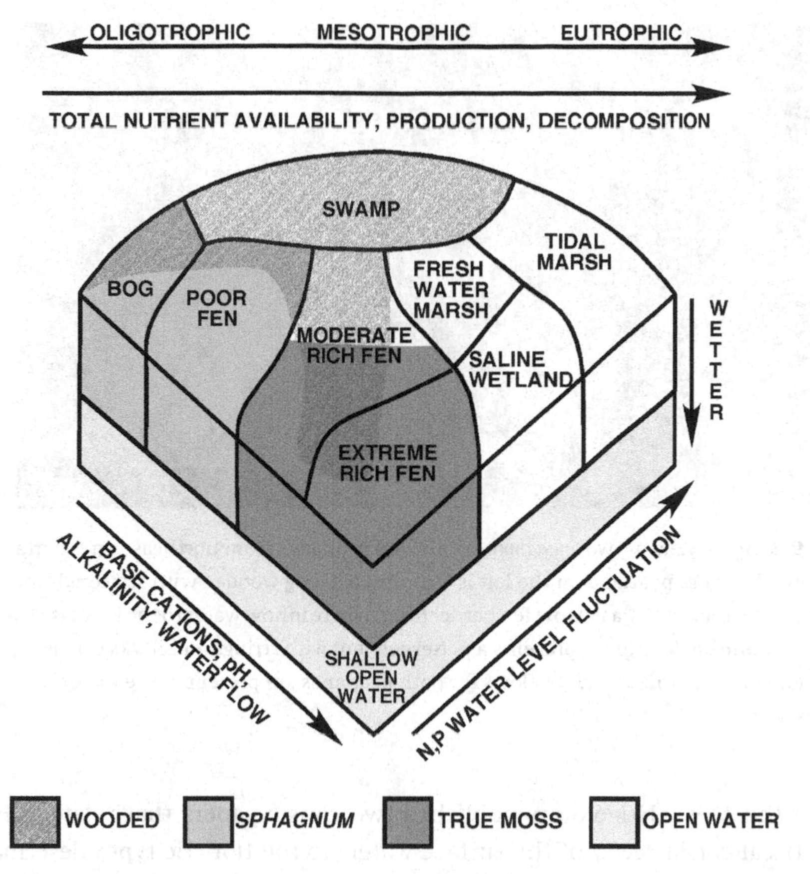

Fig. 9.1. Ternary diagram showing five wetland classes in relation to hydrology, chemistry, and vegetation. Modified from Vitt (1994) and Zoltai & Vitt (1995).

classification, then the primary division of these ecosystems is ombrogenous bogs and geogenous fens. Bogs can be viewed as ecosystems that are oligotrophic with ombrotrophic vegetation and fens are ecosystems that may be either oligotrophic or mesotrophic and are dominated by minerotrophic vegetation.

9.2.2 *Chemistry*

In the 1940s, Einar DuReitz recognized that Scandinavian peatlands could be divided into several types based on vegetation structure and floristic species composition (DuReitz 1949). He recognized that some peatlands have a large number of plant species with high fidelity to particular site conditions. The fens that were "rich" in floristic site indicators he called rich fens. Other fens had fewer species with high fidelity and he called these poor fens. He recognized that ombrotrophic bogs had no, or very few, plant species that were exclusive to bog conditions.

Fig. 9.2. Aerial view of two *Sphagnum*-dominated peatlands from northeastern Alberta, Canada. The large peatland on the left is a continental bog wooded with scattered *Picea mariana*, surrounded by a narrow fen border (lagg) where inflow water to the basin is channeled. The peatland on the right represents a patterned fen, with strings and flarks oriented perpendicular to waterflow. Outflows for both peatlands are present at the top of the photo.

In the 1950s, Hugo Sjörs published two classic papers that related pH and electrical conductivity of the surface waters to the floristic types described by DuReitz. Sjörs (1950, 1952) described pH and "corrected" conductivity gradients (the latter calculated by subtracting conductivity due to hydrogen ions and serving as a surrogate for base cation abundance) that ranged from acidic, low conductivity in bogs through somewhat less acidic poor fens to basic, high conductivity in waters of rich fens. He proposed that rich fens could be subdivided into two types: moderate- (transitional) rich fens and extreme-rich fens. Further work by a number of researchers has carefully characterized these four peatland types (bogs, poor fens, moderate-rich fens, extreme-rich fens) in terms of both chemistry and vegetation (e.g. Gorham 1956, Slack *et al.* 1980, Malmer 1986, Vitt & Chee 1990, Gorham & Janssens 1992, Vitt *et al.* 1995a).

In general, peatland surface water pH is in the range 3.0–4.2 in bogs, 4.2–5.5 in poor fens, 5.5–7.0 in moderate-rich fens, and 7.0–8.0 or higher in extreme-rich fens (reviewed in Vitt 1990). Associated with this acidity gradient is one of alkalinity, with bogs and poor fens having no alkalinity, moderate-rich fens having some alkalinity (500–1000 $\mu eq\,l^{-1}$ of $CaCO_3$) and extreme-rich fens having highly alkaline waters to the extent that $CaCO_3$ may be deposited as marl (Vitt *et al.* 1995b). Along these acidity/alkalinity gradients, base cation

(Ca^{2+}; Mg^{2+}; Na^{+}; K^{+}) concentrations in peatland surface waters increase. Typically, surface water concentrations of Ca^{2+} are less than $3\,mg\,l^{-1}$ in bogs, around $5\,mg\,l^{-1}$ in poor fens, and from 5 to $35\,mg\,l^{-1}$ or more in rich fens (Fig. 9.3). However, concentrations of potentially limiting nutrients (dissolved inorganic forms of N and P) (Walbridge & Navaratnam 2006) are highly variable and show little correlation with the defining chemical gradients of acidity, alkalinity, and base cations (Fig. 9.3). Chemically, poor fens are more similar to bogs than they are to rich fens. Thus, when surface water chemical characteristics are considered, the critical division is between systems that are acidic and possess no alkalinity (poor fens and bogs) and systems that are neutral to basic and alkaline (rich fens).

At the regional scale, surface water sampling of over 100 peatlands reveals a clearly bimodal pH distribution: bogs and poor fens with pH less than about 5.0 form one group and rich fens with pH greater than about 6.0 form a second group (Fig. 9.4). Peatlands with intermediate surface water pH values between 5.2 and 5.7 can be rare on the northern landscape (e.g. Glaser *et al.* 2004); however, there are other regions where this gap is not so apparent (Glaser 1992b). As surface water pH increases within this narrow pH range, alkalinity becomes established as a key chemical characteristic (Vitt *et al.* 1995a).

9.2.3 *Vegetation and flora*

In terms of physiognomy, peatlands vary considerably. Bogs are relatively dry and have a relatively thick acrotelm (aerobic layer), and a large percentage of their area is covered by hummocks; they may be wooded, shrub-dominated, or completely without trees (open) (Glaser & Janssens 1986, Belland & Vitt 1995). Generally, maritime bogs are open, often contain a sedge component on lawns (see Table 9.1 for definitions), and contain pools of water (which may be arranged in reticulate or parallel patterns), whereas continental bogs are wooded, have almost no sedges, and have no open water (Damman 1979, Glaser & Janssens 1986, Davis & Anderson 1991, Vitt *et al.* 1994). Bogs are always dominated by *Sphagnum* mosses (or feather mosses and lichens), almost exclusively lack a sedge component, and often have abundant shrubs (Glaser 1992a, Belland & Vitt 1995). Poor and rich fens are relatively wet, have a thin acrotelm, and have a higher percentage of their area covered by lawns and carpets (Vitt 1990). Pools are sometimes present (Vitt *et al.* 1975). These fens may exhibit surface patterning of reticulate or parallel arrangements of raised, dry, elongate strings separated by pools (flarks) of water often filled with carpets of moss (Fig. 9.2) (Foster *et al.* 1983, Halsey *et al.* 1997b). Fens are usually sedge-dominated. They may be wooded, shrubby, or open (Vitt & Chee 1990, Halsey *et al.* 1997b). Poor fens are dominated by *Sphagnum* moss and ericaceous shrubs may be

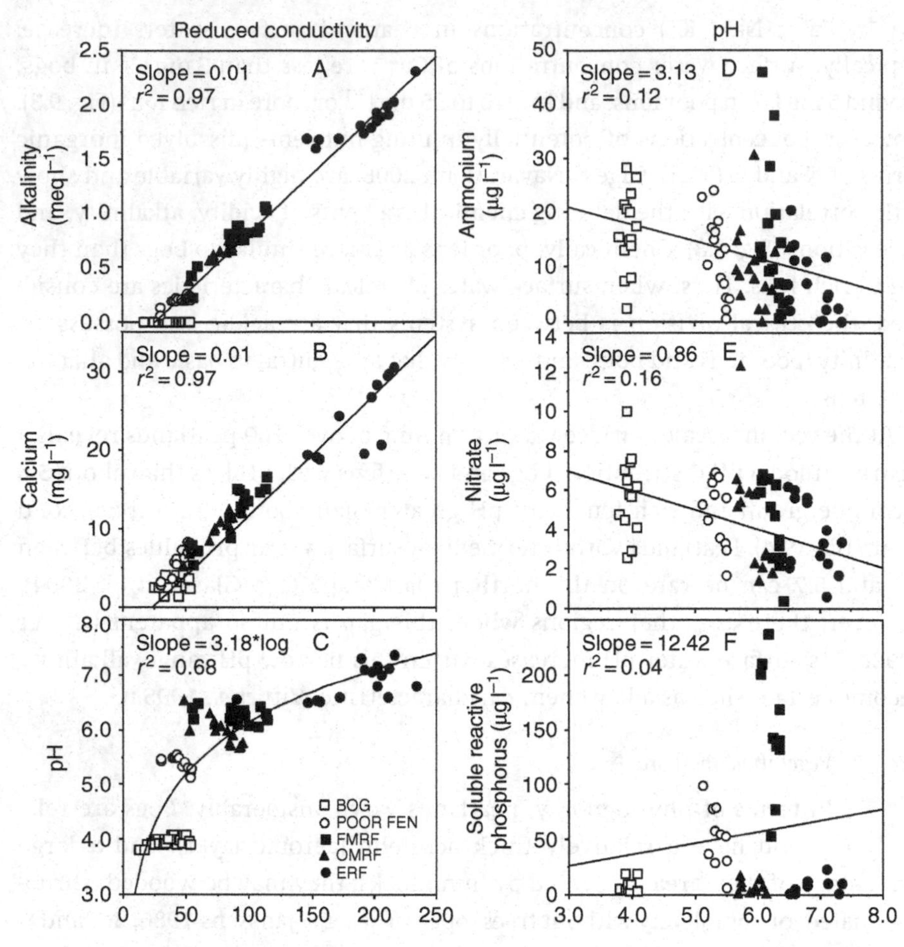

Fig. 9.3. Relationship between reduced conductivity (µS cm^{-1}) and (A) calcium, (B) alkalinity, and (C) pH, and between pH and the nutrients (D) ammonium, (E) nitrate, and (F) soluble reactive phosphorus for surface waters along the bog–rich fen gradient. Open symbols include bogs and poor fens; closed symbols include several rich fens. Modified from Vitt *et al.* (1995a). Reduced conductivity (often termed corrected conductivity) is the total electrical conductivity minus that supplied by H^+ (Sjörs 1952). (Reprinted with permission from the National Research Council Press/*Canadian Journal of Fisheries and Aquatic Sciences*, Vol. 52, 1995, Seasonal variation in water chemistry over a bog-rich fen gradient in continental western Canada, by Dale H. Vitt, Suzanne E. Bayley, and Tai-Long Jin, Figs. 16 and 17, p. 602.)

present; furthermore, they differ from bogs by the appearance of a few rare fen indicators with high fidelity to poor fens (i.e. species of *Juncus* and *Carex*). "Brown mosses" dominate moderate- and extreme-rich fens, and ericaceous shrubs are sparse or absent (Vitt 1990); rich fens have relatively high numbers of indicator species. Vascular plant indicator species of these peatland types are regionally

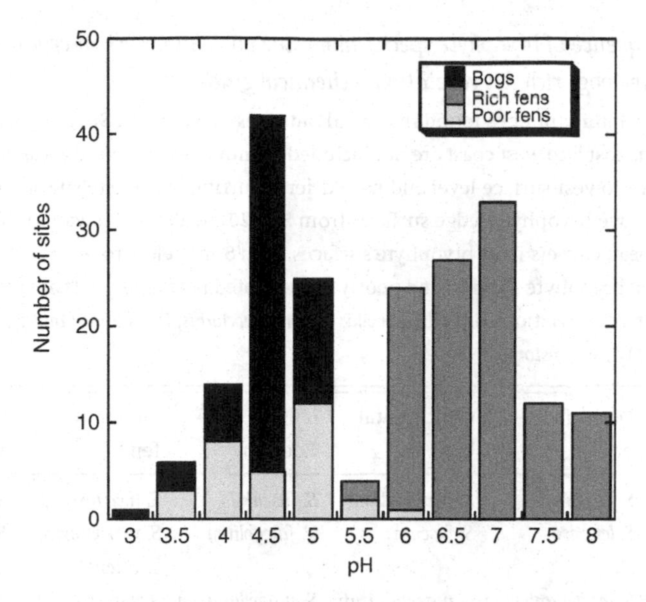

Fig. 9.4. Histogram of pH and peatland type. There are relatively few sites that have surface water pH between 5.2 and 5.7, when alkalinity values approach zero. Data for sites ($n = 100$) from continental western Canada.

based; however, the broad ranges of bryophytes coupled with their sensitivity to nutrient, acidity/alkalinity gradients, and water levels make bryophytes nearly perfect sensitive indicators of peatland conditions (Table 9.1).

In summary, when bogs, poor fens, and rich fens are compared floristically through multivariate techniques (e.g. Nicholson *et al.* 1996, Gignac *et al.* 1998), bogs and poor fens are generally more similar to each other than are poor fens and rich fens. Any of these three peatland types may be wooded, shrubby, or totally without trees (Glaser 1992a, Belland & Vitt 1995, Halsey *et al.* 1997b). Bogs, especially continental ones, differ from fens in their general lack of sedges (Glaser 1992a, Belland & Vitt 1995). Bogs and poor fens are *Sphagnum*-dominated; rich fens are brown moss-dominated. In oceanic areas, bogs may be patterned, whereas fens tend to be patterned more frequently in more continental areas. Bogs have water flowing away from their centers, but fens have water flowing through the system. Bogs have a well-developed acrotelm and are drier, whereas fens of all types have a poorly developed acrotelm and are wetter (Vitt *et al.* 1994). The dominance of *Sphagnum* in poor fens and bogs and lack of it in rich fens correlates well with acidity/alkalinity criteria (compare open symbols: *Sphagnum*-dominated peatlands, with closed symbols: brown moss-dominated peatlands of Fig. 9.3). Thus, if chemical, vegetational, and floristic criteria are used to categorize peatlands, then poor fens and bogs should be grouped together as "*Sphagnum*-dominated peatlands" (or "acidic peatlands") versus rich

Table 9.1 *Sequence of bryophyte species along the pool–hummock microtopographic gradient for the bog–rich fen vegetation – chemical gradients*

Species are the dominant ones found in central* and western Canada. Species found in oceanic peatlands of the east and west coast are not included. Hummocks are raised areas 20 to more than 50 cm above the lowest surface level and have drier occurring bryophyte species, sedges, and shrubs; lawns have bryophyte/sedge surfaces from 5 to 20 cm above the water table and well-consolidated peat; carpets have bryophyte surfaces from 5 cm below to 5 cm above the water table (emergent bryophyte species) and poorly consolidated peat; and pools are areas with open, standing water. Abbreviations *C., Calliergonella; D., Drepanocladus; H., Hamatocaulis; S., Sphagnum; T., Tomenthypnum; W., Warnstorfia.*

	Permafrost bog	Continental bog	Poor fen	Moderate-rich fen	Extreme-rich fen
Hummock: Top	*S. fuscum* *S. lenense*	*S. capillifolium* *S. fuscum*	*S. fuscum* *T. falcifolium*	*S. fuscum* *S. warnstorfii* *T. nitens*	*S. fuscum* *S. warnstorfii* *T. nitens*
Hummock: Side	*S. magellanicum*	*S. magellanicum*	*S. magellanicum*	*S. warnstorfii* *T. nitens*	*S. warnstorfii* *T. nitens*
Lawn	*S. angustifolium*	*S. angustifolium*	*S. angustifolium*	*H. vernicosus*	*Campylium stellatum*
	S. balticum	*S. rubellum**	*S. papillosum**	*S. teres*	
Carpet	*S. jensenii*	*S. lindbergii*	*S. jensenii*	*C. cuspidata*	*Scorpidium revolvens*
	S. majus *S. riparium*		*S. riparium*	*S. subsecundum*	
Pool	*W. fluitans*	*S. cuspidatum**	*W. exannulata*	*D. aduncus*	*Scorpidium scorpioides*
				H. lapponicus	

Data derived from numerous sources, including Slack *et al.* (1980), Crum (1988), Gignac and Vitt (1990), Gignac *et al.* (1991), Vitt *et al.* (1995b). Modified from Vitt (1994).

fens that should be called "brown moss-dominated peatlands" (or "alkaline peatlands"). However if hydrological criteria are used, ombrogenous bogs stand alone and can be contrasted with geogenous rich and poor fens.

9.3 Function and ecological importance of the moss layer

A 90%–100% cover of mosses dominates the ground layer in peatlands; although several species of hepatics occur in peatlands, only *Leiomylia (Mylia) anomala* is ever abundant. Functioning of the peatland ecosystem is highly dependent on this moss layer, and both production and decomposition, as well as community development, are all influenced by this layer of mosses. In particular, the moss layer influences peatland function in several ways. Especially noteworthy are: (1) nutrient uptake, (2) water-holding abilities, (3) decomposition, and (4) acidification.

9.3.1 *Nutrient uptake and the consequences of atmospheric deposition*

Some of the earliest studies in Great Britain suggested that elevated atmospheric nitrogen deposition could inhibit the growth of *Sphagnum* and also alter the habitats of some species. These responses of *Sphagnum* species were implicated as causally related to the loss of some *Sphagnum* species from British peatlands (Press & Lee 1982, Woodin *et al.* 1985, Press *et al.* 1986). In North America, however, subsequent experiments revealed that increasing nitrogen input to bogs (Rochefort *et al.* 1990) and rich fens (Rochefort & Vitt 1988) resulted in increases in moss growth, suggesting that nitrogen was limiting to *Sphagnum* growth. Other factors, such as precipitation (Bayley 1993) and phosphorus availability (Aerts *et al.* 1992), have been shown to influence plant growth. At high nitrogen deposition sites, *Sphagnum* growth has been shown to be phosphorus limited (Bragazza *et al.* 2004, Limpens *et al.* 2004).

When isotopically labeled nitrogen was experimentally added to peatlands (a bog and a rich fen) in simulated precipitation at sites with low nitrogen deposition, 98% of the nitrogen was recovered in the top 12 cm of peat after one year. After two years, less than 2% of the added nitrogen was found in the vascular plants. In both cases, the mosses increased in growth but the vascular plants did not (Li & Vitt 1997). This particular study did not distinguish whether the added nitrogen that was recovered in the ground layer and in the peat was sequestered in micro-organisms, within moss cell walls, or contained in the moss living cell cytoplasm. Under low nitrogen deposition, it appears that nitrogen is quickly sequestered by the moss layer and subsequent movement to vascular plants is dependent on release of the nitrogen from the moss and mineralization rates within the moss layer. *Sphagnum* tightly holds much of this added nitrogen and in the second year 19% of the first year's nitrogen was found in the new second year's growth, indicating that *Sphagnum* can translocate nitrogen upward (Li & Vitt 1997). However, under moderate and high deposition, nitrogen may pass through the *Sphagnum* layer directly to vascular plants and micro-organisms.

In summary, it appears that in *Sphagnum* (and probably all mosses), both the growth response and the sequestration of nitrogen is triphasic. First, under pristine boreal conditions, nitrogen limits the growth of *Sphagnum* (and true mosses in rich fens as well) and the *Sphagnum* layer is able to sequester nearly all of the deposited atmospheric nitrogen for its own use; moss growth occurs without a measurable change in moss tissue nitrogen concentration. Second, with increasing deposition nitrogen no longer limits *Sphagnum* growth, but is taken up by the growing mosses and is sequestered in the living plant tissues (thus concentrations of nitrogen in the *Sphagnum* plants increase). Third, at even higher nitrogen deposition, the living *Sphagnum* layer becomes nitrogen-saturated and some of the newly deposited nitrogen bypasses the living *Sphagnum* and is leached downward in the upper peat column, becoming available for either vascular plant or micro-organism growth (Lamers *et al.* 2000, Berendse *et al.* 2001, Limpens & Berendse 2003, Limpens *et al.* 2003). *Sphagnum* responses in the early studies in Great Britain appear to reflect this third phase, whereas *Sphagnum* responses in the later North American studies appear to reflect phase 1. Thus, under phase 1 and 2 conditions, *Sphagnum* is able to scavenge nearly all of the atmospheric nitrogen and act as a "gatekeeper" for accessibility of nitrogen into peatlands. However, under high nitrogen deposition regimes (phase 3), this gatekeeper role is overrun (Wieder 2006). Vitt *et al.* (2003) suggested that the boundary between phases one and two may be reached at atmospheric nitrogen deposition rates of around 16 kg ha^{-1} yr^{-1}.

A recent synthesis of ground layer net primary production suggests that at the regional scale in continental western Canada the four peatland types do not differ (Campbell *et al.* 2000). Overall, variability in ground layer production is high both spatially and temporally, but in general, the ground layer produces about 41% of the total annual plant production (Table 9.2). In a global review of *Sphagnum* production, Gunnarsson (2005) reported an overall mean of 259 g m^{-2} yr^{-1} with large variation (standard deviation + 206). Thus, mosses sequester nutrients efficiently, and through release and mineralization the moss layer can effectively control subsequent nutrient (at least nitrogen) availability, and hence plant production. However, changes in atmospheric nitrogen deposition can have serious consequences in ground layer functioning. A more thorough review of peatland production is provided in Wieder (2006).

9.3.2 *Water-holding capacity*

In comparison to almost all vascular plants, which are drought-avoiding and unable to survive water deficit, most mosses are drought-tolerant and can survive water deficit (Wood 2005). A few mosses are desiccation-tolerant and can survive even severe water deficit (Wood 2007; see also Chapters 6 and 7, this volume). Mosses are photosynthetically active when they are wet, have the ability to become inactive when dry, and can revitalize when rewetted (Bewley 1979, Proctor 1979, 1984, Proctor *et al.* 2007). Whereas mosses occurring in dry habitats (e.g. *Grimmia, Orthotrichum, Syntrichia*) have evolved to be able to survive in the face

Table 9.2 *Summary of net primary production pooled state/province means by layer for wetland types*

Ground layer in peatlands is moss-dominated. Peatland and northern wetland means are pooled by wetland type and location. Standard deviations are shown in brackets for those layers where original published data did not include pooled means. For layers containing pooled means no standard deviations could be calculated. Symbol x, layers that are not present for the particular peatland type. Amounts given are in g m^{-2} yr^{-1}. For details see Campbell *et al.* (2000).

	Tree	Shrub	Herb	Ground	Total
Bog	106 (192)	247 (104)	13	156 (157)	449 (215)
Wooded fen	44	108	34	74	358
Shrubby fen	x	63	125	118	263
Open fen	x	x	365 (458)	163	268 (34)
Peatlands	**88 (68)**	**210 (136)**	**166 (298)**	**139 (106)**	**337 (142)**
Wooded swamp	542 (279)	31 (29)	62	x	654 (197)
Shrubby swamp	x	480 (260)	727 (667)	x	1232 (405)
Marsh	x	x	999 (529)	x	934 (518)
Northern wetlands	**542 (279)**	**255 (296)**	**820 (592)**	**x**	**970 (467)**

of daily cycles of drying and rewetting, peatland mosses tolerate frequent drought to a much lesser extent (Glime & Vitt 1984). Instead, peatland mosses have developed morphological adaptations to retain moisture, allowing longer photosynthetically active periods and thus greater growth (Stålfelt 1937). Pool and carpet species such as *Scorpidium scorpioides, Hamatocaulis lapponicus, Warnstorfia (Drepanocladus) exannulata*, and *Sphagnum cuspidatum* live submerged or form poorly consolidated, emergent carpets (Vitt & Chee 1990). These species appear to have limited physiological abilities to live for extended time after drying out (but little information is available on this topic). Among the rich fen species, hummock species such as *Tomenthypnum nitens* have abundant stem tomentum and numerous side branches apparently facilitating water uptake through capillarity. In addition, the dense canopy structure of *Tomenthypnum* plant communities may diminish evaporative water losses. These adaptations can also be seen in other peatland species such as *Aulacomnium palustre, Catoscopium nigritum, Dicranum undulatum, Polytrichum strictum*, and *Tomenthypnum falcifolium.* Interestingly, among these mosses and their dense canopies, it is common to find less dought-tolerant species of hepatics, with species of *Cephalozia, Calypogeia*, and *Lophozia* occurring as mini-lianas among the larger mosses.

Although dead hyaline leaf cells that contain pores and hold large amounts of water occur sporadically throughout mosses (Proctor 1984), they are

Fig. 9.5. *Sphagnum papillosum*, with its complex canopy of dense capitula and turgid spreading branches.

particularly well developed in the genus *Sphagnum*. In species of this genus, the leaves consist of alternating, large, hyaline cells enclosing smaller, living, green cells, in a 1:2 ratio, an arrangement that is unique among plants. Additionally, the stems and branches are often encased in one or more layers of dead enlarged hyaline cells, forming a hyalodermis. As hyaline cells develop, they lose their living cell contents, resulting in a relatively high C/N ratio of the dead cells that compose peat, with implications for peat decomposition rates. However, the large hyaline cells, reinforced with internal cell wall thickenings (termed fibrils), partly to entirely enclose photosynthetically active green cells. Thus, through internal hyaline cell water-holding capacity and through a complex canopy of dense capitula (Rydin & McDonald 1985), spreading and hanging branches, and concave leaves (Fig. 9.5), *Sphagnum* produces a plant morphology suggestive of drought avoidance, not drought tolerance.

Communities of hummock species such as *S. fuscum* and *S. capillifolium* have the most highly developed canopies that protect the living green cells from drying out; communities of lawn species such as *S. angustifolium* have a looser canopy structure and less well developed water-retention abilities. *Sphagnum* species occurring on the highest hummocks occur there not because they are physiologically drought-tolerant, but because they are morphologically drought-avoidant (Titus & Wagner 1984). Lawn species, however, occurring in wetter but more variable conditions, are in fact physiologically more drought-insensitive and lack the highly developed canopy modifications of the true hummock-formers (Titus & Wagner 1984). These interpretations have been corroborated experimentally through establishment experiments showing that the lawn species *S. angustifolium* established efficiently on bare peat without the protection of a developed canopy, whereas *S. fuscum* and *S. magellanicum*, both hummock-formers, did not establish efficiently on bare peat (Li & Vitt 1994).

The high water-holding capacities of *Sphagnum* plant communities can result in local water table raising that facilitates the lateral expansion of peatlands into adjacent upland areas. This swamping, or paludification, of neighboring habitats is a major factor in increasing the amount of organic terrain in northern landscapes (Vitt & Kuhry 1992).

In summary, the dead hyaline cells allow *Sphagnum* species to maintain hydrated conditions of the living cells for extended periods of time, lengthening photosynthetically active periods and promoting net primary production and vertical plant growth. Hyaline cells of individual *Sphagnum* plants, along with the dense packing of *Sphagnum* species into communities, result in the upward movement of water from the water table, elevating the peatland water table and apparently facilitating lateral peatland expansion. At the same time, the high water-holding capacity promotes the development of anaerobic conditions in microsites above the peatland water table, where dissolved oxygen consumption rates exceed oxygen diffusion rates into wet peat. In addition, the water-holding capabilities of moss plants and moss communities buffer peatland ground-layer temperatures against changes in air temperature at the moss–atmosphere interface, conferring some degree of protection of individual moss plants against evaporative stress and diminishing evaporative losses from peatlands as a whole. Together these conditions promote the accumulation of peat by limiting decomposition.

9.3.3 *Decomposition*

Bryophytes are small organisms that appear to have difficulties tolerating large water level fluctuations (Zoltai & Vitt 1990). In wetlands such as swamps and marshes where water level fluctuations are seasonally quite variable, bryophytes are not dominant (Vitt 1994). These non-peat-forming wetlands, dominated by vascular plants that produce copious litter, have relatively high rates of decomposition (Mitch & Gosselink 1993). Although it has been commonly argued that one of the factors attributing to slow decomposition rates in peatlands is acidity, in fact both acidic and basic peatlands accumulate large amounts of peat. Both true-moss and *Sphagnum*-dominated peatlands accumulate large amounts of peat. In continental boreal Canada, an analysis of 341 peatland cores clearly indicates that rich fens accumulate peat to depths similar to those found in poor fens and bogs (Fig. 9.6). Although the basal portions of some cores may largely be detrital without recognizable plant parts, in many of these cores

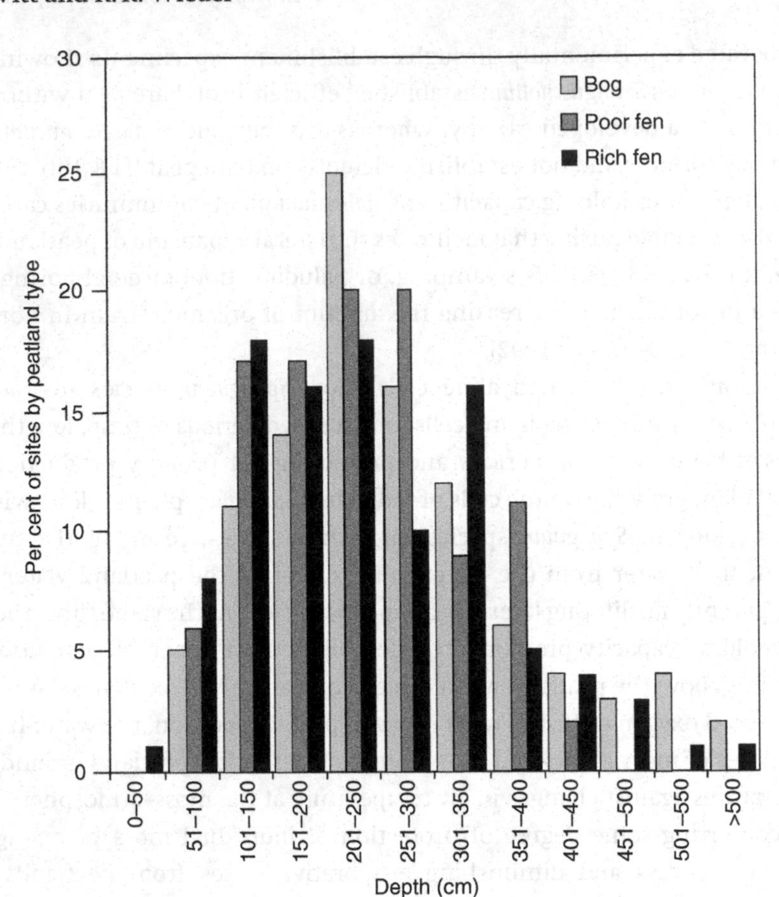

Fig. 9.6. Peatland depths partitioned by per cent of bogs, poor fens, and rich fens (as determined by present-day surface vegetation). Number of sites: bog, 129; poor fen, 66; rich fen, 146. All sites are from continental western Canada (Alberta, Saskatchewan, and Manitoba). Data are from Zoltai *et al.* 1999.

the major component of the accumulated peat is bryophytic; *Sphagnum* in poor fens and bogs, and brown mosses in rich fens (including species of the genera *Aulacomnium, Catoscopium, Campylium, Drepanocladus* [sensu lato], *Meesia, Scorpidium,* and *Tomenthypnum*). Mass losses in the upper 30 cm of true moss-dominated rich fens and *Sphagnum*-dominated poor fens are slightly (but significantly) greater than in bogs, with fens averaging 6.4% of total peat mass lost compared with 5.3% in bogs over a 26 month period of time (Fig. 9.7, unpublished data). However, higher initial inputs (NPP) to the peat column of fens cause substantial quantities of peat to be deposited in both peatland types.

Comparison of hummock decomposition by using litter bags (placed just beneath the peatland surface) in rich fens (*Tomenthypnum nitens*) compared

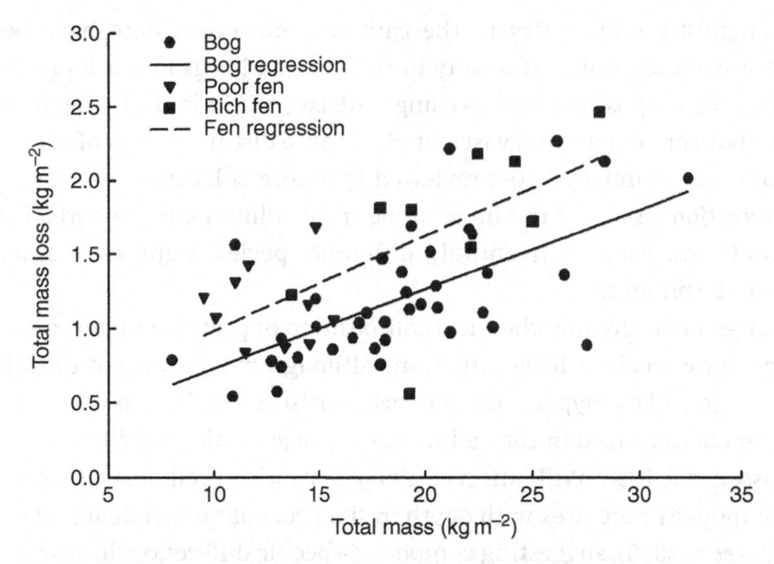

Fig. 9.7. Relation between the total mass and total mass loss of peat in the upper 30 cm of the peat column in fens and bogs in continental western Canada. Details of methods in Benscoter 2007. Slopes for bog ($r^2 = 0.40$, $p = 0.0001$) and fen ($r^2 = 0.486$, $p = 0.0009$) are significantly different (ANCOVA $p = 0.561$; however, elevations of the slopes are significantly different at $p = 0.0025$). Y (% lost over a 24 month period) = 5.3% for bog and 6.4% for fen.

with bogs (*Sphagnum fuscum*) indicates that over the short term (16 months) significantly more decomposition occurred in hummocks composed of *Tomenthypnum nitens* (Vitt 1990, Li & Vitt 1997). Within bogs along the hummock–hollow gradient, hummock species retain about 86% of their initial dry mass after three years, but hollow species retain only 74% after a similar time period (Rochefort *et al.* 1990). In addition, bryophyte material decomposes at slower rates than vascular plant material, and moss net primary production (NPP) in hummocks is about half that of bog hollows (carpets and lawns) (Vitt 1990, Gunnarsson 2005) whereas moss NPP of fen hummocks is greater than or equal to that of hollows (reviewed in Vitt 1990). Although few data exist on true moss production it appears that hummocks can develop higher than hollows above the water table in both bogs and rich fens, owing to higher rates of moss NPP (fens) and/or less decomposition in hummocks compared with hollows (bogs).

When *Sphagnum*-dominated hummocks were compared with brown moss-dominated hummocks, both with well-developed acrotelms, the *Sphagnum* system decomposed 11% less than the brown moss hummocks after two years (Li & Vitt 1997). These data suggest that there should be a fundamental difference in plant chemistry upon entry to the catotelm; however, there have been no comparative studies of peat quality in rich fens and bogs to our knowledge. In summary, decomposition studies using litter bag results indicate that individual species (both bryophytic and vascular plant) have distinctive and often different rates of decay (Turetsky 2004, reviewed in Moore & Basiliko 2006) and when decomposition losses for the intact upper peat column are compared between bogs and fens, each with entirely different species, significant differences appear to be present.

Changes in the organic chemical composition of peat as it undergoes decomposition have received little attention. Although mosses do not contain true lignin, a lignin-like polyphenolic network consisting of *p*-hydroxyphenyl groups has been characterized in the cell walls of *Sphagnum* plants (Wilson *et al.* 1989, Rasmussen *et al.* 1995, Williams *et al.* 1998). The ratio of cellulose to these lignin-like compounds decreases with depth in the peat column (Williams *et al.* 1998, Turetsky *et al.* 2000), suggesting compound-specific differences in susceptibility to decay.

An empirical modeling approach based on ^{210}Pb dating of peat cores was used to estimate annual net primary production and depth-dependent decomposition rates within the upper 30–40 cm of bog peat (Wieder 2001). We have expanded on this modeling approach to estimate that "lignin" and holocellulose constitute 27%–32% and 57%–60%, respectively, of annual net primary production in three Alberta bogs (as compared with measured values of 18%–21% and 59%–62%, respectively, in these same bogs; Turetsky *et al.* 2000). The holocellulose fraction is composed mainly of α-cellulose (70%–85%), with a smaller contribution from hemicellulose (15%–22%), as determined from both the empirical modeling approach and measurements of peat at the surface of the peat deposit (Turetsky *et al.* 2000).

The empirical modeling approach to simulating bulk peat decomposition (Wieder 2001) can also be applied to the decomposition of cellulose and lignin fractions of newly formed peat over a 100 year period (Fig. 9.8). Using this approach, we predicted that after 100 years of decomposition, only 14%–27% of the initial mass (i.e. material produced at the surface in one year's net primary production), 14%–22% of the initial holocellulose, 12%–16% of the initial α-cellulose, 4%–16% of the initial hemicellulose, and 13%–32% of the initial lignin would remain. Correspondingly, the lignin : holocellulose ratio would change from about 0.4:1 in the peat produced from net primary production at the surface of the peat column to about 0.8:1 after 100 years of decomposition. Although cellulosic components of peat decompose more rapidly than lignin-like components, it is not the case that cellulosic components disappear, with lignin-like components showing great recalcitrance to decomposition; rather,

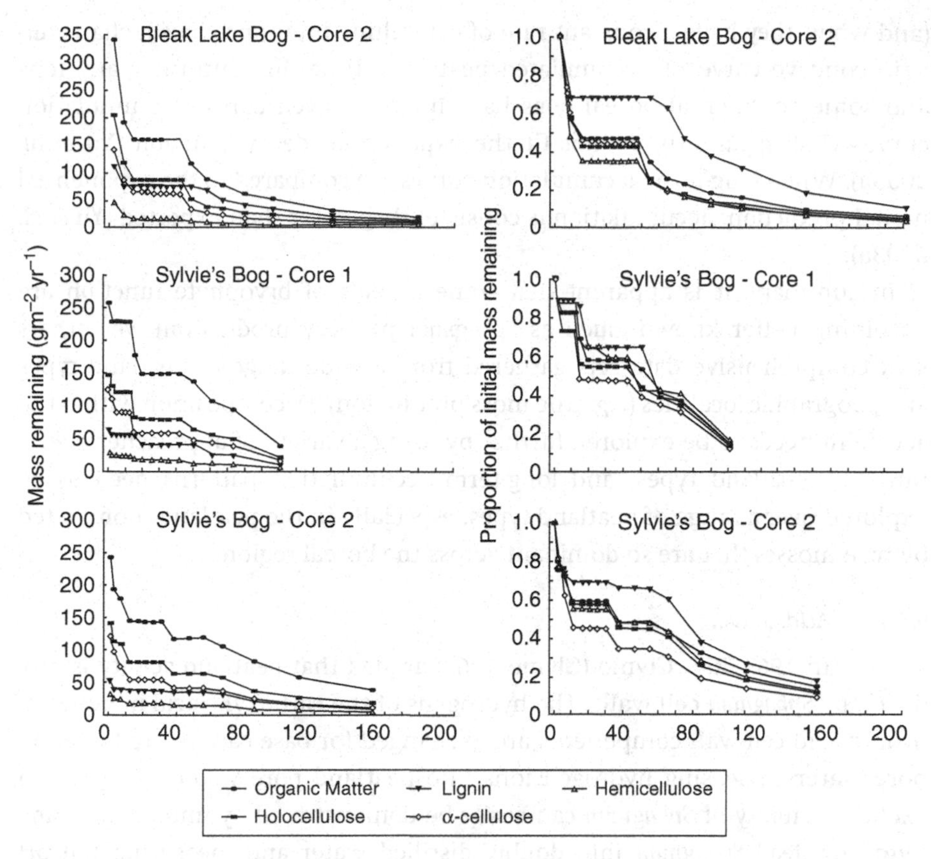

Fig. 9.8. Fate of a single year's cohort of organic matter and its components estimated by using the ^{210}Pb-based empirical modeling approach described by Wieder (2001) and the quantitative characterization of organic matter components as a function of depth in ^{210}Pb-dated peat cores, following the procedure of Wieder & Starr (1998).

both cellulose and lignin decay in concert with a slow, but progressive, enrichment in lignin-like compounds in the remaining peat. It may be that the lignin-like compounds are intimately associated with the cellulosic compounds in *Sphagnum* cell walls (Rasmussen *et al.* 1995, Wilson *et al.* 1989, Williams *et al.* 1998) in a way that confers some degree of overall resistance to decay.

Peat decomposition continues, albeit at a considerably slower pace, throughout the deeper, permanently water-saturated peat column, the catotelm. For example, in some *Sphagnum*-dominated bogs, the anaerobic catotelm receives from the acrotelm a relatively large amount (although probably less than 20% of the original mass) of undecomposed material with low bulk density. Once in the catotelm, deposition is considerably less, and has been modeled following simple exponential decay functions (Clymo 1984) that when plotted as a graph (and when they have a constant rate of catotelmic input) produce a characteristic concave curve of accumulated peat over time. In contrast, many fens and some continental boreal bogs have linear or even convex accumulation curves that apparently do not fit the exponential decay function (Yu *et al.* 2003a). When long-term accumulation curves are compared to the exponential model prediction, accumulation is consistently greater than expected (Yu *et al.* 2003a).

In summary, it is apparent that some aspects of bryophyte function are becoming better known (such as *Sphagnum* primary production), yet others lack comprehensive data sets gathered from a wide range of peatland types and geographic localities (e.g. true moss production). Decomposition within the acrotelm needs to be explored further by using a variety of approaches over a range of peatland types, and long-term accumulation patterns need to be explored over a range of peatland types, especially in the peatlands dominated by true mosses that are so dominant across the boreal region.

9.3.4 *Acidification*

In 1963, R. S. Clymo (Clymo 1963) argued that peatland acidity is produced by *Sphagnum* cell walls. The hydrogens of the carboxylic acid moieties of uronic acid cell wall components are exchanged for base cations contained in pore waters, releasing hydrogen ions into peatland pore waters. This cation exchange ability of *Sphagnum* can easily be demonstrated by immersing some living or dead *Sphagnum* into doubly distilled water and measuring the pH change and then by adding common table salt to the same solution and again measuring the pH change. In the former case, no pH change is evident; in the latter, the pH will immediately decrease by 2–3 pH units. In 1980, Harry Hemond argued that although this process of cation exchange by *Sphagnum* undoubtedly occurs, it is not sufficient to produce the acidity that is present in bogs (i.e. pH 3.0–3.7). He concluded that bog acidity is largely due to decomposition and the production of humic acids present in pore water as DOC, and that acidity in bogs is a result of hydrogen release through decomposition of organic molecules that in turn are dissolved in the pore water as organic carbon. Under the former scenario, the anions are attached to the *Sphagnum* cell walls (Richter & Dainty 1989), whereas in the latter case the anions are dissolved in the pore water as DOC (Hemond 1980) and thus create the predominant "brown water" characteristic of bogs.

Regional landscape analyses of natural lake acidity support this pattern. Halsey *et al.* (1997a) showed that across 29 watersheds, watershed bog cover and lake DOC concentrations are positively correlated. Watersheds with a high percentage of poor fen cover also tend to have acidic lakes (Fig. 9.9). Importantly,

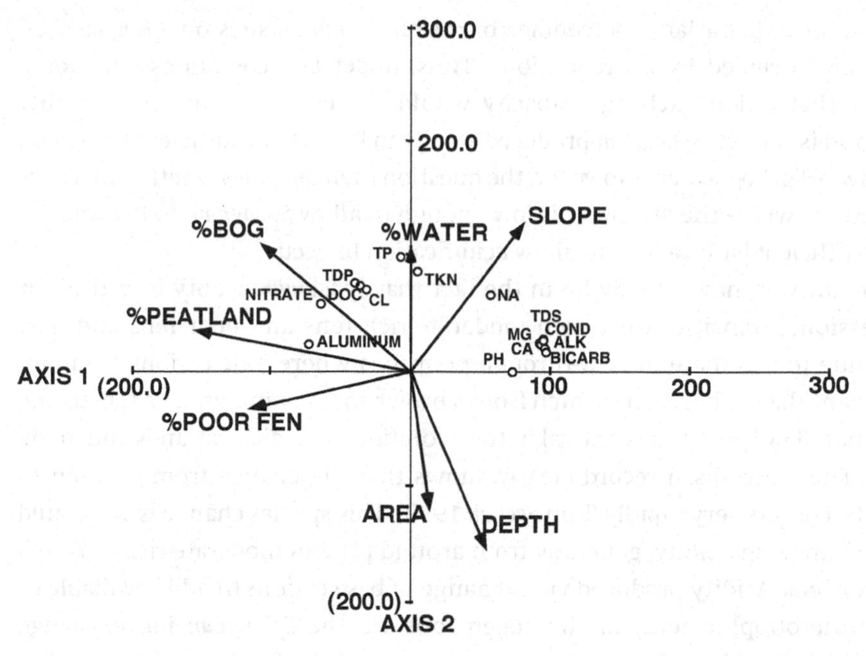

Fig. 9.9. Biplot of the chemical parameters measured in water from 29 boreal lakes in northeastern Alberta, Canada. Watershed variables that explain a significant amount of the variation on the first two axes are shown by directed arrows. Abbreviations: %WATER, per cent of open water in the watershed; ALK, alkalinity; BICARB, bicarbonates; CA, calcium; CL, chlorine; COND, reduced conductivity; DOC, dissolved organic carbon; MG, magnesium; NA, sodium; PH, pH; TDP, total dissovled phosphorus; TDS, total dissolved solids; TKN, total Kjehldahl nitrogen; TP, total potassium. DEPTH and AREA refer to lake and area depth; SLOPE is the regional watershed slope. (Reprinted with permission from Kluwer Academic Publishers/ *Water, Air, and Soil Pollution*, Vol. 96, 1997, Influence of peatlands on the acidity of lakes in northeastern Alberta, Canada, by Linda A. Halsey and Dale H. Vitt, Fig. 5, p. 33.)

this study indicates that lakes occurring in watersheds with greater than 30% cover of poor fens and bogs are nearly always acidic, while those without extensive acidic peatland cover have higher pH (all lakes of this study were situated on glacial deposits over acidic shales). In addition, outputs from poor fens appear to have more influence on downstream acidity than do stagnant bogs.

Sphagnum cation exchange activity is dependent on the presence of free cations in the surface waters of a peatland. Rich fens have high concentrations of free cations (Vitt *et al.* 1995a); however, the charge of these free base cations – largely Ca^{2+} – is balanced, mainly by HCO_3^- (Vitt & Chee 1990). Gignac (1994) has shown that *Sphagnum* quickly dies when grown in water having any amount of alkalinity. Bogs, on the other hand, have low concentrations of free base cations (Vitt *et al.*

1995a), such that a large percentage of cation exchange sites on *Sphagnum* cell walls are occupied by hydrogen ions. Thus, under bog conditions one would expect that cation exchange capacity would not be base-saturated, and that Hemond is correct: *Sphagnum*-produced acidity in bogs is not sufficient to explain the low pH of bog waters. However, the question then becomes whether there are conditions where the alkalinity is low enough to allow *Sphagnum* to live and yet have sufficient base cations to allow acidification to occur.

The answer may actually lie in the fact that *Sphagnum* acidity is critical for successional transitions between moderate-rich fens and poor fens and may continue to dominate in flow-through poor fens where base cations are more abundant than in bogs. The switch from a brown moss-dominated system to one dominated by *Sphagnum* is critical in the evolution of acidic peatlands and many bogs. The macrofossil record clearly shows that the change from rich fen to poor fen occurs very rapidly (Kuhry *et al.* 1993). This species change is associated with changes in acidity, generally from around pH 7 in moderate-rich fens to 5 in poor fens. Acidity produced via exchange of base cations (readily available in the minerotrophic fens) for hydrogen ions via the *Sphagnum* ion exchange system will be relatively more important in accounting for the acidity at higher pH (4–7). At low pH (3–4), where much higher concentrations of hydrogen ions are required and where base cations are low due to ombrotrophic conditions, acidity is due more to humic acid decomposition. This acidity would in many cases produce waters with high concentrations of DOC and these would be more heavily colored. Thus, a corollary is that *Sphagnum*-produced acidity may more strongly influence external downstream chemistry, while decomposition acidity may be more internally influential, especially in stagnant bogs. The relative importance of these two types of acidity needs further study, especially in poor fens.

The available evidence suggests that rich fens may persist without change for thousands of years, with fens being just as deep as bogs (Fig. 9.6). However, if alkalinity concentrations allow establishment of *Sphagnum*, then rapid acidification via *Sphagnum* cation exchange may result in the development of poor fen vegetation within 100–200 years (Vitt & Kuhry 1992). This rapid change at pH around 5.5 appears to be responsible for the rarity of peatlands of this transitional nature in many areas of the boreal landscape (cf. Fig. 9.4).

9.4 Responses to environmental change and disturbance

Bryophytes are small plants closely tied to their substrate. This is especially true in peatlands, where changes in height above water table and in water chemistry affect both the structure of bryophyte communities and the functioning of these moss-dominated peatland ecosystems. Although most of the bryophyte species that dominate the ground layers of peatlands have broad geographical ranges, they have extremely narrow habitats. For example, *Sphagnum fuscum* and *Scorpidium scorpioides* are two species that have circumboreal ranges across northern Eurasia and North America, and both are clear and abundant indicators of specific water level and water chemistry conditions in northern peatlands (*S. fuscum* on hummocks in bogs and *S. scorpioides* in carpets and pools of extreme-rich fens). Thus responses of bryophytes to environmental changes are useful across the hemisphere, as opposed to most peatland vascular plant species that are confined to narrower geographic ranges (Vitt 2006). The abundance of bryophytes in northern peatlands, and their narrow habitat tolerances yet broad geographic ranges, make bryophytes useful as key indicators of both community changes and ecosystem responses to disturbance and environmental change.

Disturbances across the boreal forest are tightly coupled to both the activities of humans and changes in climate: both natural climatic cycles and global climate change. Turetsky *et al.* (2002a) examined cumulative effects of disturbance on peatlands of western Canada and estimated that current disturbances reduce carbon uptake in continental peatlands by about 85% when compared with a non-disturbance scenario. These authors concluded that wildfire was by far the major disturbance, followed by peat extraction, permafrost melt (having a positive influence on carbon uptake), then reservoirs, and mining of oil sands. The two most important disturbances, fire and permafrost melt, affect peatland community structure as well as ecosystem function in contrasting ways (Robinson & Moore 2000).

9.4.1 *Permafrost melt*

Near the southern boundary of discontinuous permafrost, permafrost is confined to wooded peatlands. In these wooded peatlands, permafrost occurs as sporadic lenses of ice, creating distinctive landscape features termed frost mounds (Beilman *et al.* 2001), generally believed to have aggraded during the past climatic cold spell (the Little Ice Age). Frost mounds are most often found in bogs and differ from unfrozen bog areas by having more dense populations of trees that are taller and of larger diameter and a ground layer with more abundant feather mosses (*Hylocomium splendens* and *Pleurozium schreberi*) and reindeer lichens (*Cladina mitis* and *Cladonia uncialis*). Comparatively, unfrozen bogs have a less well-developed tree layer and greater abundance of *Sphagnum fuscum*, *S. magellanicum*, and *S. angustifolium*. Warming climate over the past 100–150 years has resulted in permafrost gradually melting, and the southern boundary of permafrost has migrated

Fig. 9.10. Melting of permafrost at the southern edge of discontinuous permafrost, (see Vitt *et al.* 1994 for details). (A) Area of peat collapse due to localized permafrost melt at the edge of a frost mound in a continental bog. Emergent mounds of moss are *Pleurozium schreberi* that in the previous year formed hummocks 0.5–1.0 m high on the frost mound. Photo taken in northern Alberta. (B) Internal lawn formed by collapse of frost mound. Leaning snags of dead *Picea mariana* trees are visible. Vegetation is sedge- and *Sphagnum*-dominated, especially *S. riparium*.

northward some 200 km in western Canada (Vitt *et al.* 1994, Halsey *et al.* 1995, Camill 1999, Halsey *et al.* 2000).

At the local scale, the results of this melting are spectacular. Frost mounds, elevated one meter or more above the water table, collapse, thereby submerging the previously dry surface of feather mosses (Fig. 9.10a). Submergence continues and the newly created internal lawn is quickly vegetated by sedges (*Carex* spp.) and a variety of carpet-dwelling sphagna (Fig. 9.10b), including *Sphagnum riparium*, *S. jensenii*, *S. majus*, and more rarely, *S. obtusum* and *S. lindbergii*. Since the original feather moss surface layer is identifiable in the peat column as a dense sylvic [=woody]-feather moss horizon, modern ^{210}Pb dating methods can determine the age of the melt and the rates of peat accumulation since the melt, compared with both neighboring frost mound and unfrozen bog.

Bogs, frost mounds, and internal lawns differ not only in terms of moss and vascular plant communities, but also in rates of recent net peat accumulation. In particular, peat accumulation in internal lawns is substantially greater than in the frost mound features, whose degradation through permafrost melt lead to internal lawn creation (Robinson & Moore 2000, Turetsky *et al.* 2002b). However, internal lawns undergo autogenic succession over time, with progressive development toward bogs (Beilman *et al.* 2001, Camill *et al.* 2001) and concurrent decreases in peat accumulation rates. Through ^{210}Pb dating of internal lawn peat, Turetsky *et al.* (2007) estimated that within about 70 years following initial permafrost melt, peat accumulation rates in internal lawns become stable and similar to peat accumulation rates in bogs.

9.4.2 *Wildfire*

When peatlands are affected by fire, peat (and stored carbon) is lost both directly (from the aboveground vegetation and surface peat) as well as indirectly (organic matter losses through ongoing peat decomposition with little or no plant production, at least early on after fire). Turetsky *et al.* (2002a) estimated that across continental western Canada, 75% of the organic matter lost as a result of wildfire was from the direct action of the fire itself, and 25% was from indirect effects of post-fire mineralization of organic matter.

Re-establishment of the ground layer in peatlands is critical for the return of long-term ecosystem processes, especially those that lead to the formation of peat and accumulation of carbon in the long-term peatland sink. Until recently, the time frame and trajectories of vegetation succession were unknown. Detailed examination of vegetation recovery along a 102-year bog chronosequence in Alberta, Canada, indicates that post-fire ground layer recovery takes place in three phases. True mosses are dominant in early succession during the first 20 years post-fire, and of these mosses *Polytrichum strictum* and *Aulacomnium palustre* are especially important (Benscoter 2007). Initial populations of these mosses begin in the wet depressions and expand outward (Benscoter 2006). *Sphagnum* recruits, especially *S. angustifolium*, establish within the true moss populations and aggressively expand, and at about 20 years post-fire a continuous ground layer of *Sphagnum* has become established. From 20 to 80 years post-fire, *Sphagnum* hummock and hollow topography becomes established and persists. Gradually in late succession, as tree canopy density increases feather mosses become more abundant (Benscoter 2007, Fig. 9.11). Whereas continuous ground layer cover is established at about 20 years post-fire, tree layer density, canopy cover, and biomass continue to increase, thus lagging behind development of the ground layer.

These structural changes in the bog ground layer are reflected in temporal changes in net ecosystem exchange (net CO_2 flux) after fire. Immediately after fire, bogs function as a net C source to the atmosphere, with a rate of 8.9 ± 8.4 mol C m^{-2} yr^{-1}. Bogs switch from C sources to C sinks at about 13 years after fire, as the moss and shrub layers become re-established. The strength of the bog C sink peaks at about 18.4 mol C m^{-2} yr^{-1} at about 74 years after fire, concomitant with the peak in aboveground net primary production of black spruce trees. At 100 years after fire, the bog C sink reaches a fairly stable value of about 10 mol C m^{-2} yr^{-1} (Fig. 9.12; Wieder *et al.* 2009). Projecting the carbon balance recovery trajectory across the 2280 km^2 of bogs in the well studied Wabasca region of north-central Alberta, with a bog fire return interval of 120 yr, the regional bog C sink is about 171 ± 61 Gg C yr^{-1}. However, two likely consequences of

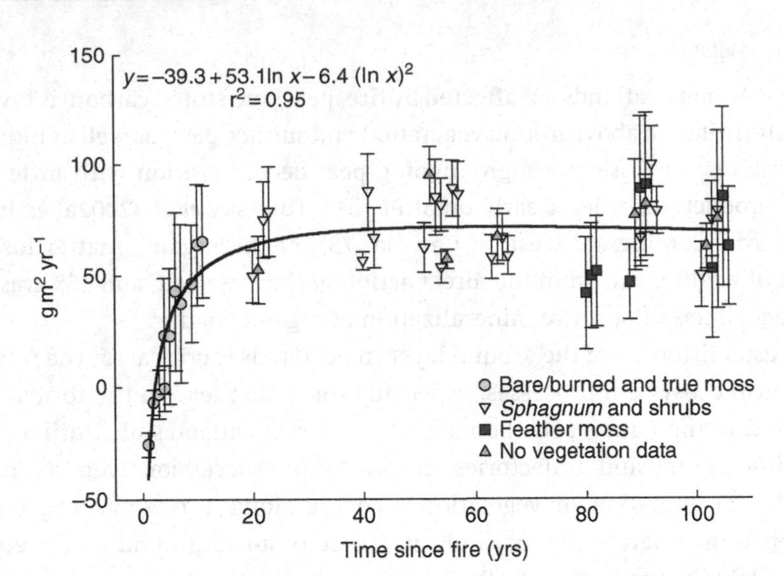

Fig. 9.11. The response for net primary production (NPP) of the ground layer (dominated by species of *Sphagnum, Cladina*, and feather mosses) along a chronosequence of bog sites after fire. Data for 2003–2006 from Benscoter (2007). Species groups based on cluster and indictor species analyses. Mean ± SD, $n=5$.

ongoing climate change are a shortening of the fire return interval and a temperature-driven enhancement of peatland respiratory carbon losses, either or both of which will diminish the regional peatland carbon sink across continental western Canada (Wieder *et al.* 2009).

9.4.3 *Climatic cycles*

In the examples of environmental disturbance discussed above, the influences of change were sufficiently severe to modify the vegetation, with ecosystem responses tied closely to wholesale plant community changes that were driven by successional patterns of the ground layer, and structure was tied closely to function. However, ecosystem processes can change without evidence of community change, and even though key species may remain unchanged, changes in the rates of important processes may take place and modify the overall functioning of the ecosystem. Such a situation appears to be evident in some fens in western Canada over a large part of the past 10 000 years.

One of the unique properties of peatlands is that the peat is deposited *in situ*, and plant parts are incorporated into the peat column exactly where they grew. Analysis of these plant macrofossils provides insight into patterns of local plant succession. One of the important conclusions that can be reached from the study of numerous paleo-records of a core is that in fens and bogs the

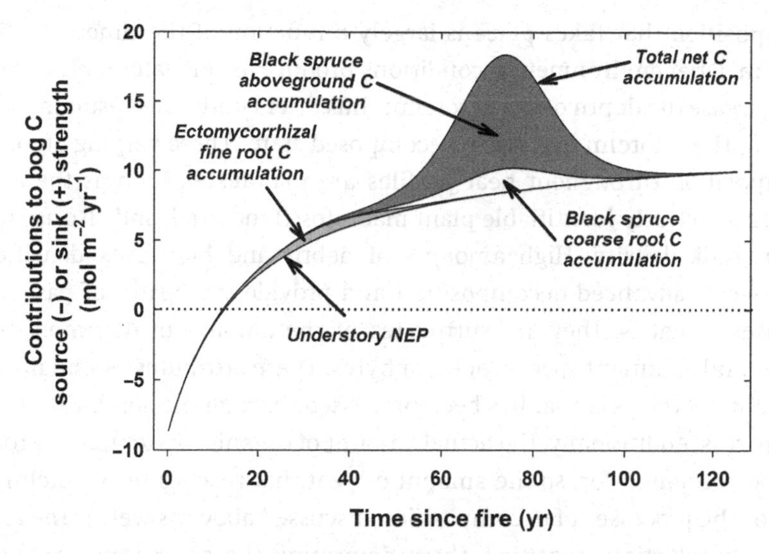

Fig. 9.12. Changes in net ecosystem production (NEP) of C accumulation along a time-since-fire chronosequence of 10 bog peatlands in central Alberta, Canada (Wieder *et al.* 2007). Annual NEP from static chambers is based on measurements of CO_2 at different intensities of photosynthetically active radiation (PAR; from full sun to full dark), fitting a rectangular equation to the resulting NEP versus PAR measurements collected over a 3-year period, applying hourly measurements for PAR and air temperature (from a weather station installed at one of the bog sites) to the fitted equation, and summing the hourly estimates of NEP over a full year. Accumulation of C in black spruce aboveground tissues and roots was calculated from the derivative of the best fit equations describing changes over time in aboveground and root biomass at the 10 chronosequence sites.

key and most abundant species may exist for hundreds and even thousands of years unchanged in composition (Kuhry *et al.* 1993). However, fine-scale paleo-analysis of long (2–5 m) peat cores from sites that have a continuous, single species occurrence suggest that functional long-term changes clearly have occurred. A case in point is Upper Pinto Fen in western Alberta (Yu *et al.* 2003b) (see below).

Interpretation of these data requires some additional background. Above-ground organic matter is deposited at the peatland surface from vascular plant litter and combines with moss plants and belowground vascular plant roots to form the peat column. Decomposition is relatively rapid in the aerobic acrotelm; after spending some time in the acrotelm, peat arrives at the anaerobic catotelm, where decomposition rates are relatively uniform and very low (Clymo 1984). So, in a given peatland and for a given species, the amount of decomposition that takes place is largely a function of the time spent in the acrotelm. Drier environmental conditions produce lower water tables, which in turn increase the depth of the acrotelm; thus drier conditions result in material reaching the catotelm in a more decomposed state. These varying amounts of decomposition throughout peat profiles are manifested both as the amount of debris vs. clearly identifiable plant macrofossil material, and also as changes in peat bulk density. High amounts of debris and high bulk densities are indicative of advanced decomposition and provide surrogates of past surface processes - that is, they are surrogates of the amount of decomposition of specific and dominant species of bryophytes. These attributes essentially assess the quality of the peat that has been processed through the aerobic decomposition process. Additionally, the actual amount of organic matter that is produced must be accounted for, so the amount of peat that reaches the catotelm is the result of the processes of decomposition discussed above as well as the amount of plant production; together, these determine the gross input to the peat column. Bryophytes are desiccation-tolerant, and as such photosynthesize as long as they are wet and turgid. Rates of photosynthesis decrease as the mosses become drier; thus high water tables throughout the growing season increase the bryophyte production and increase the gross organic matter inputs to peatlands.

In conclusion, dry climatic conditions and the resultant lowered peatland water table affect the accumulating peat surface in two ways: bryophytic production decreases because the growing apices are subjected to an increased frequency of desiccation, and total acrotelmic decomposition increases not because of enhanced rates but because lower water table effectively extends the aerobic acrotelm deeper into the peat column. These functional changes are manifested in the peat column as changes in vertical peat accumulation ($g\,cm^{-1}$), the ratio of debris to recognizable bryophyte macofossils, and peat bulk density ($g\,cm^{-3}$).

Upper Pinto Fen (UPF), Western Alberta, Canada

At this site, Yu *et al.* (2003b) found that both ash-free bulk density and debris did not increase with depth (indicating that substantial decomposition was not occurring in the catotelm); however, both ash-free bulk density and per cent debris as well as per cent *Scorpidium scorpioides* (in the macrofossil analysis) exhibited extremely variable profiles (Fig. 9.13). All three of these parameters showed periodicities at both millennial (1500–2190 yr with a mean of 1795 yr) and century scales (386 and 667 yr). Especially significant are three periods of 200–600 yr duration at 4000, 5500, and 6900 cal years BP of high rates of peat accumulation. In this peatland that was dominated by one bryophyte species

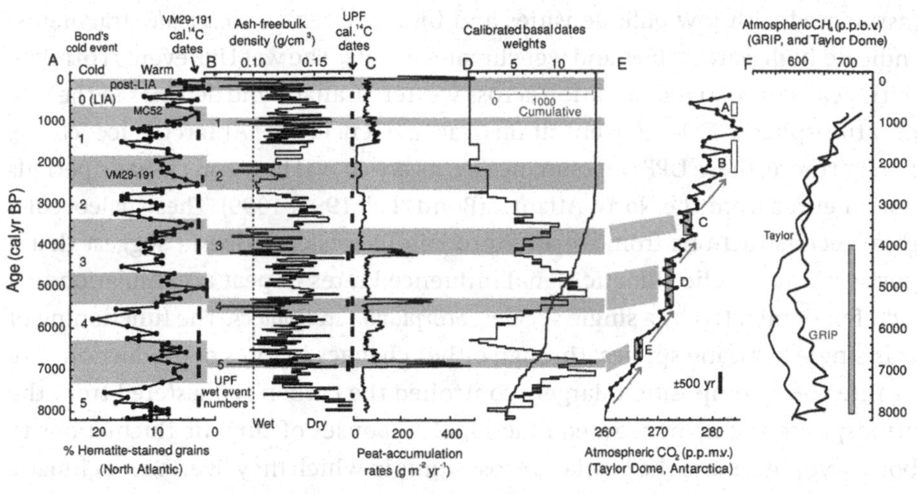

Fig. 9.13. Climate, peatland and carbon-cycle correlation. (A) Percentage of hematite-stained grains from the North Atlantic. Numbers show the cold events for the North Atlantic region, together with the "Little Ice Age" (LIA). Rectangles on the right are locations and 2 standard deviation ranges (180–480 years, with a mean of 300 years) of 12 calibrated ^{14}C dates. There might be an additional age error of $\pm$ 200 years owing to variable reservoir correction. The correlation with the Upper Pinto Fen (UPF) record (shaded in B) is suggestive within the dating uncertainty. (B) Ash-free bulk density from the UPF core in central Alberta. The UPF wet events (shaded bands) were defined as the lowest ash-free density values (shaded in B). Rectangles on the right are locations and 2 s.d. ranges (45–335 years, with a mean of 190 years) of calibrated ^{14}C dates. (C) Peat-accumulation rates from the UPF core. (D) Weight of calibrated basal peat dates from 79 paludified peatlands in Continental western Canada as a measure of probability of peatland initiation. Cumulative curve can be used as a proxy for the increase of new peatland areas. See Halsey *et al.* (1997b) for detailed information on location, peat depth and reference of each basal date. (E) Atmospheric CO_2 concentrations from Taylor Dome in Antarctica. Rectangles indicate periods of decreases in the CO_2 rising rate, especially during the CO_2 plateaus E, D, and C shortly after the peat accumulation and peatland initiation peaks, and during the phase of rapid increase in new peatland area (C, D). The suggestive correlation with the UPF and western interior Canadian peatlands is shown as shaded bands. (F) Atmospheric CH_4 concentrations from GRIP core in Greenland and from Taylor Dome core in Antarctica. Legend shortened from Yu *et al.* 2003b.

continually for 5200 years (between 6500 and 1300 cal yr BP), few species changes are evident in the paleoecological record and past environmental change produced no detectable species changes. However, peat accumulation rates varied considerably (Fig. 9.13) from a norm of less than 100 g m^{-2} yr^{-1} to peaks of more than 400 g m^{-2} yr^{-1}. These peaks in peat accumulation (associated with low bulk densities and high per cent *S. scorpioides* fragments) indicate high water tables and wet climatic events. The wet UPF events correlate with peaks in peatland initiation across western Canada and declines in the rate of atmospheric CO_2 concentrations (as evidenced in Antarctic ice cores). Furthermore, these UPF wet climatic events are coeval to warm climatic periods documented from the North Atlantic (Bond *et al.* 1997, 1999). These paleoecological reconstructions from Upper Pinto Fen in western Alberta suggest that a pervasive and cyclic climatic signal influenced rates of peat accumulation in a rich fen dominated by a single species, *Scorpidium scorpioides*. The functioning of this single keystone species, through either changes in rates of production and/or rates of decomposition, largely controlled the carbon sequestered from the atmosphere and demonstrated functional responses of climatic fluctuations to both bryophytes and the peatland ecosystems in which they live. Future climatic changes may also influence function dynamics of individual keystone bryophytes and may affect carbon sequestration of boreal peatlands.

9.5 Conclusions

In many northern areas of the world with cool climates and short growing seasons, bryophyte-dominated peatlands form a substantial part of the landscape. These ecosystems have expanded over the past 6000 to 10 000 years, and have sequestered large amounts of carbon. The functioning of these northern peatlands is strongly influenced by bryophytes, and our understanding of the nutrient flow, diversity, and carbon sequestering of these ecosystems can only be advanced by thorough knowledge of the bryophytes that dominate in these ecosystems.

Acknowledgments

This chapter is based on research made possible through research funded by the National Science Foundation (U.S.) and also by the Natural Science and Engineering Research Council (Canada). Support for individual projects includes funding from the Province of Alberta, Sun Gro Horticulture (through Tony Cable); Networks of Centers of Excellence in Sustainable Forest Management; Canadian Forest Service; and the University of Alberta. We are especially grateful to the efforts of Sandi Vitt for technical expertise. These ideas and the data on which they are founded have accumulated through the years by interactions with students and colleagues; in particular D. Beilman, B. Benscoter, D. Gignac, L. Halsey, P. Kuhry, Y. Li, B. Nicholson, L. Rochefort, K. Scott, M. Turetsky, M. Vile, A. Wood, B. Xu, Z. Yu, and S. Zoltai.

References

Aerts, R., Wallen, B. & Malmer, N. (1992). Growth-limiting nutrients in *Sphagnum*-dominated bogs subject to low and high atmospheric nitrogen supply. *Journal of Ecology*, **80**, 131–40.

Bayley, S. E. (1993). Mineralization of nitrogen in bogs and fens of western Canada. In *Proceedings of the American Society of Limnology and Oceanography and the Society of Wetland Scientists Combined Annual Meetings, Edmonton, Alberta*, p. 88. Edmonton, Canada: University of Alberta.

Beilman, D. W., Vitt, D. H. & Halsey, L. A. (2001) Localized permafrost peatlands in western Canada: Definition, distributions, and degradation. *Arctic and Antarctic Alpine Research*, **33**, 70–7.

Belland, R. J. & Vitt, D. H. (1995). Bryophyte vegetation patterns along environmental gradients in continental bogs. *Ecoscience*, **2**, 395–407.

Benscoter, B. W. (2006). Post-fire bryophyte establishment in a continental bog. *Journal of Vegetation Science*, **17**, 647–52.

Benscoter, B. W. (2007). Post-fire compositional and functional recovery of boreal bogs. Ph.D. dissertation, Southern Illinois University, Carbondale, IL.

Berendse, F., Van Breemen, N., Rydin, H. *et al.* (2001). Raised atmospheric CO_2 levels and increased N deposition cause shifts in plant species composition and production in *Sphagnum* bogs. *Global Change Biology*, **7**, 591–8.

Bewley, J. D. (1979). Physiological aspects of desiccation tolerance. *Annual Review of Plant Physiology*, **30**, 195–238.

Bond, G. C., Showers, W., Cheseby, M. *et al.* (1997). A pervasive millennial-scale cycle in north Atlantic holocene and glacial climates. *Science*, **278**, 1257–66.

Bond, G. C., Showers, W., Elliot, M. *et al.* (1999). The North Atlantic's 1–2 kyr climate rhythm: Relation to Heinrich Events, Dansgaard/Oeschger Cycles and the Little Ice Age. *Geophysical Monograph – American Geophysical Union*, **112**, 35–8.

Bragazza, L., Tahvanainen, T., Kutnar, L. *et al.* (2004). Nutritional constraints in ombrotrophic *Sphagnum* plants under increasing atmospheric nitrogen deposition in Europe. *New Phytologist*, **163**, 609–16.

Bridgham, S. D., Updegraff, K. & Pastor, J. (1998). Carbon, nitrogen, and phosphorus mineralization in northern wetlands. *Ecology*, **79**, 1545–61.

Camill, P. (1999). Patterns of boreal permafrost peatland vegetation across environmental gradients sensitive to climate warming. *Canadian Journal of Botany*, **77**, 721–33.

Camill, P., Lynch, J. A., Clark, J. S., Adams, J. B. & Jordan, B. (2001). Changes in biomass, aboveground net primary production, and peat accumulation following permafrost thaw in the boreal peatlands of Manitoba, Canada. *Ecosystems*, **4**, 461–78.

Campbell, C., Vitt, D. H., Halsey, L. A. *et al.* (2000). Net primary production and standing biomass in northern continental wetlands. Northern Forestry Centre Information Report NOX -X-369. Ottawa: Canadian Forest Service, 57 pp.

Clymo, R. S. (1963). Ion exchange in *Sphagnum* and its relation to bog ecology. *Annals of Botany*, **27**, 309–24.

Clymo, R. S. (1984). The limits of peat growth. *Proceedings of the Royal Society of London*, B**303**, 605–54.

Damman, A. W. H. (1979). Geographic patterns in peatland development in eastern North America. In *Classification of Peat and Peatlands, Proceedings of the International Peat Society Symposium, Hyytälä, Finland*, ed E. Kivinen, L. Heikurainen & P. Pakarinen, pp. 42–57. Helsinki: International Peat Society.

Davis, R. B. & Anderson, D. S. (1991). The eccentric bogs of Maine: a rare wetland type in the United States. Maine State Planning Office Critical Areas Program Planning Report 93.

DuReitz, G. E. (1949). Huvidenheter och huvidgranser i Svensk myrvegetation. (Summary: Main units and main limits in Swedish mire vegetation.) *Svensk Botanisk Tidskrift*, **43**, 274–309.

Foster, D. R., King, G. A., Glaser, P. H. & Wright, Jr., H. E. (1983). Origin of string patterns in boreal peatlands. *Nature*, **306**, 256–8.

Gignac, L. D. (1994). Habitat limitations and ecotope structure of mire *Sphagnum* in western Canada. Ph.D. dissertation, University of Alberta, Edmonton, Canada.

Gignac, L. D. & Vitt, D. H. (1990). Habitat limitations of *Sphagnum* along climatic, chemical and physical gradients in mires of western Canada. *Bryologist*, **93**, 7–22.

Gignac, L. D., Nicholson, B. J. & Bayley, S. E. (1998). The utilization of bryophytes in bioclimatic modeling: predicted northward migration of peatlands in the MacKenzie River basin, Canada, as a result of global warming. *Bryologist*, **101**, 572–87.

Glaser, P. H. (1992a). Raised bogs in eastern North America – regional controls for species richness and floristic assemblages. *Journal of Ecology*, **64**, 535–54.

Glaser, P. H. (1992b). Vegetation and water chemistry. In *Patterned Peatlands of Northern Minnesota*, ed. H. E. Wright, Jr. & B. A. Coffin, pp. 15–26. Minneapolis, MN: University of Minnesota Press.

Glaser, P. H. & Janssens, J. A. (1986). Raised bogs in eastern North America: transitions in landforms and gross stratigraphy. *Canadian Journal of Botany*, **64**, 395–415.

Glaser, P. H., Siegel, D. I., Reeve, S. I., Janssens, J. A. & Janecky, D. R. (2004). Tectonic drivers for vegetation patterning and landscape evolution in the Albany River region of the Hudson Bay Lowlands. *Journal of Ecology*, **92**, 1054–70.

Glime, J. M. & Vitt, D. H. (1984). The physiological adaptations of aquatic Musci. *Lindbergia*, **10**, 41–52.

Gorham, E. (1956). The ionic composition of some bog and fen waters in the English lake district. *Journal of Ecology*, **44**, 142–52.

Gorham, E. & Janssens, J. A. (1992). Concepts of fen and bog re-examined in relation to bryophyte cover and the acidity of surface waters. *Acta Societatis Botanicorum Poloniae*, **61**, 7–20.

Gunnarsson, U. (2005). Global patterns of *Sphagnum* productivity. *Journal of Bryology*, **27**, 267–77.

Halsey, L. A., Vitt, D. H. & Zoltai, S. C. (1995). Disequilibrium response of permafrost in boreal continental western Canada to climate change. *Climatic Change*, **30**, 57–73.

Halsey, L. A., Vitt, D. H. & Trew, D. (1997a). Influence of peatlands on the acidity of lakes in northeastern Alberta, Canada. *Water, Air and Soil Pollution*, **96**, 17–38.

Halsey, L. A., Vitt, D. H. & Zoltai, S. C. (1997b). Climatic and physiographic controls on wetland type and distribution in Manitoba, Canada. *Wetlands*, **17**, 243–62.

Halsey, L. A., Vitt, D. H. & Zoltai, S. C. (2000). The changing landscape of Canada's western boreal forest: the dynamics of permafrost. *Canadian Journal of Forest Research*, **30**, 283–7.

Hemond, H. F. (1980). Biogeochemistry of Thoreau's Bog, Concord, Massachusetts. *Ecological Monographs*, **50**, 507–26.

Ingram, H. A. P. (1978). Soil layers in mires: function and terminology. *Journal of Soil Science*, **29**, 224–7.

Kuhry, P. & Vitt, D. H. (1996). Fossil carbon/nitrogen ratios as a measure of peat decomposition. *Ecology*, **77**, 271–5.

Kuhry, P., Nicholson, B. J., Gignac, L. D., Vitt, D. H. & Bayley, S. E. (1993). Development of *Sphagnum*-dominated peatlands in boreal continental Canada. *Canadian Journal of Botany*, **71**, 10–22.

Lamers, L. P. M., Bobbink, R. & Roelofs, J. G. M. (2000). Natural nitrogen filter fails in raised bogs. *Global Change Biology*, **6**, 583–6.

Li, Y. & Vitt, D. H. (1994). The dynamics of moss establishment: temporal responses to nutrient gradients. *Bryologist*, **97**, 357–64.

Li, Y. & Vitt, D. H. (1997). Patterns of retention and utilization of aerially deposited nitrogen in boreal peatlands. *Ecoscience*, **4**, 106–16.

Limpens, J. & Berendse, F. (2003). Growth reduction of *Sphagnum magellanicum* subjected to high nitrogen deposition: The role of amino acid nitrogen concentration. *Oecologia*, **135**, 339–45.

Limpens, J., Berendse, F. & Klees, H. (2003). N deposition affects N availability in interstitial water, growth of *Sphagnum* and invasion of vascular plants in bog vegetation. *New Phytologist*, **157**, 339–47.

Limpens, J., Berendse, F. & Klees, H. (2004). How phosphorus availability affects the impact of nitrogen deposition on *Sphagnum* and vascular plants in bogs. *Ecosystems*, **7**, 793–804.

Malmer, N. (1962). Studies of mire vegetation in the archaean area of southwestern Götaland (south Sweden). II. Distribution and seasonal variation in elementary constituents on some mire sites. *Opera Botanica*, **7**, 1–67.

Malmer, N. (1986). Vegetational gradients in relation to environmental conditions in northwestern European mires. *Canadian Journal of Botany*, **82**, 899–910.

Malmer, N., Horton, D. G. & Vitt, D. H. (1992). Elemental concentrations in mosses and surface waters of western Canadian mires relative to precipitation chemistry and hydrology. *Ecography*, **15**, 114–28.

Mitch, W. J. & Gosselink, J. G. (1993). *Wetlands*, 2nd edn. New York: Van Nostrand Reinhold.

Moore, T. R. (1989). Plant production, decomposition, and carbon efflux in a subarctic patterned fen. *Arctic and Alpine Research*, **21**, 156–62.

Moore, T. R. & Basiliko, N. (2006). Decomposition in boreal peatlands. In *Boreal Peatland Ecosystems*, ed. R. K. Wieder & D. H. Vitt, pp. 331–58. Berlin: Springer.

National Wetlands Working Group (1988). *Wetlands of Canada*. Ecological Land Classification Series, No. 24. Ottawa, Ontario: Sustainable Development Branch, Environment Canada, and Montreal, Quebec: Polyscience Publications.

Nicholson, B., Gignac, L. D. & Bayley, S. E. (1996). Peatland distribution along a north-south transect in the MacKenzie River Basin in relation to climatic and environmental gradients. *Vegetatio*, **126**, 119–33.

Press, M. C. & Lee, J. A. (1982). Nitrate reductase activity of *Sphagnum* species in the south Pennines. *New Phytologist*, **92**, 487–92.

Press, M. C., Woodin, S. J. & Lee, J. A. (1986). The potential importance of an increased atmospheric nitrogen supply to the growth of ombrotrophic *Sphagnum* species. *New Phytologist*, **103**, 45–55.

Proctor, M. C. F. (1979). Structure and eco-physiological adaptation in bryophytes. In *Bryophyte Systematics*, ed. G. C. S. Clarke & J. G. Duckett, pp. 479–509. London: Academic Press.

Proctor, M. C. F. (1984). Structure and ecological adaptation. In *The Experimental Biology of Bryophytes*, ed. A. F. Dyer & J. G. Duckett, pp. 9–37. London: Academic Press.

Proctor, M. C. F., Oliver, M. J., Wood, A. J. *et al.* (2007). Desiccation tolerance in bryophytes: a review. *Bryologist*, **110**, 595–621.

Rasmussen, S., Wolff, C. & Rudolph, H. (1995). Compartmentalization of phenolic constituents in *Sphagnum*. *Phytochemistry*, **38**, 35–9.

Richter, C. & Dainty, J. (1989). Ion behaviour in plant cell walls. I. Characterization of the *Sphagnum russowii* cell wall ion exchanger. *Canadian Journal of Botany*, **67**, 451–9.

Robinson, S. D. & Moore, T. R. (2000). The influence of permafrost and fire upon carbon accumulation in high boreal peatlands, Northwest Territories, Canada. *Arctic, Antarctic and Alpine Research*, **32**, 155–66.

Rochefort, L. & Vitt, D. H. (1988). Effects of simulated acid rain on *Tomenthypnum nitens* and *Scorpidium scorpioides* in a rich fen. *Bryologist*, **91**, 121–9.

Rochefort, L., Vitt, D. H. & Bayley, S. E. (1990). Growth, production and decomposition dynamics of *Sphagnum* under natural and experimentally acidified conditions. *Ecology*, **71**, 1986–2000.

Rydin, H. & McDonald, A. J. S. (1985). Tolerance of *Sphagnum* to water level. *Journal of Bryology*, **13**, 571–8.

Siegel, D. I. & Glaser, P. H. (1987). Groundwater flow in a bog/fen complex, Lost River peatland, northern Minnesota. *Journal of Ecology*, **75**, 743–54.

Sjörs, H. (1950). Regional studies in north Swedish mire vegetation. *Botaniska Notiser*, **1950**, 174–221.

Sjörs, H. (1952). On the relation between vegetation and electrolytes in north Swedish mire waters. *Oikos*, **2**, 242–58.

Slack, N. G., Vitt, D. H. & Horton, D. G. (1980). Vegetation gradients of minerotrophically rich fens in western Alberta. *Canadian Journal of Botany*, **58**, 330–50.

Stålfelt, M. G. (1937). Der Gasaustausch der Moose. *Planta*, **27**, 30–60.

Thormann, M. N. & Bayley, S. E. (1997). Decomposition along a moderate-rich fen-marsh peatland gradient in Boreal Alberta, Canada. *Wetlands*, **17**, 123–37.

Thormann, M. N., Szumigalski, A. R. & Bayley, S. E. (1999). Aboveground peat and carbon accumulation potentials along a bog-fen-marsh peatland gradient in southern boreal Alberta, Canada. *Wetlands*, **19**, 305–17.

Titus, J. E. & Wagner, D. J. (1984). Carbon balance for two *Sphagnum* mosses: Water balance resolves a physiological paradox. *Ecology*, **65**, 1765–74.

Turetsky, M. R. (2004). Decomposition and organic matter quality in continental peatland: the ghost of permafrost past. *Ecosystems* **7**, 740–50.

Turetsky, M. R., Wieder, R. K., Williams, C. & Vitt, D. H. (2000). Organic matter accumulation, peat chemistry, and permafrost melting in peatlands of boreal Alberta. *Ecoscience*, **7**, 379–92.

Turetsky, M. R., Wieder, R. K., Halsey, L. & Vitt, D. H. (2002a). Current disturbance and the diminishing peatland carbon sink. *Geophysical Research L* 29(11)10.1029/2001GLO14000, 12 June 2002.

Turetsky, M. R., Wieder, R. K. & Vitt, D. H. (2002b). Boreal peatland C fluxes under varying permafrost regimes. *Soil Biology and Biochemistry*, **34**, 907–12.

Turetsky, M. R., Wieder, R. K., Vitt, D. H., Evans, R. J. & Scott, K. D. (2007). The disappearance of relict permafrost in boreal North America: effects on peatland carbon storage and fluxes. *Global Change Biology*, **13**, 1922–34.

Vitt, D. H. (1990). Growth and production dynamics of boreal mosses over climatic, chemical, and topographic gradients. *Botanical Journal of the Linnean Society*, **104**, 35–59.

Vitt, D. H. (1994). An overview of factors that influence the development of Canadian peatlands. *Memoirs of the Entomological Society of Canada*, **169**, 7–20.

Vitt, D. H. (2006). Functional characteristics and indicators of boreal peatlands. In *Boreal Peatland Ecology*, ed. R. K. Wieder & D. H. Vitt, pp. 9–24. Berlin: Springer-Verlag.

Vitt, D. H. & Chee, W. L. (1990). The relationships of vegetation to surface water chemistry and peat chemistry in fens of Alberta, Canada. *Vegetatio*, **89**, 87–106.

Vitt, D. H. & Kuhry, P. (1992). Changes in moss-dominated wetland ecosystems. In *Bryophytes and Lichens in a Changing Environment*, ed. J. W. Bates & A. M. Farmer, pp. 178–210. Oxford: Clarendon Press.

Vitt, D. H., Achuff, P. & Andrus, R. E. (1975). The vegetation and chemical properties of patterned fens in the Swan Hills, north central Alberta. *Canadian Journal of Botany*, **53**, 2776–95.

Vitt, D. H., Bayley, S. E. & Jin, T.-L. (1995a). Seasonal variation in water chemistry over a bog-rich fen gradient in continental western Canada. *Canadian Journal of Fisheries and Aquatic Sciences*, **52**, 587–606.

Vitt, D. H., Halsey, L. A. & Zoltai, S. C. (1994). The bog landforms of continental western Canada, relative to climate and permafrost patterns. *Arctic and Alpine Research*, **26**, 1–13.

Vitt, D. H., Halsey, L. A., Thormann, M. N. & Martin, T. (1995b). *Peatland Inventory of Alberta*. Edmonton, Alberta: Alberta Peat Task Force, University of Alberta.

Vitt, D. H., Horton, D. G., Slack, N. G. & Malmer, N. (1990). *Sphagnum*-dominated peatlands of the hyperoceanic British Columbia coast: patterns in surface water chemistry and vegetation. *Canadian Journal of Forest Research*, **20**, 696–711.

Vitt, D. H., Halsey, L. A., Wieder, K. & Turetsky, M. (2003). Response of *Sphagnum fuscum* to nitrogen deposition: A case study of ombrogenous peatlands in Alberta, Canada. *Bryologist*, **106**, 235–45.

Wagner, D. J. & Titus, J. E. (1984). Comparative desiccation tolerance of two *Sphagnum* mosses. *Oecologia*, **62**, 182–7.

Walbridge, M. R. & Navaratnam, J. A. (2006). Phosphorus in boreal peatlands. In *Boreal Peatland Ecosystems*, ed. R. K. Wieder & D. H. Vitt, pp. 231–58. Berlin: Springer.

Wieder, R. K. (2001). Past, present and future peatland carbon balance - an empirical model based on 210Pb-dated cores. *Ecological Applications*, **7**, 321–36.

Wieder, R. W. (2006). Primary production in boreal peatlands. In *Boreal Peatland Ecosystems*, ed. R. K. Wieder & D. H. Vitt, pp. 145–64. Berlin: Springer.

Wieder, R. W., Scott, K. D., Kamminga, K. *et al.* (2009). Post-fire carbon balance recovery in boreal bogs of Alberta, Canada. *Global Change Biology*.

Wieder, R. K. & Starr, S. T. (1998). Quantitative determination of organic fractions in highly organic, *Sphagnum* peat soils. *Communications in Soil Science and Plant Analysis*, **29**, 847–57.

Wieder, R. K., Yavitt, J. B. & Lang, G. E. (1990). Methane production and sulphate reduction in two Appalachian peatlands. *Biogeochemistry*, **10**, 81–104.

Williams, C. J., Yavitt, J. B., Wieder, R. K. & Cleavitt, N. L. (1998). Cupric oxidation products of northern peat and peat-forming plants. *Canadian Journal of Botany*, **76**, 51–62.

Wilson, M. A., Sawyer, J., Hatcher, P. G. & Lerch, H. E. III. (1989). 1-3-5-hydroxybenzene structures in mosses. *Phytochemistry*, **28**, 1395–400.

Wood, A. J. (2005). Eco-physiological adaptations to limited water environments. In *Plant Abiotic Stress*, ed. M. A. Jenks & P. M. Hasegawaand, pp. 1–13. Oxford: Blackwell Publishing.

Wood, A. J. (2007). Frontiers in Bryological and Lichenological Research. The nature and distribution of vegetative desiccation tolerance in hornworts, liverworts and mosses. *Bryologist*, **110**, 163–77.

Woodin, S. J., Press, M. C. & Lee, J. A. (1985). Nitrate reductase activity in *Sphagnum fuscum* in relation to wet deposition of nitrate from the atmosphere. *New Phytologist*, **99**, 381–8.

Yu, Z. C., Campbell, I. D., Campbell, C. *et al.* (2003a). Carbon sequestration in western Canadian peat highly sensitive to Holocene wet-dry climate cycles at millennial timescales. *Holocene*, **13**, 801–8.

Yu, Z. C., Vitt, D. H., Campbell, I. D. & Apps, M. J. (2003b). Understanding Holocene peat accumulation pattern of continental fens in western Canada. *Canadian Journal of Botany*, **81**, 267–82.

Zoltai, S. C. & Vitt, D. H. (1990). Holocene climatic change and the distribution of peatlands in western interior Canada. *Quaternary Research*, **33**, 231–40.

Zoltai, S. C. & Vitt, D. H. (1995). Canadian wetlands: environmental gradients and classification. *Vegetatio*, **118**, 131–7.

Zoltai, S. C., Siltanen, R. M. & Johnson, J. D. (1999). A wetland environmental data base. Canadian Forest Service NOX Publication Series. Edmonton, Alberta: Northern Forestry Centre.

10

Population and community ecology of bryophytes

HAKAN RYDIN

10.1 Introduction

Modern textbooks in general ecology contain very few, if any, bryophyte examples of patterns and processes such as population and metapopulation dynamics, dispersal, competition, herbivory, and species richness variation. A question then arises: can we freely adapt theories developed from studies of vascular plants or even animals and apply them to bryophytes? In this chapter I give examples of population and community-level processes based on bryophyte studies, and discuss how life history, morphology and physiology of bryophytes can help us to understand population dynamics, community diversity, and species composition.

Bryophytes are important in terms of species richness and cover in many habitats, and also for ecosystem functions. Most obvious is the role of *Sphagnum* as peat former. A calculation based on an average peat depth of 2 m indicates that the amount of carbon in northern hemisphere peatlands is 320 Gt, about 44% of the amount held in the atmosphere as carbon dioxide (Rydin & Jeglum 2006; see Chapter 9, this volume). Another example is the finding that nitrogen-fixation by the cyanobacterium *Nostoc* associated with *Pleurozium schreberi* contributes substantially to the nutrient budget of boreal forests (DeLuca *et al.* 2002). An illustration of the community importance of bryophytes is their contribution to biodiversity in many ecosystems at northern latitudes. As an example, Sweden hosts *c.* 0.8% of the world's vascular plant species, and 7.5% of the bryophytes (Table 10.1). Using bryophytes as model organisms in population and community studies has the advantage that many species, especially the dominant ones, have very wide distributions. Detailed studies of, for example, niche relations among peat mosses or forest feather mosses in boreal Europe and North America are highly comparable down to species level.

Bryophyte Biology: Second Edition, ed. B. Goffinet & A. J. Shaw. Published by Cambridge University Press.

Table 10.1 *Bryophytes (mosses, liverworts, and hornworts) constitute a large share of the plant diversity at northern latitudes*
Data from Sweden (Gustafsson & Ahlén 1996) are used as an example.

	World	Sweden	Proportion of world's species found in Sweden
Bryophyte species	14 000	1050	7.5%
Vascular plant species	261 000	1972	0.8%
Ratio bryophyte : vascular	1 : 19	1 : 1.9	—

Even though most ecological theories apply to bryophytes as well as to vascular plants (Steel *et al.* 2004), a sound interpretation of population and community processes must take into account that several morphological, physiological and life-history attributes are basically different in bryophytes. Most obvious are the simpler morphology with absence of roots, stomata, and in many species, of conducting tissues, and the peculiar life cycle (Table 10.2). Most bryophytes are modular and clonal, a fact with a number of ecological consequences (Svensson *et al.* 2005), and particularly important is the ability to regenerate vegetatively from almost any part of the gametophyte after fragmentation (During 1990) and from the specialized asexual propagules produced by many species of liverworts and acrocarpous mosses. Bryophytes are able to tolerate a range of habitat conditions owing to a considerable phenotypic plasticity in morphological or physiological attributes. Finally, dioicy is much more common in bryophytes than in vascular plants.

Grazing animals, parasitic fungi, and mycorrhiza exert strong influences on populations and communities of vascular plants, but in most circumstances these can be ignored in studies of bryophytes. With their low nutrient content, and in many cases peculiar biochemistry, bryophytes are generally avoided by grazers. However, in some northern habitats mosses may be heavily grazed. Examples are several goose species in the Arctic, and mice, voles and lemmings in alpine heath ecosystems and boreal forests (listed in Prins 1981) where much grazing can occur under snow cover during the winter. In years, when their populations are large, lemmings may severely reduce moss biomass in alpine snow beds (Virtanen *et al.* 1997, and references therein) and boreal forests (Ericson 1977), and strongly influence the species richness and composition. In other ecosystems the most obvious effect of grazing is to reduce the light competition from vascular plants, and as secondary effects to decrease the amount of litter covering the bryophytes and to produce small-scale gaps for colonization. Additional effects could be trampling and fertilizing by grazers. Parasitic fungi may cause necrosis, but lacking roots,

Table 10.2 *Some features that are unique to bryophytes, or more typical for bryophytes than for vascular plants, and that affect patterns and processes in populations and communities*

Attribute	Ecological consequence
Morphology	
No roots	• Photosynthesis, water and nutrient uptake in same tissue (leaves). • Dependent on environment in immediate contact with the green shoot.
No stomata in leaves	• No control of water losses (poikilohydric) or of gas (CO_2, O_2) exchange.
Most species lack conducting tissue	• Transport of water in external capillary network (ectohydric). Some internal nutrient translocation via cell-to-cell connections (plasmodesmata). Specialized conducting tissue in few (e.g. *Polytrichum*), but even these (endohydric) have mostly external water transport.
Many species grow at apex and decay at base	• No accumulation of respiring tissue. • Photosynthesizing green part always young; no senescence.
Physiology	
Tolerate desiccation and freezing	• Quick recovery after dry and cold periods during vegetation season.
Less control of growth phenology than in vascular plants	• Can grow any part of the year as soon as weather permits.
Life history	
Haploid dominant (vegetative) phase	• No "sheltering" of recessive alleles in dominant life stage; all alleles exposed to environment.
Sexual dispersal by spores	• Dispersal by large numbers covering long distances but with low probability of establishment. • Wide distribution of many species.
Male gametes transported in water to female archegonium	• Fertilization usually only between shoots within a cm–dm distance.

bryophytes are less susceptible to soil pathogens than vascular plants. The full range of interactions with micro-organisms that occur among bryophytes was reviewed by During & van Tooren (1990).

10.2 Population patterns and processes

The basis for understanding population processes in a species is its life history. Life-history traits are characteristics that affect the transition between different stages of the life cycle, such as birth, reproduction, dispersal, and death. Life-history traits evolve as the organism faces a trade-off in the use of limited resources, so that, for instance, in disturbed habitats a genotype with a large number of spores has the highest fitness, whereas in other habitats fewer but larger vegetative propagules is a more viable strategy. Life-history strategies are co-evolved integrated combinations of traits, and the most common scheme used to group bryophyte species based on life-history strategies is that by During (1979, 1992; Fig. 10.1). His classification is drawn on variation in lifespan (annual, few years, long-lived) and spore trade-off (many small or few large) in response to habitat duration and distances among suitable habitats (see Chapter 8, this volume). A more elaborate classification can include breeding system (monoicous, dioicous, and various intermediates), gametophyte size and longevity, presence of asexual propagules, features of the capsule that affect dispersal, and spore size and number (Longton 1997). In the following sections we will see how life histories affect population processes. Further details on life histories and their evolution in bryophytes are found in Longton (1997). For a thorough review of the reproductive biology of bryophytes, see Longton & Schuster (1984).

10.2.1 *Spore production*

Sexual reproduction is a process that spans many months and is vulnerable to environmental stress at all stages. A tremendous variation in spore production between years can be caused by weather variation. For example, in some boreal *Sphagnum* species gametangia are formed in the late summer, and if this period is dry the next year's sporophyte production will be reduced. In the next stage, spring precipitation has an effect on fertilization rates, and finally large numbers of developing sporophytes abort in dry summers (Sundberg 2002). The total result is that, even in the generally wet peatland habitat, spore output in some species varies between years by four orders of magnitude because of desiccation (and may even totally fail in some years). Similar large effects of variation in wetness has been demonstrated in such contrasting species as the epixylic *Buxbaumia viridis* (Wiklund 2002) and the forest floor feather moss *Hylocomium splendens* (Rydgren *et al.* 2006).

Monoicous species (potentially selfing) produce spores more frequently than dioicous ones (Cronberg 1993), which may explain why fewer monoicous mosses (compared with dioicous ones) are rare (Longton 1992, Laaka-Lindberg *et al.* 2000, Söderström & During 2005). In dioicous species the distance between male and female shoots may limit fertilization, and variation in spore capsule density will depend on the distribution and ratio of male and female shoots (Pujos 1994). Sexual reproduction may be completely lacking owing to the absence of male shoots in the population (Longton & Greene 1979). Rydgren *et al.* (2006) found that the probability that a female shoot in the dioicous

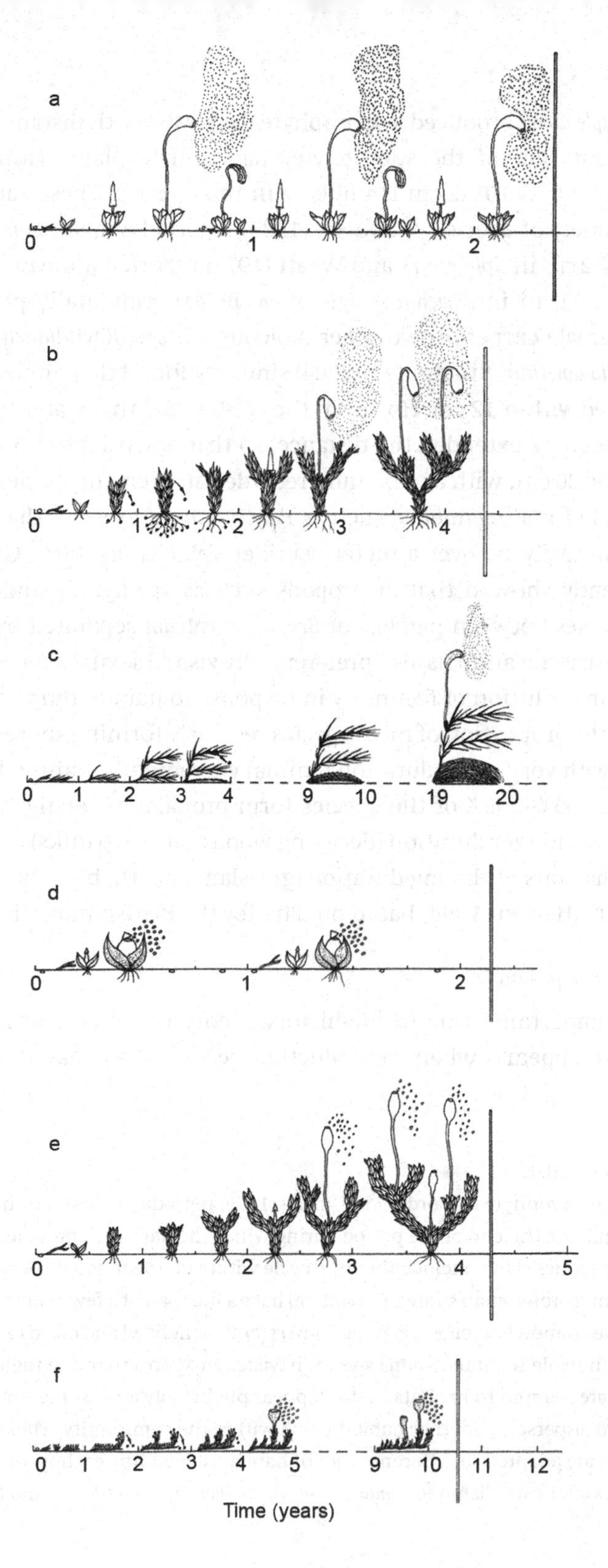

a
0
1
2
b
0
1
2
3
4
c
0
1
2
3
4
9
10
19
20
d
0
1
2
e
0
1
2
3
4
5
f
0
1
2
3
4
5
9
10
11
12
Time (years)

Hylocomium splendens produced a sporophyte decreased with distance to the nearest male plant: 85% of the sporophytes had a male plant within 5 cm, and fertilization beyond 10–12 cm is unlikely in this species. These ranges seem to hold for a variety of species: McQueen (1985) reported a mean gamete dispersal distance of 2.2 cm in *Sphagnum* and Wyatt (1977) reported a mean of 2.1 cm and maximum of 11 cm in *Atrichum angustatum*. By experimentally placing a male shoot in a female carpet of two other dioicous mosses, *Rhytidiadelphus triquetrus* and *Abietinella abietina*, Bisang *et al.* (2004) similarly found that almost all fertilizations occurred within 12 cm. However, they also noted that water flow governed by microtopograpy extended the distance, so that downslope (>5°) fertilization easily reached 20 cm, with a maximum recorded at 34 cm. In species with splash-cup dispersal of male gametes, such as *Polytrichum*, it appears that fertilization distances can easily be over a meter (van der Velde *et al.* 2001). Cronberg *et al.* (2006a) recently showed that arthropods such as springtails and mites could transfer gametes between patches of *Bryum argenteum* separated by some centimeters, and that the animals also preferentially visited sexual shoots of the moss.

There is an evolution of fecundity in response to habitat duration and this is reflected in the proportion of moss species regularly forming spores or gemmae. In habitats with very short duration (animal excrement) or annual disturbance (arable fields) 95%–100% of the species form propagules regularly. In habitats with somewhat longer duration (decaying wood and tree trunks) 60%–70% do so, whereas in habitats with long duration (grassland, heath, bog) the proportion is only 25%–30% (Herben 1994; based on data for the British moss flora).

10.2.2 *Cost of reproduction*

An important issue in life-history theory is the cost of reproduction. Such a cost appears when reproduction leads to increased mortality or

---

Caption for Fig. 10.1.

Life strategies of bryophytes according to During (1979; reproduced with permission). The vertical bars indicate the end of the period during which the habitat is suitable for the species. *Fugitives* (a) are species that conclude their life cycle within one year and disperse to ephemeral habitats with numerous small spores. *Colonists* (b) have a life span of a few years and are found in habitats that last somewhat longer. *Perennial stayers* (c) are longlived and often rely on vegetative reproduction in stable habitats. *Shuttle* species have larger spores than the fugitives and colonists and are adapted to habitats that disappear predictably after some time, so that the species have to disperse to another habitat patch within the community. There are three categories that are adapted to different types of habitats depending on how quickly they produce spores after establishment: *annual shuttles* (d), *short-lived shuttles* (e) and *long-lived shuttles* (f).

reduced growth rate such that ultimately reproduction in the future will decrease. A cost of reproduction has only recently been demonstrated in bryophytes. In *Dicranum polysetum* growth was lower in shoots that developed sporophytes than in those where sporophyte formation was aborted or experimentally terminated (Ehrlen *et al.* 2000), and in *Anastrophyllum hellerianum* there appeared to be a higher mortality in fertile female shoots than in males (Pohjamo & Laaka-Lindberg 2003). In many bryophytes the sporophytes are minute compared with the gametophyte, and intuitively a reproductive cost seems unlikely. However, in *D. polysetum* as much as 75% of the year's growth could be allocated to reproduction (Ehrlen *et al.* 2000). Furthermore, nitrogen, phosphorus, or potassium, which are limiting nutrients, occur in high concentrations in the sporophyte. It is possible that the reproductive effort in terms of these nutrients is large and even leads to a significant cost of reproduction for the gametophyte. Several types of cost appear in fertile female shoots of the dioicous *Hylocomium splendens*: reduced number of growing points, higher risk of shoot termination, and reduced daughter segment size (Rydgren & Økland 2003). As a result, a non-sporulating shoot population has a higher population growth rate. Since successful spore establishment has not been observed in the natural habitat of this species it seems unlikely that the sporulating shoots can reach equal fitness under "equilibrium" conditions (Rydgren & Økland 2002b), but instead they may be superior in colonizing new substrates.

It is not always that the sporophyte is the most costly part of reproduction. In the desert dioicous moss *Syntrichia caninervis*, the reproductive effort in male organs was larger than in female ones, and this has been proposed as one factor behind the strongly female-biased sex ratio in this species (Stark *et al.* 2000).

10.2.3 *Dispersal*

Several experiments have been made in which a source colony with a known number of capsules and spores has been set up, and spores collected on sticky surfaces at different distances. Because the density of spore deposition decreases rapidly with distance, such experiments can only reveal the shape of the dispersal curve up to a few meters. In *Sphagnum*, spore density fitted well to an inverse power function, i.e., a linear relationship after log-transforming spore density and distance (Sundberg 2005). Despite the fact that the spore density was quickly reduced to undetectable levels, a majority of the spores (60%–90%) traveled beyond the sampled distance of 3.2 m. The proportion of spores that disappear beyond a few meters is somewhat dependent on spore size (Miles & Longton 1992), but even in species with quite large spores most of the spores traveled beyond 2 m (Söderström & Jonsson 1989). The wood-inhabiting liverwort *Anastrophyllum hellerianum* has small spores (about 10 μm) and in an experiment more than 50% dispersed further than 10 m (Pohjamo *et al.* 2006). Whereas the gradient of the power function for *Sphagnum* was around −2 (Sundberg 2005; implying that a tenfold increase in distance from source reduced the spore density by a factor of 100), the gradient for *Anastrophyllum* was close to −1 (a tenfold increase in distance leads to a tenfold decrease in density). *Anastrophyllum* also produces gemmae of the same size as its spores, and they can be dispersed equally well (Pohjamo *et al.* 2006).

These experiments give some understanding of within-community dispersal and how species can fill small vegetation gaps created after disturbances. However, since they predict near-zero spore density at distances beyond a few meters, extrapolations to between-community dispersal or species migrations should be made with caution. For epiphytes (which start their dispersal at some height above ground) a somewhat more realistic appreciation of spore densities at longer distances can be achieved with functions other than the commonly used inverse power or negative exponential. In particular, the log-normal function is considered useful and realistic, and gives a "fatter" tail (i.e. higher densities at larger distances). Dispersal experiments indicate that a single patch has a great impact on spore deposition at close range, but for dispersal at a larger scale (colonization of more isolated sites), spores produced by numerous sources further away play an increasingly important role.

At a larger geographic scale, colonizations depend on events that may occur with very low probability but over a much longer time. To study this, spore trapping is not a feasible method. Instead dispersal can be inferred from distribution patterns, and especially useful are habitats that are of known age and/ or of known distance from dispersal sources. In the northern part of the Baltic Sea, the land is rising at a rate of 5–9 mm yr^{-1}, and the height of an island is a measure of its age. Colonization patterns on islands can thus be related to age, size, and distance from the mainland. In a study of *Hylocomium splendens* molecular markers were used to identify clones (Cronberg 2002). There was a linear relation between number of clones and island age, increasing from on average 12.5 clones on an island 100 yr old to 27.5 at 300 yr, but no indications of isolation by distance. With increasing island age the moss patches tended to be multiclonal, indicating repeated recruitment to the island. This species is dioicous, and the probability that male and female shoots colonize the same patch is low. Therefore only the oldest island had spore capsules and it may take as much as three centuries before any dispersal can take place within or between islands.

Species of *Sphagnum* also colonize the uplift islands in the Baltic Sea to form patches in rock crevices. Island species richness correlated positively with island area, but not with distance from the mainland (up to 40 km) or island age (Sundberg *et al.* 2006). The species were differently successful in reaching the islands. The highest colonization rate was found in species with a high regional spore output, which was estimated by the product of regional abundance, sporophyte frequency, and number of spores per capsule. Similar to *Hylocomium*, the rarity of spore capsules in most species on the islands indicates the mainland as a source for colonization rather than dispersal among islands. Another system indicating the effectiveness of long-distance dispersal in *Sphagnum* is peat pits, in which block-cut peat extraction has ceased and left a bare peat surface. After 50 years, abandoned pits in eastern Sweden contained species that are not observed in natural peatlands in the region, such as *S. lindbergii*, *S. aongstroemii*, and *S. molle*, as evidence for spore dispersal over tens of kilometres (Soro *et al.* 1999). The ecological importance of *Sphagnum* spores have been questioned, since their germination seems to be strongly phosphorus-limited in most peatland habitats. However, recent experiments show that they germinate readily in the presence of decaying vascular plant litter or animal faeces (Sundberg & Rydin 2002), and the studies on islands and in peat pits show that spores are indeed important for establishment in disturbed and newly formed wetlands. Similarly, Miller & McDaniel (2004) could demonstrate dispersal over at least 5 km of calcicole species reaching mortared (i.e. calcium-rich) roads built 65 years ago in an area with acidic rocks in New York State, U.S.A.

One way to test the importance of dispersal vs. habitat limitation is to experimentally introduce diaspores. Lloret (1994) compared the success of three dominant forest floor mosses to colonize experimental gaps 1 m^2. All three species colonized after experimental planting; this result shows that the environmental conditions were not limiting. The experiment indicated that *Dicranum scoparium* and *Hylocomium splendens* were dispersal-limited, whereas *Pleurozium schreberi* was not, and this species probably colonized effectively from the adjacent carpet. Such short-distance dispersal may often be by vegetative fragments or specialized asexual diaspores produced in many species. For such diaspores it is generally concluded that effective dispersal is in the centimeter range (Laaka-Lindberg *et al.* 2003), with the above-mentioned small gemmae in *Anastrophyllum hellerianum* as exceptions.

We normally assume that bryophytes are wind-dispersed, but other vectors may be involved. In wetlands and along rivers, fragments and spores could easily be water-dispersed (Dalen & Söderström 1999). The adaptation to dispersal by flies in Splachnaceae is discussed below. Breil & Moyle (1976) found 65 species of bryophyte in birds' nests in Virginia, U.S.A.; this finding indicates that birds may assist in the dispersal of fragments as they forage (for instance in bark crevices) and collect nesting material.

A quickly developing research field is to use genetic markers and infer dispersal from the distribution of haplotypes or from the relationship between genetic similarity and distance (e.g. Skotnicki *et al.* 2000, Cronberg 2002, Snäll *et al.* 2004a). In the metapopulation section below, an approach in which dispersal distances are modeled from distribution patterns is discussed.

10.2.4 *Germination and establishment*

The germinability of spores is often high. Hassel & Söderström (1999) reported that 97% of the spores in *Pogonatum dentatum* germinated in the laboratory, and they found similar values for other species in the literature. Establishment in the field occurs with much lower probability; even when spores are spread out experimentally in what appears to be a suitable habitat, establishment probabilities seem to be of the order of 10^{-4} to 10^{-3} (Hassel & Söderström 1999) or even lower. *Sphagnum* spores hardly germinate at all on the peat where the adult plants grow; some nutrient additions from decaying litter seems to be required (Sundberg & Rydin 2002). Spore germination is of course dependent on moisture, and an example of the highly specific requirements during spore germination comes from studies of *Neckera pennata* (epiphyte on tree trunks) and *Buxbaumia viridis* (on decaying wood). The germination depends on an interaction between moisture and pH so that high water availability facilitates germination at suboptimal pH, and vice versa (Wiklund & Rydin 2004b). For *Neckera*, which normally occupies rather desiccation-prone tree trunks, this may explain its preference for host trees with high pH where it can germinate quickly and therefore exploit the short windows of opportunity with wet bark after rain events. Not only is spore germination difficult, but there is probably also high mortality in the protonema stage (Lloret 1991) as an effect of desiccation (Thomas *et al.* 1994).

Establishment from vegetative fragments has a much higher probability of success, but is a habitat-sensitive process, too (Cleavitt 2001). A very practical example are the methods developed for re-establishing *Sphagnum* on peatlands after peat harvest (summarized by Rochefort *et al.* 2003). Here fragments are used, and for success the hydrology must be controlled to produce a wet peat surface and the fragments need initial protection by a layer of straw mulch. Hence, the technique highlights the fact that surface desiccation is the critical factor for the establishment of bryophyte fragments.

10.2.5 *Diaspore banks*

The presence of spores as a diaspore bank in the soil has been demonstrated for many species (During 2001), and it also appears that gemmae can enter a state of dormancy (Laaka-Lindberg & Heino 2001). The ecological importance of the diaspore bank is difficult to assess, but it suggests at least a potential

Table 10.3 *The relative contribution (%) of bryophytes with different life-histories in the diaspore bank and in the vegetation in Dutch chalk grassland sites*

Data are for an average of three sites (During & ter Horst 1983) and Swedish boreal forest (Jonsson 1993). The numbers relate to abundance of the different strategies, not to number of species.

	Chalk grassland		Boreal forest	
Life history	Diaspore bank	Vegetation	Diaspore bank	Vegetation
Colonists	76	27	32	<1
Annual shuttle species	1	4	—	—
Short-lived shuttle species	12	14	23	<1
Long-lived shuttle species	—	—	15	15
Perennials	10	56	18	84

for secure and rapid colonization after disturbance. In Swiss arable fields 15 species germinated from soil samples. Five of these were not present in the surface vegetation, and only four species in the vegetation were absent from the diaspore bank (Bisang 1996). From a mixed forest in New Brunswick, Canada, Ross-Davis & Frego (2004) list 29 species from the diaspore bank, of which 13 did not appear in the extant community. From peat samples down to 30 cm, regeneration of three *Sphagnum* and eleven liverwort species was observed (Clymo & Duckett 1986, Duckett & Clymo 1988). Many of the shoots came from buried stems, indicating that these may retain a regenerating capacity up to perhaps 60 years. Some plants probably also developed from spores, and experiments with buried *Sphagnum* capsules show that spores may survive for several decades, perhaps even a century (Sundberg & Rydin 2000), an observation that will increase the significance of the spore bank.

As in flowering plants, the representation of species in the diaspore bank is strongly related to life-history attributes. Relative to their abundance in the vegetation cover, colonists are considerably more common than perennials in the diaspore bank (Table 10.3). As expected, monoicous species with frequent production of spores or gemmae are often found in the diaspore bank, but contrary to observations in seed banks (where small seeds are often numerous), Jonsson (1993) noted that the species found in the diaspore bank had on average larger spores than those found in the vegetation cover.

10.2.6 *Clonal expansion and population persistence*

As noted above, many bryophytes have the capacity to expand and disperse by vegetative fragments or specialized propagules. Of particular ecological importance is clonal expansion by branching, since the new shoots benefit from being physiologically integrated with the mother plant and hence have a higher chance of survival than detached propagules. Clonal species may be very persistent (During 1990), and *Sphagnum* individuals can probably survive for centuries (Rydin & Barber 2001) as they slowly expand clonally and at the same time avoid a respiratory burden by losing old tissue to peat formation.

A distinction is sometimes made between "guerrilla" and "phalanx" strategies of clonal growth. Among bryophytes the typical phalanx species are the acrocarps forming dense cushions, whereas guerrilla species are weft- or mat-forming pleurocarps (Cronberg *et al.* 2006b). Even in the phalanx type clones tend to mix with time since the mixing is not merely dependent on the type of clonal growth but also on longevity, spore dispersal, etc. For example, a square decimeter of *Hylocomium splendens* often has 2–3 genotypes, and sometimes even 5 (Cronberg 2004). The presence of guerrilla and phalanx species has several community consequences (review in Svensson *et al.* 2005). Phalanx growth leads to aggregation of individuals of the same species which diminishes the interspecific competition, whereas guerrilla species evade intraspecific competition and encounter more interspecific contacts. Expanding as a physiologically integrated front, phalanx species are generally considered strong resource competitors in an undisturbed community. Instead guerrilla ramets carry fewer resources from the mother plant, but they could capture new space effectively, and be good at pre-emptive competition.

Clonal growth potentially gives the plant an opportunity to explore the habitat, thereby reaching positions with favorable conditions. There are not many tests of such "foraging" in bryophytes, but experiments by Rincon & Grime (1989) indicate that species with high growth potential (*Brachythecium praelongum* and *Thuidium tamariscinum*) may have some ability to expand laterally from dark to light patches.

10.2.7 *Density-dependence in bryophyte populations*

Intraspecific competition in vascular plants is often described by the negative effects (such as decreased growth or increased mortality) that follow from increasing shoot density. For example, if the reduced growth per individual exactly compensates for increase in density, the total biomass produced per unit area will be independent of sowing density as described by the *law of constant final yield*. When mean shoot mass is plotted against density in a log–log diagram a slope of approximately −1 is observed (i.e. a tenfold increase in density results in a tenfold decrease in shoot size, Fig. 10.2). In many cases, an increase in density is also followed by an increase in mortality. This is referred to as self-thinning, and it has often been found that the average size of the surviving

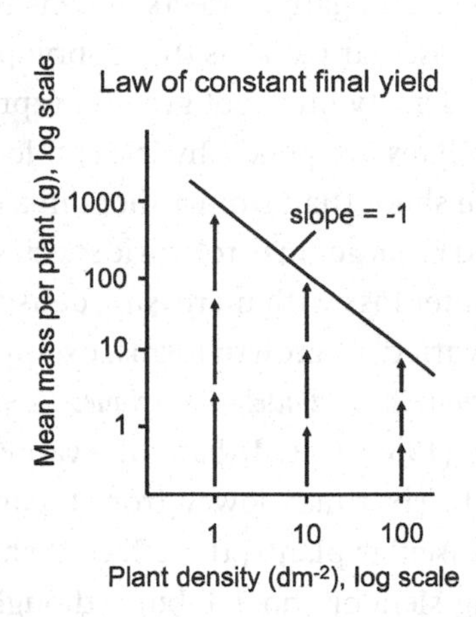

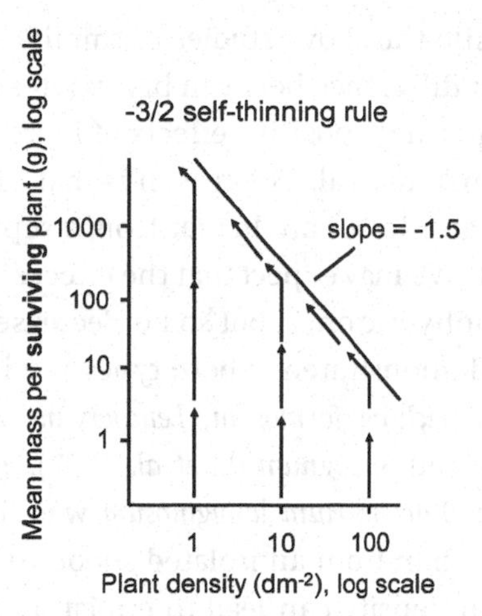

Fig. 10.2. Conceptual models of density-dependence in plant populations showing the log–log relation between density and individual plant size. Plants are sown at different densities, and the growth of the individuals per unit time (e.g. per month) is lower at higher density (as indicated by the shorter arrows). If the population follows the *law of constant final yield* the total biomass produced per unit area will be independent of sowing density. Dense populations often suffer from earlier and higher mortality (as indicated by the arrows showing a reduced density as the plants grow), and many natural populations fit a line with a slope of approximately −1.5; hence the *−3/2 self-thinning rule*.

individuals increases more rapidly than density decreases. The result is that many natural populations fit a line with a slope of approximately -1.5, a pattern that is referred to as the *−3/2 self-thinning rule* (Fig. 10.2; Begon *et al.* 2006).

Collins (1976) found that log mass–log density slopes for *Polytrichum alpestre* in the maritime Antarctic were in the range -0.66 to -0.82. This indicates a weaker intraspecific competition than predicted by the law of constant final yield, so that increased density led to increased total yield. Although there was a considerable flux in the populations (up to 37% of the population lost, and equally many gained over a year), this turnover was not related to density, which remained constant. Scandrett & Gimingham (1989) assessed intraspecific competition by growing monocultures of *Pleurozium schreberi*, *Hylocomium splendens*, and *Hypnum jutlandicum* from fragments that were spread out at densities of 0.8 and 8 mg cm^{-2} (dry mass). Yield was in most cases reduced by more than 50% at the higher density, indicating a stronger density-effect on individual size than predicted from the law of constant final yield. An example in which density-dependence was manifested as self-thinning was given by Lloret (1991), who observed high mortality rates among sporophytes in dense populations of the coprophytic *Tayloria tenuis*.

While the above studies exemplify expected negative effects of density, the largest difference between bryophytes and vascular plants is that in bryophytes we often note positive effects of increased density on shoot growth, reproduction and survival. The reason is that bryophytes are poikilohydric (Table 10.2), and water losses are higher from a separate shoot than from a shoot in a dense carpet. We may expect that the effect should be larger in ectohydric species than in endohydric ones, but so far decreased water loss with increasing density has been demonstrated in both types and in a variety of growth forms, exemplified by *Polytrichum formosum*, *Leucobryum glaucum*, *Rhytidiadelphus triquetrus* (Filzer 1933), and *Sphagnum* (Li *et al.* 1992). Tallis (1959) stated that the evaporation rate in *Racomitrium lanuginosum* was four to six times lower from a compact carpet than from an isolated shoot. As in vascular plants, the effect of shading at high density can lead to etiolation (long slender shoots), but although this plastic response leads to smaller shoots it need not result in self-thinning. Bates (1988) grew *Rhytidiadelphus triquetrus* at a range of shoot spacing and noted that individual growth increased with increasing density. There was no mortality, and the production of offshoots was not restricted by initial high density. Hence, total growth was highest in dense populations. Watson (1979) studied several coexisting species of the Polytrichaceae that could be aged through their distinct annual growth increments. She noted a positive relation between shoot density and mean age in the populations, indicating that increased density led to increased shoot longevity. Water and light availability will determine how biomass growth is affected by density (Pedersen *et al.* 2001), but overall it appears that in most cases the positive effects on moisture outweigh the negative effects of shading (van der Hoeven 1999).

In experiments with *Sphagnum* some negative effects of density on net recruitment (number of new shoots minus mortality), were observed (Rydin 1995). In most cases this negative density-dependence was weak, so that shoot density increased during the experiment even in samples that started with a density similar to natural ones. This reflects the phenotypic plasticity in *Sphagnum*; although there is an upper size limit for a shoot, there is virtually no limit to how slender a *Sphagnum* shoot can be. In *S. tenellum*, for example, the natural density is about 10 capitula cm^{-2}, but occasionally populations can be three times as dense (Rydin 1995). As demonstrated by Hayward & Clymo (1983) *Sphagnum* carpets are controlled by an intriguing balance between shading and desiccation. Shoots growing more rapidly in length than their neighbors will be more prone to desiccation, which leads to diminished growth. In contrast, shoots that accumulate less biomass will escape burial as long as they can form slender but tall shoots through etiolation. In the extreme case they will of course be overtopped, but since shoot size is very plastic they can keep their apex at the surface even if the shoot becomes very slender. This explains how *Sphagnum* mats maintain a very smooth surface by forces acting on individuals. Shoot size variation can be even larger in other bryophytes. In *Hylocomium splendens*, shoot size can vary by a factor of 1000 (Økland 1995, 2000) indicating a considerable potential to avoid burial. Thinning and disturbance can also lead to increased size variation when light suddenly reaches farther down and activates sprouting from basal parts (van der Hoeven & During 1997).

10.2.8 *Population dynamics in* Hylocomium splendens: *a case study using matrix modeling*

The use of matrix models has been a standard method in the study of population dynamics in animals and vascular plants for some time (Caswell 2001), especially for structured populations. Such a population can be divided in discrete classes (according to size, age or stage, or a combination) and the fate of each individual is followed from one census to the next, most often from one year to the next. Using these data a matrix is produced with transition probabilities between stages. The number of individuals in each stage in the next year is calculated by multiplying the transition matrix by the vector of number of individuals in each stage class, and by doing this repeatedly, the long-term population growth rate, λ, can be calculated ($\lambda > 1$ in a population that increases in numbers, $\lambda = 1$ in a stable population and $\lambda < 1$ in a decreasing population). Matrix models may thus be used to project future population sizes. As for all models certain assumptions are made; for the basic matrix models (the deterministic ones) the most important one is that biotic and abiotic conditions recorded at the time data were collected will not change. Other outputs we get from this type of analysis are stable age–size–stage structures, reproductive values (the expected number of offspring produced by individuals in different life-stage class per time interval), and different measures of sensitivity of population change to variation in transition probabilities. The most commonly used matrix models also assume that the transitions are independent of history, that is, the fate of an individual depends only on its present stage, not on its previous history. For bryophytes this is probably a reasonable assumption.

The practical difficulties in marking and relocating very small plants have restrained most bryologists from using matrix models. An extremely thorough study of population dynamics in bryophytes is that by Rune Økland and associates on *Hylocomium splendens* (Økland 1995, Økland & Økland 1996, Økland 1997, Rydgren *et al.* 1998, Økland 2000, Rydgren *et al.* 2001, Rydgren & Økland 2001). They use colored plastic rings to tag individual shoots and are thus able to follow the fate of these over many years. Some results from their papers are compiled in the following section, to illustrate population processes in bryophytes and

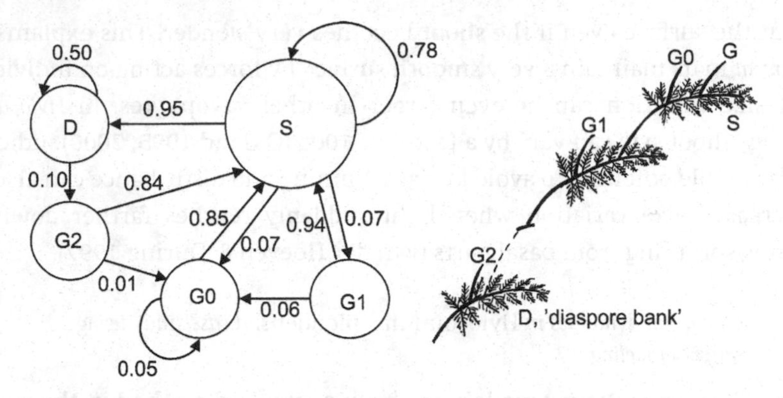

Fig. 10.3. Life-cycle graph and transition probabilities in *Hylocomium splendens* from a study in SE Norway, simplified from Rydgren & Økland (2002a) by merging some stage classes. Circles represent the different stages in the life cycle, and arrows are the transitions from one stage to another. Most often the current year's vegetative segment (S) produces a growing point (G) forming next year's segment (this is the S → S loop with probability 0.78). Sometimes an extra growing point develops on the current (G0) or last year's (G1) segment. Growing points (G2) can also develop from older segments or on detached segments of unknown age (in the model such segments are treated as a diaspore bank, D). Peculiar to this life-cycle is that spores and spore bank are not included: earlier studies had implied that they were unimportant in mature stands of *Hylocomium*.

also to indicate the usefulness of the method. A methodological summary was given by Rydgren & Økland (2002a) in which they suggest that this method can be applied to a range of bryophytes.

The first and critical point is to identify the life-cycle stages in a way that is biologically meaningful and reproducible. With its peculiar annually reiterating growth form, *Hylocomium splendens* is a classic example (Fig. 10.3). From each segment ("frond") a new growing point normally develops to form next year's segment. In some shoots an extra growing point develops, and sometimes growing points emerge also from older segments. The population size does not refer to number of physiologically independent units, but instead reflects the total number of active growing points. Which study units will be appropriate depend on the structure of the plant; in non-clonal plants the individual shoots can be used, but in clonal plants number of ramets, modules, or meristems must normally be used as a measure of population size.

The life cycle in *Hylocomium splendens* is (as in clonal plants in general) characterized by shoot persistence: a large majority of the current year's segments form a growing point (G) to build a new segment next year (S → S loop with probability 0.78 in Fig. 10.3). Extra growing points (G0), and new growing points in one-year-old segments (G1) are much rarer, but once formed they have a very high probability of developing into a new segment, which leads to branching of the moss: an increase in population size. Even regenerations by growing points developing on older segments or on detached segments of unknown age (G2; acting as a "diaspore bank") have a high probability of growing into a mature new segment. Matrix models not only predict the growth or decline of the population and changes in numbers in the different life-stages, they also suggest which stages and transitions have the highest influence on changes in population size. This is referred to as elasticity, which measures the proportional change in λ as a function of a proportional change in the transition probability (elasticities for all transitions sum to 100%). In *Hylocomium* it seems that elasticity for the S $\rightarrow$ S loop (persistence via a normal growing point) is normally in the range 60%–70%, whereas all other transitions in general have values $<$10%. This is probably quite typical for long-lived perennials.

Detailed monitoring of natural populations and thinning experiments have revealed both negative and positive density-effects. One negative effect was the reduced regenerations from old segments (G2) with increasing density. Another was the increased risk that shoots were overtopped and buried by neighbors. By contrast, density and segment size are positively correlated, which can be attributed to the more favorable moisture regime in denser carpets. Larger segments have a higher probability of forming new growing points, larger daughter segments, and lower risk of termination. Density effects on reproduction are central in population studies of vascular plants, but studies on bryophytes are rare. As indicated in the section on spore production above, density may also have both negative and positive effects on reproduction. Reduced density stimulated sporophyte production, probably through reduced shading, but the increased distances between male and female shoots may be problematic at too low a density: all in all, it appears that the direct effects of density on population regulation are relatively weak, and the indirect effect of segment size is stronger.

While the basic use of matrix models – to project future changes in population size – assumes that the transition probabilities are the same from year to year and that they are not affected by population density, the effects of external factors can be studied by comparing the transition probabilities among years and sites. Variation in population dynamics among sites is expected and related to microclimate and other local factors. More interesting is that some of the between-year variation is synchronous in different localities: favorable (wetter) years result in larger segments with higher probability of successful production of next year's segment (survival of growing point G; Fig. 10.3). The variation in λ between years (0.87–1.22) in *Hylocomium* is similar to values reported for vascular clonal plants in these forest communities. Even short wet periods in a critical period of the year can lead to bursts of regeneration from lower and detached segments (growing point G2), and in years with large lemming and vole populations, grazing under snow can have large impacts on the demographic structure.

10.3 Metapopulation patterns and processes

Many bryophyte species grow on substrates that are patchily distributed, such as boulders, dead wood, and tree trunks. Lacking a root system that could "explore" the surface and deeper soils, bryophytes are more strongly linked to substrate patches than are vascular plants. For these reasons, metapopulation theory is particularly applicable to bryophytes. A *metapopulation* is a set of populations linked by dispersal. A *patch* is the place with conditions that are suitable for the species, for instance the tree trunk for an epiphytic species. Where the species is present in a patch it is referred to as a *local population. Local dynamics* are caused by population processes and interactions with other species. The local population may eventually become extinct – *local extinction* – in which case it can only re-occur if it is successfully dispersed from another local population in the metapopulation. The *occupancy* – the proportion of patches that is occupied by the studied species – is determined by the balance between local extinctions and colonizations.

Several types of metapopulation model have been developed over the past decades, and they somewhat differently describe the factors affecting local extinctions and colonizations. A common assumption is that local population size fluctuates with time because of demographic and environmental stochasticity, and the larger the population, the smaller the extinction risk. Since population size is often correlated with patch area, many models use patch area as a predictor for extinction probability (the larger the area, the lower the extinction risk). In early metapopulation models the immigration (recolonization) probability was the same for all patches, but in most modern models immigration probabilities depend on patch *connectivity*, which is a central concept in metapopulation theory. It is inversely related to isolation and should reflect the expected number of immigrants arriving at the patch per unit time (Hanski 1999). A patch has high connectivity if it is surrounded by many other patches, if the distances to these patches are small and if they host large populations of the studied species (see Fig. 10.4 for details). If a statistical effect of connectivity on the occupancy pattern can be established, it is taken as an indication of dispersal limitation. The fitted slope parameter of the connectivity function (α) indicates how quickly the influence of a source patch declines with distance, and hence gives an indication of effective dispersal distances. More

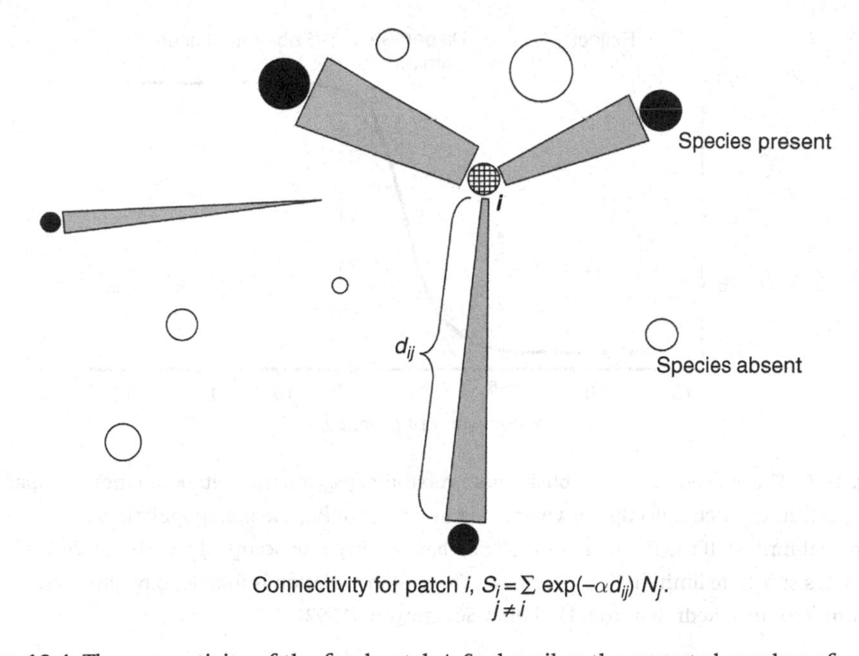

Fig. 10.4. The connectivity of the focal patch i, S_i, describes the expected number of immigrants per unit time from the surrounding patches j (Hanski 1999). In this graphic model diaspore density is shown by the width of the grey wedges. The surrounding patches are at different distances (d_{ij}) from the focal patch, have different areas (circle size), and are occupied (filled) or unoccupied (open). In the equation, N_j represents the potential diaspore output from patch j, most realistically measured by the population size. Since population sizes are often unknown, area is commonly used as an indicator of population size ($N_j = 0$ if the species is absent and $N_j = A_j$ if it is present). More simply, but less realistically, N_j could represent presence ($N_j = 1$) or absence ($N_j = 0$) of the study species. In the simplest form all patches are equal ($N_j = 1$). The parameter α determines how quickly the diaspore rain decreases with increasing distance from the source patch according to the negative exponential function. In some cases, e.g. for epiphytes, a log-normal function is often considered more realistic than the negative exponential (see, for example, Snäll *et al.* 2005 for details).

rarely, experimentally measured dispersal curves (e.g. Söderström & Jonsson 1989) could be inserted in the connectivity function.

The connectivity measure illustrates an important aspect of metapopulations, namely source–sink dynamics. A local population in a patch with negative population growth rate is doomed to extinction (sink population), but it may reappear because other patches with positive growth rates (source populations) deliver a surplus of migrants. The patches thus have unequal influence on the metapopulation, and the exporting patches would be the ones that are large or

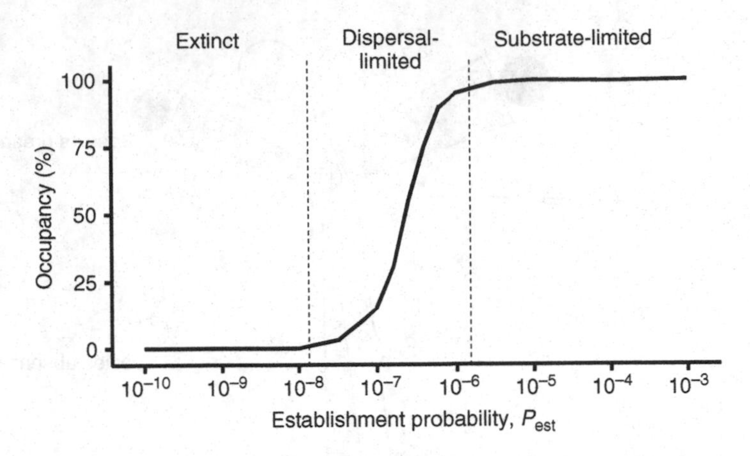

Fig. 10.5. The effect of spore establishment probability (P_{est}) on the metapopulation occupancy (proportion of patches occupied). Over a narrow range of P_{est} the metapopulation is dispersal-limited. If the P_{est} increases, all patches will become occupied and the species becomes substrate-limited. If P_{est} decreases the metapopulation is unable to persist, and becomes extinct. Redrawn from Herben & Söderström (1992).

have high habitat quality. By incorporating the area or population size of surrounding patches in the connectivity measure (Fig. 10.4), the metapopulation model will capture the essence of source–sink dynamics.

In metapopulation models "colonization" summarizes two processes that are in practice quite difficult to separate: dispersal and establishment. A crucial parameter is the probability that a diaspore establishes once it has reached the patch. With a simulation model based on field data for the fugitive species *Orthodontium lineare* growing on decaying wood patches, Herben *et al.* (1991) suggested that the range of values under which the establishment probability affects the metapopulation size is rather narrow. Within this range the metapopulation is dispersal-limited (Fig. 10.5). With increasing establishment success all patches will become occupied, and the metapopulation size becomes substrate limited (Herben & Söderström 1992). For lower probabilities the metapopulation will become extinct because the colonization rate cannot match the extinction rate.

Metapopulation theory pictures the landscape as consisting of hospitable patches and an inhospitable matrix. This applies well to species that are confined to such patches, but makes the theory unsuitable for, for example, facultative epiphytes. Even for species that are truly confined to the particular substrate type, a notorious difficulty is to determine whether an unoccupied patch is in fact hospitable. It is often necessary to use rather rough criteria to define the "suitable" patches in a study; for epiphytes one may, for example, assume that all trunks of a host tree species are suitable patches even though many suitable patches of the same tree species are never colonized for unknown reasons.

Traditionally, studies of bryophytes focused on substrate factors (Chapter 8, this volume) to explain the presence and composition of species, whereas metapopulation studies focus on colonization–extinction dynamics, dispersal limitation, and the spatial configuration of the patches in the landscape. Recent studies combine the approaches by trying to disentangle the relative importance of substrate quality and colonization–extinction dynamics for the occupancy pattern. Another recent development is models that account for the fact that the patches are also dynamic. They may change in quality over time, and more importantly they may have a limited duration. Boulders are permanently available patches, tree trunks last for decades to centuries, and dung patches for a couple of years (and are open for new colonizations only when the dung is fresh). It is therefore necessary to include in metapopulation models the fact that extinctions can happen as a deterministic effect of patch destruction. New patches arise at different spots, so the connectivity pattern also changes over time. How important the dynamics of patches is depends on the patch turnover rate. At one extreme, classical metapopulation models assume that patches are invariable in quality and position, and all extinctions are caused by internal population dynamics. At the other end of the spectrum is a system in which the local populations are extremely persistent and the only factor that causes local extinction is the destruction of the patch. In the latter case the species must have the capacity to disperse to new patches as these appear, and this can be referred to as a patch-tracking metapopulation (Snäll *et al.* 2003). In the following, bryophyte metapopulation processes are exemplified for species growing on dung patches and for epiphytes on tree stems, but the ideas apply also to other systems, such as epiphyllous bryophytes (Zartman 2003, Zartman & Shaw 2006, Zartman & Nascimento 2006) and decaying logs (see Söderström & Herben 1997 for a general review).

10.3.1 *Bryophytes on dung: patch quality, local interactions, and metapopulation processes*

Several species of Splachnaceae growing on droppings of large mammals exemplify the dynamics of mosses confined to ephemeral, patchy habitats. The moss spores are dispersed from one dropping to another by flies, which also are dependent on the dung for their breeding. Both the moss and the insect have to conclude their life cycle before the dung patch becomes inhospitable. The moss species attract somewhat different sets of flies (Marino 1991a), and each moss species is dispersed by 10–17 fly species. The adaptation is remarkable:

several species have a long seta with the hypophysis under the capsule bright red or yellow and even umbrella-shaped. Most important is probably that entomophilous species have sticky spores and produce volatile attractants (Pyysalo *et al.* 1983). The following discussion of the processes affecting coexistence and persistence of *Splachnum ampullaceum*, *S. luteum*, *Tetraplodon angustatus*, and *T. mnioides* growing on droppings of large mammals is based on the thorough investigation by Marino (1991a,b,c, 1997).

Marino noted that unoccupied droppings were rare, and that the mosses often occupied the whole dung patch. This contrasts with many epiphytic systems and indicates that local interspecific competition is superimposed on the metapopulation processes and may strongly affect local extinctions and spore output from the patches. Each species produced more spores when growing alone than when co-occurring in a mixture with another species, an observation that indicates resource competition. However, the competitive ability was dependent on habitat. In wet habitats *Splachnum* species had a competitive advantage, produced more spores, and even eliminated *Tetraplodon*, and in dry sites the relations were reversed. The fundamental niches seem largely to be overlapping, and in absence of competitors the species could occupy both habitat types (Marino 1991b). The altered competitive relations may be mediated by habitat effects on patch chemistry, and this causes the infrequency of co-occurrence between species of *Splachnum* and *Tetraplodon* so that *Splachnum* was found in peatlands with *Sphagnum* and brown mosses and *Tetraplodon* in dry upland sites with *Cladonia* lichens and feather mosses. Competition between congeners was rather symmetric, but from laboratory experiments it appears that *Splachnum ampullaceum* was slightly competitively inferior to *S. luteum* in wet habitats (Marino 1991b). Balancing this, flies captured on *S. ampullaceum* carried more than twice as many spores as those on *S. luteum*, and several of the flies specializing in *S. ampullaceum* were also strongly related to wetland habitats, indicating some degree of colonization–competition trade-off.

Theoretical models, for instance lottery models, suggest that competitive species may coexist if they have equal chance to be the first settler as sites become available for colonization. Of particular interest is that coexistence of competitive equivalents should be facilitated by phenological separation (Fagerström & Ågren 1979) if patches are renewed randomly in time. The first diaspore to arrive has an advantage so that the phenological separation may lead to pre-emptive space competition among the dung mosses. Spores of *Tetraplodon angustatus* were dispersed in May (Marino 1991a) and *T. mnioides* and the *Splachnum* species in the summer from mid-June. The power of pre-emption should be especially strong in this case since the dung patch is suitable for the insects for a very short time, with almost all visits taking place during the first day. For this reason, the metapopulation of *T. angustatus* is probably rather independent of the other species since it is based on a different set of patches: those deposited in spring.

Dung patches have a more rapid turnover than most other patch types inhabited by bryophytes. To understand the metapopulation one must account for the short time window for colonization, and that local extinctions may be caused by interspecific competition. We do not have a spatially explicit model for Splachnaceae, but Marino (1991c) has attempted to model the relative importance of different factors for the *Splachnum* species. The model implies that the number of patches, interspecific competition and aggregation within species were all important in explaining the persistence time of the metapopulations. Aggregation may follow, for instance, from short dispersal distances, and its effect would be to decrease the degree of interspecific contacts and hence promote coexistence by reducing the risk of competitive exclusion (Bengtsson *et al.* 1994).

10.3.2 *Epiphytes: local environment, connectivity and tree dynamics*

Bryophytes confined to deciduous trees that occur as scattered individuals in an otherwise inhospitable conifer forest matrix are useful study systems. Here it is possible to test the relative importance of local factors, patch connectivity and tree dynamics for the occupancy pattern. A starting point is to verify that the study species is really confined to one or several tree species, so that all suitable trees (hosts) can be mapped and checked for occurrence of the epiphyte. Such epiphytes generally do not appear on saplings of the host, and therefore a lower diameter limit can be set for the host definition. The next step is to measure relevant local environmental variables. The selection is difficult and must be based on previous knowledge on the biology of epiphytic bryophytes in general and the focal species in particular. Tree species (if several hosts are used) together with age and/or diameter are obvious variables, and substrate factors of interest are depth of bark crevices (indicating microenvironmental variation) and bark pH. The local environment also includes microclimatic measures of light, temperature, and humidity. These can be measured directly, but more commonly indicated by crown cover, stem density, shading, local soil moisture, etc.

Heegaard (2000) partitioned the variance in the distribution of *Ulota crispa* in mixed forests in Norway. A regression model could explain 55% of the variation of number of moss cushions per forest plot (2.5 m × 2.5 m). Of this 20% could be accounted for by metapopulation processes (i.e. related to number of host trees per plot and spatial co-ordinates) and 20% by substrate factors. Since tree density influences both the environment (shading) and dispersal distance, the remaining explained variance could not be split into substrate and patch dynamics. A tentative conclusion was then that dispersal and habitat factors were approximately equally important.

Intuitively, spatial aggregation of the epiphyte would indicate dispersal limitation, but such aggregation could also follow from an aggregation of host trees or from spatial variation in environmental factors (Hedenås *et al.* 2003). A useful strategy is instead to test how well the occupancy pattern can be predicted from patch connectivity, given that the environmental factors have been accounted for. An example of this approach is the study of *Neckera pennata* growing on broadleaved trees in a conifer-dominated landscape in eastern Sweden. The species occupied 30% of 1050 studied trees in three forest stands. The probability of finding *N. pennata* on a particular host tree increased with tree diameter and depth of bark crevices, and it differed among tree species, with *Acer platanoides* having the highest, and *Populus tremula* the lowest occupancy (Snäll *et al.* 2004b). Tree diameter, crevices and tree species were approximately equally important, but, in comparison, connectivity had a much stronger effect (Fig. 10.6) which resulted in a spatial aggregation of the *Neckera* in these forests.

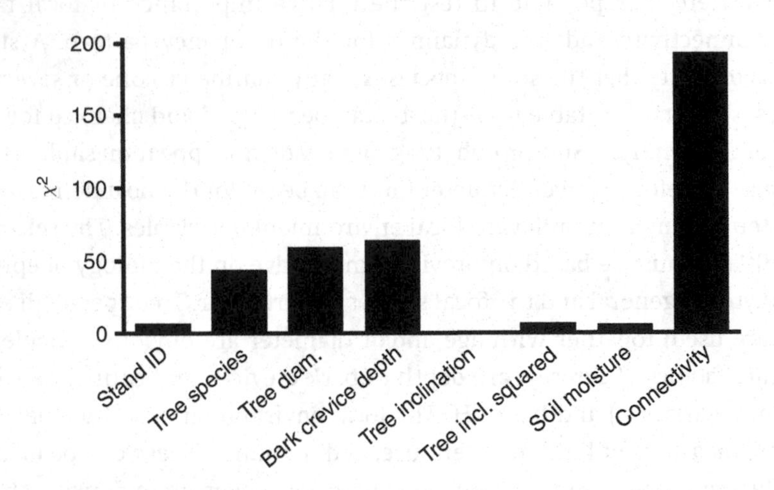

Fig. 10.6. The effects of environmental variables and connectivity on the probability of occurrence of *Neckera pennata* on deciduous host trees in coniferous forests in eastern Sweden. Connectivity was here calculated according to a log-normal function based on population size of *N. pennata* on potential source trees. The χ^2 for each predictor variable indicates its importance for explaining the occurrence of *N. pennata*. Occupancy differed among sites and tree species. Tree stem diameter and depth of bark crevices had positive effects on occupancy, but connectivity appeared to be more important than all the environmental variables together. Based on data in Snäll *et al.* (2004b).

Patch area is obviously an important variable for metapopulation processes, and tree diameter is often a good predictor for occupancy and cover of many epiphytes (McGee & Kimmerer 2002). However, the causal relationship is obscure since tree diameter affects the occupancy in several ways, as follows. (1) The area *per se* effect, by which larger areas have larger populations and therefore lower extinction risks. (2) Larger trees are older and have had longer time to catch diaspores, which for *Neckera* seemed to be the most important mechanism (Snäll *et al.* 2005). (3) Larger trees have had longer times to develop spore-producing colonies, and dispersal within the tree can increase population size and decrease the extinction risk. In *Neckera pennata*, for example, it was estimated that first spore production occurs when the moss patch is 20–30 years old (Wiklund & Rydin 2004a). (4) The bark structure, and probably also its chemistry, changes with tree age. A complete separation of habitat factors and dispersal limitation is obviously difficult.

Trees die and new trees are established, and an understanding of the effect of the dynamics of the patches requires repeated surveys. In the *Neckera* project, the sites were re-analyzed after 2 and 4 years, and even such a large effort yielded only a few colonizations and extinctions; hence the conclusions are somewhat tentative. Starting with 280 of 831 trees being occupied in year 0, there were 38 colonizations, 3 extinctions on standing trees, and 5 extinctions caused by tree death in the first year (Snäll *et al.* 2005). Thus, indications are that "deterministic extinctions" caused by the death of host trees are at least as important as "metapopulation extinctions", but even though a considerable number of trees fell, the metapopulation is growing. Once established, epiphytic bryophyte populations seem quite resistant to the extinction causes depicted in classical metapopulation theory. The fact that many extinctions are caused by patch destruction means that extinctions may often occur when the local population is large, rather than striking small populations (as in the classical metapopulation model).

There is also a less dramatic form of dynamics in the patch structure with a successional sequence of bryophytes occupying the tree as it grows and attains a different bark structure. Some early successional species experience a decreasing patch quality as the tree grows, and the trunk becomes inhospitable long before the tree falls (Studlar 1982), unless the bryophyte can move to younger branches of the tree.

The metapopulation concept can be scaled up to model presence/absence of the species in separate forest stands in a larger landscape. In the *Neckera* study, it was possible to define 128 "suitable stands", i.e. forest patches that contained at least some of the deciduous host trees in a 2565 ha conifer-dominated landscape (Snäll *et al.* 2004b). The probability of occurrence of *N. pennata* in a stand was most strongly affected by the densities and stem diameters of the main hosts (*Acer platanoides* and *Fraxinus excelsior*). Also at this scale dispersal limitations were indicated: occupancy increased with increasing connectivity to other stands with *N. pennata*, and it was later found that the occupancy of a number of other epiphytic bryophytes was also affected by host tree densities and diameters as well as by landscape connectivity (Löbel *et al.* 2006a). Logging operations have changed the landscape structure, and to some extent a connectivity measure based on the analysis of 25-year-old air photos could be used as predictor for today's occupancy of *N. pennata* and other epiphytes (Snäll *et al.* 2004b, Löbel *et al.* 2006b). This again indicates that the metapopulations are quite resistant, but also that there is a time lag to extinction, so that these species may be approaching a new equilibrium with lower metapopulation sizes as a result of the decreasing landscape connectivity. Forest landscapes are dynamic also in absence of forestry: in the boreal forests many epiphytes depend on deciduous trees that appear as successional stages after storms and fire.

The understanding of population processes in epiphytes has increased with the use of metapopulation models and spatial statistics. Experimental studies are fewer, but a promising tool is to install tree branches differing in size, species, bark structure, and isolation to test how environment and connectivity affect colonizations. Using an experimental approach in Douglas fir stands in Oregon, Sillett *et al.* (2000) could show that rates of colonization for bryophytes were higher in old growth forests than in young ones (particularly for *Antitrichia curtipendula*).

10.3.3 *Bryophyte metapopulations: a synthesis*

In general terms we have seen that patch duration, patch size, patch configuration, patch quality, and internal dynamics (population dynamics, competition, and small-scale disturbances) can all affect colonizations and extinctions in bryophyte metapopulations. An attempt can be made to compare the importance of different factors that affect colonizations and extinctions and thereby determine the occupancy patterns in different types of patchy substrates (see also discussion in Herben 1994). If the patches are permanent, for instance boulders, extinctions are caused by demographic and environmental stochasticity (as depicted in classical metapopulation models), small-scale within-patch dynamics, or local competition. Deterministic extinctions caused by patch destruction becomes relatively more important with decreasing patch duration from tree stems via fallen logs and smaller parts of dead wood to dung patches and leaves inhabited by epiphyllous bryophytes, and the inhabitants will have to disperse for successful patch-tracking. In most metapopulation models colonizations are limited by dispersal, and this should be relatively more important in patches that do not differ much in habitat quality or local environmental condition. Finally, life-history characteristics have evolved in response to patch dynamics (Southwood 1977); early reproduction is required to cope with rapid patch turnover, and species with efficient dispersal can persist in systems with low connectivity.

10.4 Community patterns and processes

Studies in community ecology focus on processes affecting species composition and richness at local scale. Classical competition theory, founded on Lotka–Volterra dynamics and Gause's principle, states that species coexistence must be based on niche separation, otherwise competitive exclusion will take place and after some time an equilibrium will be reached when the competitively weak species are ousted. While niche separation undoubtedly promotes coexistence, the modern view is that niche-overlapping species often coexist because the competitive equilibrium is not reached. Habitat conditions always change (as in a sequence of wet and dry years, for instance) and this may alter the competitive relations among species. Physical disturbances could open up gaps in the carpet of competitive dominants and promote species with a trade-off to colonization rather than competition. In this section we will look at niche differentiation, disturbance, and competition in bryophyte communities.

10.4.1 *Niche differentiation and coexistence patterns*

Niche separation among bryophyte species has been demonstrated in many cases (Slack 1997), by using techniques such as regression models to demonstrate species response curves (e.g. Gignac *et al.* 1991), calculation of niche breadth and niche separation along one or several habitat gradients (for methods see Soro *et al.* 1999), separation of species in multivariate analysis of environmental factors (Hedderson & Brassard 1990, Nordbakken 1996) and analysis of positive and negative associations of species in small sample plots (Økland 1994). Niche relations are often discussed in terms of the physiological and morphological attributes of the species, but as shown in the metapopulation section, a complete understanding of distribution patterns requires that life-history attributes are also taken into account. Such an approach was taken by Vanderpoorten and Engels (2002). They found that the presence of many bryophytes could be predicted by soil variables (by using logistic regression), and that the predictability of the species was correlated with life-history attributes, such as spore size.

If competition is a strong structuring force, one would expect that bryophyte species should not be able to coexist in close contact. However, Økland (1994)

Table 10.4 *Positive and negative associations between pairs of bryophyte species in a Norwegian boreal spruce forest*

The values for positive associations refer to the proportion (of all potential associations) of species pairs that co-occur in sample plots more often than expected if they were independently distributed, and negative associations indicate the proportion of species pair that co-occur more seldom than expected.

Plot size (m^2)	Positive associations (%)	Negative associations (%)
1	13.1	0.3
1/4	9.6	0.9
1/16	9.8	2.7
1/64	15.8	2.1
1/256	7.8	2.1

Data from Økland (1994).

observed that it is much more common that species are positively than negatively associated, i.e. they co-occur in sample square plots ranging in size from 6.25 cm × 6.25 cm to 1 m × 1 m more often than would be expected if they were distributed independently (Table 10.4). The rarity of negative associations indicates that interspecific competition was not an important structuring force at the scales studied. For several reasons more positive associations are to be expected as the plot size increases. (1) Larger plots will have more environmental variation, enabling species with different niches to inhabit the plot. (2) Facilitation by assisted water balance is discussed above, and enables interstitial species that require a moss matrix as substrate to inhabit the plot. Such positive interaction may counteract the effect of competition (During & Lloret 2001). (3) Resource competition in bryophytes can only occur through direct contacts between gametophytes, and larger plots include shoots that do not interact. The zone of interaction is within one or a few centimeters, in contrast with vascular plants where shading and root interactions can span over several decimeters. Studying plots as small as 13 mm × 13 mm (a relevant size for individual interactions in bryophytes), Wilson *et al.* (1995) found a large number of negative associations indicating that species interactions may affect individual shoots even if it does not structure the community at larger scale.

10.4.2 *Regeneration processes and the role of disturbance*

Since all plants basically need the same resources, it has been difficult to explain all coexistences by niche differentiation. Grubb (1977) introduced the concept of regeneration niche, including all stages of reproduction, dispersal, and establishment. This opens up a whole range of coexistence-promoting mechanisms. Of particular interest is the trade-off between attributes that make a species persistent and competitive in the habitat and attributes that confer dispersal ability. An example that outlines the mechanisms of these relationships is the co-habitation of *Dicranum flagellare* and *Tetraphis pellucida* on decaying logs. In *D. flagellare*, high reproductive allocation to asexual brood branches with high establishment potential gives a successful short-range dispersal, whereas *T. pellucida* puts most of its reproductive effort into spores that enable the species to colonize a larger proportion of more isolated logs (Kimmerer 1994). Kimmerer (1993) used a matrix model to describe the dynamics of *Tetraphis pellucida* (Fig. 10.7). When it occupies an open patch, it

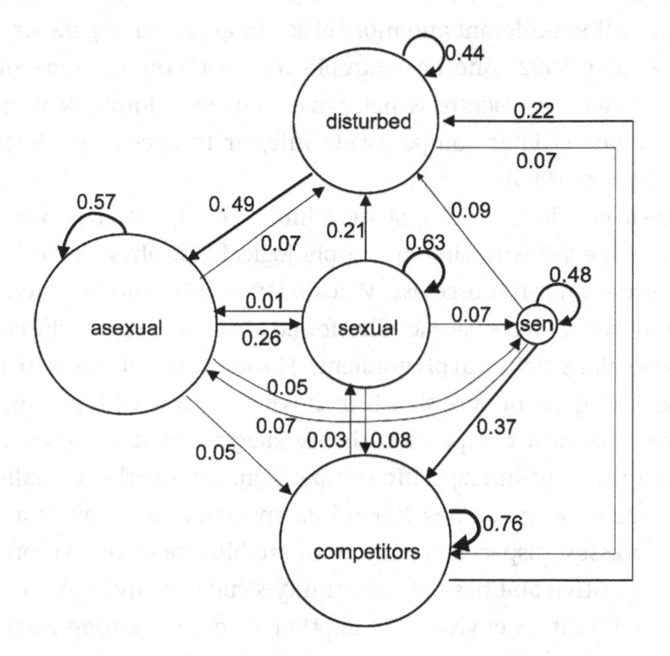

Fig. 10.7. The dynamics in patches of decaying wood in the Adirondack Mountains, New York State, showing the stages of change in the dominant species *Tetraphis pellucida*. Disturbed (vegetation-free) patches are colonized by *T. pellucida*. The colonization starts with an asexual stage that later can develop into sexual and also decline into senescence. All three stages, but particularly the senescent one, can be invaded by competing species, and disturbance can open up for recolonization. The probabilities of annual transition between the stages are shown (also indicated by line thickness), and the proportion of patches of each type (stable stage distribution) is shown by the diameter of the circles. The proportions are strongly dependent on disturbance regime. If disturbances are reduced by 50%, the proportion of patches with competitors will increase from 26% to 47%. Based on data in Kimmerer (1993).

often starts as an asexual colony, which gradually becomes sexual. Such colonies may be quite persistent, but may also be overcome by competitors or wiped out by disturbance. Recurrent disturbances enables the species to maintain dominance in the system, even though it cannot resist local competitive exclusion. It also turns out that the brood branches produced by *D. flagellare* are much more desiccation-tolerant than the gemmae produced by *T. pellucida*. This leads to a niche differentiation that may facilitate species coexistence: *D. flagellare* can colonize the drier parts of log tops, whereas *T. pellucida* is primarily found on moister sides of the logs (Kimmerer & Young 1996).

There are probably many cases in which coexistence of species can be explained by disturbances breaking the competitive dominance. One example is *Conocephalum conicum* and *Fissidens obtusifolius* in riparian habitats. On frequently flooded cliffs *C. conicum* was overgrowing *F. obtusifolius*, but the latter species is more flood-tolerant and more efficient at colonizing disturbed patches (Kimmerer & Allen 1982). Another example of such a "competition–colonization" trade-off is *Aulacomnium palustre*, which can effectively colonize peat by vegetative reproduction, but is later competitively inferior to species of *Polytrichum* and *Sphagnum* (Li & Vitt 1995).

According to classical theories the most intense competition is expected among closely related species with similar morphological and physiological characteristics, and hence similar resource use. Watson (1980, 1981) investigated six coexisting species of the Polytrichaceae. She found that the species differed in their realized niches along light and pH gradients. However, the niches were not broader when the species grew in pure stands than when it grew with potential competitors. Whereas classical competition theory suggests that niche overlap should decrease as a result of interspecific competition, the overlap actually increased with increasing contacts among *Polytrichum* species. Again, this is a case where stochastic processes, dispersal, and order of establishment seem more important than the competitive abilities for community structure, and a strict competitive hierarchy is difficult to envisage. In another case with strong dominance and almost complete cover of bryophytes, the boreal forest floor, niches are also highly overlapping, and it is difficult to explain the spatial occupancy of different species by habitat partitioning (Frego & Carleton 1995a). Similarities in response to habitat conditions leads to symmetric competition among established plants and if all species suffer equally they are unable to outcompete each other. The results are more compatible with the interpretation that local non-equilibrium and stochastic processes rather than competition are structuring the community.

The "intermediate disturbance hypothesis" predicts that species richness in a community should peak at intermediate frequency or severity of disturbance. Without disturbances, a few species tend to dominate, and with too much disturbance only the most effective colonizers will persist. In the boreal forests, dominance of a few species (notably feather mosses) leads to low bryophyte diversity. In small gaps the colonization will reflect the abundance of the adjacent vegetation, and in experiments with gaps of 10 cm diameter in black spruce forest in Ontario, the forest floor dominant *Pleurozium schreberi* was the most successful colonizer by virtue of its high abundance of propagules (Frego 1996). It appears that the gaps must be of certain size to promote diversity. Tree fall, for instance, causes larger gaps with exposed soil where some early-successional species can colonize (Jonsson & Esseen 1990). Such gaps will contribute to overall diversity in the forest, even if dispersal limitation means that each gap or tree fall mound only support few colonizing species (Heinken & Zippel 2004, Kimmerer 2005). During & van Tooren (1988) followed colonization and succession in experimentally cleared plots in Dutch forests. Early colonists were mosses growing in adjacent vegetation which had short-range dispersal by gemmae and spores, and liverworts and perennial mosses followed after the second year. There were no indications that the early-arriving species inhibited the possibilities for the forest floor dominants to regain control of the area.

10.4.3 *Competition studies*

Evidence for interspecific competition in bryophytes comes from (1) studies of niche separation and spatial segregation of species; (2) peat stratigraphy; (3) monitoring of permanent plots during succession or assumed equilibrium conditions; (4) observations of dead remnants of species under the living ones; (5) reciprocal transplant experiments; and (6) experimental assemblages, such as replacement series starting from plant fragments. The evidence from descriptive methods is somewhat circumstantial; experiments are required to convincingly demonstrate the role of competition.

The two main mechanisms for competition among vascular plants are resource depletion in the root zone and shading by foliage. In contrast, bryophytes typically form a monolayer, and the effect of competition is that shoots can be buried. If shoot number is used to monitor interspecific competition it should be combined with a measure of shoot size, since we have seen above that a declining population can survive as a constant number of very small etiolated shoots. In experiments with vascular plants biomass is often used as a response, but in bryophytes the border between living and dead parts is often obscure, sometimes to the point of making "biomass" a useless concept. Growth rate (length or biomass) can be used, but one should be cautious with the use of relative growth rate. It uses initial mass or length as denominator, and this makes it a somewhat dubious concept in bryophytes: a shoot that grows 1 cm (or x mg) in a year will probably do so regardless of whether it was cut to 2 or 4 cm at the start, but the relative growth rate will be very different in the two cases. A practical way to monitor the effects of competition on the community is often to measure the change in area covered by the competing species. Competition is then manifested as a struggle for space (see Crowley *et al.* 2005 for a modeling approach of overgrowth competition), even if the underlying mechanism is a resource competition (light, water, or nutrients).

Detailed niche studies backed up by experiments have been made on *Sphagnum* and other peatland species along gradients of water level and pH (Rydin & Jeglum 2006, Chapter 9, this volume, and references therein). *Sphagnum* offers a system in which interspecific contacts are tight and therefore competition may at least potentially acts as a structuring force. Ten species of *Sphagnum* may coexist in a floating soft carpet community even with little variation in topography (Vitt & Slack 1975), and up to five species in a plot 4 cm in diameter (Rydin 1986). To some degree, differences in fundamental niches and competitive relations along the gradients can explain the distribution patterns (for instance, hollow species cannot grow high up on a hummock because they will dry out and die), but the hummock-former *S. fuscum* seems able to grow also in sites closer to the water table (Rydin 1993a, see also Mulligan & Gignac 2001). However, it appears that several species are competitive equivalents. As discussed above, competition might still be very intense, but it may not lead to competitive exclusion, or at least the competitive replacement may take a very long time (Rydin & Barber 2001). It is likely that colonization order and pre-emptive competition is important for the distribution patterns (Rydin 1993b).

Results from a number of other experimental investigations on interspecific competition are summarized in Table 10.5. Overall, it appears that many communities dominated by bryophytes are characterized by high niche overlap with no clear competitive hierarchy among the community dominants. The bryophyte cover often shows a close mixing of several species. Competitive replacements occur very slowly, if at all, in the closed bryophyte cover. Instead, the species composition is often affected by small-scale disturbances. These are invaded and held by pre-emptive competition by the same species that dominate the closed community (founder control), but when the disturbance is large or drastically alters the substrate (for instance after fire) a different set of species will colonize and gradually be outcompeted by the dominants (dominance control). Anthropogenic changes, notably nitrogen deposition, can shift the competitive relations and lead to dominance of one or a few species, but sometimes the change in environment is more important than competition.

10.4.4 *Interactions with vascular plants*

Being small and lacking roots, bryophytes have a disadvantage in exploitation competition with vascular plants, even though this is partly

Table 10.5 *Examples of experimental studies of interspecific competition in bryophytes*

Habitat (reference)	Approach; species	Result
Grassland (van der Hoeven 1999)	Transplants; *Calliergonella cuspidata, Ctenidium molluscum*	Differences in growth rate among species, but no clear effects of competition.
Calcareous grassland (Zamfir & Goldberg 2000)	Replacement series from fragments; two acrocarps and seven pleurocarps	Differences in biomass growth indicate a competitive hierarchy. Species composition, but not evenness, affected by interactions. No competitive exclusions.
Boreal forest (Frego & Carleton 1995b)	Transplants; feather mosses	Highly overlapping tolerance ranges; differences in growth rate among species, but no clear community effects of competition.
Subalpine forests (Cleavitt 2004)	Fragments and patches of *Mnium arizonicum* transplanted into mats of *Hylocomium splendens*	Dominance control: *Mnium* established well in gaps but was overgrown by *Hylocomium*.
Bog (Rydin 1993a,b, 1997)	Transplants, permanent plots, growth of assemblages in different microhabitats; *Sphagnum* spp.	No competitive exclusion and no clear competitive hierarchy. Some species limited by microhabitat conditions.
Poor fen (Mulligan & Gignac 2001, 2002)	Transplants, effect on the growth of a phytometer (*Aulacomnium palustre*); *Sphagnum* spp. and feather mosses	Feather mosses may be limited by habitat; indications of a competitive hierarchy among *Sphagnum* species.
Bog (Lütke Twenhöven, 1992)	Growth rate and cover with different nitrogen doses; *Sphagnum fallax, S. magellanicum*	In general shoot growth rate increased in *S. fallax* when N was added, but no increase in cover could be detected.
Fen (Kooijman 1993, Kooijman & Kanne 1993, Kooijman & Bakker 1995)	Transplants and replacement series; *Scorpidium scorpioides, Calliergonella cuspidata, Sphagnum subnitens, S. squarrosum*	*Scorpidium* transplanted into *Sphagnum squarrosum* was overgrown after 8 months. However, competitive relations (length growth in monoculture vs. mixed stands) cannot fully explain recent decrease in *Scorpidium scorpioides* and *Sphagnum subnitens* in the Netherlands: the species are largely restricted by the chemical environment, rather than competition.
Decaying wood (McAlister 1995)	Replacement series from fragments; *Anomodon rostratus, Leucobryum albidum, Platygyrium repens*	Biomass growth not affected by interspecific competition.
Peatlands (Li *et al.* 1993)	Pure and mixed cultures; *Sphagnum*	No difference in length or biomass growth between pure and mixed cultures under different phosphate concentrations.
Shore cliffs (Kimmerer & Allen 1982)	Replacement series in greenhouse, permanent plots in field; *Conocephalum conicum, Fissidens obtusifolius*	*C. conicum* showed increased cover in experimental mixture and *F. obtusifolius* decreased relative to pure stands, but this did not lead to competitive replacement in the field.
Oceanic heath (Scandrett & Gimingham 1989)	Replacement series; *Pleurozium schreberi, Hylocomium splendens* and *Hypnum jutlandicum*	In mixtures *H. jutlandicum* was the most successful and *Pleurozium schreberi* the least successful (biomass growth). Mixture yield probably also depends on the relative ability to establish from fragments, and thus reflect pre-emption as well as resource competition.

balanced by their generally lower demands for light and nutrients. The light competition is generally completely asymmetric (not affecting the vascular plant), and often amplified by the accumulation of litter (Chapman & Rose 1991).

In most plant communities it is doubtful whether there is any direct nutrient competition since the uptake zones of roots and gametophytes hardly overlap. In ecosystems with nutrient-poor soils (such as leached podsols) or peat, the situation is different. Here a considerable share of the nutrients enters through precipitation (in bogs, *all* nutrients) or become available in throughfall from trees or from decomposing litter (Oechel & van Cleve 1986, Bayley *et al.* 1987). The nutrients are trapped in the thick mat of feather mosses or *Sphagnum*, and the vascular plants have to rely on nutrients released from decaying basal parts of the mosses (Malmer *et al.* 1994, Svensson 1995). An intriguing balance between light and nutrient interactions was demonstrated in Tamm's classical study of *Hylocomium splendens*: the maximum growth of the moss was at the edge of the spruce crown where shading was not too strong and water and nutrients from crown drip were most readily available (Tamm 1953).

Other cases where nutrient limitations lead to bryophyte dominance can be mentioned. In heathland fire successions Gloaguen (1993) observed *Polytrichum commune* invading and covering the grass *Agrostis curtisii*, and the constant struggle by gardeners trying to keep their lawns free from mosses such as *Rhytidiadelphus squarrosus* and *Pseudoscleropodium purum* is a classic demonstration that bryophytes are not always competitively weaker than vascular plants. These relations become completely changed with increased anthropogenic nitrogen deposition or fertilization. At low doses the bryophytes will increase their growth, and some cover of vascular plants may also reduce desiccation and facilitate the growth of bryophytes (Ingerpuu *et al.* 2005). With higher nutrient doses more nitrogen will reach the root zone, and a circle of positive feedbacks is initiated. The growth of the vascular plants is promoted, and by the increasing shading the bryophytes' growth and capacity to retain nutrients will decline and even more nutrients pass down to the roots. The decrease in bryophyte abundance or richness at high cover of vascular plants has been documented in many cases: bog vegetation with *Sphagnum* (see Rydin & Jeglum 2006 for discussion and further references), fens (Bergamini *et al.* 2001), grasslands (Aude & Ejrnaes 2005), and Tasmanian eucalypt woodlands (Pharo *et al.* 2005). When boreal forests are fertilized to increase timber production, an expansion of grasses is coupled with a drastic decrease in dominant feather mosses, but other bryophytes such as *Brachythecium starkei*, *B. reflexum*, and *Plagiothecium denticulatum* may increase (Strengbom *et al.* 2001).

Direct competition for nutrients may occur between *Sphagnum* and vascular plants with very shallow root systems, such as *Drosera* spp. They could easily be overgrown, and with higher *Sphagnum* growth the vascular plant will have to invest more assimilate into stem biomass (Thum 1988) to keep pace with the moss carpet. Redbo-Torstensson (1994) found that mortality in *Drosera rotundifolia* increased as nitrogen was added, suggesting that the decrease in light availability following the increase in *Sphagnum* growth was more important than nitrogen competition.

Bryophytes often have a negative effect on seed germination and establishment. Examples are *Campylopus introflexus* hampering germination of *Calluna vulgaris* (Equihua & Usher 1993) in British heaths, and bryophyte mats inhibiting germination of non-native species in Australian grasslands (Morgan 2006). Sometimes seedling mortality after germination is the main problem. Scots pine (*Pinus sylvestris*) germinate well in *Sphagnum*, but many seedlings are overgrown by the moss. The seedlings may initially be distributed at random, but surviving pines are only found in bog hummocks where *Sphagnum* height growth is low and aeration is sufficient for the pine roots (Gunnarsson & Rydin 1998). In calcareous grasslands, Zamfir (2000) showed that seedling mortality by burial was high in a moss carpet, except in the grass *Festuca ovina*, which develops a cotyledon that quickly penetrates the moss layer. In the thick feather moss mat in boreal forests, conifer seedlings face the opposite problem. In a study in Sweden, 99% of the spruce seedlings died within a year (Leemans 1991). The most common cause of death was desiccation because the seedlings were unable to grow a primary root sufficiently long to reach the mineral soil beneath the moss carpet, and seedling establishment can increase after heavy grazing in years with peak populations of voles and lemmings (Ericson 1977).

In summary, the interactions between bryophytes and vascular plants are complex, but the bryophyte features listed in Table 10.2 can help to explain under what circumstances bryophytes are likely to be competitive. Especially important for success are the low resource requirements, the long growing season, the capture of nutrients directly into photosynthesizing tissue, and the ability to form thick carpets that monopolize the ground. Further examples of negative and positive interactions are described in Longton (1988).

10.5 Species richness on patchy substrates and islands

Bryophytes on patchy substrates have been used to investigate factors affecting species richness. Boulders, host trees, decaying logs, and forest stands can be viewed as "habitat islands" with conditions contrasting to the surrounding matrix, and the theory of island biogeography is often applied to predict a positive species–area relationship in habitat islands. In several ways this analogy can be questioned. (1) Island biogeography assumes that there is a "mainland": an area in which all species in the system occur and from which they disperse to the islands. In reality it is more common that dispersal occurs between the patches and the patch inhabitants are better described as a set of metapopulations, i.e., a metacommunity (Holyoak *et al.* 2005). (2) Both the island biogeography and the metapopulation theories are applicable for the obligate patch species for which the matrix is inhospitable, but many species have the matrix as their main habitat and appear in the patch only because they expand clonally or from spores from the immediately adjacent matrix. Without a continuous inflow of propagules they would not appear in the patches. This can be referred to as a mass effect (Shmida & Ellner 1984), and is clearly related to the source–sink dynamics discussed above. Since so many bryophytes depend on patchy substrates, the diversity at a larger scale, such as a whole forest stand, also depends on the amount and configuration of suitable substrates within the stand (Berglund & Jonsson 2001, Pharo *et al.* 2004, Mills & Macdonald 2004).

10.5.1 *Species richness on true islands*

Studies of true islands invariably demonstrate a strong species–area relationship, and bryophytes are no exception. According to island biogeography the reason is that larger islands have larger populations of each species, and therefore the extinction risks are lower. However, it is also a universal phenomenon that larger islands have more habitat types and therefore should hold more species, and to separate the effect of area from that of habitat diversity is notoriously difficult. For example, Nakanishi (2001) studied bryophyte species richness in islands off Japan ranging in area from 11 m^2 to 12 000 km^2. The species–area relationship was very strong ($R^2 = 0.98$ for the log–log function). The islands ranged in elevation between 10 and 217 m and there was a strong correlation between elevation and log area ($r = 0.87$), and hence also a strong relation between species richness and elevation ($R^2 = 0.84$). Therefore the effects of habitat diversity (as reflected by elevation) cannot be separated from that of area. The islands ranged in distance from mainland between 50 m and 10 km, but there were no indications of any effects of isolation on species richness. Among vascular plants, it is common that isolation distances of the order of kilometers do not affect species richness (Rydin & Borgegård 1988), and *Sphagnum* diversity on islands in the Baltic Sea seemed unaffected by distances to mainland up to 40 km (Sundberg *et al.* 2006). In an attempt to isolate the effect of island area, Tangney *et al.* (1990) studied bryophyte species richness in sample plots on lake islands in New Zealand. They found a strong effect of island area that demonstrated an effect that was independent of habitat diversity.

Therefore the results are in agreement with island biogeography theory, but again there was almost no effect of isolation. A tentative conclusion from island studies is that species richness of bryophytes as well as vascular plants will almost always depend both on area and habitat diversity, and (at least within kilometer-scale ranges) be independent of distance to mainland.

10.5.2 *Species richness in epiphytes*

A question remains, namely whether it is possible to understand species richness in patchy substrates as simply the sum of the occurrences determined by the metapopulation processes of the different species. The answer would probably be "no" in saturated communities where colonizations and extinctions may depend on the number and identity of other species in the patch, since interspecific competition would modify the metapopulation processes. There may also be facilitation, for instance if bryophyte mats provide water storage and sheltered conditions for desiccation-sensitive species (Ellyson & Sillett 2003). In the sites where *Neckera pennata* was studied, however, large proportions of each tree stem were not covered by epiphytes, and it seems reasonable to try to separate the importance of habitat factors and dispersal for patch species richness in much the same way as described above for the occupancy of *N. pennata*. In this case a main predictor of species richness was tree diameter, but spatial aggregation (indicating dispersal limitation and metapopulation processes) was equally important (Löbel *et al.* 2006b). Total species richness is composed of both obligate and facultative epiphytes, and to gain further understanding of the processes it is necessary to divide the total diversity into functional groups (Fig. 10.8). For facultative epiphytes (generalists in Fig. 10.8), tree diameter and other habitat factors were important, whereas the diversity of obligate epiphytes differed considerably among forest sites and was also influenced by tree species. Most interestingly, among the obligate epiphytes the spatial processes were more important for asexually dispersed species than for those mostly dispersed by spores. It appears that species dispersed by small spores are less limited by dispersal distances, but instead more demanding in the establishment phase: host tree species was the most important diversity predictor for this group. Conversely, vegetative diaspores are larger, and epiphytes dispersed by gemmae or gemma-like branchlets showed stronger spatial aggregation (dispersal limitation), but were less dependent on specific host trees for their establishment.

Following island biogeography theory a strong effect of area on species richness is expected, and also because patch area is an important predictor for occupancy in metapopulations of individual species. However, in this study system tree diameter explained only 12% of the variation in bryophyte species

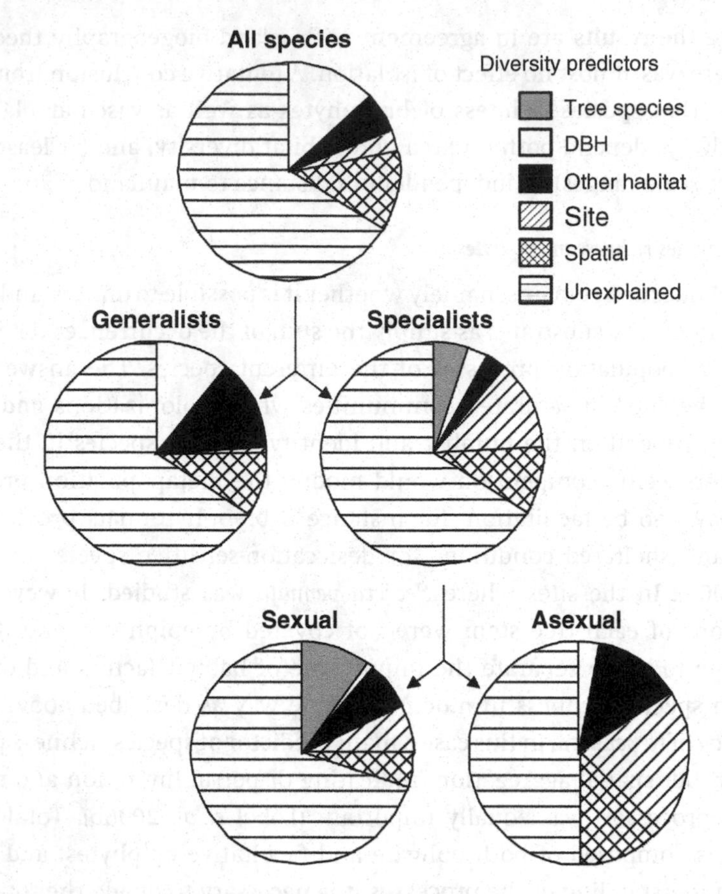

Fig. 10.8. The relative importance of tree species, local habitat factors and spatial factors (related to dispersal) for species richness of bryophytes growing as epiphytes on deciduous tree in a coniferous landscape in eastern Sweden. For generalists (facultative epiphytes), tree diameter at breast height (DBH) and a range of habitat factors were important. In sexual (mostly spore-dispersed) specialists (obligate epiphytes), local species richness differed among tree species, whereas for asexual species (mostly dispersed by gemmae) the importance of spatial factors indicates dispersal limitations. Based on data in Löbel *et al.* (2006b).

richness (Löbel *et al.* 2006b). At a larger scale, this changes drastically: species richness of obligate epiphytes in separate forest stands was strongly related to forest stand area ($R^2 = 0.41$) and even stronger when the number of host trees per stand was used a measure of area ($R^2 = 0.57$) (Löbel *et al.* 2006a). A similar study of epiphytes on *Populus tremula* in the boreal coniferous landscape was made in Finland and adjacent parts of Russia. At local scale, the presence of several species was here favored by larger tree diameter and denser tree cover (probably through the effect on microclimate), and at regional scale the stand diversity increased with increasing abundance of *Populus* (Ojala *et al.* 2000).

10.5.3 *Species richness on boulders*

In a study of bryophyte species richness on boulders in eastern Sweden, Weibull & Rydin (2005) noted a strong species–area relationship. This is consistent with island biogeography theory. However, in addition to (or alternative to) a direct effect of area, the relationship can also be caused by a positive correlation between island area and habitat diversity. In this case species richness was higher on boulders with more variation in microhabitats, e.g. variation in leaf litter cover, smoothness, shape, and amount of fissures. An important factor was also the species identity of the tree above the boulder, with lowest richness under *Picea abies*, intermediate under *Betula pendula* and *Quercus robur*, and highest under *Acer platanoides*, *Ulmus glabra*, and *Fraxinus excelsior* (twice as many species on boulders under the latter three trees compared with *Picea*). The ranking indicates that chemical base saturation of throughfall is important. Hence there were three main determinants of species richness: area *per se*, habitat diversity, and local environment. In a similar study in the northeastern U.S.A. (Kimmerer & Driscoll 2000), there was some effect of microhabitat diversity on species richness, but more important were small-scale disturbances. Small moss patches often fall off, or are washed away. This prevents the dominance of competitive species and creates open spaces for colonization. Thereby, the equilibrium between area-dependent extinctions and isolation-dependent immigrations assumed in island biogeography becomes less important than the disturbance regime. Species richness seems to peak at some intermediate level of disturbance (Weibull & Rydin 2005), and therefore illustrate the intermediate disturbance hypothesis: too frequent or intense levels of disturbance would allow only the most efficient colonists to grow on the boulder, but intermediate levels allow the coexistence of competitive dominants and fugitive species. Boulders often contain many of the ground-dwelling forest floor species, and therefore connectivity effects on species richness are not to be expected. Virtanen & Oksanen (2007) studied cryptogams (bryophytes, lichens and ferns) on erratic calcareous boulders in an area with acidic soils. This is a particularly useful study system for metacommunity patterns, and for the calcicolous species (not growing in the matrix) the species richness was positively affected both by boulder area and connectivity.

10.5.4 *Species richness on decaying wood*

Decaying wood is an important substrate for many bryophytes. Even though many of the species also grow on other substrates (McAlister 1997) there are a number of wood specialists for which the habitat island or metacommunity concepts are applicable. As an example, Andersson & Hytteborn (1991) found 16 epixylic specialists in an old-growth coniferous forest in eastern Sweden but only five in an adjacent managed site. The old-growth site had more dead wood, larger pieces of wood, and more substrate in suitable stages of decay than the managed stand. These variables describe the patch network, but the more humid local climate in old-growth forests may be equally important for many bryophytes (Söderström 1988a). Fallen logs become hospitable when the bark has disappeared and the wood has softened to achieve some water-holding capacity, and there is a clear successional sequence with facultative epiphytes as first colonizers, followed by species that could be classified as early and late epixylics. At a very late stage of decay the wood specialists become overgrown by forest floor mosses (Söderström 1988b), and fallen logs are suitable patches for only a few decades (Hytteborn & Packham 1987, Söderström 1988b). Considering that it takes considerable time for some of the inhabitants to start to reproduce, for instance about nine years in *Ptilidium pulcherrimum* (Jonsson & Söderström 1988), the time to build up high diversity is short.

10.6 Species composition and richness at different temporal and spatial scales

The discussion so far indicates that species composition and richness in the community not only depend on local processes but are also affected by processes operating between communities (e.g. Zobel 1997) as described by the theories of metapopulation biology and island biogeography. We can use the conceptual model in Fig. 10.9 for a discussion on the role of scale-dependent processes for the composition in bryophyte assemblages, and how they appear to differ from those of vascular plants.

With their numerous small spores, bryophytes appear less dispersal-limited than vascular plants at the continental scale. This is witnessed by the large similarity in species composition over continental distances mentioned initially. This also indicates that bryophyte distributions may be less governed by climatic factors, which makes sense considering their tolerance to hazards such as desiccation and frost. The filter between continental and regional scales appears quite coarse.

At the interface between regional and local scales we have seen that the theories of metapopulation biology and island biogeography apply to bryophytes in much the same way as to vascular plants, and dispersal limitations at this scale have been repeatedly demonstrated. Species richness in the community is constrained by the fact that the species do not have the dispersal

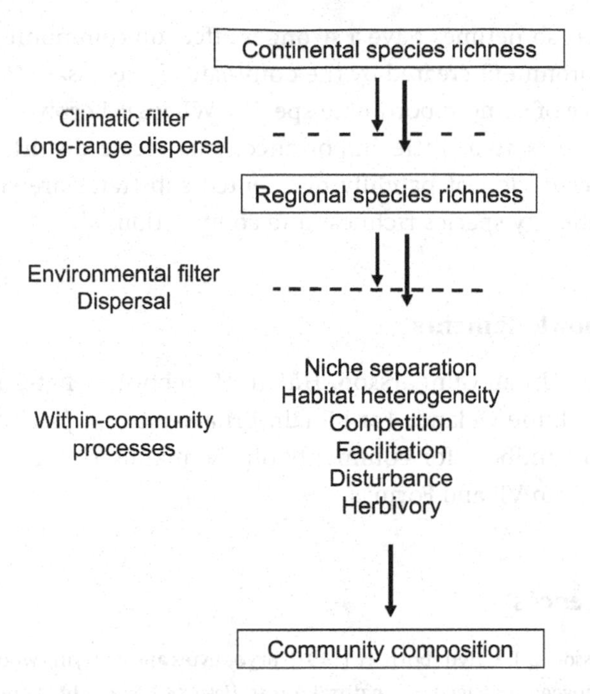

Fig. 10.9. Species composition and diversity depends on processes acting at different temporal and spatial scale. Long-range dispersal over the continental scale can operate over centuries and millennia and explain, for example, species migration after the last glaciation. The local community is affected by other communities in the region as described by metapopulation dynamics and island biogeography. Species are filtered out from the continental to the regional and down to the local scale. Classical equilibrium theories assume that the local species richness is further reduced by competition, unless there is some heterogeneity in habitat conditions that matches a niche differentiation among species. According to non-equilibrium theories, species can be rescued by several mechanisms that prevent or slow down competitive exclusion. Herbivory or fine-scale disturbance may break the dominance of strong competitors and favor competitively weak species with good local dispersal and establishment.

capacity required to counteract local extinctions. The tight connection between the bryophytes and their substrate (in the absence of a spatially integrating root system) makes the environmental filter quite fine-meshed for many species.

Entering into the community level, we have seen cases of coexistence by niche differentiation and a separation among species along microscale environmental gradients. More striking are the many ecosystems dominated by bryophytes in which several species with similar morphology and life history closely coexist. They may well compete intensively but the symmetric nature of the interactions makes competitive exclusion a very slow process. In fact, facilitation may sometimes have a stronger effect on community composition: the moist environment created by the dominants is necessary for the presence and persistence of some subordinate species. Whereas herbivory in most bryophyte assemblages is of little importance, small-scale disturbances and the continuous generation of patchily distributed substrates are strong determinants of community species richness and composition.

Acknowledgments

I thank Urban Gunnarsson, Håkan Hytteborn, Bengt-Gunnar Jonsson, Swantje Löbel, Rune Økland, Tord Snäll, Brita Svensson, Joachim Strengbom, and Sebastian Sundberg for comments on the manuscript. Financial support was obtained from VR and Formas.

References

Andersson, L. I. & Hytteborn, H. (1991). Bryophytes and decaying wood – a comparison between managed and natural forest. *Holarctic Ecology*, **14**, 121–30.

Aude, E. & Ejrnaes, R. (2005). Bryophyte colonisation in experimental microcosms: the role of nutrients, defoliation and vascular vegetation. *Oikos*, **109**, 323–30.

Bates, J. W. (1988). The effect of shoot spacing on the growth and branch development of the moss *Rhytidiadelphus triquetrus*. *New Phytologist*, **109**, 499–504.

Bayley, S. E., Vitt, D. H., Newbury, R. W. *et al.* (1987). Experimental acidification of a *Sphagnum*-dominated peatland: first year results. *Canadian Journal of Fisheries and Aquatic Sciences*, **44** (Suppl. 1), 194–205.

Begon, M., Townsend, C. R. & Harper, J. L. (2006). *Ecology. From Individuals to Ecosystems*, 4th edn. Oxford: Blackwell Science.

Bengtsson, J., Fagerström, T. & Rydin, H. (1994). Competition and coexistence in plant communities. *Trends in Ecology and Evolution*, **9**, 246–50.

Bergamini, A., Pauli, D., Peintinger, M. & Schmid, B. (2001). Relationships between productivity, number of shoots and number of species in bryophytes and vascular plants. *Journal of Ecology*, **89**, 920–9.

Berglund, H. & Jonsson, B. G. (2001). Predictability of plant and fungal species richness of old-growth boreal forest islands. *Journal of Vegetation Science*, **12**, 857–66.

Bisang, I. (1996). Quantitative analysis of the diaspore banks of bryophytes and ferns in cultivated fields in Switzerland. *Lindbergia*, **21**, 9–20.

Bisang, I., Ehrlén, J. & Hedenäs, L. (2004). Mate limited reproductive success in two dioicous mosses. *Oikos*, **104**, 291–8.

Breil, D. A. & Moyle, S. M. (1976). Bryophytes used in construction of bird nests. *Bryologist*, **79**, 95–8.

Caswell, H. (2001). *Matrix Population Models: Construction, Analyses, and Interpretation*, 2nd edn. Sunderland, MA: Sinauer.

Chapman, S. B. & Rose, R. J. (1991). Changes in the vegetation at Coom Rigg Moss National Nature Reserve within the period 1958–86. *Journal of Applied Ecology*, **28**, 140–53.

Cleavitt, N. L. (2001). Disentangling moss species limitations: the role of physiologically based substrate specificity for six species occurring on substrates with varying pH and percent organic matter. *Bryologist*, **104**, 59–68.

Cleavitt, N. L. (2004). Controls on the distribution of *Mnium arizonicum* along an elevation gradient in the Front Ranges of the Rocky Mountains, Alberta. *Journal of the Torrey Botanical Society*, **131**, 150–60.

Clymo, R. S. & Duckett, J. G. (1986). Regeneration of *Sphagnum*. *New Phytologist*, **102**, 589–612.

Collins, N. J. (1976). Growth and population dynamics of the moss *Polytrichum alpestre* in the maritime Antarctic. Strategies of growth and population dynamics of tundra plants 2. *Oikos*, **27**, 389–401.

Cronberg, N. (1993). Reproductive biology of *Sphagnum*. *Lindbergia*, **17**, 69–82.

Cronberg, N. (2002). Colonization dynamics of the clonal moss *Hylocomium splendens* on islands in the Baltic land uplift area: reproduction, genet distribution and genetic variation. *Journal of Ecology*, **90**, 925–35.

Cronberg, N. (2004). Genetic differentiation between populations of the moss *Hylocomium splendens* (Hedw.) Schimp. from low versus high elevation in the Scandinavian mountain range. *Lindbergia*, **29**, 64–72.

Cronberg, N., Natcheva, R. & Hedlund, K. (2006a). Microarthropods mediate sperm transfer in mosses. *Science*, **313**, 1255.

Cronberg, N., Rydgren, K. & Økland, R. H. (2006b). Clonal structure and genet-level sex ratios suggest different roles of vegetative and sexual reproduction in the clonal moss *Hylocomium splendens*. *Ecography*, **29**, 95–103.

Crowley, P. H., Davis, H. M., Ensminger, A. L. *et al.* (2005). A general model of local competition for space. *Ecology Letters*, **8**, 176–88.

Dalen, L. & Söderström, L. (1999). Survival ability of moss diaspores in water – an experimental study. *Lindbergia*, **24**, 49–58.

DeLuca, T. H., Zackrisson, O., Nilsson, A.-C. & Sellstedt, A. (2002). Quantifying nitrogen-fixation in feather moss carpets of boreal forests. *Nature*, **419**, 917–20.

Duckett, J. G. & Clymo, R. S. (1988). Regeneration of bog liverworts. *New Phytologist*, **110**, 119–27.

During, H. J. (1979). Life strategies of bryophytes: a preliminary review. *Lindbergia*, **5**, 2–18.

During, H. J. (1990). Clonal growth patterns among bryophytes. In *Clonal Growth in Plants: Regulation and Function*, ed. J. van Groenendael & H. de Kroon, pp. 153–76. The Hague: SPB Academic Publishing.

During, H. J. (1992). Ecological classification of bryophytes and lichens. In *Bryophytes and Lichens in a Changing Environment*, ed. J. W. Bates & A. M. Farmer, pp. 1–31. Oxford: Oxford University Press.

During, H. J. (2001). Diaspore banks. *Bryologist*, **104**, 92–7.

During, H. J. & Lloret, F. (2001). The species-pool hypothesis from a bryological perspective. *Folia Geobotanica*, **36**, 63–70.

During, H. J. & ter Horst, B. (1983). The diaspore bank of bryophytes and ferns in chalk grassland. *Lindbergia*, **9**, 57–64.

During, H. J. & van Tooren, B. F. (1988). Patterns and dynamics in the bryophyte layer of a chalk grassland. In *Diversity and Pattern in Plant Communities*, 1st edn, ed. H. J. During, M. J. Werger & J. H. Willems, pp. 195–208. The Hague: SPB Academic Publishing.

During, H. J. & van Tooren, B. F. (1990). Bryophyte interactions with other plants. *Botanical Journal of the Linnean Society*, **104**, 79–98.

Ehrlen, J., Bisang, I. & Hedenäs, L. (2000). Costs for sporophyte production in the moss *Dicranum polysetum*. *Plant Ecology*, **149**, 207–17.

Ellyson, W. J. T. & Sillett, S. C. (2003). Epiphyte communities on Sitka spruce in an old-growth redwood forest. *Bryologist*, **106**, 197–211.

Equihua, M. & Usher, M. B. (1993). Impacts of carpets of the invasive moss *Campylopus introflexus* on *Calluna vulgaris* regeneration. *Journal of Ecology*, **81**, 359–65.

Ericson, L. (1977). The influence of voles and lemmings on the vegetation in a coniferous forest during a 4-year period in northern Sweden. *Wahlenbergia*, **4**, 1–114.

Fagerström, T. & Ågren, G. I. (1979). Theory for coexistence of species differing in regeneration properties. *Oikos*, **33**, 1–10.

Filzer, P. (1933). Experimentelle Beiträge zur Synökologie der Pflanzen. I. *Jahrbücher für Wissenschaftliche Botanik*, **79**, 9–130.

Frego, K. A. (1996). Regeneration of four boreal bryophytes: colonization of experimental gaps by naturally occurring propagules. *Canadian Journal of Botany*, **74**, 1937–42.

Frego, K. A. & Carleton, T. J. (1995a). Microsite conditions and spatial pattern in a boreal bryophyte community. *Canadian Journal of Botany*, **73**, 544–51.

Frego, K. A. & Carleton, T. J. (1995b). Microsite tolerance of four bryophytes in a mature black spruce stand: reciprocal transplants. *Bryologist*, **98**, 452–8.

Gignac, L. D., Vitt, D. H. & Bayley, S. E. (1991). Bryophyte response surfaces along ecological and climatic gradients. *Vegetatio*, **93**, 29–45.

Gloaguen, J. C. (1993). Spatio-temporal patterns in post-burn succession on Brittany heathlands. *Journal of Vegetation Science*, **4**, 561–6.

Grubb, P. J. (1977). The maintenance of species richness in plant communities: the importance of the regeneration niche. *Biological Reviews*, **52**, 107–45.

Gunnarsson, U. & Rydin, H. (1998). Demography and recruitment of Scots pine on raised bogs in eastern Sweden and relationships to microhabitat differentiation. *Wetlands*, **18**, 133–41.

Gustafsson, L. & Ahlén, I. (1996). *The National Atlas of Sweden*. Part 16. *Geography of Plants and Animals*. [Also in Swedish: *Växter och djur*.] Stockholm: SNA.

Hanski, I. (1999). *Metapopulation Ecology*. Oxford: Oxford University Press.

Hassel, K. & Söderström, L. (1999). Spore germination in the laboratory and spore establishment in the field in *Pogonatum dentatum* (Brid.) Brid. *Lindbergia*, **24**, 3–10.

Hayward, P. M. & Clymo, R. S. (1983). The growth of *Sphagnum*: experiments on, and simulation of, some effects of light flux and water-table depth. *Journal of Ecology*, **71**, 845–63.

Hedderson, T. A. & Brassard, G. R. (1990). Microhabitat relationships of five co-occurring saxicolous mosses on cliffs and scree slopes in eastern Newfoundland. *Holarctic Ecology*, **13**, 134–42.

Hedenås, H., Bolyukh, V. O. & Jonsson, B. G. (2003). Spatial distribution of epiphytes on *Populus tremula* in relation to dispersal mode. *Journal of Vegetation Science*, **14**, 233–42.

Heegaard, E. (2000). Patch dynamics and/or the species-environmental relationship in conservation bryology. *Lindbergia*, **25**, 85–8.

Heinken, T. & Zippel, E. (2004). Natural re-colonization of experimental gaps by terricolous bryophytes in Central European pine forests. *Nova Hedwigia*, **79**, 329–51.

Herben, T. (1994). The role of reproduction for persistence of bryophyte populations in transient and stable habitats. *Journal of the Hattori Botanical Laboratory*, **76**, 115–26.

Herben, T., Rydin, H. & Söderström, L. (1991). Spore establishment probability and the persistence of the fugitive invading moss, *Orthodontium lineare*: a spatial simulation model. *Oikos*, **60**, 215–21.

Herben, T. & Söderström, L. (1992). Which habitat parameters are most important for the persistence of a bryophyte species on patchy, temporary substrates? *Biological Conservation*, **59**, 121–6.

Holyoak, M., Leibold, M. A. & Holt, R. D. (eds.). (2005). *Metacommunities. Spatial Dynamics in Ecological Communities.* Chicago, IL: University of Chicago Press.

Hytteborn, H. & Packham, J. R. (1987). Decay rate of *Picea abies* logs and the storm gap theory: a re-examination of Sernander plot III, Fiby urskog, central Sweden. *Arboricultural Journal*, **11**, 299–311.

Ingerpuu, N., Liira, J. & Pärtel, M. (2005). Vascular plants facilitated bryophytes in a grassland experiment. *Plant Ecology*, **180**, 69–75.

Jonsson, B. G. (1993). The bryophyte diaspore bank and its role after small-scale disturbance in a boreal forest. *Journal of Vegetation Science*, **4**, 819–26.

Jonsson, B. G. & Esseen, P.-A. (1990). Treefall disturbance maintains high bryophyte diversity in boreal spruce forest. *Journal of Ecology*, **78**, 924–36.

Jonsson, B. G. & Söderström, L. (1988). Growth and reproduction in the leafy hepatic *Ptilidium pulcherrimum* (G. Web.) Vainio during a 4-year period. *Journal of Bryology*, **15**, 315–25.

Kimmerer, R. W. (1993). Disturbance and dominance in *Tetraphis pellucida*: a model of disturbance frequency and reproductive mode. *Bryologist*, **96**, 73–9.

Kimmerer, R. W. (1994). Ecological consequences of sexual versus asexual reproduction in *Dicranum flagellare* and *Tetraphis pellucida*. *Bryologist*, **97**, 20–5.

Kimmerer, R. W. (2005). Patterns of dispersal and establishment of bryophytes colonizing natural and experimental treefall mounds in northern hardwood forests. *Bryologist*, **108**, 391–401.

Kimmerer, R. W. & Allen, T. F. H. (1982). The role of disturbance in the pattern of a riparian bryophyte community. *American Midland Naturalist*, **107**, 370–83.

Kimmerer, R. W. & Driscoll, M. J. L. (2000). Bryophyte species richness on insular boulder habitats: the effects of area, isolation, and microsite diversity. *Bryologist*, **103**, 748–56.

Kimmerer, R. W. & Young, C. C. (1996). Effect of gap size and regeneration niche on species coexistence in bryophyte communities. *Bulletin of the Torrey Botanical Club*, **123**, 16–24.

Kooijman, A. M. (1993). On the ecological amplitude of four mire bryophytes; a reciprocal transplant experiment. *Lindbergia*, **18**, 19–24.

Kooijman, A. M. & Bakker, C. (1995). Species replacement in the bryophyte layer in mires: the role of water type, nutrient supply and interspecific interactions. *Journal of Ecology*, **83**, 1–8.

Kooijman, A. M. & Kanne, D. M. (1993). Effects of water chemistry, nutrient supply and interspecific interactions on the replacement of *Sphagnum subnitens* by *S. fallax* in fens. *Journal of Bryology*, **17**, 431–8.

Laaka-Lindberg, S., Hedderson, T. A. & Longton, R. E. (2000). Rarity and reproductive characteristics in the British hepatic flora. *Lindbergia*, **25**, 78–84.

Laaka-Lindberg, S. & Heino, M. (2001). Clonal dynamics and evolution of dormancy in the leafy hepatic *Lophozia silvicola*. *Oikos*, **94**, 525–32.

Laaka-Lindberg, S., Korpelainen, H. & Pohjamo, M. (2003). Dispersal of asexual propagules in bryophytes. *Journal of the Hattori Botanical Laboratory*, **93**, 319–30.

Leemans, R. (1991). Canopy gaps and establishment patterns of spruce (*Picea abies* (L.) Karst.) in two old-growth coniferous forests in central Sweden. *Vegetatio*, **93**, 157–65.

Li, Y., Glime, J. M. & Liao, C. (1992). Responses of two interacting *Sphagnum* species to water level. *Journal of Bryology*, **17**, 59–70.

Li, Y., Glime, J. M. & Drummer, T. D. (1993). Effects of phosphorus on the growth of *Sphagnum magellanicum* Brid. and *S. papillosum* Lindb. *Lindbergia*, **18**, 25–30.

Li, Y. & Vitt, D. H. (1995). The dynamics of moss establishment: temporal response to a moisture gradient. *Journal of Bryology*, **18**, 677–87.

Lloret, F. (1991). Population dynamics of the coprophilous moss *Tayloria tenuis* in a Pyrenean forest. *Holarctic Ecology*, **14**, 1–8.

Lloret, F. (1994). Gap colonization by mosses on a forest floor: an experimental approach. *Lindbergia*, **19**, 122–8.

Löbel, S., Snäll, T. & Rydin, H. (2006a). Metapopulation processes in epiphytes inferred from patterns of regional distribution and local abundance in fragmented forest landscapes. *Journal of Ecology*, **94**, 856–68.

Löbel, S., Snäll, T. & Rydin, H. (2006b). Species richness patterns and metapopulation processes – evidence from epiphyte communities in boreo-nemoral forests. *Ecography*, **29**, 169–82.

Longton, R. E. (1988). *Biology of Polar Bryophytes and Lichens*, 1st edn. Cambridge: Cambridge University Press.

Longton, R. E. (1992). Reproduction and rarity in British mosses. *Biological Conservation*, **59**, 89–98.

Longton, R. E. (1997). Reproductive biology and life-history strategies. *Advances in Bryology*, **6**, 65–101.

Longton, R. E. & Greene, S. W. (1979). Experimental studies of growth and reproduction in the moss *Pleurozium schreberi* (Brid.) Mitt. *Journal of Bryology*, **10**, 321–38.

Longton, R. E. & Schuster, R. M. (1984). Reproductive biology. In *New Manual of Bryology*, ed. R. M. Schuster, pp. 386–462. Nichinan: The Hattori Botanical Laboratory.

Lütke Twenhöven, F. (1992). Competition between two *Sphagnum* species under different deposition levels. *Journal of Bryology*, **17**, 71–80.

Malmer, N., Svensson, B. M. & Wallén, B. (1994). Interactions between *Sphagnum* mosses and field layer vascular plants in the development of peat-forming systems. *Folia Geobotanica et Phytotaxonomica*, **29**, 483–96.

Marino, P. C. (1991a). Dispersal and coexistence of mosses (Splachnaceae) in patchy habitats. *Journal of Ecology*, **79**, 1047–60.

Marino, P. C. (1991b). Competition between mosses (Splachnaceae) in patchy habitats. *Journal of Ecology*, **79**, 1031–46.

Marino, P. C. (1991c). The influence of varying degrees of spore aggregation on the coexistence of the mosses *Splachnum ampullaceum* and *S. luteum*: a simulation study. *Ecological Modelling*, **58**, 333–45.

Marino, P. C. (1997). Competition, dispersal and coexistence of Splachnaceae in patchy habitats. *Advances in Bryology*, **6**, 241–63.

McAlister, S. (1995). Species interactions and substrate specificity among log-inhabiting bryophyte species. *Ecology*, **76**, 2184–95.

McAlister, S. (1997). Cryptogam communities on fallen logs in the Duke Forest, North Carolina. *Journal of Vegetation Science*, **8**, 115–24.

McGee, G. G. & Kimmerer, R. W. (2002). Forest age and management effects on epiphytic bryophyte communities in Adirondack northern hardwood forests, New York, U.S.A. *Canadian Journal of Forest Research*, **32**, 1562–76.

McQueen, C. B. (1985). Spatial pattern and gene flow distances in *Sphagnum subtile*. *Bryologist*, **88**, 333–6.

Miles, C. J. & Longton, R. E. (1992). Deposition of moss spores in relation to distance from parent gametophytes. *Journal of Bryology*, **17**, 355–68.

Miller, N. G. & McDaniel, S. F. (2004). Bryophyte dispersal inferred from colonization of an introduced substratum on Whiteface Mountain, New York. *American Journal of Botany*, **91**, 1173–82.

Mills, S. E. & Macdonald, S. E. (2004). Predictors of moss and liverwort species diversity on microsites in conifer-dominated boreal forests. *Journal of Vegetation Science*, **15**, 189–98.

Morgan, J. W. (2006). Bryophyte mats inhibit germination of non-native species in burnt temperate native grassland remnants. *Biological Invasions*, **8**, 159–68.

Mulligan, R. C. & Gignac, L. D. (2001). Bryophyte community structure in a boreal poor fen: reciprocal transplants. *Canadian Journal of Botany*, **79**, 404–11.

Mulligan, R. C. & Gignac, L. D. (2002). Bryophyte community structure in a boreal poor fen II: interspecific competition among five mosses. *Canadian Journal of Botany*, **80**, 330–9.

Nakanishi, K. (2001). Floristic diversity of bryophyte vegetation in relation to island area. *Journal of the Hattori Botanical Laboratory*, **91**, 301–16.

Nordbakken, J.-F. (1996). Fine-scale patterns of vegetation and environmental factors on an ombrotrophic mire expanse: a numerical approach. *Norwegian Journal of Botany*, **16**, 197–209.

Oechel, W. C. & van Cleve, K. (1986). The role of bryophytes in nutrient cycling in the Taiga. In *Forest Ecosystems in the Alaskan Taiga*, 1st edn, ed. K. van Cleve, F. S. Chapin III, P. W. Flanagan, L. A. Viereck & C. T. Dyrness, pp. 121–37. New York: Springer-Verlag.

Ojala, E., Mönkkönen, M. & Inkeröinen, J. (2000). Epiphytic bryophytes on European aspen *Populus tremula* in old-growth forests in northeastern Finland and in adjacent sites in Russia. *Canadian Journal of Botany*, **78**, 529–36.

Økland, R. H. (1994). Patterns of bryophyte association at different scales in a Norwegian boreal spruce forest. *Journal of Vegetation Science*, **5**, 127–38.

Økland, R. H. (1995). Population biology of the clonal moss *Hylocomium splendens* in Norwegian boreal spruce forests. I. Demography. *Journal of Ecology*, **83**, 697–712.

Økland, R. H. (1997). Population biology of the clonal moss *Hylocomium splendens* in Norwegian boreal spruce forests. III. Six-year demographic variation in two areas. *Lindbergia*, **22**, 49–68.

Økland, R. H. (2000). Population biology of the clonal moss *Hylocomium splendens* in Norwegian boreal spruce forests. 5. Vertical dynamics of individual shoot segments. *Oikos*, **88**, 449–69.

Økland, R. H. & Økland, T. (1996). Population biology of the clonal moss *Hylocomium splendens* in Norwegian boreal spruce forests. II. Effects of density. *Journal of Ecology*, **84**, 63–9.

Pedersen, B., Hanslin, H. M. & Bakken, S. (2001). Testing for positive density-dependent performance in four bryophyte species. *Ecology*, **82**, 70–88.

Pharo, E. J., Kirkpatrick, J. B., Gilfedder, L., Mendel, L. & Turner, P. A. M. (2005). Predicting bryophyte diversity in grassland and eucalypt-dominated remnants in subhumid Tasmania. *Journal of Biogeography*, **32**, 2015–24.

Pharo, E. J., Lindenmayer, D. B. & Taws, N. (2004). The effects of large-scale fragmentation on bryophytes in temperate forests. *Journal of Applied Ecology*, **41**, 910–21.

Pohjamo, M. & Laaka-Lindberg, S. (2003). Reproductive modes in the epixylic hepatic *Anastrophyllum hellerianum*. *Perspectives in Plant Ecology, Evolution and Systematics*, **6**, 159–68.

Pohjamo, M., Laaka-Lindberg, S., Ovaskainen, O. & Korpelainen, H. (2006). Dispersal potential of spores and asexual propagules in the epixylic hepatic *Anastrophyllum hellerianum*. *Evolutionary Ecology*, **20**, 415–30.

Prins, H. H. T. (1981). Why are mosses eaten in cold environments only? *Oikos*, **38**, 374–80.

Pujos, J. (1994). Systémes de croisement et fécondité chez le *Sphagnum*. *Canadian Journal of Botany*, **72**, 1528–34.

Pyysalo, H., Koponen, A. & Koponen, T. (1983). Studies on entomophily in Splachnaceae (Musci). II. Volatile compounds in the hypophysis. *Annales Botanici Fennici*, **20**, 335–8.

Redbo-Torstensson, P. (1994). The demographic consequences of nitrogen fertilization of a population of sundew, *Drosera rotundifolia*. *Acta Botanica Neerlandica*, **43**, 175–88.

Rincon, E. & Grime, J. P. (1989). Plasticity and light interception by six bryophytes of contrasted ecology. *Journal of Ecology*, **77**, 439–46.

Rochefort, L., Quinty, F., Campeau, S., Johnson, K. & Malterer, T. J. (2003). North American approach to the restoration of *Sphagnum* dominated peatlands. *Wetlands Ecology and Management*, **11**, 3–20.

Ross-Davis, A. L. & Frego, K. A. (2004). Propagule sources of forest floor bryophytes: spatiotemporal compositional patterns. *Bryologist*, **107**, 88–97.

Rydgren, K. & Økland, R. H. (2001). Sporophyte production in the clonal moss *Hylocomium splendens*: the importance of shoot density. *Journal of Bryology*, **23**, 91–6.

Rydgren, K. & Økland, R. H. (2002a). Life-cycle graphs and matrix modelling of bryophyte populations. *Lindbergia*, **27**, 81–9.

Rydgren, K. & Økland, R. H. (2002b). Ultimate costs of sporophyte production in the clonal moss *Hylocomium splendens*. *Ecology*, **83**, 1573–9.

Rydgren, K. & Økland, R. H. (2003). Short-term costs of sexual reproduction in the clonal moss *Hylocomium splendens*. *Bryologist*, **106**, 212–20.

Rydgren, K., Økland, R. H. & Økland, T. (1998). Population biology of the clonal moss *Hylocomium splendens* in Norwegian boreal spruce forests. 4. Effects of experimental fine-scale disturbance. *Oikos*, **82**, 5–19.

Rydgren, K., de Kroon, H., Økland, R. H. & van Groenendael, J. (2001). Effects of fine-scale disturbances on the demography and population dynamics of the clonal moss *Hylocomium splendens*. *Journal of Ecology*, **89**, 395–405.

Rydgren, K., Cronberg, N. & Økland, R. H. (2006). Factors influencing reproductive success in the clonal moss, *Hylocomium splendens*. *Oecologia*, **147**, 445–54.

Rydin, H. (1986). Competition and niche separation in *Sphagnum*. *Canadian Journal of Botany*, **64**, 1817–24.

Rydin, H. (1993a). Mechanisms of interactions among *Sphagnum* species along water-level gradients. *Advances in Bryology*, **5**, 153–85.

Rydin, H. (1993b). Interspecific competition among *Sphagnum* mosses on a raised bog. *Oikos*, **66**, 413–23.

Rydin, H. (1995). Effects of density and water level on recruitment, mortality and shoot size in *Sphagnum* populations. *Journal of Bryology*, **18**, 439–53.

Rydin, H. (1997). Competition between *Sphagnum* species under controlled conditions. *Bryologist*, **100**, 302–7.

Rydin, H. & Barber, K. E. (2001). Long-term and fine-scale co-existence of closely related species. *Folia Geobotanica*, **36**, 53–62.

Rydin, H. & Borgegård, S.-O. (1988). Plant species richness on islands over a century of primary succession: Lake Hjälmaren. *Ecology*, **69**, 916–27.

Rydin, H. & Jeglum, J. K. (2006). *The Biology of Peatlands*. Oxford: Oxford University Press.

Scandrett, E. & Gimingham, C. H. (1989). Experimental investigation of bryophyte interactions on a dry heathland. *Journal of Ecology*, **77**, 838–52.

Shmida, A. & Ellner, S. (1984). Coexistence of plant species with similar niches. *Vegetatio*, **58**, 29–55.

Sillett, S. C., McCune, B., Peck, J. E. & Rambo, T. R. (2000). Four years of epiphyte colonization in Douglas-fir forest canopies. *Bryologist*, **103**, 661–9.

Skotnicki, M. L., Ninham, J. A. & Selkirk, P. M. (2000). Genetic diversity, mutagenesis and dispersal of Antarctic mosses – a review of progress with molecular studies. *Antarctic Science*, **12**, 363–73.

Slack, N. G. (1997). Niche theory and practice: bryophyte studies. *Advances in Bryology*, **6**, 169–204.

Snäll, T., Ribeiro Jr, P. J. & Rydin, H. (2003). Spatial occurrence and colonizations in patch-tracking metapopulations: local conditions versus dispersal. *Oikos*, **103**, 566–78.

Snäll, T., Fogelquist, J., Ribeiro Jr, P. J. & Lascoux, M. (2004a). Spatial genetic structure in two congeneric epiphytes with different dispersal strategies analysed by three different methods. *Molecular Ecology*, **13**, 2109–19.

Snäll, T., Hagström, A., Rudolphi, J. & Rydin, H. (2004b). Distribution pattern of the epiphyte *Neckera pennata* on three spatial scales – importance of past landscape structure, connectivity and local conditions. *Ecography*, **27**, 757–66.

Snäll, T., Ehrlén, J. & Rydin, H. (2005). Colonization-extinction dynamics of an epiphyte metapopulation in a dynamic landscape. *Ecology*, **86**, 106–15.

Söderström, L. (1988a). The occurrence of epixylic bryophyte and lichen species in an old natural and a managed forest stand in northeast Sweden. *Biological Conservation*, **45**, 169–78.

Söderström, L. (1988b). Sequence of bryophytes and lichens in relation to substrate variables of decaying coniferous wood in northern Sweden. *Nordic Journal of Botany*, **8**, 89–97.

Söderström, L. & During, H. J. (2005). Bryophyte rarity viewed from the perspectives of life history strategy and metapopulation dynamics. *Journal of Bryology*, **27**, 261–8.

Söderström, L. & Herben, T. (1997). Dynamics of bryophyte metapopulations. *Advances in Bryology*, **6**, 205–40.

Söderström, L. & Jonsson, B. G. (1989). Spatial pattern and dispersal in the leafy hepatic *Ptilidium pulcherrimum*. *Journal of Bryology*, **15**, 793–802.

Soro, A., Sundberg, S. & Rydin, H. (1999). Species diversity, niche width and species associations in harvested and undisturbed bogs. *Journal of Vegetation Science*, **10**, 549–60.

Southwood, T. R. E. (1977). Habitat, the templet for ecological strategies? *Journal of Animal Ecology*, **46**, 337–65.

Stark, L. R., Mishler, B. D. & McLetchie, D. N. (2000). The cost of realized sexual reproduction: assessing patterns of reproductive allocation and sporophyte abortion in a desert moss. *American Journal of Botany*, **87**, 1599–608.

Steel, J. B., Wilson, J. B., Anderson, B. J., Lodge, H. E. & Tangney, R. S. (2004). Are bryophyte communities different from higher-plant communities? Abundance relations. *Oikos*, **104**, 479–86.

Strengbom, J., Nordin, A., Näsholm, T. & Ericson, L. (2001). Slow recovery of boreal forest ecosystem following decreased nitrogen input. *Functional Ecology*, **15**, 451–7.

Studlar, S. M. (1982). Succession of epiphytic bryophytes near Mountain Lake, Virginia. *Bryologist*, **85**, 51–63.

Sundberg, S. (2002). Sporophyte production and spore dispersal phenology in *Sphagnum*: the importance of summer moisture and patch characteristics. *Canadian Journal of Botany*, **80**, 543–56.

Sundberg, S. (2005). Larger capsules enhance short-range spore dispersal in *Sphagnum*, but what happens further away? *Oikos*, **108**, 115–24.

Sundberg, S. & Rydin, H. (2000). Experimental evidence for a persistent spore bank in *Sphagnum*. *New Phytologist*, **148**, 105–16.

Sundberg, S. & Rydin, H. (2002). Habitat requirements for establishment of *Sphagnum* from spores. *Journal of Ecology*, **90**, 268–78.

Sundberg, S., Hansson, J. & Rydin, H. (2006). Colonisation of *Sphagnum* on land uplift islands in the Baltic Sea: time, area, distance and life history. *Journal of Biogeography*, **33**, 1479–91.

Svensson, B. M. (1995). Competition between *Sphagnum fuscum* and *Drosera rotundifolia*: a case of ecosystem engineering. *Oikos*, **74**, 205–12.

Svensson, B. M., Rydin, H. & Carlsson, B. Å. (2005). Clonal plants in the community. In *Vegetation Ecology*, ed. E. van der Maarel, pp. 129–46. Oxford: Blackwell Science.

Tallis, J. H. (1959). Studies in the biology and ecology of *Rhacomitrium lanuginosum* Brid. II. Growth, reproduction and physiology. *Journal of Ecology*, **47**, 325–50.

Tamm, C. O. (1953). Growth, yield and nutrition in carpets of a forest moss (*Hylocomium splendens*). *Meddelanden från Statens Skogsforskningsinstitut*, **43**, 1–140.

Tangney, R. S., Wilson, J. B. & Mark, A. F. (1990). Bryophyte island biogeography: a study in Lake Manapouri, New Zealand. *Oikos*, **59**, 21–6.

Thomas, P. A., Proctor, M. C. F. & Maltby, E. (1994). The ecology of severe moorland fire on the North York Moors: chemical and physical constraints on moss establishment from spores. *Journal of Ecology*, **82**, 457–74.

Thum, M. (1988). The significance of carnivory for the fitness of *Drosera* in its natural habitat. 1. The reactions of *Drosera intermedia* and *D. rotundifolia* to supplementary feeding. *Oecologia*, **75**, 472–80.

van der Hoeven, E. C. (1999). Reciprocal transplantation of three chalk grassland bryophytes in the field. *Lindbergia*, **24**, 23–8.

van der Hoeven, E. C. & During, H. J. (1997). The effect of density on size frequency distributions in chalk grassland bryophyte populations. *Oikos*, **80**, 533–9.

van der Velde, M., During, H. J., van de Zande, L. & Bijlsma, R. (2001). The reproductive biology of *Polytrichum formosum*: clonal structure and paternity revealed by microsatellites. *Molecular Ecology*, **10**, 2423–34.

Vanderpoorten, A. & Engels, P. (2002). The effects of environmental variation on bryophytes at a regional scale. *Ecography*, **25**, 513–22.

Virtanen, R. & Oksanen, J. (2007). The effects of habitat connectivity on cryptogam richness in a boulder metacommunity system. *Biological Conservation*, **135**, 415–22.

Virtanen, R., Henttonen, H. & Laine, K. (1997). Lemming grazing and structure of a snowbed plant community – a long-term experiment at Kilpisjärvi, Finnish Lapland. *Oikos*, **79**, 155–66.

Vitt, D. H. & Slack, N. G. (1975). An analysis of the vegetation of *Sphagnum*-dominated kettle-hole bogs in relation to environmental gradients. *Canadian Journal of Botany*, **53**, 332–59.

Watson, M. A. (1979). Age structure and mortality within a group of closely related mosses. *Ecology*, **60**, 988–97.

Watson, M. A. (1980). Shifts in patterns of microhabitat occupation by six closely related species of mosses along a complex altitudinal gradient. *Oecologia*, **47**, 46–55.

Watson, M. A. (1981). Patterns of microhabitat occupation of six closely related species of mosses along a complex altitudinal gradient. *Ecology*, **62**, 1067–78.

Weibull, H. & Rydin, H. (2005). Bryophyte species richness on boulders: effects of area, habitat diversity and covering tree species. *Biological Conservation*, **122**, 71–9.

Wiklund, K. (2002). Substratum preference, spore output and temporal variation in sporophyte production in the epixylic moss *Buxbaumia viridis*. *Journal of Bryology*, **24**, 187–95.

Wiklund, K. & Rydin, H. (2004a). Colony expansion of *Neckera pennata*: Modelled growth rate and effect of microhabitat, competition, and precipitation. *Bryologist*, **107**, 293–301.

Wiklund, K. & Rydin, H. (2004b). Ecophysiological constraints on spore establishment in bryophytes. *Functional Ecology*, **18**, 907–13.

Wilson, J. B., Newman, J. E. & Tangney, R. S. (1995). Are bryophyte communities different? *Journal of Bryology*, **18**, 689–705.

Wyatt, R. (1977). Spatial pattern and gamete dispersal distances in *Atrichum angustatum*, a dioicous moss. *Bryologist*, **80**, 284–91.

Zamfir, M. (2000). Effects of bryophytes and lichens on seedling emergence of alvar plants: evidence from greenhouse experiments. *Oikos*, **88**, 603–11.

Zamfir, M. & Goldberg, D. E. (2000). The effect of initial density on interactions between bryophytes at individual and community levels. *Journal of Ecology*, **88**, 243–55.

Zartman, C. E. (2003). Habitat fragmentation impacts on epiphyllous bryophyte communities in central Amazonia. *Ecology*, **84**, 948–54.

Zartman, C. E. & Nascimento, H. E. M. (2006). Are habitat-tracking metacommunities dispersal limited? Inferences from abundance-occupancy patterns of epiphylls in Amazonian forest management. *Biological Conservation*, **127**, 46–54.

Zartman, C. E. & Shaw, A. J. (2006). Metapopulation extinction thresholds in rain forest remnants. *American Naturalist*, **167**, 177–89.

Zobel, M. (1997). The relative role of species pools in determining plant species richness: an alternative explanation of species coexistence? *Trends in Ecology and Evolution*, **12**, 266–9.

11

Bryophyte species and speciation

A. JONATHAN SHAW

11.1 Introduction

The three lineages of bryophytes, mosses, liverworts, and hornworts, compose successful groups of early embryophytes. The mosses are estimated to include some 12 700 species (Crosby *et al.* 2000), the liverworts approximately 6000–8000 extant species (Crandall-Stotler & Stotler 2000, Chapter 1, this volume), and the hornworts about 100–150 species (Chapter 3, this volume). Mosses are comparable in species richness to the monilophytes, which are estimated to include about 11 500 species (Pryer *et al.* 2004). Among the extant land plants, therefore, only the angiosperms are currently more species-rich than are the bryophytes.

It is often stated that bryophytes are most diverse in the tropics and fit the general pattern found in many groups of organisms, with increasing species richness toward the equator (Rosenzweig 1995). However, a quantitative analysis of latitudinal diversity patterns in the mosses failed to detect any such latitudinal gradient, except perhaps a weak one in the Americas (Shaw *et al.* 2005a). It appears that liverwort diversity is highest at moderate to high latitudes of the Southern Hemisphere, although one family, the Lejeuneaceae, is hyperdiverse in wet tropical forests of both the New and Old Worlds (Gradstein 1979).

The fossil record for mosses, liverworts, and hornworts is too incomplete to assess whether these groups were more or less diverse in the geological past (Miller 1984, Oostendorp 1987). Many early Tertiary or even older fossils look quite similar to extant taxa (e.g. Janssens *et al.* 1979), and this interpretation has contributed to the view held by some (e.g. Anderson 1963, Crum 1972) that many or most bryophytes have changed little over vast amounts of time (measured in tens of millions of years, at least), and as such, should be viewed as

Bryophyte Biology: Second Edition, ed. B. Goffinet & A. J. Shaw. Published by Cambridge University Press.

"living fossils". One case of morphological stasis over tens of millions of years is provided by Frahm's (2005) report of plants seemingly identical to *Hypnodontopsis mexicana* from Baltic amber of Eocene age (45–58 million years old). Crum's (1972) claim that mosses represent "unchanging, unmoving sphinxes of the past" that are "evolutionary failures nonetheless well adapted to a modest role in nature" is one of the most oft-quoted speculations in the literature of bryophyte evolutionary biology. Most quotations of Crum's statements are given as introduction to evidence that many bryophytes are in fact highly diverse at the genetic level, in contradiction to the situation expected if they are indeed unmoving, unchanging, living fossils characterized by very low rates of evolution. Moreover, recent studies have documented ecotypic differentiation among populations of some bryophyte species (Shaw 2001), as well as complex interspecific evolutionary patterns that include cryptic speciation, hybridization, and allopolyploidy (see below, and previous reviews by Wyatt *et al.* 1989, Stoneburner *et al.* 1991, Bischler & Boisselier-DuBayle 1997, Shaw 2000a). Interestingly, Stenøien (unpublished) compared nucleotide substitution rates in mosses and seed plants using relative rates tests applied to phylogenies for the groups and found that substitution rates were indeed lower in mosses than in seed plants. The difference held true for analyses of nuclear (26*S*), chloroplast (*rps*4), and mitochondrial (*nad*5) substitution rates.

Nevertheless, there is systematic and phylogenetic evidence that some groups of bryophytes have undergone periods of rapid diversification. Gradstein (1979, 1994, 1997) speculated that the liverwort family Lejeuneaceae, which includes at least 1000 species, has undergone relatively recent spurts of diversification and is actively diversifying in tropical regions. In the first attempt to put a time frame on evolutionary rates in bryophytes, Wall (2005) estimated a rate of 0.56 ($\pm$0.004) new lineages per million years in the moss genus *Mitthyridium*, which is fast even in comparison to estimates for rapidly radiating angiosperm groups (e.g. Baldwin & Sanderson 1998). Shaw *et al.* (2003b) used phylogenetic tree shape to infer relative rates of net diversification (speciation minus extinction) in pleurocarpous mosses and found evidence for an increase in diversification rate associated with the origin of hypnalian pleurocarpous mosses. Newton *et al.* (2007), however, did not detect evidence for an early period of rapid diversification in their study of pleurocarp phylogeny.

Based on a phylogeny including bryophytes and seed plants and calibrated by a date for the origin of embryophytes estimated from fossils, Newton *et al.* (2007) came up with a date for the origin of pleurocarpous mosses of 194–161 million years ago (mya). They estimated the diversification of major pleurocarpous lineages (mainly families) at 165–131 mya. Their calibrated phylogeny suggested that many moss families may have originated in the Cretaceous, but that some families, at least, appear to have diversified more recently. The Hypnodendraceae and Racopilaceae, for example, appear to have originated in the Cretaceous (>100 mya) but *Racopilum* itself may have diversified within the past 20–40 my. Similarly, an origin for Ptychomniaceae was dated at 172–140 mya, but the divergence between *Garovaglia* and *Euptychium* might have been as recent as 28–16 mya. Accuracy of dating estimates for bryophytes is very much limited by the paucity of good calibration by fossils. Wall (2005) used oceanic islands whose geologic history is understood to provide maximum age estimates for endemics.

A similar analysis of liverwort diversification was recently published by Heinrichs *et al.* (2007). They estimated that the Jungermanniopsida (leafy plus simple thalloid liverworts) diverged from the Marchantiopsida (complex thalloid liverworts) approximately 370 mya in the late Devonian period, and that the leafy and simple thalloid lineages diverged about 310 mya, in the late Carboniferous. They further suggested that many of the genera of leafy liverworts diverged as early as the late Cretaceous or early Tertiary. Based on fossils preserved in amber, Hartmann *et al.* (2006) dated the origin of the genus *Bryopteris* (Lejeuneaceae) as minimally 50 mya, during the Cretaceous Period. Using that date and an estimate of the mutation rate for the nuclear ribosomal ITS region, these authors suggested that clades of *Bryopteris* (*B. diffusa* versus *B. filicina* plus *B. gaudichaudii*) diverged in the Miocene, about 15 mya.

These estimates would suggest that many families of bryophytes are quite old, although we still know little about the ages of extant species. Morphologically cryptic species and allopolyploids are now known to be common in mosses and liverworts and we might think of these cases as evidence of recent divergence and speciation, but there is no *a priori* reason to assume that their origins were recent; genetic patterns in some allopolyploids suggest ancient origins (see below).

11.2 Species concepts

If we wish to study the origin of species, we must first ask the question: what are species? The conceptual basis for defining and delimiting species has been the subject of voluminous discussion since (and even before) Darwin (1859) argued that species differ in no fundamental way from infraspecific taxa such as varieties and subspecies. Hey (2001) gleaned from the literature 24 different attempts to define species. One thing that most systematists agree on is that species should on some level be "units of evolution".

11.2.1 *Morphological definitions*

In a survey of taxonomic literature, McDade (1995) noted that most monographers do not discuss species concepts at all. As in most groups of organisms, the most commonly applied approach to delimiting species of bryophytes is morphological: species are groups of individuals or populations that are morphologically distinguishable from other groups. Bryologists rarely have any information beyond "these recurring morphotypes are distinguishable", and therefore represent separate species. Geographic and ecological information can provide useful supplementary data to the extent that distinct distributions and/or ecological characteristics suggest that morphologically defined species have biological meaning as evolutionary entities. Indeed, the observation that two morphotypes differ in their geographic distributions strongly suggests that they have unique evolutionary histories as separate lineages. Morphological variation within and among species results from environmental and genetic differences, and too few studies have employed experimental approaches to determine the extent to which morphological variation is genetically based (but see, for example, Mishler 1985, Såstad 1999, Buryova & Shaw 2005).

11.2.2 *Biological and phylogenetic species concepts*

A flurry of papers discussing alternative species concepts were published during the 1980s and 1990s (e.g. Mishler & Donoghue 1982, Nixon & Wheeler 1990, Baum & Donoghue 1995, McDade 1995, Olmstead 1995, Davis 1997). A main issue surrounds the conceptual (dis)advantages of the so-called biological species concept (BSC, Mayr 1965), versus various phylogenetic concepts. According to the BSC, primacy is given to reproductive isolation as the most important criterion for delimiting species. That is, species comprise gene pools that are reproductively isolated from other gene pools (species). There are a number of methodological and conceptual problems with the BSC. These include the fact that reproductive information is lacking for the vast majority of species on the planet, and it is difficult or impossible to apply the BSC to asexual or nearly asexual taxa (Wheeler & Meier 2000). In addition, reproductive isolation is in many groups a matter of degree rather than all or nothing, and species defined by reproductive isolation sometimes result in taxa that are not congruent with delimitations based on morphology or phylogenetic history.

The development of cladistic methodologies, the ability to easily gather molecular data, and the conceptual framework of phylogenetic systematics, has yielded a number of (phylogenetic) tree-based species concepts. Proponents of phylogenetic species agree about the use of phylogenetic methods and reconstructing ancestor–descendant relationships for "discovering" the units we call species, but differ in exactly what criterion is most appropriate for delineating species in practice. An important point of disagreement among those supporting some form of phylogenetic species concept is whether species have to be monophyletic.

Aside from the conceptual controversies surrounding which variant of the phylogenetic species concept is most appropriate, there is every reason to expect that many "good" species in nature may not be monophyletic. The basic coalescence model is a relatively new paradigm in population genetics focused on reconstructing the ancestry of allele copies sampled from a population. Coalescence theory tells us that in the absence of mitigating processes such as changes in population size, natural selection, mutation, or population structure, coalescence of allele copies backward in time to the most recent common ancestor (MRCA) of all extant copies through the process of genetic drift (the so-called coalescence time) takes $2N$ generations in a diploid, where N is the population size (Hudson 1990). Thus, coalescence times are proportional to the population size and the absolute time to coalescence depends on generation time. It is significant that the time to coalescence is faster in a small population than in a large population. A population that has gone through a bottleneck may be a "small" population in terms of genetic characteristics, even if the census population size subsequently rebounded.

Molecular data gathered for the purpose of delineating species consist of allele copies sampled from individuals representing putative species. Resolution of monophyletic groups of allele copies (which are sequenced to represent the species) depends on the amount of time since the species have exchanged genes and the effective population sizes of the two divergent taxa. For any particular gene, allele copies from a population (or species) should coalesce to the MRCA after $2N$ generations in the case of a diploid, or $1N$ generation for a locus with uniparental inheritance (i.e. chloroplast or mitochondrial genes).

Rosenberg (2003) found that we should expect allele trees sampled from recently diverged taxa to show polyphyly until about $1.3N$ generations after isolation, paraphyly from $1.3N$ to $1.6N$ generations, and only after $1.6N$ generations should we expect an allele tree to indicate reciprocal monophyly. Those numbers are for a randomly selected gene, but studies nowadays generally utilize sequence data from multiple genes to infer species status. Reciprocal monophyly for 99% of sampled genes is expected after approximately $5.3N$ generations (Rosenberg 2003). As time since divergence increases, the required sample size (number of individuals or populations) required to demonstrate monophyly decreases. Two species that have been reproductively isolated for many generations and represent independently evolving lineages may not display reciprocal monophyly for genes sampled in a systematic study. In such cases, allele trees may not reflect the underlying species trees. Rosenberg &

Nordberg (2002) provided a readable overview of coalescence theory as it relates to the sorts of questions frequently addressed by systematists.

The likelihood of resolving reciprocally monophyletic species in a phylogenetic analysis depends in part on the mode and timing of speciation. In the case of a long-distance founder, monophyly of the derived taxon might occur relatively quickly because of the population size bottleneck through which it passes, but phylogenetically it would be seen as nested within the ancestral, larger population. The ancestral (source) population would appear paraphyletic until it accumulates enough mutations to be resolved as monophyletic relative to the derived founder population. If speciation through peripheral isolation or long-distance dispersal is common, we might expect to find many paraphyletic species in molecular systematic analyses. When speciation occurs because an ancestral taxon is divided into relatively large population fragments, coalescence times would be long and neither derived taxon might appear monophyletic from analyses of gene trees. As per inferences made by Rosenberg (2003, see above), such species would initially be seen as polyphyletic in terms of individual gene trees, then, over time, paraphyletic, and only later, reciprocally monophyletic. Factors such as population structure in the widespread ancestral species, selective sweeps, or changes in population size since isolation can also affect the coalescence process.

These considerations suggest that many species would appear to be non-monophyletic when studied with a limited sample of genes and during one slice of time. Based on a survey of published species-level phylogenetic analyses for plants, Crisp & Chandler (1996) found that more than 30% of commonly recognized species appear to be paraphyletic (see also Funk & Omland 2003). Paraphyletic bryophyte species were resolved by Shaw (2000a), and Shaw & Allen (2000).

It is important to keep in mind that molecular evidence that two taxa are *not* genetically differentiated (or monophyletic) is negative evidence, and it can always be argued that additional data would demonstrate them to be differentiated. Taxonomic decisions based on short sequences from just one genomic region should be viewed cautiously. So-called DNA bar-coding has become popular in recent years and shows much promise for rapid species identification in some groups. Bar-coding is based on comparing short sequences from an unknown to a database of homologous sequences from known taxa. As an identification tool, bar-code sequences have value, but they are of limited utility and can be positively misleading for discovering new species and for making evolutionary inferences because of the stochastic nature of the coalescent process and the limited information content of short sequences.

11.3 Bryophyte species delimitation based on molecular markers

A variety of molecular tools have been applied to evolutionary and systematic problems in bryology. Most such studies have focused on deeper relationships among genera and families, but genetic data have also been applied to species-level problems. Isozyme markers and DNA-based methods provide valuable information that is complementary to – but not a good substitute for – careful morphological studies. Isozyme-based research is briefly summarized below and papers aimed at resolving species-level systematic problems are listed in Table 11.1. Since DNA-based work on resolving species has not been previously summarized, these studies are discussed in somewhat more detail, but are also listed in Table 11.1. Papers documenting allopolyploid speciation in bryophytes are discussed separately and listed in Table 11.2.

11.3.1 *Isozyme-based studies*

Isozymes, defined broadly as allelic variants of enzyme-encoding genes, have been and continue to be important sources of information about infraspecific population structure and genetic relationships among species. Isozymes have the important characteristic of being co-dominant markers (both alleles are expressed in heterozygotes), making them valuable for studies of hybridization and allopolyploidy. They are also relatively inexpensive and once the protocols are worked out for a given taxon, they are highly reproducible. Disadvantages of isozymes relative to DNA-based methods is that isozyme analyses require living material, and levels of variation may be lower than with some DNA-based fingerprinting methods.

Table 11.1 provides a list of studies that have addressed species-level taxonomic problems using isozyme markers. The table does not include papers focused on population structure and geographic patterns within individual species rather than the genetic delimitation of species. Nor does Table 11.1 list papers that resolve relationships among species with no or only limited sampling within species. Earlier isozyme studies were reviewed by Wyatt *et al.* (1989), Stoneburner *et al.* (1991), and Bischler & Boisselier-DuBayle (1997).

Isozyme data are not amenable to phylogenetic analyses of ancestor-descendant relationships, but phenograms provide insight into genetic differences among species and conspecific populations. Several generalities have emerged from isozyme analyses of bryophytes. One is that morphological variation does not always reveal the underlying genetic structure of species. Morphologically cryptic or nearly cryptic species have been documented in

Table 11.1 *Published studies that have utilized isozymes and DNA-based information to address problems of species delimitation in mosses and liverworts*

Taxonomic group	Marker	Reference(s)
Amblystegium	DNA sequences	Vanderpoorten *et al.* 2001, 2004
Anacolia menziesii – A. webbii	ISSRs, DNA sequences	Werner *et al.* 2003
Aneura pinguis	isozymes	Szweykowski & Odrzykoski 1990
Calypogeia	isozymes	Buczkowska 2004, Buczkowska *et al.* 2004
Campylopus pilifer, C. introflexus	DNA sequences	Stech & Dohrmann 2004
Cinclidotus	isozymes	Ahmed & Frahm 2003
Climacium americanum – C. dendroides	isozymes	Shaw *et al.* 1994
Climacium americanum – C. kindbergii	isozymes	Shaw *et al.* 1987
Conocephalum conicum	isozymes	Akiyama & Hiraoka 1994, Itouga *et al.* 1999, Odrzykoski 1987, Odrzykoski *et al.* 1981, Odrzykoski & Szweykowski 1991, Szweykowski *et al.* 1981b, Szweykowski & Krzakowa 1979
Dicranoloma	DNA sequences	Stech *et al.* 2006a
Eurhynchium crassinervium	DNA sequences	Frahm *et al.* 2000
Eurhynchium, Rhynchostegiella, Rhynchostegium	DNA sequences	Stech & Frahm 1999b
Fontinalis antipyretica group	DNA sequences	Shaw & Allen 2000
Herbertus	DNA sequences	Feldberg & Heinrichs 2006
Herbertus borealis	DNA sequences	Feldberg & Heinrichs 2005
Herbertus sendtneri	DNA sequences	Feldberg *et al.* 2004
Hymenophyton	DNA sequences	Pfeiffer 2000b
Hypnobartlettia fontana	DNA sequences	Stech *et al.* 1999
Hypopterygium "rosulatum"	DNA sequences	Pfeiffer 2000a
Hypopterygium tamarisci complex	DNA sequences	Pfeiffer *et al.* 2000
Isothecium	DNA sequences	Ryall *et al.* 2005
Jensenia	DNA sequences	Schaumann *et al.* 2004
Leucobryum	DNA sequences	Vanderpoorten *et al.* 2003a
Leucobryum albidium – L. glaucum	PCR-RFLP	Patterson *et al.* 1998
Leucodon	isozymes	Akiyama 2004
Lunularia	isozymes	Boisselier-Dubayle *et al.* 1995a
Marchantia polymorpha	isozymes, PCR-RFLP, RAPDs	Boisselier-Dubayle *et al.* 1995b
Mielichhoferia elongata – M. mielichhoferiana	isozymes, DNA sequences	Shaw 1994, 1998, 2000b
Mnium orientale– M. hornum	isozymes	Wyatt *et al.* 1997
Mitthyridium	DNA sequences	Wall 2005

Table 11.1 (*cont.*)

Taxonomic group	Marker	Reference(s)
Neckera	isozymes	Appelgren & Cronberg 1999
Orthotrichum freyanum	DNA sequences	Goffinet *et al.* 2007
Pellia	DNA sequences	Fiedorow *et al.* 2001
Pellia endiviifolia	isozymes	Szweykowski 1984, Szweykowski *et al.* 1981b, 1995, Zielinski 1987
Pellia epiphylla complex	isozymes	Pacek *et al.* 1998, Pacek & Szweykowska-Kulinska 2003, Zielinski 1987, Zielinski *et al.* 1985
Philonotis	isozymes	Buryova 2004
Plagiochila	DNA sequences	Groth *et al.* 2003, 2004, Heinrichs *et al.* 2002a,b, 2003, 2005
Plagiochila carringtonii	DNA sequences	Renker *et al.* 2002
Plagiochila cucullifolia var. *anomala*	DNA sequences	Heinrichs *et al.* 2003
Plagiochila detecta	RAPDs	So & Grolle 2000
Plagiochila virginica	DNA sequences	Heinrichs *et al.* 2002a
Platyhypnidium riparioides – *P. mutatum*	DNA sequences	Stech & Frahm 1999a
Polytrichum	microsatellites	van der Velde & Bijlsma 2000
Polytrichum	RAPDs	Zouhair *et al.* 2000
Polytrichum commune	isozymes	Bijlsma *et al.* 2000
Polytrichum commune – *P. uliginosum*	microsatellites	van der Velde & Bijlsma 2004
Porella	isozymes, RAPDs	Boisselier-Dubayle & Bischler 1994
Porella	isozymes	Boisselier-Dubayle *et al.* 1998a, Bischler *et al.* 2006
Porella platyphylla – *P. platyphylloidea*	isozymes	Therrien *et al.* 1998
Preissia quadrata	isozymes	Boisselier-Dubayle & Bischler 1997
Ptilidium	isozymes	Adamczak *et al.* 2005
Pyrrhobryum mnioides	DNA sequences	McDaniel & Shaw 2003
Racopilum	isozymes	Vries *et al.* 1983
Reboulia	isozymes	Boisselier-Dubayle *et al.* 1998b
Rhynchostegium, Rynchostegiella	DNA sequences	Stech & Frahm 1999b
Rhytidiadelphus	ISSRs, DNA sequences	Vanderpoorten *et al.* 2003b
Riccia	isozymes	Dewey 1989
Schizymenium shevockii	PCR-RFLP	Shaw 2000c
Sphagnum cuspidatum, S. viride	isozymes	Hanssen *et al.* 2000
Sphagnum rubellum – *S. capillifolium*	isozymes	Cronberg 1987, 1989, 1998
Sphagnum subsecundum complex	isozymes	Melosik *et al.* 2005

Table 11.1 (*cont.*)

Taxonomic group	Marker	Reference(s)
Sphagnum (3 species)	isozymes, RAPDs	Stenøien & Såstad 1999
Sphagnum capillifolium – *S. quinquefarium*	isozymes	Cronberg & Natcheva 2002
Sphagnum ehyalinum	DNA sequences	Shaw & Goffinet 2000
Sphagnum macrophyllum – *S. cribrosum*	DNA sequences	Zhou & Shaw 2008
Sphagnum mirum – *S. tundrae*	isozymes	Flatberg & Thingsgaard 2003
Sphagnum pylaesii	DNA sequences	Shaw *et al.* 2004
Sphagnum recurvum complex	isozymes	Såstad *et al.* 1999b
Sphagnum sect. *Acutifolia*	DNA sequences	Shaw *et al.* 2005b
Sphagnum sect. *Subsecunda*	DNA sequences	Shaw *et al.* 2005c
Symphogyna	DNA sequences	Schaumann *et al.* 2003
Syntrichia laevipila complex	ISSRs	Gallego *et al.* 2005
Targionia	isozymes	Boisselier-Dubayle & Bischler 1999
Timmia	DNA sequences	Budke & Goffinet 2006
Tortula subulata complex	DNA sequences	Cano *et al.* 2005
Weissia wimmeriana	ISSRs	Werner *et al.* 2004
Weymouthia	DNA sequences	Quandt *et al.* 2001

both moss and liverwort genera (Shaw 2001). Early studies by Polish workers (e.g. Szweykowski & Krzakowa 1979, Szweykowski *et al.* 1981b) demonstrated that widespread species of the liverwort genera *Conocephalum* and *Pellia* in fact consist of morphologically cryptic or nearly cryptic taxa that appear to represent non-recombining gene pools, and subsequent examples of cryptic species have been documented in *Calypogeia, Marchantia, Riccia*, and *Reboulia*, among others. Morphologically cryptic or nearly cryptic species resolved by isozymes have also been discovered in mosses (e.g. *Climacium* [Shaw *et al.* 1987, Shaw *et al.* 1994], *Neckera* [Appelgren & Cronberg 1999], *Polytrichum* [Bijlsma *et al.* 2000]).

A second general observation from accumulating isozyme studies is that congeneric species tend to be more genetically divergent from one another than is typical of seed plants (Gottlieb 1981, Wyatt *et al.* 1989). Whereas the mean genetic identity (Nei 1972) between congeneric seed plants is 0.67, many congeneric bryophyte species are more genetically differentiated. Indeed, even some pairs of morphologically cryptic species exhibit genetic identities of less than 0.50 (Wyatt *et al.* 1989, and papers listed in Table 11.1). Reasons for this may be several, including possibly older ages for many bryophyte species compared with species of seed plant genera. It should also be kept in mind that the taxonomic level

Table 11.2 *Putative allopolyploid species of bryophytes*

Species names in bold indicate the plastid DNA parent, presumed to be maternal.

Genus	Polyploid name	Technique	Parent(s)	Origins	Reference(s)
Calypogeia	*C. azurea, C. muelleriana, C. sphagnicola*	isozymes	unknown		Buczkowska *et al.* 2004
Corsinia	*C. coriandrina*	isozymes	*C. coriandrina* × unknown	>1	Boisselier-Dubayle & Bischler 1998
Pellia	*P. borealis*	isozymes, DNA sequences	***P. epiphylla*** "N" × *P. epiphylla* "S"	>2	Odrzykoski *et al.* 1996, Fiedorow *et al.* 2001, Pacek & Szweykowska-Kulinska 2003
Plagiochasma	*P. rupestre*	isozymes		?	Boisselier-Dubayle *et al.* 1996
Plagiomnium	*P. medium*	isozymes	***P. insigne*** × *P. ellipticum*	>3	Wyatt *et al.* 1988, 1992
Plagiomnium	*P. cuspidatum*	isozymes	*P. acutum* × unknown	?	Wyatt & Odrzykoski 1998
Plagiomnium	*P. curvatulum*	isozymes	*P. ellipticum* × *P. elatum?*	?	Wyatt *et al.* 1993a
Polytrichastrum	*P. pallidisetum, P. sexulare, P. ohioense*	isozymes	unknown	?	Derda & Wyatt 2000
Polytrichum	*P. longisetum*	isozymes, microsatellites	*P. formosum?* × unknown	?	van der Velde & Bijlsma 2001
Porella	*P. baueri*	isozymes	***P. cordaeana*** × *P. platyphylla*	>1	Boisselier-Dubayle *et al.* 1998a, Jankowiak & Szweykowska-Kulinska 2004
Reboulia	*R. queenslandica*	isozymes	*R. hemisphaerica* v. *hemisphaerica* × *R. h.* v. *orientalis*	1?	Boisselier-Dubayle *et al.* 1998b
Rhizomnium	*R. pseudopunctatum*	isozymes	***R. magnifolium*** × *R. gracile*	?	Wyatt *et al.* 1993b
Sphagnum	*S. russowii*	isozymes, DNA sequences	*S. girgensohnii* × *S. rubellum* or *S. quinquefarium (S. warnstorfii)*	>1	Cronberg 1996, Shaw *et al.* 2005b
Sphagnum	*S. majus*	isozymes	***S. cuspidatum*** × *S. ? annulatum*	?	Såstad *et al.* 2000
Sphagnum	*S. troendelagicum*	isozymes	*S. tenellum* × *S. balticum*	?	Såstad *et al.* 2001
Sphagnum	*S. auriculatum, S. carolinianum, S. inundatum, S. lescurii*	DNA sequences, microsatellites	*S. subsecundum* × unknown	>1	A.J. Shaw, unpublished
Sphagnum	*S. jensenii*	isozymes	*S. balticum* × *S. annulatum*	?	Såstad *et al.* 1999a
Targionia	*T. lorbeeriana*	isozymes	*T. hypophylla* × unknown / unknown × unknown	>2	Boisselier-Dubayle & Bischler 1999

we call genus is a human construct and "genera" really are not comparable in bryophytes and seed plants.

11.3.2 *DNA-based studies*

DNA-based methods include nucleotide sequencing as well as various "DNA fingerprinting" approaches. Fingerprinting methods that have been used to address systematic problems in the bryophytes include random amplified polymorphic DNA (RAPDs, Welsh & McClelland 1990), amplified fragment length polymorphism (AFLPs, Vos *et al.* 1995), inter-simple sequence repeats (ISSRs, Zietkiewicz *et al.* 1994), and microsatellites (also known as simple sequence repeats). One advantage of DNA-based methods over isozymes is that dried herbarium collections can be used to more easily sample from across the ranges of taxa without conducting additional and often expensive fieldwork.

RAPD markers were developed in the 1990s and have been used extensively in plant and animal breeding, as well as in studies of population genetics. Much concern has been expressed, however, about the reproducibility of RAPD data. The journal *Molecular Ecology*, for example, cautions that RAPD-based studies are discouraged for submission because of such concerns (www.blackwellpublishing.com). AFLP markers are generally reproducible, but artefacts can arise, as with RAPDs (Stevens *et al.* 2007), because of contamination of samples by microorganisms such as fungi. Recent work (e.g. Davis *et al.* 2003) has shown that liverworts contain multiple species of fungi living endosymbiotically within the gametophytes, and the same is true for mosses (A. J. Shaw, unpublished). It has been suggested that high levels of apparent genetic variation detected in some mosses could result at least in part from external and/or endophytic fungi (Stevens *et al.* 2007).

Microsatellite primers, in contrast, are specifically designed for the organisms under study and are highly unlikely to amplify endophytes or other contaminating organisms. Also, unlike RAPDs, AFLPs, and ISSRs, microsatellites are codominant markers so heterozygotes can be distinguished from homozygotes. This feature is especially valuable to studies of mating patterns, hybridization, and polyploidy. Korpelainen *et al.* (2007) described a relatively simple (and therefore economical) method for developing microsatellite markers. RAPDs, AFLPs, and ISSRs provide highly polymorphic data because the number of loci that can be investigated is virtually unlimited. Much fewer microsatellite loci are typically investigated, but high resolving power is gained because each locus is multi-allelic. In a group of closely related *Sphagnum* species, we (B. Shaw & A. J. Shaw, unpublished) resolve up to ten alleles at some loci. One potential problem with microsatellite loci is the possibility of size homoplasy: the independent evolution of indistinguishable alleles because of high microsatellite mutation rates (Estoup *et al.* 2002).

DNA sequence data are more expensive than fingerprint data to gather, but have been used for species-level research in both bryophytes and vascular plants. Researchers have three genomes from which to choose markers: mitochondrial (mtDNA), chloroplast (cpDNA), and nuclear (nDNA). Nacheva & Cronberg (2007b) recently confirmed that both mitochondrial and chloroplast DNA are inherited through the maternal parent in *Sphagnum*. Mitochondrial genes tend to be relatively conserved and are rarely useful at the population and species levels in plants (in contrast to animals, where mtDNA is the preferred marker for such studies). No species-level studies based on mtDNA sequences has been published, although mtDNA genes have been utilized for phylogenetic problems involving genera, families, and orders (e.g. Cox *et al.* 2004). cpDNA exhibits moderate levels of variation, and can be useful for separating species and even for resolving phylogenetic patterns within species. The two most common regions are *trn*L–*trn*F, which includes both intron and non-coding spacer sequences as well as a small portion of coding region (Quandt & Stech 2004) and *trn*G, which includes non-coding intron sequences (Pacak & Szweykowska 2003). Coding genes such as *rbc*L are generally too conserved to be useful at the intraspecific level. The most common nuclear region for species- and population-level research is the internal transcribed spacer (ITS) of the ribosomal RNA repeat (nrDNA) (Vanderpoorten *et al.* 2006). Other nuclear genes (or, more commonly, introns within them) that have been used for species-level bryophyte studies include glyceraldehyde 3-phosphate dehydrogenase (*gpd*: Wall 2005), adenosine kinase (*adk*: Vanderpoorten *et al.* 2004, McDaniel & Shaw 2005), phytochrome (*phy*: McDaniel & Shaw 2005), glyceraldehyde 3-phosphate dehydrogenase (*GapC*: Szövényi *et al.* 2006), and *LEAFY/FLO* (Shaw *et al.* 2004, 2005a). Shaw *et al.* (2003a) developed anonymous nuclear regions from which sequence data have proven informative for resolving relationships between and within *Sphagnum* species. These regions were developed specifically for *Sphagnum*; however, the protocol outlined in Shaw *et al.* (2003a) could be applied to any other genus.

Following up on research based on isozymes, Pacek *et al.* (1998) found that RAPD markers clearly resolve Polish populations of two morphologically cryptic forms *of Pellia epiphylla*. The two forms are less differentiated from one another than between *P. epiphylla* and *P. neesiana*, which are morphologically distinguishable, but appear to represent different gene pools.

Stenøien & Såstad (1999) used RAPD markers, along with isozymes, in a study of *Sphagnum angustifolium, S. lindbergii, S. fallax*, and *S. isoviitae*. Of the four species, only *S. angustifolium* varied at isozyme loci, and this species was the most variable for RAPD markers. *Sphagnum fallax* and *S. isoviitae* could not be distinguished by either marker type, although populations within *S. fallax* were polymorphic and differentiated for RAPDs. The authors considered *S. isoviitae* and *S. fallax* to be conspecific based on their results.

Såstad *et al.* (1999b) used isozyme markers and RAPDs to examine genetic relationships among species in the so-called *Sphagnum recurvum* complex. Both types of data resolved two groups, one characterized by brown and the other by yellow spores. The two populations each of *S. angustifolium*, *S. flexuosum*, and *S. recurvum* (brown spores) grouped together, but no such grouping could be discerned in a complex that includes *S. fallax*, *S. brevifolium*, and *S. isoviitae* (yellow spores). The authors consequently questioned whether the latter represent different species.

Werner *et al.* (2004) used ISSR markers to compare the common, widespread species *Weissia controversa* and the rare congener *W. wimmeriana*. The two species are clearly differentiated, and the rare *W. wimmeriana* was relatively depauperate in genetic variability.

Patterson *et al.* (1998) used a PCR–RFLP technique, in which the ITS region is amplified and then cut with restriction enzymes, for a genetic study of *Leucobryum albidum* and *L. glaucum* in a mixed population where colonies of these two clump-forming taxa seemed to grade from one species to another. The main difference between the two species is size, and in particular, the length of the leaves. Although the distribution of leaf length measurements taken from plants was continuous with no hint of bimodality, two ITS genotypes were found at the site. A comparison of morphological and molecular results showed that plants with one ITS genotype always had leaves greater than 5 mm long, whereas the other had leaves less than 5 mm. Despite morphological continuity, at least two distinguishable genetic types are present, and they differed, on average, morphologically. Data from additional markers are needed to determine whether the two species are reproductively isolated, since one locus is insufficient to address that question.

Vanderpoorten *et al.* (2003b) used the same PCR–RFLP technique to compare species of *Rhytidiadelphus*. ISSRs distinguished the four putative species they sampled but ITS haplotypes only distinguished *R. japonicus*, *R. loreus*, and *R. triquetrus* from *R. squarrosus* plus *R. subpinnatus*. Morphological differences were correlated with ISSR–ITS haplotypes. This is an example of a circumboreal group where intercontinental sampling is necessary to thoroughly assess the taxonomic status of the putative species because the study was based on a total of 16 samples from a limited geographic area.

Stech *et al.* (2006b) used *trnL* intron and *trnL–trnF* spacer sequences in a study of the liverwort genus *Tylimanthus*. They resolved that two endemic Macaronesian taxa are sister species and appear to be closely related to a species that occurs in Reunion and is also disjunct in the Neotropics.

Shaw *et al.* (2005b) gathered sequence data from seven cpDNA and nuclear genes representing about 20 species of *Sphagnum* section *Acutifolia*. Most morphologically defined species were distinguishable, but there was also some evidence of interspecific gene flow. *Sphagnum subtile* could not be distinguished from *S. capillifolium*, and *S. andersonianum* could not be separated from *S. rubellum*. Surprisingly, *S. tenerum*, which Crum (1984) considered a variety of *S. capillifolium*, turned out to be unambiguously monophyletic, and one of the most genetically divergent species in the section.

Zhou & Shaw (2008) found that the nearly cryptic species *Sphagnum macrophyllum* and *S. cribrosum* (section *Subsecunda*) are reciprocally monophyletic based on multilocus sequence data. Fixed nucleotide differences suggest that interspecific gene flow is limited or non-existent. A morphologically aberrant morphotype of *S. cribrosum*, known informally as the "wave-form" (L. E. Anderson, pers. comm.), is known only from two Carolina Bays (shallow lakes) in the North Carolina Coastal Plain. The wave-form is remarkably distinctive, characterized by sparsely forking stems with no branch fascicles or capitula, but it is anatomically similar to normal *S. cribrosum*. It looks much more like a *Fontinalis* than a *Sphagnum*. As the name suggests, it had been assumed that the wave-form is a non-genetic habitat expression, but molecular data revealed that it differs from normal *S. cribrosum*, including plants growing in the same lake, in a number of nucleotide substitutions. Normal *S. cribrosum* growing at the site with the wave-form also has a 25 base-pair insertion that is not shared with sympatric wave-form plants.

European and American populations of *Sphagnum pylaesii* are reciprocally monophyletic based on sequence data, but plants from Newfoundland and South America were barely different (Shaw *et al.* 2004). Shaw *et al.* (2005c) found that haploid and polyploid species in the *S. subsecundum* complex do not sort out on the basis of sequences from eight nuclear and cpDNA loci. This turns out in part to be a result of reticulate evolution involving polyploidy, and geographically correlated cryptic speciation that was undetected at the time (A. J. Shaw, unpublished).

Using isozymes, Shaw & Schneider (1995) found two groups of populations in the rare "copper moss" *Mielichhoferia elongata*, and resolved the same two groups in a sequence-based (ITS) study of nearly the same set of populations (Shaw 2000b). Based on rooting provided by a congeneric outgroup species, *M. mielichhoferiana* was resolved as paraphyletic, within which was nested both (morphologically cryptic) groups of *M. elongata*. The two groups of populations within *M. elongata* have partially allopatric geographic ranges but occur in the same general area of Colorado, in the Rocky Mountains. Inter-group mixtures were not observed within individual colonies.

Discriminant analysis of morphological characters scored from 76 herbarium specimens representing four putative species in the *Tortula subulata* complex resolved four morphotypes (with some overlap) (Cano *et al.* 2005). Sequence data from the ITS region indicated that morphological types attributable to *T. mucronifolia* and *T. schimperi* are monophyletic. Two clades that differ in 17 indel mutations were resolved within *T. mucronifolia*. The two clades are partly allopatric along a north–south gradient in Europe; one found in Spain and Italy east to Ukraine, and the other found in Sweden, Greenland, Russia, and also Canada and Alaska. Cano *et al.* (2005) interpreted these two clades as cryptic species. In contrast to *T. mucronifolia* and *T. schimperi*, *T. inermis* was more weakly supported as monophyletic and may be nested within a paraphyletic *T. subulata*. Taxonomic varieties of *T. subulata* were not resolved as reciprocally monophyletic, but support in that part of the tree was weak so monophyly probably could not be rejected (and the authors did not explicitly test this). ITS sequences from North America and Europe, respectively, were very similar in both *T. mucronifolia* and *T. schimperi*.

Stech & Dohrmann (2004) sequenced the nuclear ITS region and the cpDNA *atp*B–*rbc*L spacer in 22 species of *Campylopus*, with a focus on *C. pilifer* and *C. introflexus*. *Campylopus introflexus* was resolved as monophyletic, with little infraspecific variation, whereas *C. pilifer* was paraphyletic because *C. introflexus* was nested within it. Molecular data suggested divergence between New and Old World populations of *C. pilifer*, although the New World populations were paraphyletic. The Old World populations of *C. pilifer* were nested within the New World populations (suggesting possible dispersal from the Old to the New World), and *C. introflexus* was in turn nested within the Old World populations.

Using nucleotide sequences from the nuclear (ITS) and chloroplast (*trnL–trnF*) genomes, Shaw & Allen (2000) found that groups of species defined by leaf morphology within the aquatic moss genus *Fontinalis* are non-monophyletic. *Fontinalis antipyretica*, a species found in both North America and Europe, is paraphyletic because European populations are more closely related to another European species (*F. squamosa*) than they are to North American populations of *F. antipyretica*. Similarly, North American populations of *F. antipyretica* are more closely related to other North American species (*F. gigantea* and *F. chrysophylla*) than they are to the European populations of *F. antipyretica*. These observations suggest that speciation in temperate *Fontinalis* species has been allopatric and that widespread morphotypes may be old and non-monophyletic.

Several authors have used isozymes and DNA-based information to support the description of new taxa. Flatberg & Thingsgaard (2003) showed that isozyme patterns corroborate morphological observations when they described *Sphagnum tundrae* from Svalbard. So & Grolle (2000) provided evidence from

RAPD markers when they described *Plagiochila detecta*. Shaw & Goffinet (2000) showed that *Sphagnum ehyalinum* is an intersectional hybrid when they described it as new from Chile. The new species has cpDNA of section *Subsecunda* and nuclear DNA of section *Cuspidata*. Shaw (2000c) found that a new species of *Schizymenium* from California occurred sympatrically with morphologically similar plants of *Mielichhoferia elongata*. The two species have very similar gametophytes, differing mainly in sporophyte morphology. Primers were designed that would only amplify the ribosomal DNA of *M. elongata* in order to survey and identify ambiguous sterile plants from sites in the Sierra Nevada Mountains.

Pfeiffer *et al.* (2000) found that five putative species in the "*Hypopterygium tamarisci* complex", a group that is widespread in the Southern Hemisphere, could not be distinguished by about 300 nucleotides of cpDNA $trnL_{UAA}$ sequence. A similar lack of differentiation between Patagonian and New Zealand populations of *Weymouthia* led Quandt *et al.* (2001) to conclude that putatively allopatric species in this group should be synonymized. Their data was based on sequences from the ITS2 region of nuclear DNA and longer cpDNA sequences that included the *trnL* intron as well as the non-coding spacer between the *trnL* and *trnF* genes.

Goffinet *et al.* (2007) compared cpDNA sequences from two loci (*trnL–trnF* and *rps4*) in a putative new species of *Orthotrichum* from Chile with those from the morphologically similar Northern Hemisphere species *O. alpestre*. Although the Chilean species (described as *O. freyanum*) is very similar in morphology to *O. alpestre*, it turned out to be more closely related to a sympatric South American species, *O. assimile*, which occurs in proximity to the new species on *Nothofagus* bark.

Stech & Frahm (1999a) compared 375 nucleotides of the *trnL* intron and 521 nucleotides of the ITS nuclear region in two specimens collected at the same site in Germany. One specimen was typical *Platyhypnidium riparioides* and the other was *P. mutatum*, a species recently described from plants collected at the site and known only from there. Stech and Frahm found that the two plants had identical *trnL* intron sequences and differed by only one substitution in the ITS region, and concluded that they were conspecific.

Budke & Goffinet (2006) sequenced one nuclear and two cpDNA regions to test species concepts in the moss genus *Timmia*. Based on 27 samples, they concluded that *T. austriaca*, *T. megapolitana* subsp. *megapolitana*, and *T. megapolitana* subsp. *bavarica* are demonstrably monophyletic, whereas the varieties of *T. norvegica* were not, in part owing to the nested position of *T. siberica*; consequently they combined the two species taxonomically.

Feldberg & Heinrichs (2005, 2006) used the cpDNA *trnL–trnF* region and nrITS for an analysis of Neotropical *Herbertus* species. This study is a very nice example of a combined taxonomic revision, with keys, illustrations, descriptions, etc., and a molecular analysis of relationships within and among species.

Partly on the basis of their molecular results, Feldman and Heinrichs combined several previously recognized South American binomials under a single species, *H. juniperoideus*. Their phylogenetic analysis demonstrated multiple dispersals between the New and Old Worlds, involving plants presently found in Europe, Africa, and South America. Groth *et al.* (2003, 2004) and Heinrichs *et al.* (2003, 2005, 2006) also demonstrated multiple dispersals in the leafy liverwort genus *Plagiochila* between tropical America and Africa and/or Asia. Long-distance dispersal appears to be infrequent and speciation has occurred in disjunct regions to produce geographically restricted species groups, but it is difficult to interpret the phylogeny of *Plagiochila* without hypothesizing occasional long-distance range expansions, along with extinction in some areas. ITS sequences of *P. cambuena* from Madagascar were very similar to those of *P. corrugata* from Brazil, prompting the authors to synonymize them (Heinrichs *et al.* 2003). Heinrichs *et al.* (2005) also found little divergence between South American and African plants of *P. boryana*, *P. punctata*, and *P. stricta*, which they interpreted as evidence of long-distance intercontinental dispersal. Based on their phylogenetic analyses, it appears that the direction of dispersal was from South America to Africa.

11.4 Speciation mechanisms in bryophytes

Hybridization in natural populations can provide a window into evolutionary processes relevant to speciation; a thorough review of natural bryophyte hybrids was provided by Natcheva & Cronberg (2004). Premolecular studies of naturally occurring moss hybrids and mechanisms of reproductive isolation were reviewed by Anderson & Snider (1982). Relatively few studies of hybridization in natural bryophyte populations based on genetic data have been published.

Shaw (1994, 1998) documented viable recombinant gametophytes derived from hybrids between *Mielichhoferia elongata* and *M. mielichhoferiana*, and found that most recombinants were genetically closer to *M. mielichhoferiana* than to *M. elongata*. Natcheva & Cronberg (2007a) found that interspecific recombinant gametophytes derived from hybrid sporophytes involving *Sphagnum capillifolium* and *S. quinquefarium* consistently had the maternal cpDNA of *S. quinquefarium*, but were closer to *S. capillifolium* in terms of nuclear ISSR markers. The asymmetric nature of interspecific recombinants in terms of nuclear DNA, observed by Shaw in *Mielichhoferia* and by Natcheva and Cronberg in *Sphagnum*, could reflect back-crossing to one of the parental species, or lower fitness of recombinants that have a more even contribution of genetic material from the two parents. Recombinant sphagna grown from spores exhibited the same asymmetry of parental contributions, supporting the latter interpretation (Natcheva & Cronberg 2007a). These observations suggest that genic interactions, perhaps involving organellar and nuclear loci, may be important in determining the result of interspecific hybridization and, conversely, the evolution of reproductive isolation. In the case of *S. capillifolium–S. quinquefarium* hybrids, experimentally grown recombinants were viable and performed well under a range of environments (Natcheva 2006). These were the recombinants that contained a preponderance of *S. capillifolium* nuclear markers.

Using microsatellite markers, van der Velde & Bijlsma (2004) found no evidence of established interspecific recombinant gametophytes between *Polytrichum commune* and *P. uliginosum*, which appear to be reproductively isolated. Within a sympatric population, however, hybrid, albeit abortive, sporophytes were found attached to *P. uliginosum* female gametophytes, but not on *P. commune* females. Hybrid sporophytes produced few if any viable spores. It appears that the mechanisms of reproductive isolation between these two species is asymmetric: prezygotic or very early postzygotic when *P. commune* is the female parent (i.e. hybrid sporophytes do not form or at least do not develop to a visible stage) and postzygotic when *P. uliginosum* is the female parent (hybrid sporophytes begin development but abort).

A first attempt to get at the genetic basis of hybrid breakdown was conducted by McDaniel (2005) using a QTL (quantitative trait loci) approach. McDaniel found that reduced protonemal growth in a cross between genetically divergent populations of *Ceratodon purpureus* could be traced to multiple unlinked loci. He also demonstrated that non-additive interactions among loci (i.e. epistasis) contributed to reduced protonemal growth of inter-racial hybrids. McDaniel's analyses thus provide direct evidence of genic interactions affecting the outcome of mating between genetically differentiated plants.

New species of bryophytes undoubtedly originate in a variety of ways including the geographic subdivision of ancestral ranges (classic allopatric speciation), through founder events associated with infrequent long-distance dispersal, and through cytological mechanisms such as polyploidization. Many bryophytes make good experimental organisms, yet the pioneering work of von Wettstein (reviews: 1928, 1932) has not been followed up in recent years.

So what do we know about speciation mechanisms in bryophytes? Unfortunately, not much. Anderson (1963) and Crum (1972) argued mainly from biogeographical observations that many or most bryophyte species are ancient, and that the broad geographic distributions characterizing many taxa result from vicariance associated with continental drift. Crum (1972) argued extensively that long-distance dispersal as a general explanation for the common intercontinental geographic distributions of many bryophytes is highly unlikely. It is now clear from experimental (van Zanten 1978), atmospheric (Muñoz *et al.* 2004), and phylogenetic (see above) data that intercontinental long-distance dispersal, while not common, has played an important role in generating the geographic distributions of bryophyte species. Molecular analyses have corroborated taxonomic conclusions that many tropical species are disjunct across several continents and such disjunctions appear to be especially common between the Neotropics and Africa. It should nevertheless always be kept in mind that finding a lack of divergence between disjunct populations is a negative result and does not preclude the possibility that additional markers will reveal differentiation. In the Northern Hemisphere temperate and boreal zones, it is fair to generalize that most species have continuous or discontinuous intercontinental ranges and there is little evidence from molecular markers of significant intercontinental divergence. (See, however, isozyme evidence provided by Cronberg (1998) of divergence between British and Scandinavian populations of *Sphagnum rubellum*.)

So the question remains, how do we explain the broad geographic distributions of many bryophyte species without substantial morphological differentiation among widely disjunct populations, and in many cases, without obvious genetic differentiation as well? Either there is sufficient intercontinental gene flow to prevent differentiation, or divergence is slow indeed, as Crum (1972) and others have suggested. Neither a level of gene flow sufficient to cause genetic homogenization, nor such slow evolution that populations have not diverged by genetic drift over millions of years, seem very likely, yet there appears to be no other explanation! Stenøien & Såstad (1999) interpreted the lack of genetic differentiation between North American and European populations of *Sphagnum angustifolium* as slow evolution, due to very large effective population sizes that make genetic drift negligible. Whether this is true, is hard to test. It seems clear that bryophytes are able to disperse effectively over both short and long distances and such changes in distribution erase historical information that might be used to make inferences about speciation modes. For this reason, bryophytes are not always very good organisms for formulating and testing hypotheses about speciation mechanisms thought to be common, such as allopatric divergence caused by drift or natural selection.

The most studied mode of speciation in bryophytes is through allopolyploid formation. Wyatt *et al.* (1988) first demonstrated allopolyploidy in bryophytes, and since then allopolyploids have been documented in at least 12 genera (Table 11.2). Allopolyploids involve hybridization followed by chromosome doubling (polyploidization). Polyploidization in bryophytes was assumed by early authors to result primarily from apospory: the regeneration of diploid gametophytes from immature sporophyte tissues (e.g. Anderson

1980, Wyatt & Anderson 1984). Apospory has been reported in both mosses and liverworts (Lal 1984). However, recent work suggests that diplospory - the production of unreduced spores - might be as or more important a mechanism, especially in the generation of stable allopolyploid species (Såstad 2005, Flatberg *et al.* 2006). Såstad (2005) estimated that approximately 5%–10% of liverwort species and 6%–19% of moss species are polyploids, and thus genome duplication has been an important process in bryophyte evolution. Såstad (2005) argued that no well-established examples of autopolyploidy exist; all polyploid species that have been studied genetically appear to be allopolyploids. Infraspecific variation in chromosome numbers is common as well (Fritsch 1991), and autopolyploidy may also be important in bryophyte evolution. There has been little study of genetic and phylogenetic relationships among polyploid "races" within individual, morphologically defined species. Consequently, Såstad's (2005) rejection of autopolyploidy as an important feature of bryophyte evolution may have been premature.

The primary evidence in favor of polyploid bryophytes being of hybrid origin is the observation of fixed heterozygosity for molecular markers. That is, all sampled gametophytes are heterozygous for codominant markers, suggesting that the alleles do not segregate at meiosis to yield both homozygotes and heterozygotes. Fixed heterozygosity occurs when the two haploid genomes present in a diploid gametophyte are sufficiently differentiated that chromosomes do not pair properly and segregate. It is worth noting that the absence of homozygotes in a population survey, while suggestive, provides only indirect evidence that heterozygosity is fixed. Studies involving growth of gametophytes from spores in order to test for segregation directly is the only way to demonstrate unequivocally that heterozygosity is really fixed. Inferences based on population surveys require sufficient sample sizes to adequately demonstrate that homozygotes were not simply missed.

Important questions addressed by studies on polyploid bryophyte species are as follows. (1) Is the polyploid an allo- or autopolyploid? (2) What (is) are the parental species? (3) If allopolyploid, what is the maternal parent and is that species always the maternal parent? (4) How many times has the polyploid originated? (5) Is (are) the origin(s) recent or ancient? Once these questions have been answered, then polyploids are valuable species for addressing questions about molecular evolution following genome duplication (Wendel 2000).

11.4.1 *Allopolyploidy in liverworts*

The European species of *Pellia* are perhaps the most thoroughly studied group of taxa, through a variety of molecular techniques. Isozymes and RAPDs have shown that both *P. endiviifolia* and *P. epiphylla* consist of at least two morphologically cryptic taxa (Szweykowski *et al.* 1981b, Zielinski 1984, 1987, Szweykowski & Odrzykoski 1990). Moreover, the polyploid species, *P. borealis*, is an allopolyploid with *P. epiphylla* "cryptic species N" and *P. epiphylla* "cryptic species S" as the progenitors. Fixed heterozygosity for isozyme alleles in *P. borealis* show that the cryptic species differ in significant genomic features such that homeologous chromosome pairing and independent assortment are precluded (Odrzykoski *et al.* 1996). Fiedorow *et al.* (2001) used a PCR–RFLP technique to compare $tRNA^{LEU}$ genes in the haploids and polyploids and found an additive pattern in *P. borealis*. Sequences from the same region indicated that *P. borealis* differs from the two progenitors only slightly. *Pellia borealis* is one of only a few bryophyte allopolyploids in which male and female parentage has been ascertained; *P. epiphylla* "cryptic species N" contributed the mt- and cpDNA for all of 14 Polish populations of *P. borealis* sampled.

Reboulia hemisphaerica is traditionally considered to be the only species in this genus, but Boisselier-Dubayle *et al.* (1998b) found that it consists of three genetically differentiated but morphologically cryptic haploid taxa (pairwise Nei's $I = 0.325$–0.550). In addition, plants from New Zealand and Australia proved to be polyploid. Only one isozyme locus, AAT, was heterozygous in the polyploids, but the same heterozygous pattern was found in the two polyploid samples analyzed, one from New Zealand and one from Australia. One of the alleles present in the heterozygotes was also detected in the Japanese cryptic haploid species of *R. hemisphaerica* (*R. hemisphaerica* var. *japonica*) and the other occurred in European populations of *R. hemisphaerica* s. str. (*R. hemisphaerica* var. *hemisphaerica*). The polyploid was interpreted as an allopolyploid involving these two parents. Polyploid colonies in New Zealand and Australia are genetically closer to the European haploids but are morphologically more similar to the Japanese haploids. Additional markers are needed to test the hypothesis that the whole genome, rather than just the locus coding for AAT, is duplicated.

Boisselier-Dubayle & Bischler (1998) found that the complex thalloid liverwort genus *Corsinia*, previously considered monospecific, consists of at least three morphologically cryptic taxa, one haploid and two diploid. *Corsinia coriandrina* (s. l.) is widespread though sporadic in southern Europe and Micronesia and is also found in the U.S.A. and South America. One sample from Texas proved to be diploid, as were some samples from the Mediterranean region. New World and Old World diploids appear to have originated independently, and were highly divergent. European diploids exhibited fixed heterozygosity at six of eight enzyme systems assayed and some alleles could have been provided by the sampled haploid form, but other alleles could not be accounted for, so an additional unsampled parent is implied.

Porella baueri, with a chromosome number of $N=16$, is the only reported polyploid in that genus (Boisselier-Dubayle *et al.* 1998a). The polyploid exhibits fixed heterozygosity at four of 13 isozyme loci and alleles present at these and the remaining homozygous loci occurred in either or both of the putative haploid parents, *P. cordeana* and *P. platyphylla*. Populations of the polyploid, *P. baueri*, from Western Europe tended to be morphologically and genetically closer to *P. cordaeana*, whereas eastern European populations were closer to the other parent, *P. platyphylla*. Jankowiak and Szwekowska-Kulinska (2004) recently showed that *P. cordaeana* was the maternal parent, based on cpDNA and mtDNA sequences. At least two origins of the polyploid were inferred from the observation that two different alleles of phosphatase were detected in each of the putative haploid parents, and both turned up in different polyploid individuals. *Porella baueri* appears to be sexually fertile, and some preliminary indication of recombination between sympatric haploids and diploids was detected.

A more complex history was inferred for triploids in the thallose liverwort genus *Targionia* (Boisselier-Dubayle & Bischler 1999). *Targionia hypophylla* ($N=9$) and triploids known as *T. lorbeeriana* ($N=27$) are widespread in both the New and Old Worlds. European and Macaronesian populations include both haploids and triploids and the two cytotypes occur sympatrically and sometimes close together, although apparently not in mixed colonies. Haploids appear to be more common in northern Europe and triploids more common in southern Europe. Triploids exhibited fixed heterozygosity at all of the seven isozyme loci investigated. European triploids contain two sets of alleles from the European haploids, while a third set of alleles was not detected in any haploid population and the authors hypothesized that the other parent is extinct. They excluded other allopatric species of *Targionia* as potential parents based on morphological considerations. Based on limited sampling, African and Australasian triploids had a highly divergent allelic profile so Boisselier-Dubayle & Bischler (1999) interpreted them as independently derived. In total, a minimum of three origins were hypothesized for triploid *Targionia*; two in Europe and one in Africa/Australasia. Boisselier-Dubayle & Bischler (1999) hypothesized that triploids originated by hybridization and chromosome doubling, followed by meiotic non-disjunction.

Southern Hemisphere gametophytes of *Plagiochasma rupestre* are haploid ($N=9$), but European samples are gametophytically diploid (Boisselier-Dubayle *et al.* 1996). Although Boisselier-Dubayle *et al.* initially assumed the diploids were derived by autopolyploidy, they assayed AAT isozymes from sporelings germinated from experimentally crossed plants and found no segregation of alleles. Tetrasomic inheritance would have been expected if the autopolyploid hypothesis were correct, so these authors suggested that *P. rupestre* is an allopolyploid. This is the only published study in bryophytes that actually shows, experimentally, that heterozygotes are fixed rather than segregating.

Buczkowska *et al.* (2004) assayed four enzyme systems in 223 Polish samples of the leafy liverwort *Calypogeia*, representing six species. Only nine multilocus haplotypes were detected, but three species, *C. azurea*, *C. muelleriana*, and *C. sphagnicola*, were shown to be polyploid based on chromosome counts, and exhibited apparent fixed heterozygosity for one or two enzymes. Although the authors acknowledged that additional work is necessary to corroborate their interpretation, they considered the latter three species to be allopolyploids. For the enzyme TPI, variation in the allelic composition among polyploid species suggests independent origins.

11.4.2 *Allopolyploidy in mosses*

Three species of *Polytrichastrum*, namely *P. pallidisetum*, *P. sexangulare*, and *P. ohioense*, are allopolyploids based on fixed heterozygosity at five or six of eleven isozyme loci screened from some 7000 shoots representing 304 populations (Derda & Wyatt 2000). Isozyme profiles from the polyploids were sufficiently differentiated from any haploid species sampled that parents could not be identified. It appears that the hybrids are derived from crosses between species in different genera of Polytrichaceae, possibly involving *Polytrichum*, *Polytrichastrum*, and *Pogonatum*. Monophyly of and phylogenetic relationships among these genera are, however, incompletely resolved at present (Hyvönen *et al.* 2004). Haploid species including *Polytrichastrum appalachianum* and *Polytrichum commune* have isozyme alleles that might indicate that they were involved in the parentage of one or more of the polyploids, but other species could not be eliminated. Derda & Wyatt (2000) speculated that polyploidization might have occurred so long ago that precise parentage may never be uncovered.

Polytrichum longisetum also appears to be an allopolyploid based on fixed heterozygosity of four isozyme and 12 microsatellite loci (van der Velde & Bijlsma 2001). Forty-three percent of the microsatellite loci assayed for *P. longisetum* exhibited a single band, and therefore appeared to be homozygous. These authors, however, argued that because microsatellite loci have very high mutation rates, these seemingly homozygous loci were unlikely to result from indistinguishable alleles in the two progenitor species. Rather, they thought the single banded patterns reflected a failure to amplify one of the two alleles. They concluded from their analyses that the haploid species, *P. formosum*, or a taxon very similar to it, was one of the parents, but they could not identify the other parent. As in the case of *Polytrichastrum* studied by Derda & Wyatt (2000), these results suggest that the allopolyploid does not have a recent origin.

Rhizomnium pseudopunctatum is an allopolyploid of *R. gracile* of western North America and northern Asia, and *R. magnifolium* of Europe (Wyatt *et al.* 1993b). Although the two putative progenitor species are currently allopatric and their ranges could not be much more disjunct, the genetic evidence is strong that they gave rise to *R. pseudopunctatum*. If true, this observation provides a statement about how current distributional ranges may be misleading in the formulation of hypotheses about bryophyte speciation. Jankowiak *et al.* (2005) demonstrated from cp- and mtDNA sequences sampled from the two parents and the allopolyploid that *R. magnifolium* is the maternal parent, assuming that organellar inheritance is only through the female.

Plagiomnium medium is an allopolyploid derivative of *P. ellipticum* and *P. insigne* (Wyatt *et al.* 1988, 1992). Isozyme alleles found in the allopolyploid were detected in different individuals of the two parents, indicating multiple origins. In addition, more than 30 different multilocus isozyme genotypes were detected in *P. medium*, suggesting recombination among allopolyploid individuals subsequent to their origin(s). Restriction digests of cpDNA from the two parents and *P. medium* indicate that *P. insigne* is the maternal parent. However, RFLP analysis of whole chloroplast genomes is too crude to determine whether there is variation at the cpDNA sequence level within the polyploid or its parents.

Plagiomnium cuspidatum also appears to be an allopolyploid based on fixed or nearly fixed heterozygosity at eight allozyme loci (Wyatt & Odrzykoski 1998). The east-Asian *P. acutum* appears to be one parent but the other could not be identified. During the course of their investigations, Wyatt & Odrzykoski (1998) also uncovered genetic evidence of previously unrecognized species, one of which was subsequently described (Wyatt *et al.* 1997).

Only two chromosome numbers have been reported in the genus *Sphagnum*, $N = 19$ and 38 (Fritsch 1991). Cronberg (1996) presented isozyme evidence that *S. russowii* ($N = 38$) is an allopolyploid with *S. girgensohnii* and either *S. quinquefarium* or *S. rubellum* as progenitors. Shaw *et al.* (2005b) reported corroborating evidence from nuclear and chloroplast DNA sequences, also suggesting that *S. russowii* has originated multiple times, and one origin might have involved *S. warnstorfii* as well. Flatberg *et al.* (2006) described hybrid sporophytes in a mixed population of *S. girgensohnii* and *S. russowii*, derived from inter-ploidal backcrossing. The hybrid sporophytes were found only on female *S. girgensohnii* gametophytes. Less than 5% of the spores in hybrid capsules germinated but triploid plants were successfully reared to at least a juvenile gametophyte stage. Hybrid sporophytes on female *S. girgensohnii* gametophytes were reported by the authors from scattered Scandinavian sites, and were also found at the same Norwegian site in multiple years. The extent to which triploid hybrids between *S. girgensohnii* and *S. russowii* persist in nature, or cross with either of the progenitors, is unknown.

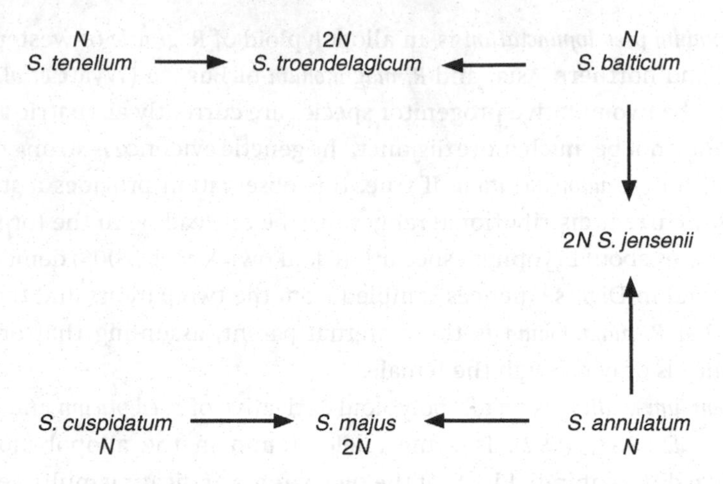

Fig. 11.1. Hypothesized ancestry for allopolyploid species of *Sphagnum* section *Cuspidata*. Four (gametophytically) haploid species (*N*), *S. tenellum*, *S. balticum*, *S. cuspidatum*, and *S. annulatum*, have given rise to three allopolyploid species (2*N*), *S. jensenii*, *S. majus*, and *S. troendelagicum* with overlapping parentage. Summarized from results of Såstad *et al.* (1999a, 2000, 2001).

Flatberg *et al.* pointed out that because of the huge numbers of spores produced by hybrid sporophytes, 5% germination could yield thousands of viable offspring.

Several allopolyploid species have been documented in *Sphagnum* section *Cuspidata* (Fig. 11.1). Såstad *et al.* (2000) found that the boreal species, *S. majus*, exhibits fixed heterozygosity at three out of nine isozyme loci assayed, while two other loci were homo- or heterozygous in different individuals. Both alleles detected at the loci characterized by fixed heterozygosity were found also in the haploid species *S. cuspidatum* and *S. annulatum*. Six alleles found in the polyploid were not detected in either putative parent. These authors suggested that a relatively high frequency of orphan (or silenced) alleles in *S. majus* might indicate an ancient origin (and subsequent divergence from the ancestral allopolyploid(s)). In contrast, all alleles of another allopolyploid in section *Cuspidata*, *S. jensenii*, were detected in one or both putative parents (*S. balticum* and *S. annulatum*), suggesting a more recent origin (Såstad *et al.* 1999a). *Sphagnum jensenii* exhibited fixed heterozygosity at four out of nine isozyme loci assayed. Different polyploid individuals had common MNR-1 alleles found in different plants of *S. balticum*, suggesting at least two independent origins.

In contrast to the broad geographic ranges of *Sphagnum majus* and *S. jensenii*, another allopolyploid *Sphagnum*, *S. troendelagicum*, is endemic to a relatively small area of central Norway. Isozyme and RAPD data support an origin for *S. troendelagicum* through hybridization between *S. tenellum* and *S. balticum* (Såstad *et al.* 2001). Fixed heterozygosity was observed at two loci in *S. troendelagicum* and both alleles at each locus were also found in *S. balticum* and/or *S. tenellum*. The allopolyploid exhibited limited polymorphism at isozyme loci but was highly polymorphic for RAPD markers (Stenøien & Flatberg 2000, Såstad *et al.* 2001). Surprisingly, there was little or no linkage disequilibrium among RAPD markers in three intensively sampled populations of *S. troendelagicum* (Stenøien & Flatberg 2000). Sexual reproduction has never been observed in *S. troendelagicum*, so even if the polyploid originated several times, strong linkage disequilibrium would be expected. Recent, unpublished cpDNA sequences indicate that *S. tenellum* is the chloroplast parent (Stengrunet *et al.*, unpublished data).

Sphagnum balticum appears to have participated in the origin of at least two allopolyploid species in section *Cuspidata* (*S. jensenii* and *S. troendelagicum*). Similarly, *S. annulatum* appears to be one parent of both *S. majus* and *S. jensenii*. These observations clearly indicate that species of section *Cuspidata* are able to hybridize with related species and it may be that the high levels of phenotypic variation characteristic of these species reflect hybridization in natural populations.

Allopolyploidy also appears to be common in *Sphagnum* section *Subsecunda*. Of the species found in Europe and eastern North America, *S. contortum, S. platyphyllum*, and *S. subsecundum* appear to be consistently haploid whereas *S. auriculatum, S. carolinianum, S. inundatum*, and *S. lescurii* are polyploid (Fritsch 1991, Melosik *et al.* 2005, A. J. Shaw, unpublished data). Here the evolutionary patterns are complex and raise challenging taxonomic problems; many of the data are currently unpublished but well-supported aspects of the story are briefly summarized. Sequence data from the chloroplast and nuclear genomes, and microsatellites, show that the North American polyploid species *S. lescurii* is distinct from the European polyploid *S. auriculatum*. Furthermore, European *S. inundatum* originated independently of North American *S. inundatum* and the two are differentiated for both nuclear and chloroplast markers. On the other hand, the North American polyploids *S. carolinianum, S. lescurii*, and (American populations of) *S. inundatum* cannot be distinguished by cpDNA or nuclear sequences and differ only very slightly in microsatellite allele frequencies. Similarly, European *S. inundatum* and *S. auriculatum* are undifferentiated for either sequence-based or microsatellite markers. *Sphagnum contortum* and *S. platyphyllum* were eliminated as haploid parents; only *S. subsecundum* is implicated by both cpDNA sequences and microsatellite markers. The other parent may be an unsampled race of *S. subsecundum* or some other, perhaps extinct, species.

11.5 Tempo and mode of allopolyploid evolution

We do not have any way to know with confidence how old allopolyploid bryophyte species are. In the absence of direct methods for dating allopolyploid origins, we can infer that if all alleles found in the allopolyploid species can also be detected in the parental haploids, the origin may have been relatively recent. Examples include *Sphagnum jensenii* and *S. troendelagicum*. The latter species is highly restricted in geographic distribution and presently occurs sympatrically with the two parental haploids. It may be very recent. The allopolyploid *Pellia borealis* may be a relatively recent derivative of two European, morphologically cryptic haploid species. Similarly, the allelic profile of polyploid *Porella baueri* can be completely accounted for by alleles found in two haploid species, *P. cordeana* and *P. platyphylla*. Polyploid species of *Polytrichastrum*, in contrast, may have originated so long ago that we cannot even identify the parental haploids, which may be extinct. In *Rhizomnium*, the parental haploids can be identified, but presently have widely disjunct allopatric distributions. Clearly the allopolyploid is old enough that major range changes have occurred since the haploids had opportunities to hybridize. When sufficient data have been collected, it appears that some or perhaps most allopolyploid bryophyte species originated multiple times. This is in keeping with what we know about allopolyploid origins in vascular plants (Soltis & Soltis 1999).

11.6 Reconciling evolutionary inferences from molecular data with species concepts

To some extent the controversy over biological versus phylogenetic species concepts is an artificial one. Phylogenetic concepts that define species as the least inclusive group of populations/individuals that are hierarchically related (Nixon & Wheeler 1990) are based on the fact that relationships below the species level are reticulate rather than hierarchical because of recombination. Moreover, reciprocal monophyly of related species can only occur when they have been reproductively isolated for sufficient time for allele coalescence to occur within species. At least partial reproductive isolation is necessary, even if not sufficient, for speciation to proceed. Thus, evolutionary biologists focused on speciation mechanisms (e.g. Coyne & Orr 2004) tend to adopt a biological species concept whereas those focused on defining and delimiting species prefer phylogenetic approaches.

Allopolyploids present special problems for taxonomists. It is now known that most allopolyploid "species" have originated multiple times and in some cases there is genetic evidence of more than ten origins for a single taxon (Soltis & Soltis 1999). Such species are thus demonstrably polyphyletic; what is a taxonomist to do? Inferences that polyphyletic allopolyploids go on to function as biologically meaningful species come from evidence of genetic recombination among independently derived plants and animals (e.g. Wyatt *et al.* 1988,

1992, Doyle *et al.* 1999, Espinoza & Noor 2002). It is standard practice to recognize allopolyploids as species, even when known to be polyphyletic. Yet the practice goes contrary to any phylogenetic species concept that requires monophyly. The alternative is to recognize two, three, ... ten or more monophyletic species that cannot be distinguished morphologically, and which in some cases at least appear to function together as "evolutionarily significant units". Molecular approaches to study polyploid formation have clarified some aspects of bryophyte evolution, but have perhaps muddied the waters of bryophyte taxonomy.

Acknowledgment

Preparation of this chapter was supported by NSF grant DEB-0515749.

References

Adamczak, M., Buczkowska, K., Baczkiewicz, A. & Wachowiak, W. (2005). Comparison of allozyme variability in Polish populations of two species of *Ptilidium* Nees (Hepaticae) with contrasting degrees of sexual reproduction. *Cryptogamie Bryologie*, **26**, 151–65.

Ahmed, J. & Frahm, J. P. (2003). Isozyme variability among Central European species of the aquatic moss *Cinclidotus*. *Cryptogamie Bryologie*, **24**, 147–54.

Akiyama, H. (2004). Allozyme variability within and among populations of the epiphytic moss *Leucodon* (Leucondontaceae: Musci). *American Journal of Botany*, **81**, 1280–7.

Akiyama, H. & Hiraoka, T. (1994). Allozyme variability within and among divergent populations of liverwort *Conocephalum conicum* (Marchantiales: Hepaticae). *Japanese Journal of Plant Research*, **107**, 307–20.

Anderson, L. E. (1963). Modern species concepts: mosses. *Bryologist*, **66**, 107–19.

Anderson, L. E. (1980). Cytology and reproductive biology of mosses. In *The Mosses of North America*, ed. R. J. Taylor & A. E. Leviton, pp. 37–76. San Francisco, CA: Pacific Division of the American Association for the Advancement of Sciences.

Anderson, L. E. & Snider, J. A. (1982). Cytological and genetic barriers in mosses. *Journal of the Hattori Botanical Laboratory*, **52**, 241–54.

Appelgren, L. & Cronberg, N. (1999). Genetic and morphological variation in the rare epiphytic moss *Neckera pennata* Hedw. *Journal of Bryology*, **21**, 97–107.

Baldwin, B. G. & Sanderson, M. J. (1998). Age and rate of diversification of the Hawaiian silversword alliance (Compositae). *Proceedings of the National Academy of Sciences, U.S.A.*, **95**, 9402–6.

Baum, D. A. & Donoghue, M. J. (1995). Choosing among alternative "phylogenetic" species concepts. *Systematic Botany*, **20**, 560–73.

Bijlsma, R., van der Welde, M., van de Zande, L., Boerema, A. C. & van Zanten, B. O. (2000). Molecular markers reveal cryptic species within *Polytrichum commune* (common hair-cap moss). *Plant Biology*, **2**, 408–14.

Bischler, H. & Boisselier-Dubayle, M.-C. (1997). Population genetics and variation in liverworts. *Advances in Bryology*, **6**, 1–34.

Bischler, H., Boisselier-Dubayle, M.-C., Fontinha, S. & Lambourdiére, J. (2006). Species boundaries in European and Macaronesian *Porella L.* (Jungermanniales, Porellaceae). *Cryptogamie Bryologie*, **27**, 35–57.

Boisselier-Dubayle, M.-C. & Bischler, H. (1994). A combination of molecular and morphological characters for delimitation of taxa in European *Porella*. *Journal of Bryology*, **18**, 1–11.

Boisselier-Dubayle, M.-C. & Bischler, H. (1997). Enzyme polymorphism in *Preissia quadrata* (Hepaticae, Marchantiaceae). *Plant Systematics and Evolution*, **205**, 73–84.

Boisselier-Dubayle, M.-C. & Bischler, H. (1998). Allopolyploidy in the thalloid liverwort *Corsinia* (Marchantiales). *Botanica Acta*, **111**, 490–6.

Boisselier-Dubayle, M.-C. & Bischler, H. (1999). Genetic relationships between haploid and triploid *Targionia* (Targioniaceae, Hepaticae). *International Journal of Plant Sciences*, **160**, 1163–9.

Boisselier-Dubayle, M.-C., De Chaldee, M., Lambourdiere, J. & Bischler, H. (1995a). Genetic variability in western European *Lunularia*. *Fragmenta Floristica et Geobotanica*, **40**, 379–91.

Boisselier-Dubayle, M.-C., Jubier, M. F., Lejeune, B. & Bischler, H. (1995b). Genetic variability in three subspecies of *Marchantia polymorpha*: isozymes, RFLP, and RAPD markers. *Taxon*, **44**, 363–76.

Boisselier-Dubayle, M.-C., Lambourdiere, J. & Bischler, H. (1996). Progeny analysis by isozyme markers in the polyploid liverwort *Plagiochasma rupestre*. *Canadian Journal of Botany*, **74**, 521–7.

Boisselier-Dubayle, M.-C., Lambourdiere, J. & Bischler, H. (1998a). The leafy liverwort *Porella baueri* (Porellaceae) is an allopolyploid. *Plant Systematics and Evolution*, **210**, 175–97.

Boisselier-Dubayle, M.-C., Lambourdiere, J. & Bischler, H. (1998b). Taxa delimitation in *Reboulia* investigated with morphological, cytological, and isozyme markers. *Bryologist*, **101**, 61–9.

Buczkowska, K. (2004). Genetic differentiation of *Calypogeia fissa* Raddi (Hepaticae, Jungermanniales) in Poland. *Plant Systematics and Evolution*, **247**, 187–201.

Buczkowska, K., Odrzykoski, I. J. & Chudzinska, E. (2004). Delimitation of some European species of *Calypogeia* Raddi (Jungermanniales, Hepaticae) based on cytological characters and multienzyme phenotype. *Nova Hedwigia*, **78**, 147–63.

Budke, J. M. & Goffinet, B. (2006). Phylogenetic analyses of Timmiaceae (Bryophyta : Musci) based on nuclear and chloroplast sequence data. *Systematic Botany*, **31**, 633–41.

Buryova, B. (2004). Genetic variation in two closely related species of *Philonotis* based on isozymes. *Bryologist*, **107**, 316–27.

Buryova, B. & Shaw, A. J. (2005). Phenotypic plasticity in *Philonotis fontana* (Bryopsida : Bartramiaceae). *Journal of Bryology*, **27**, 13–22.

Cano, M., Werner, O. & Guerra, J. (2005). A morphometric and molecular study in *Tortula subulata* complex (Pottiaceae, Bryophyta). *Biological Journal of the Linnean Society*, **149**, 333–50.

Cox, C. J., Goffinet, B., Shaw, A. J. & Boles, S. B. (2004). Phylogenetic relationships among the mosses based on heterogeneous Bayesian analysis of multiple genes from multiple genomic compartments. *Systematic Botany*, **29**, 234–50.

Coyne, J. & Orr, H. A. (2004). *Speciation*. Sunderland, Massachusetts: Sinauer Associates.

Crandall-Stotler, B. & Stotler, R. E. (2000). Morphology and classification of the Marchantiophyta. In *Bryophyte Biology*, ed. A. J. Shaw & B. Goffinet, pp. 21–70. Cambridge: Cambridge University Press.

Crisp, M. D. & Chandler, G. T. (1996). Paraphyletic species. *Telopea*, **6**, 813–44.

Cronberg, N. (1987). Genotypic differentiation between the two related peat mosses, *Sphagnum rubellum* and *S. capillifolium* in northern Europe. *Journal of Bryology* **19**, 715–29.

Cronberg, N. (1989). Patterns of variation in morphological characters and isoenzymes in populations of *Sphagnum capillifolium* (Ehrh.) Hedw. and *S. rubellum* Wils. from two bogs in southern Sweden. *Journal of Bryology*, **15**, 683–96.

Cronberg, N. (1996). Isozyme evidence of relationships within *Sphagnum* section *Acutifolia* (Sphagnaceae, Bryophyta). *Plant Systematics and Evolution*, **203**, 41–64.

Cronberg, N. (1998). Population structure and interspecific differentiation of the peat moss sister species *Sphagnum rubellum* and *S. capillifolium* (Sphagnaceae) in northern Europe. *Plant Systematics and Evolution*, **209**, 139–58.

Cronberg, N. & Natcheva, R. (2002). Hybridization between the peat mosses, *Sphagnum capillifolium* and *S. quinquefarium* (Sphagnaceae, Bryophyta) as inferred by morphological characters and isozyme markers. *Plant Systematics and Evolution*, **234**, 53–70.

Crosby, M. R., Magill, R. E., Allen, B. & He, S. (2000). A checklist of the mosses. Missouri Botanical Garden, St. Louis, MO, USA. www.mobot.org/MOBOT/tropicos/most/checklist.shtml.

Crum, H. A. (1972). The geographic origins of the mosses of North America's eastern deciduous forest. *Journal of the Hattori Botanical Laboratory*, **35**, 269–98.

Crum, H. A. (1984). Sphagnopsida, Sphagnaceae. *North American Flora*, ser. II, Part **11**, 1–180.

Darwin, C. (1859). *On the Origin of Species*. A Facsimile of the First Edition. Cambridge, MA: Harvard University Press.

Davis, J. I. (1997). Evolution, evidence, and the role of species concepts in phylogenetics. *Systematic Botany*, **22**, 373–403.

Davis, E. C., Franklin, J. B., Shaw, A. J. & Vilgalys, R. (2003). Endophytic *Xylaria* (Xylariaceae) among liverworts and angiosperms: phylogenetics, distribution, and symbiosis. *American Journal of Botany*, **90**, 1661–7.

De Queiroz, K. & Donoghue, M. J. (1988). Phylogenetic systematics and the species problem. *Cladistics*, **4**, 317–38.

Derda, G. S. & Wyatt, R. (2000). Isoenzyme evidence regarding the origin of three allopolyploid species of *Polytrichastrum* (Polytrichaceae, Bryophyta). *Plant Systematics and Evolution*, **220**, 37–53.

Dewey, R. M. (1989). Genetic variation in the liverwort *Riccia dictyospora* (Ricciaceae, Hepaticopsia. *Systematic Botany*, **15**, 155–67.

Doyle, J. J., Doyle, J. J. & Brown, A. H. D. (1999). Origins, colonization, and lineage recombination in a widespread perennial soybean polyploidy complex. *Proceedings of the National Academy of Sciences, U.S.A.*, **96**, 10741–5.

Espinoza, N. R. & Noor, M. A. F. (2002). Population genetics of a polyploidy: is there hybridization between lineages of *Hyla versicolor? Journal of Heredity*, **93**, 81–5.

Estoup, A., Jarne, P. & Cornuet, J.-M. (2002). Homoplasy and mutation model at microsatellite loci and their consequences for population genetics analysis. *Molecular Ecology*, **11**, 1591–604.

Feldberg, K. & Heinrichs, J. (2005). On the identity of *Herbertus borealis* (Jungermanniopsida: Herbertaceae) with notes on the possible origin of *H. sendtneri*. *Journal of Bryology*, **27**, 343–50.

Feldberg, K. & Heinrichs, J. (2006). A taxonomic revision of *Herbertus* (Jungermanniidae: Herbertaceae) in the Neotropics based on nuclear and chloroplast DNA and morphology. *Biological Journal of the Linnean Society*, **151**, 309–32.

Feldberg, K., Groth, H., Wilson, R., Schafer-Verwimp, A. & Heinrichs, J. (2004). Cryptic speciation in *Herbertus* (Herbertaceae, Jungermanniopsida): range and morphology of *Herbertus sendtneri* inferred from nrITS sequences. *Plant Systematics and Evolution*, **249**, 247–61.

Fiedorow, P., Odrzkoski, I., Szweykowski, J. & Szweykowska-Kulinska, Z. (2001). Phylogeny of the European species of the genus *Pellia* (Hepaticae; Metzgeriales) based on molecular data from nuclear tRNA(CAA)(LEU) intergenic spacers. *Gene*, **262**, 309–15.

Flatberg, K. I. & Thingsgaard, K. (2003). Taxonomy and geography of *Sphagnum tundrae* with a description of *S. mirum*, sp. nov. (Sphagnaceae, sect. *Squarrosa*). *Bryologist*, **106**, 501–15.

Flatberg, K. I., Thingsgaard, K. & Såstad, S. M. (2006). Interploidal gene flow and introgression in bryophytes: *Sphagnum girgensohnii* × *S. russowii*, a case of spontaneous neotriploidy. *Journal of Bryology*, **28**, 27–37.

Frahm, J. P. (2005). The genus *Hypnodontopsis* (Bryophyta, Rhachitheciaceae) in Baltic and Saxon amber. *Bryologist*, **108**, 228–35.

Frahm, J. P., Müller, K. & Stech, M. (2000). The taxonomic status of *Eurhynchium crassinervium* from river banks based on ITS sequence data. *Journal of Bryology*, **22**, 291–2.

Fritsch, R. (1991). Index to bryophyte chromosome counts. *Bryophytorum Bibliotheca*, **40**, 19–20.

Funk, D. J. & Omland, K. E. (2003). Species-level paraphyly and polyphyly: frequency, causes, and consequences, with insights from animal mitochondrial DNA. *Annual Review of Ecology and Systematics*, **34**, 397–423.

Gallego, M. T., Werner, O., Sergio, C. & Guerra, J. (2005). A morphological and molecular study of the *Syntrichia laevipila* complex (Pottiaceae) in Portugal. *Nova Hedwigia*, **80**, 301–22.

Goffinet, B., Buck, W. R. & Wall, M. A. (2007). *Orthotrichum freyanum* (Orthotrichaceae, Bryophyta), a new epiphytic species from Chile. *Beiheft zur Nova Hedwigia*, in press.

Gottlieb, L. D. (1981). Electrophoretic evidence and plant populations. *Progress in Phytochemistry*, **7**, 1–46.

Gradstein, S. R. (1979). The genera of the Lejeuneaceae: past and present. In *Bryophyte Systematics*, ed. G. C. S. Clarke & J. G. Duckett, pp. 83–107. London: Academic Press.

Gradstein, S. R. (1994). Lejeuneaceae: Ptychantheae. Brachiolejeuneae. *Flora Neotropica Monograph*, **62**, 1–225.

Gradstein, S. R. (1997). The taxonomic diversity of epiphyllous bryophytes. *Abstracta Botanica*, **21**, 15–19.

Groth, H., Linder, M., & Heinrichs, J. (2004). Phylogeny and biogeography of *Plagiochila* based on nuclear and chloroplast DNA sequences. *Monographs in Systematic Botany from the Missouri Botanical Garden*, **98**, 365–87.

Groth, H., Linder, M., Wilson, R. *et al.* (2003). Biogeography of *Plagiochila* (Hepaticae): natural species groups span several floral kingdoms. *Journal of Biogeography*, **30**, 965–78.

Hanssen, L., Såstad, S. M. & Flatberg, K. I. (2000). Population structure and taxonomy of *Sphagnum cuspidatum* and *S. viride*. *Bryologist*, **103**, 93–103.

Hartmann, F. A., Wilson, R., Gradstein, S. R., Schneider, H. & Heinrichs, J. (2006). Testing hypotheses on species delimitation and disjunctions in the liverwort *Bryopteris* (Jungermanniopsida: Lejeuneaceae). *International Journal of Plant Science*, **167**, 1205–14.

Heinrichs, J., Proschold, T., Renker, C., Groth, H. & Rycroft, D. S. (2002a). *Plagiochila virginica* A. Evans rather than *Plagiochila dubia* Lindenb. & Gottsche occurs in Macaronesia; placement in sect. *Contiguae* Carl is supported by ITS sequences of nuclear ribosomal DNA. *Plant Systematics and Evolution*, **230**, 221–30.

Heinrichs, J., Groth, H., Holz, I. *et al.* (2002b). The systematic position of *Plagiochila moritziana*, *P. trichostoma*, and *P. deflexa* based on ITS sequence variation of nuclear ribosomal DNA, morphology, and lipophylic secondary metabolites. *Bryologist*, **105**, 189–203.

Heinrichs, J., Gradstein, S. R., Groth, H. & Lindner, M. (2003). *Plagiochila cucullifolia* var. *anomala* var. nov. from Ecuador, with notes on discordant molecular and morphological variation in *Plagiochila*. *Plant Systematics and Evolution*, **242**, 205–16.

Heinrichs, J., Lindner, M., Gradstein, S. R. *et al.* (2005). Origin and subdivision of *Plagiochila* (Jungermanniidae: Plagiochilaceae) in tropical Africa based on evidence from nuclear and chloroplast DNA sequences and morphology. *Taxon*, **54**, 317–33.

Heinrichs, J., Lindner, M., Groth, H. *et al.* (2006). Goodbye or welcome Gondwana? – Insights into the phylogenetic biogeography of the leafy liverwort *Plagiochila* with a description of *Proskauera*, *gen. nov.* (Plagiochilaceae, Jungermanniales). *Plant Systematics and Evolution*, **258**, 227–50.

Heinrichs, J., Hentschel, J., Wilson, R., Feldberg, K. & Schneider, H. (2007). Evolution of leafy liverworts (Jungermanniidae, Marchantiophyta): estimating divergence times from chloroplast DNA sequences using penalized likelihood with integrated fossil evidence. *Taxon*, **56**, 31–44.

Hey, J. (2001). The mind of the species problem. *Trends in Ecology and Evolution*, **16**, 326–9.

Hudson, R. R. (1990). Gene genealogies and the coalescent process. In *Oxford Surveys in Evolutionary Biology*, vol. 7, ed. D. J. Futuyma & J. Antonovics, pp. 1–44. Oxford: Oxford University Press.

Hyvönen, J., Koskinen, S., Smith Merrill, G. L., Hedderson, T. A. & Stenroos, S. (2004). Phylogeny of the Polytrichales (Bryophyta) based on simultaneous analysis of molecular and morphological data. *Molecular Phylogenetics and Evolution*, **31**, 915–28.

Itouga, M., Yamagusci, T. & Deguchi, H. (1999). Allozyme variability within and among populations in the liverwort *Conocephalum japonicum* (Marchantiales, Hepaticae). *Hikobia*, **13**, 89–96.

Jankowiak, K. & Szwekowska-Kulinska, Z. (2004). Organellar inheritance in the allopolyploid liverwort species *Porella baueri* (Porellaceae): reconstructing historical events using DNA sequence analysis. *Monographs in Systematic Botany from the Missouri Botanical Garden*, **98**, 404–14.

Jankowiak, K., Rybarczyk, A., Wyatt, R. *et al.* (2005). Organellar inheritance in the allopolyploid moss *Rhizomnium pseudopunctatum*. *Taxon*, **54**, 363–88.

Janssens, J. A. P., Horton, D. G. & Bassinger, J. F. (1979). *Aulacomnium heterostichoides* sp. nov., an Eocene moss from south central British Columbia. *Canadian Journal of Botany*, **57**, 2150–61.

Korpelainen, H., Kostamo, K. & Virtanen, V. (2007). Microsatellite markers identification using genome screening and restriction-ligation. *Biotechniques*, **42**, 479–86.

Lal, M. (1984). The culture of bryophytes including apogamy, apospory, parthenogenesis and protoplasts. In *The Experimental Biology of Bryophytes*, ed. A. F. Dyer & J. G. Duckett, pp. 97–115. London: Academic Press.

Mayr, E. (1965). *Animal Species and Evolution*. Cambridge, MA: Belknap Press of Harvard University Press.

McDade, L. A. (1995). Species concepts and problems in practice: insight from botanical monographs. *Systematic Botany*, **20**, 606–22.

McDaniel, S. F. (2005). The evolutionary genetics of population divergence in *Ceratodon purpureus*. Ph.D. dissertation, Duke University, Durham, North Carolina.

McDaniel, S. F. & Shaw, A. J. (2003). Phylogeographic structure and cryptic speciation in the transantarctic moss *Pyrrhobrum mnioides*. *Evolution*, **57**, 205–15.

McDaniel, S. F. & Shaw, A. J. (2005). Selective sweeps and intercontinental migration in the cosmopolitan moss, *Ceratodon purpureus* (Hedw.) Brid. *Molecular Ecology*, **14**, 1121–32.

Melosik, I., Odrzykoski, I. J. & Sliwinska, E. (2005). Delimitation of taxa of *Sphagnum subsecundum* s.l. (Musci, Sphagnaceae) based on multienzyme phenotype and cytological characters. *Nova Hedwigia*, **80**, 397–412.

Miller, N. G. (1984). Tertiary and Quaternary fossils. In *New Manual of Bryology*, ed. R. M. Schuster, pp. 1194–232. Nichinan: The Hattori Botanical Laboratory.

Mishler, B. D. & Donoghue, M. J. (1982). Species concepts: a case for pluralism. *Systematic Zoology*, **31**, 491–503.

Mishler, B. D. (1985). Biosystematic studies of the *Tortula ruralis* complex. I. Variation of taxonomic characters in culture. *Journal of the Hattori Botanical Laboratory*, **58**, 225–53.

Muñoz, J., Felicisimo, A. M., Cabezas, F., Burgaz, A. R. & Martinez, I. (2004). Wind as a long distance dispersal vehicle in the southern hemisphere. *Science*, **304**, 1144–7.

Natcheva, R. (2006). Evolutionary processes and hybridization within the peat mosses, *Sphagnum*. Ph.D. dissertation, Lund University, Lund, Sweden.

Natcheva, R. & Cronberg, N. (2004). What do we know about hybridization among bryophytes in nature? *Canadian Journal of Botany*, **82**, 1687–704.

Natcheva, R. & Cronberg, N. (2007a). Recombination and introgression of nuclear and chloroplast genomes between the peat mosses, *Sphagnum capillifolium* and *Sphagnum quinquefarium*. *Molecular Ecology*, **16**, 811–18.

Natcheva, R. & Cronberg, N. (2007b). Maternal transmission of cytoplasmic DNA in interspecific hybrids of peat mosses, *Sphagnum* (Bryophyta). *Journal of Evolutionary Biology*, **20**, 1613–16.

Nei, M. (1972). Genetic distances among populations. *The American Naturalist*, **106**, 283–92.

Newton, A. E., Wikström, N., Bell, N., Forrest, L. L. & Ignatov, M. (2007). Dating the diversification of pleurocarpous mosses. In *Pleurocarpous Mosses: Systematics and Evolution*, ed. A. E. Newton & R. Tangney, pp. 337–66. Boca Raton, FL: Taylor & Francis.

Nixon, K. C. & Wheeler, Q. D. (1990). An amplification of the phylogenetic species concepts. *Cladistics*, **6**, 211–23.

Odrzykoski, I. J. (1987). Genetic evidence for reproductive isolation between two European "forms" of *Conocephalum conicum*. *Symposia Biologia Hungarica*, **35**, 577–87.

Odrzykoski, I. J. & Szweykowski, J. (1991). Genetic differentiation without concordant morphological divergence in the thallose liverwort *Conocephalum conicum*. *Plant Systematics and Evolution*, **178**, 135–51.

Odrzykoski, I. J., Bobowicz, M. A. & Krzakowa, M. (1981). Variation in *Conocephalum conicum* – the existence of two genetically different forms in Europe. In *New Perspectives in Bryotaxonomy and Bryogeography*, ed. J. Szweykowski, pp. 519–42. Poznan, Poland: Adam Mickiewicz University.

Odrzykoski, I., Chudzinska, J. E. & Szweykowski, J. (1996). The hybrid origin of the polyploid liverwort *Pellia borealis*. *Genetica*, **98**, 75–86.

Olmstead, R. G. (1995). Species concepts and plesiomorphic species. *Systematic Botany*, **20**, 623–30.

Oostendorp, C. (1987). *The Bryophytes of the Paleozoic and Mesozoic*. Berlin: J. Cramer.

Pacak, A. & Szweykowska, Z. (2003). Organellar inheritance in liverworts: an example of *Pellia borealis*. *Journal of Molecular Evolution*, **56**, 11–17.

Pacek, A., Fiedorow, P., Dabert, J. & Szweykowska-Kulinska, Z. (1998). RAPD technique for taxonomic studies of *Pellia epiphylla*-complex (Hepaticae, Metzgeriales). *Genetica*, **104**, 179–87.

Pacek, A. & Szweykowska-Kulinska, Z. (2003). Phylogenetic studies of liverworts from the genus *Pellia* using a new type of molecular marker. *Acta Societis Botanicorum Poloniae*, **71**, 227–34.

Patterson, E., Blake Boles, S. & Shaw, A. J. (1998). Nuclear ribosomal DNA variation in *Leucobryum glaucum* and *L. albidum* (Leucobryaceae): a preliminary investigation. *Bryologist*, **101**, 272–7.

Pfeiffer, T. (2000a). Relationships and divergence patterns in *Hypopterygium 'rosulatum'* s.l. (Hypopterygiaceae, Bryophyta) inferred from *trnL* intron sequences. Studies in austral temperate rain forest bryophytes. 7. *Edinburgh Journal of Botany*, **57**, 172–80.

Pfeiffer, T. (2000b). Molecular relationship of *Hymenophyton* species (Metzgeriaceae, Hepaticophytina) in New Zealand and Tasmania. Studies in austral temperate rain forest bryophytes. 5. *New Zealand Journal of Botany*, **38**, 415–23.

Pfeiffer, T., Kruijer, H., Frey, W. & Stech, M. (2000). Systematics of the *Hypopterygium tamarisci* complex (Hypopterygiaceae, Bryophyta): implications of molecular and morphological data. Studies in austral temperate rainforest bryophytes 9. *Journal of the Hattori Botanical Laboratory*, **89**, 55–70.

Pryer, K. P., Schuettpelz, E., Wolf, P. G. *et al.* (2004). Phylogeny and evolution of ferns (Monilophytes) with a focus on the early leptosporangiate divergences. *American Journal of Botany*, **91**, 1582–98.

Quandt, D. & Stech, M. (2004). Molecular evolution of the trnT(UGU)-trn*F*(GAA) region in bryophytes. *Plant Biology*, **6**, 545–54.

Quandt, D., Frahm, J. P. & Frey, W. (2001). Patterns of molecular divergence within the paleoaustral genus *Weymouthia* Broth. (Lembophyllaceae, Bryophyta). *Journal of Bryology*, **23**, 305–11.

Renker, C., Heinrichs, J., Pröschold, T., Groth, H. & Holtz, I. (2002). ITS sequences of nuclear ribosomal DNA support the generic placement and disjunct range of *Plagiochila (Adelanthus) carringtonii. Cryptogamie, Bryologie*, **23**, 31–9.

Rosenberg, N. A. (2003). The shapes of neutral gene phylogenies in two species: probabilities of monophyly, paraphyly, and polyphyly in a coalescent model. *Evolution*, **57**, 1465–77.

Rosenberg, N. A. (2007). Statistical tests for taxonomic distinctiveness from observations of monophyly. *Evolution*, **61**, 317–23.

Rosenberg, N. A. & Nordberg, M. (2002). Genealogical trees, coalescent theory and the analysis of genetic polymorphisms. *Nature Reviews*, **3**, 380–90.

Rosenzweig, M. L. (1995). *Species Diversity in Space and Time*. Cambridge: Cambridge University Press.

Ryall, K., Whitton, J., Schofield, W. B., Ellis, S. & Shaw, A. J. (2005). Molecular phylogenetic study of interspecific variation in the moss *Isothecium* (Brachytheciaceae). *Systematic Botany*, **30**, 242–7.

Såstad, S. M. (1999). Genetic and environmental sources of variation in leaf morphology of *Sphagnum fallax* and *Sphagnum isoviitae* (Bryopsida): comparison of experiments conducted in the field and laboratory. *Canadian Journal of Botany*, **77**, 1–10.

Såstad, S. M. (2005). Patterns and mechanisms of polyploid formation bryophytes. *Regnum Vegetabile*, **143**, 317–34.

Såstad, S. M., Flatberg, K. I. & Cronberg, N. (1999a). Electrophoretic evidence supporting a theory of allopolyploid origin of *Sphagnum jensenii*. *Nordic Journal of Botany*, **19**, 355–62.

Såstad, S. M., Stenoien, H. K. & Flatberg K. I. (1999b). Species delimitation and relationships of the *Sphagnum recurvum* complex (Bryophyta) – as revealed by isozyme and RAPD markers. *Systematic Botany*, **24**, 95–107.

Såstad, S. M., Flatberg, K. I. & Hanssen, L. (2000). Origin, taxonomy and population structure of the allopolyploid peat moss *Sphagnum majus*. *Plant Systematics and Evolution*, **225**, 73–84.

Såstad, S. M., Stenoien, H., Flatberg, K. I. & Bakken, S. (2001). The narrow endemic *Sphagnum troendelagicum* is an allopolyploid derivative of the widespread *S. balticum* and *S. tenellum*. *Systematic Botany*, **26**, 66–74.

Schaumann, F., Frey, W., Hassel de Menendez, G. & Pfeiffer, T. (2003). Geomolecular divergence in the Gondwanan dendroid *Symphyogyna* complex (Pallaviciniaceae, Hepaticophytina, Bryophyta). *Flora*, **198**, 404–12.

Schaumann, F., Pfeiffer, T. & Frey, W. (2004). Molecular divergence patterns within the Gondwanan liverwort genus *Jensenia* (Pallaviciniaceae, Hepaticophytina, Bryophyta). Studies in Austral temperate rain forest bryophytes 25. *Journal of the Hattori Botanical Laboratory*, **96**, 231–44.

Shaw, A. J. (1994). Systematics of *Mielichhoferia* (Bryaceae, Musci). III. Hybridization between *M. elongata* and *M. mielichhoferiana*. *American Journal of Botany*, **81**, 782–90.

Shaw, A. J. (1998). Genetic analysis of a hybrid zone in *Mielichhoferia* (Musci). In *Bryology for the Twenty-First Century*, ed. J. W. Bates, N. W. Ashton & J. G. Duckett, pp. 161–74. Leeds: Maney and British Bryological Society.

Shaw, A. J. (2000a). Population ecology, population genetics, and microevolution. In *Bryophyte Biology*, ed. A. J. Shaw & B. Goffinet, pp. 369–402. Cambridge: Cambridge University Press.

Shaw, A. J. (2000b). Molecular phylogeography and cryptic speciation in the mosses, *Mielichhoferia elongata* and *M. mielichhoferiana* (Bryaceae). *Molecular Ecology*, **9**, 595–608.

Shaw, A. J. (2000c). *Schizymenium shevockii* (Bryaceae), a new species of moss from California, based on morphological and molecular evidence. *Systematic Botany*, **25**, 188–95.

Shaw, A. J. (2001). Biogeographic patterns and cryptic speciation in bryophytes. *Journal of Biogeography*, **28**, 253–61.

Shaw, A. J. & Allen, B. (2000). Phylogenetic relationships, morphological incongruence, and geographic speciation in the Fontinalaceae (Bryophyta). *Molecular Phylogenetics and Evolution*, **16**, 225–37.

Shaw, A. J. & Goffinet, B. (2000). Molecular evidence of reticulate evolution in the peatmosses (*Sphagnum*), including *S. ehyalinum* sp. nov. *Bryologist*, **103**, 357–74.

Shaw, A. J. & Schneider, R. E. (1995). Genetic biogeography of the rare "copper moss," *Mielichhoferia elongata* (Bryaceae). *American Journal of Botany*, **82**, 8–17.

Shaw, A. J., Meagher, T. R. & Harley, P. (1987). Electrophoretic evidence of reproductive isolation between two varieties of the moss, *Climacium americanum*. *Heredity*, **59**, 337–43.

Shaw, A. J., Gutkin, M. S. & Bernstein, B. R. (1994). Systematics of tree mosses (*Climacium*, Musci) - genetic and morphological evidence. *Systematic Botany* **19**, 263–72.

Shaw, A. J., Cox, C. J. & Boles, S. B. (2003a). Polarity of peatmoss (*Sphagnum*) evolution: who says mosses have no roots? *American Journal of Botany*, **90**, 1777–87.

Shaw, A. J., Cox, C. J., Boles, S. B. & Goffinet, B. (2003b). Phylogenetic evidence for a rapid radiation of pleurocarpous mosses (Bryopsida). *Evolution*, **57**, 2226–41.

Shaw, S., Cox, C. J. & Boles, S. B. (2004). Phylogenetic relationships among *Sphagnum* sections, *Hemitheca, Isocladus*, and *Subsecunda*. *Bryologist*, **107**, 189–96.

Shaw, A. J., Cox, C. J. & Goffinet, B. (2005a). Global patterns of moss diversity: taxonomic and molecular inferences. *Taxon*, **54**, 337–52.

Shaw, A. J., Cox, C. J. & Boles, S. B. (2005b). Phylogeny, species delimitation, and interspecific hybridization in *Sphagnum* section *Acutifolia*. *Systematic Botany*, **30**, 16–33.

Shaw, A. J., Melosik, I., Cox, C. J. & Boles, S. B. (2005c). Divergent and reticulate evolution in closely related species of *Sphagnum* section *Subsecunda* (Bryophyta). *Bryologist*, **108**, 363–78.

So, M. L. & Grolle, R. (2000). Description of *Plagiochila detecta* sp. nov. (Hepaticae) from East Asia based on morphological and RAPD evidence. *Nova Hedwigia*, **71**, 387–93.

Soltis, D. E. & Soltis, P. S. (1999). Polyploidy: recurrent formation and genome evolution. *Trends in Ecology and Evolution*, **14**, 348–52.

Stech, M. & Dohrmann, J. (2004). Molecular relationships and biogeography of two gondwanan *Campylopus* species, *C. pilifer* and *C. introflexus* (Dicranaceae). *Monographs in Systematic Botany from the Missouri Botanical Garden*, **98**, 416–31.

Stech, M. & Frahm, J. P. (1999a). The status and systematic position of *Platyhypnidium mutatum* and the Donrichardsiaceae based on molecular data. *Journal of Bryology*, **21**, 191–5.

Stech, M. & Frahm, J. P. (1999b). Systematics of species of *Eurhynchium, Rhynchostegiella*, and *Rhynchostegium* (Brachytheciaceae, Bryopsida) based on molecular data. *Bryobrothera*, **5**, 203–11.

Stech, M., Frey, W. & Frahm, J. P. (1999). The status and systematic position of *Hypnobartlettia fontana* Ochyra and the Hypnobartlettiaceae based on molecular data. Studies on austral temperate rainforest bryophytes 4. *Lindbergia*, **24**, 97–102.

Stech, M., Osman, S., Sim-Sim, M. & Frey, W. (2006b). Molecular systematics and biogeography of the liverwort genus *Tylimantus* (Acrobolbaceae). Studies in austral temperate rain forest bryophytes 33. *Nova Hedwigia*, **83**, 17–30.

Stech, M., Pfeiffer, T. & Frey, W. (2006a). Molecular relationships and divergence of palaeoaustral *Dicranoloma* species (Dicranaceae, Bryopsida). Studies in austral temperate rain forest bryophytes 31. *Journal of the Hattori Botanical Laboratory*, **100**, 451–64.

Stenøien, H. K. & Flatberg, K. I. (2000). Genetic variability in the rare Norwegian peat moss *Sphagnum troendelagicum*. *Bryologist*, **103**, 794–801.

Stenøien, H. K. & Såstad, S. M. (1999). Genetic structure in three haploid peat mosses (*Sphagnum*). *Heredity*, **82**, 391–400.

Stevens, M. I., Hunger, S. A., Hills, S. F. K. & Gemmill, C. E. C. (2007). Phantom hitch-hikers mislead estimates of genetic variation in Antarctic mosses. *Plant Systematics and Evolution*, **263**, 191–201.

Stoneburner, A., Wyatt, R. & Odrzykoski, I. J. (1991). Applications of enzyme electrophoresis to bryophyte systematics and population biology. *Advances in Bryology*, **4**, 1–27.

Szövényi, P., Hock, Z., Urmi, E. & Schneller, J. (2006). New primers for amplifying the GapC gene in bryophytes and its ultility in infraspecific phylogenies in the genus *Sphagnum*. *Lindbergia*, **31**, 78–84.

Szweykowski, J. & Krzakowa, M. (1979). Variation in four isozyme systems in Polish populations of *Conocephalum conicum* (L.) Dum. (Hepaticae, Marchantiales). *Bulletin de l'Académie Polonaise des Sciences biologiques, classe II*, **27**, 27–41.

Szweykowski, J. & Odrzykoski, I. J. (1990). Chemical differentiation of *Aneura pinguis* (L.) Dum. (Hepaticae, Aneuraceae) in Poland and some comments on application of enzymatic markers in bryology. In *Chemotaxonomy of Bryophytes*, ed. H. D. Zinsmeister & R. Mues, pp. 437–48. New York: Academic Press.

Szweykowski, J., Odrzykoski, I. J. & Zielinski, R. (1981a). Further data on the geographic distribution of two genetically different forms of the liverwort *Conocephalum conicum* (L.) Dum.: the sympatric and allopatric regions. *Bulletin de l'Académie Polonaise des Sciences biologiques, classe II*, **28**, 437–49.

Szweykowski, J., Zielinski, R. & Mendelak, M. (1981b). Variation of peroxidase isoenzymes in central European taxa of the liverwort genus *Pellia*. *Bulletin de l'Académie Polonaise des Sciences biologiques, classe II*, **29**, 9–19.

Szweykowski, J., Zielinski, R. & Odrzykoski, I. J. (1995). Geographic distribution of *Pellia* spp. (Hepaticae, Metzgeriales) in Poland based on electrophoretic identification. *Acta Botanica Polonica*, **1**, 59–70.

Therrien, J. P., Crandall-Stotler, B. J. & Stotler, R. E. (1998). Morphological and genetic variation in *Porella platyphylla* and *P. platyphylloidea* and their systematic implications. *Bryologist*, **101**, 1–19.

Vanderpoorten, A., Boles, S. B. & Shaw, A. J. (2003a). Patterns of molecular and morphological variation in *Leucobryum albidum*, *L. glaucum*, and *L. juniperoideum* (Bryopsida). *Systematic Botany*, **28**, 651–6.

Vanderpoorten, A., Hedenäs, L. & Jacquemart, A. L. (2003b). Differentiation in DNA fingerprinting and morphology among species of the pleurocarpous moss genus, *Rhytidiadelphus* (Hylocomiaceae). *Taxon*, **52**, 229–36.

Vanderpoorten, A., Goffinet, B. & Quandt, D. (2006). Utility of the internal transcribed spacers of the 18S-5.8S-26S nuclear ribosomal DNA in land plant systematics with special emphasis on Bryophytes. In *Plant Genome: Biodiversity and Evolution*, vol. 2B,

Lower Plants, ed. A. K. Sharma & A. Sharma, pp. 385–407. Enfield, NH: Science Publishers.

Vanderpoorten, A., Shaw, A. J. & Cox, C. J. (2004). Evolution of multiple paralogous adenosine kinase genes in the moss genus *Hygroamblystegium*: phylogenetic implications. *Molecular Phylogenetics and Evolution*, **31**, 505–16.

Vanderpoorten, A., Shaw, A. J. & Goffinet, B. (2001). Testing controversial alignments in *Amblystegium* and related genera (Amblystegiaceae: Bryopsida). Evidence from rDNA ITS sequences. *Systematic Botany*, **26**, 470–9.

van der Velde, M. & Bijlsma, R. (2000). Amount and structure of intra- and interspecific genetic variation in the moss genus *Polytrichum*. *Heredity*, **85**, 328–37.

van der Velde, M. & Bijlsma, R. (2001). Genetic evidence for the allodiploid origin of the moss species *Polytrichum longisetum*. *Plant Biology*, **3**, 379–85.

van der Velde, M. & Bijlsma, R. (2004). Hybridization and asymmetric reproductive isolation between the closely related bryophyte taxa *Polytrichum commune* and *Polytrichum uliginosum*. *Molecular Ecology*, **13**, 1447–54.

Vos, P., Rogers, R., Blecker, M. *et al.* (1995). AFLP – a new technique for DNA fingerprinting. *Nucleic Acids Research*, **23**, 4407–14.

Vries, A. de, Van Zanten, B. O. & Van Dijk, H. (1983). Genetic variability within and between two species of *Racopilum* (Racopilaceae). *Lindbergia*, **9**, 73–80.

Wall, D. P. (2005). Origin and rapid diversification of a tropical moss. *Evolution*, **59**, 1413–24.

Welsh, J. & McClelland, M. (1990). Fingerprinting genomes using PCR with arbitrary primers. *Nucleic Acids Research*, **18**, 7213–18.

Wendel, J. (2000). Genome evolution in polyploids. *Plant Molecular Biology*, **42**, 225–49.

Werner, O., Ros, R. M., Guerra, J. & Cano, N. J. (2004). Inter-Simple Sequence Repeat (ISSR) markers support the species status of *Weissia wimmeriana* (Sendtn.) Bruch & Schimp. (Pottiaceae, Bryopsida). *Cryptogamie Bryologie*, **25**, 137–46.

Werner, O., Ros, R. M., Guerra, J. & Shaw, A. J. (2003). Molecular data confirm the presence of *Anacolia menziesii* (Bartramiaceae, Musci) in southern Europe and its separation from *Anacolia webbii*. *Systematic Botany*, **28**, 483–9.

Wettstein, F. von (1928). Morphologie und physiologie des Formwechsels der Moose auf genetischer Grundlage. II. *Bibliotheca Genetica*, **10**, 1–216.

Wettstein, F. von (1932). Genetik. In *Manual of Bryology*, ed. F. Verdoorn, pp. 233–72. The Hague: Martinus Nijhoff.

Wheeler, Q. & Meier, R. (eds.) 2000. *Species Concepts and Phylogenetic Theory: a Debate*. New York: Columbia University Press.

Wyatt, R. & Anderson, L. E. (1984). Breeding systems in bryophytes. In *The Experimental Biology of Bryophytes*, ed. A. F. Dyer & J. G. Duckett, pp. 39–63. London: Academic Press.

Wyatt, R. & Odrzykoski, I. J. (1998). On the origins of the allopolyploid moss *Plagiomnium cuspidatum*. *Bryologist*, **101**, 263–71.

Wyatt, R., Odrzykoski, I. J., Stoneburner, A., Bass, H. W. & Galau, G.-A. (1988). Allopolyploidy in bryophytes: Multiple origins of *Plagiomnium medium*. *Proceedings of the National Academy of Sciences, U.S.A.*, **85**, 5601–4.

Wyatt, R., Stoneburner, A. & Odrzykoski, I. J. (1989). Bryophyte isozymes: systematic and evolutionary implications. In *Isozymes in Plant Biology*, ed. D. E. Soltis & P. S. Soltis, pp. 221–40. Portland, OR: Dioscorides Press.

Wyatt, R., Odrzykoski., I. J. & Stoneburner, A. (1992). Isozyme evidence of reticulate evolution in mosses – *Plagiomnium medium* is an allopolyploid of *P. ellipticum* × *Plagiomnium insigne*. *Systematic Botany*, **17**, 532–50.

Wyatt, R., Odrzykoski, I. J. & Stoneburner, A. (1993a). Isozyme evidence regarding the origins of the allpolyploid moss *Plagiomnium curvatulum*. *Lindbergia*, **18**, 49–58.

Wyatt, R., Odrzykoski, I. J. & Stoneburner, A. (1993b). Isozyme evidence proves that *Rhizomnium pseudopunctatum* is an allopolyploid of *R. gracile* × *R. magnifolium*. *Memoirs of the Torrey Botanical Club*, **25**, 21–35.

Wyatt, R., Odrzykoski, I. J. & Koponen, T. (1997). *Mnium orientale* sp. nov. from Japan is morphologically and genetically distinct from *M. hornum* in Europe and North America. *Bryologist*, **100**, 226–36.

Zanten, B. O. van (1978). Experimental studies on trans-oceanic long range dispersal of moss spores in the southern hemisphere. *Journal of the Hattori Botanical Laboratory*, **44**, 455–82.

Zhou, P. & Shaw, A. J. (2008). Systematics and population genetics in *Sphagnum macrophyllum* and *S. cribrosum* (Sphagnaceae). *Systematic Botany* (in press).

Zielinski, R. (1984). Electrophoretic and cytological study of the *Pellia epiphylla* and *Pellia borealis* complex. *Journal of the Hattori Botanical Laboratory*, **56**, 263–69.

Zielinski, R. (1987). Interpretation of electrophoretic patterns in population-genetics of Bryophytes. 6. Genetic variation and evolution of the liverwort genus *Pellia* with special reference to central European territory. *Lindbergia*, **12**, 87–96.

Zielinski, R., Szweykowski, J. & Rutkowska, E. (1985). A further electrophoretic study peroxidase isoenzyme variation in *Pellia epiphylla* (L.) Dum. from Poland, with special reference to the status of *Pellia borealis* Lorbeer. *Monographs in Systematic Botany from the Missouri Botanical Garden*, **11**, 199–209.

Zietkiewicz, E., Rafalski, A. & Labuda, D. (1994). Genome fingerprinting by simple sequence repeat (SSR)-anchored polymerase chain reaction amplification. *Genomics*, **20**, 176–83.

Zouhair, R., Corradini, P., Defontaine, A. & Hallet, J. N. (2000). RAPD markers for genetic differentiation of species within *Polytrichum* (Polytrichaceae, Musci): a preliminary survey. *Taxon*, **49**, 217–29.

12

Conservation biology of bryophytes

ALAIN VANDERPOORTEN AND TOMAS HALLINGBÄCK

12.1 Introduction

Conservation biology is a fairly new, multidisciplinary science that has developed to deal with the crisis confronting biological diversity (Primack 1993). As a crisis discipline, conservation biology arose in response to an increasingly formulated political demand to face the dramatic loss of biodiversity and the need to take steps to anticipate, prevent, and reverse the trend (Heywood & Iriondo 2003). Subsequent ratification of the Convention on Biological Diversity at the United Nation conference held in Rio in 1992 by most of the world's governments has placed the subject of biodiversity firmly on the political agenda.

The past few years have witnessed a major evolution in our understanding of conservation. The increasing need for performing tools has rendered conservation biology a truly multidisciplinary science, feeding on a variety of other areas, including ecology, demography, population biology, population genetics, biogeography, landscape ecology, environmental management, and economics (Heywood & Iriondo 2003). Conservation interest has also been progressively enlarged to include a broad array of taxa that used to be completely overlooked. Cryptogams were, for example, the focus of only about 4% of published papers between 2000 and 2005 in leading conservation journals (Hylander & Jonsson 2007). The situation has been most recently changing and there has been an increasing awareness of the necessity to include cryptogams in general, and bryophytes in particular, in conservation programs (Hylander & Jonsson 2007).

The reasons for a late but growing interest in bryophyte conservation are manifold. Although bryophytes are rarely the most conspicuous elements in the

Bryophyte Biology: Second Edition, ed. B. Goffinet & A. J. Shaw. Published by Cambridge University Press.

landscape, they play important ecological roles in terms of water balance, erosion control, or nitrogen budget, or simply by providing habitat for other organisms (Longton 1992). Furthermore, bryophytes locally exhibit richness levels that are comparable to or even higher than those of angiosperms. In boreal forests, for example, bryophyte diversity often exceeds that of vascular plants at a scale of 0.1 ha (Berglund & Jonsson 2001). In wet sclerophyll forest of Tasmania, the ratio of the number of bryophyte to vascular plant species is often 5 : 1 (Pharo & Blanks 2000). Lastly, and perhaps most importantly, diversity patterns in bryophytes do not necessarily follow the patterns present in other, better-studied taxa (Sérgio *et al.* 2000, Pharo *et al.* 2005), so that an enlarged concept of biodiversity has become increasingly necessary.

In this chapter, we review the tools that are available for assessing threat levels in bryophytes and emphasize in particular how the IUCN classification system can be applied to the specific case of bryophytes. We then provide an overview of global threat levels and conservation needs and review the mechanisms by which bryophytes are, at least locally, severely threatened. Finally, we discuss appropriate conservation strategies for preserving and managing bryophyte diversity. We conclude by some perspectives regarding the need for and possibilities of implementation of a novel, evolutionary approach to biodiversity that may complement and, perhaps, eventually replace the traditional approach focused on threat levels and phenetic species concept.

12.2 Levels of threats and the need for conservation

12.2.1 *What to conserve? A hierarchical system of threat categories applied to bryophytes*

The IUCN classification system

Conservation, "the philosophy of managing the environment in such a way that does not despoil, exhaust, or extinguish it" (Jordan 1995), is by definition concerned with the threat of extinction of species, communities, or ecosystems due to human activities. To date, the number of species believed to be under some degree of threat makes necessary the use of a system of classification that helps categorize species according to the risk of extinction they are facing (Heywood & Iriondo 2003).

The likelihood of extinction of a species must be assessed against certain criteria. This is the purpose of a red list. The most obvious option for bryophyte species status assessment is to apply the most recent criteria and threat categories of the International Union for the Conservation of Nature and Natural Resources (hereafter, IUCN) (IUCN 2001). The IUCN criteria have the advantages that (a) they have been elaborated after much thought by a great number of

Table 12.1 *IUCN criteria of species threat categories*

Criterion	Threshold
Declining population	30–90% population decline during a time period of 10 years or 3 generations, whichever is the longest
Rarity and decline	EOO[a] <20 000 km^2 or AOO[b] <2000 km^2 and severe fragmentation, continuing decline, or extreme fluctuations
Small population size and fragmentation, decline, or fluctuations	Population size <250 reproductive individuals and continuing decline of >10% in 10 years or 3 generations
Very small population size or very restricted distribution	Number of individuals <1000 or AOO[b]<20 km^2 (or <5 locations)
Quantitative analysis of extinction risk	Population viability analysis (Gärdenfors 2000) or any other form of analysis estimating extinction probability

[a] Extent of Occurrence (EOO) is the geographical range, defined as the area contained within the shortest continuous imaginary boundary that can be drawn to encompass all the known, inferred, or projected sites of present occurrence of a taxon.
[b] Area of Occupancy (AOO) is defined as the area, calculated by summing up all grid squares with the mesh size of 2 km × 2 km that are actually occupied by a taxon, excluding cases of vagrancy.

experts; (b) they carry international weight, so that any red list using these criteria is much more powerful than a list using alternative criteria; and (c) they have a clear, repeatable methodology, applicable in a wide range of circumstances and geographical areas. The IUCN red listing system can be used at different geographic scales. In September 2003, the Species Survival Commission published the *Guidelines for Application of IUCN Red List Criteria at Regional Levels* (IUCN 2003).

Five quantitative criteria are estimated to determine whether a taxon is threatened or not (Table 12.1) and, if so, to which of the seven threat categories it belongs to (Fig. 12.1, Box 12.1).

Application of the IUCN 2001 red listing system to bryophytes

Although the IUCN criteria are mainly adapted to animals, Hallingbäck *et al.* (1998) showed how they could be used for bryophytes according to a protocol that has now been adopted officially by IUCN. Assigning bryophytes to a threat category is often associated with four main difficulties regarding the definition of an individual and of the generation length, the assessment of a fragmented distribution (see Section 12.3.2), and by the absence of proper distribution data (Hallingbäck 2007).

Box 12.1 The IUCN categories in brief

Extinct (EX): A taxon is Extinct when there is no reasonable doubt that the last individual has died. For bryophytes, a species is considered EX or RE when it has not been seen for the past 50 years throughout its entire distribution range (EX) or in part of it, in which case the taxon can be included in the "Regionally Extinct" category (RE). Alternatively, the taxon must be confined to sites that have been thoroughly surveyed without success in recent years, or where suitable habitat has disappeared. Example: *Neomacounia nitida* (Lindb.) Ireland[a]

Critically Endangered (CR) corresponds to a ≥50% risk of extinction in 10 years or 3 generations. For bryophytes, the criterion based on the distribution area (AOO) is most easy to use. The taxon must be known from a single or several severely fragmented locations[b] covering altogether <10 km^2 and experience a continuing decline. Decline in habitat size or quality is most often used as a surrogate for actual population decline in the lack of actual data on population trends. Example: *Thamnobryum angustifolium* (Holt) Crundw.[a]

Endangered (EN) corresponds to a risk of extinction ≥20% in 20 years or 5 generations. The AOO must be <500 km^2. In addition, the taxon must be known from ≤5 locations[a] (or >5 if these are severely fragmented; see IUCN 2001) and also have experienced a continuing decline. Example: *Caudalejeunea grolleana* Gradst.[a]

Vulnerable (VU) corresponds to a risk of extinction ≥10% in 100 years. The AOO must be <2000 km^2. In addition, the taxon must be known from ≤10 locations[a] (>10 if these are severely fragmented; see IUCN 2001) and have experienced a continuing decline. Example: *Hypnodontopsis apiculata* Z. Iwats. and Nog.[a]

Data Deficient (DD) is applied to taxa with insufficient data to categorize them, but which are thought likely to qualify as Extinct, Critically Endangered, Endangered or Vulnerable when they are better known.

Near Threatened (NT): A taxon is NT when it is close to qualifying for VU.

Least Concern (LC) is applied to taxa that do not qualify (and are not close to qualifying) as threatened or near threatened.

[a] See www.artdata.slu.se/guest/SSCBryo/WorldBryo.htm.

[b] The term "location" defines a geographically or ecologically distinct area in which a single threatening event can rapidly affect all individuals of the taxon present (IUCN 2001).

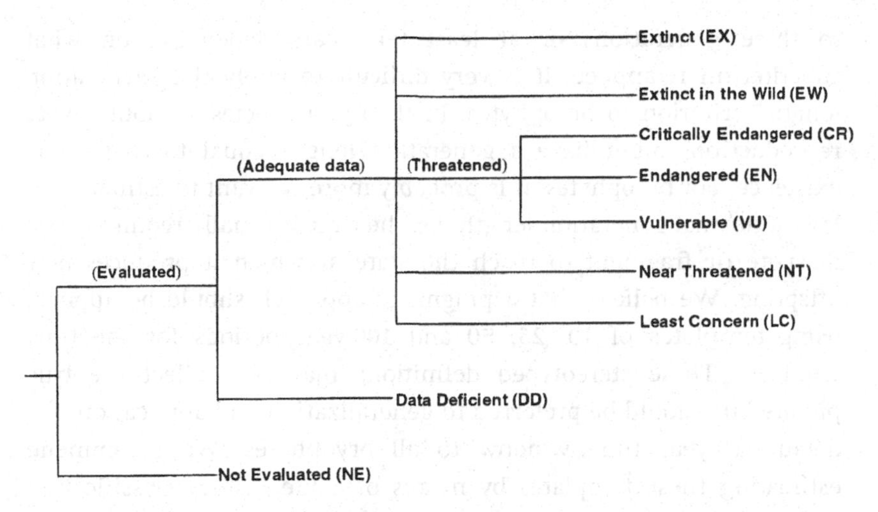

Fig. 12.1. Structure of the IUCN threat categories. See text and Box 12.1 for details.

(a) The concept of individual. The IUCN system defines an individual, hereafter termed IUCN-individual, as a distinguishable entity that is able to survive and reproduce. Although this definition is theoretically and biologically sound, practical application of this principle is impossible. One of its major drawbacks is that many bryophytes are mostly clonal, with a broad array of ramet and genet sizes, plus different survival and reproductive strategies (see Chapters 10 and 11, this volume). What is the entity that best corresponds to discrete individuals like animals? For most bryophyte species, no specific information about the number of ramets is available, which means that standardized templates have to be used to make population estimates. For practical reasons, a purely pragmatic definition is therefore used in the red list assessment process. For species that depend on discrete substrate entities (such as tree trunks or droppings), each substrate entity can be considered to contain one or two IUCN-individuals. For bryophyte species growing on ground or rocks, one IUCN-individual may be assumed to occupy a surface of 1 m^2. However, in the cases of some very small mosses (e.g. the genera *Seligeria* and *Tetrodontium*), one individual should be associated with a surface of 0.1 m^2.

(b) Generation length. IUCN has a definition of generation length, which reflects the turnover (reproductive rate) of individuals/ramets in the population. According to IUCN criteria, the population development of a species should be assessed over a time period equivalent to one up to three generations or at least ten years depending on what subcriterion is applied. It is very difficult to apply the "generation length" criterion to bryophytes. In theory, a species without sexual reproduction might have a generation length equal to that of its existence. For bryophytes, it is probably more relevant to estimate the lowest normal generation length, i.e. the time normally required for a diaspore or fragment to reach the state at which it produces new offspring. We believe that a pragmatic approach should be applied, using templates of 10, 25, 50 and 100-year periods for the time windows. These stereotyped definitions may not reflect the true picture but should be preferred to generalization and application of a default 10-year time window to all bryophytes. We recommend estimating these templates by means of a life history classification (During 1992, Söderström 2002, Söderström & During 2005). The main reason for using life history classification as a template for generation length is that the different strategy classes typically have different reproduction strategies; early (quick) vs. late (delayed) reproduction. These strategies reflect natural turnover rates. Generally, species that reproduce only after many years of growth exhibit the longest generation length, and substantial evidence indicates that individual mats or cushions may be very long-lived. A cushion may last for as long as it takes for the substratum to decay, and the life span of a terricolous moss bolster can be counted in decades and perhaps even centuries. For instance, Bates (1989) estimated from growth rates the age of large cushions of the terrestrial forest moss *Leucobryum glaucum* at about 85 years, and colonies of the strictly epiphytic moss *Neckera pennata* may last as long as 50 years (Wiklund & Rydin 2004b). There is, however, an enormous difference between the potential lifespan, achieved by very few individuals, and the actual average generation length. Therefore, a generation length of 50 years can be attributed to species with a long-lived shuttle strategy and to stayers, i.e. species such as *Leucobryum glaucum*, which often grow on stable ground or on rock. By contrast, typical colonist species, e.g. *Ceratodon purpureus*, exhibit a much shorter time span of less than 10 years. In the case of so-called short-lived shuttle species (During 1992), a group that includes most epiphytes, a period of 25 years may be used as equivalent to three generations.

(c) Absence of proper past and present distribution data. Species' extent of occurrence (EOO) is relatively accurately assessed on the basis of mapping schemes at the 2 km × 2 km level of resolution (IUCN 2001) and ecological niche modeling methods (see Section 12.4.2). Because frequency estimates are most often biased, owing to the existence of, for example, under-recorded areas and easily overlooked taxa (Urmi & Schnyder 2000), it can be useful to include an estimated level of uncertainty in the assessments. An uncertainty value of 3×, for instance, means that we estimate that only 1/3 of the true number of individuals or localities is known. Similarly, trends in frequency are fairly easy to assess when historical data from systematic surveys of species distributions are available (e.g. Bates 1995a). This is, however, rarely the case, and herbarium-based methods have been proposed to estimate, with a degree of uncertainty, past species distribution frequencies (Hedenäs *et al.* 2002, Zechmeister *et al.* 2007). However, detailed past and current distribution data are mostly only available in some European countries with a long tradition of floristic mapping. The limited information on species distribution and trends in countries where taxonomic information and floras are scarce, or even lacking, renders the implementation of the IUCN system difficult.

12.2.2 *Level of threat in the bryophyte floras*

At a worldwide scale, only 80 species (36 mosses, 43 liverworts and one hornwort) are included in the IUCN *World Red List of Bryophytes* (www.iucnredlist.org/ or http://www.artdata.slu.se/guest/SSCBryo/WorldBryo.htm) on the basis of their global threat level, occurrence in threatened habitats, and narrow distribution range. This list is not yet comprehensive and the "narrow endemic" condition precludes the inclusion of truly rare but widespread species with transoceanic distributions, which is quite a common pattern among bryophytes (Shaw 2001).

In fact, despite a tendency for occupying wide distribution ranges, the vast majority of bryophyte species are sparsely distributed, within a given area or globally (Cleavitt 2005). Species frequency distributions are typically highly skewed at a given geographic scale, with rare – and potentially threatened – species representing the bulk of the flora (Longton & Hedderson 2000). Table 12.2 shows the levels of threat at the national and regional scale in different countries. Although this exercise is mostly focused on Europe owing to the limited availability of precise data on species distributions and threat levels on other continents, it nevertheless shows that rates of extinction in most countries are already 2%–4% and that a substantial proportion of the flora is threatened in the short term.

12.2.3 *Implementation of threat levels in legislation*

In view of the vulnerability of bryophytes, a logical step beyond the recognition of highly threatened taxa is the implementation of threat levels into

Table 12.2 *Level of threat in the bryophyte flora of selected countries and areas*

	Mosses		Liverworts		
Area or country	Rates of extinction (%)[a]	% threatened species[b]	Rates of extinction (%)[a]	% threatened species[b]	Reference
Czech Republic	3.1	22.1	2.8	35.0	Kucera & Vana 2005
Estonia	2.2	3.1	0.0	4.2	Ingerpuu 1998
Finland	2.6	14.0	3.7	20.5	Ulvinen *et al.* 2002
Iberian Peninsula	1.3	9.9	0.0	11.7	Sérgio *et al.* 1994
Japan	—	9.0	—	13.0	Iwatsuki *et al.* 2000
Luxembourg	0.2	34.5	5.3	36.1	Werner 2003
New Zealand	0.0	5.3	0.0	8.6	Glenny & Fife 2005
Norway	—	14.8	—	9.9	Flatberg *et al.* 2006
Poland	—	—	0.8	15.5	Klama 2006
Rio de Janeiro state (Brazil)	—	12.1	—	19.2	Da Costa *et al.* 2005
Serbia and Montenegro	0.5	51.0	0.0	81.0	Sabovljevic *et al.* 2003
Slovakia	2.6	33.5	3.6	36.6	Kubinská *et al.* 2001
Sweden	1.9	9.3	1.2	9.3	Gärdenfors 2005
Switzerland	1.4	34.0	1.2	47.2	Schnyder *et al.* 2004
The Netherlands	3.0	29.0	2.0	31	Siebel *et al.* 2006
U.K.	2.3	13.2	0.3	11.0	Church *et al.* 2001

[a] RE *sensu* IUCN.
[b] Sensu IUCN (see Box 12.1).

the legislation. Although the level of legal protection remains very low, bryophyte species have increasingly been included in legal texts regulating the collection of selected species [e.g. the Swiss Nature and landscape protection decree, appendix 2 (www.admin.ch/ch/f/rs/451_1/index.html); the Canadian Species at Risk Act (McIntosh & Miles 2005); the European Annex V of the Habitats Directive (http://ec.europa.eu/environment/nature/nature_conservation/eu_nature_legislation/habitats_directive/index_en.htm); and Schedule 8 of the British Wildlife and Countryside Act 1981 (www.jncc.gov.uk/)]. In the European Union, for example, the habitat of 32 more or less threatened bryophyte species are protected under Annex II of the Habitats Directive, which has already led to the protection of more than 1000 localities included in the Natura 2000 network (Hallingbäck 2003). In the U.K. similarly, bryophytes may be included within a Site of Special Scientific Interest (SSSI), and guidelines have been produced to aid the selection of SSSI on the basis of the sum of species threat levels, as defined in the British Red Data Book (www.jncc.gov.uk/).

12.3 Why are bryophytes threatened?

12.3.1 *What biological properties make bryophytes vulnerable?*

As Cleavitt (2005) summarizes, rarity – and vulnerability – are linked to a series of intrinsic properties of bryophytes regarding dispersal ability, genetic potential, habitat tolerances, competitive ability, reproduction and establishment, and survival rates, whose significance for conservation are briefly reviewed below.

Dispersal

Low dispersal ability of rare species is a fundamental assumption in the metapopulation theory framework (Cleavitt 2005, see also Chapter 10, this volume). Demographic processes are especially crucial in fugitive species such as epiphytes, for which substrate availability lasts for a limited amount of time. Thus, epiphytes are critically dependent on their ability to disperse to new patches, and metapopulation models predict that habitat density is the crucial factor for their persistence (Hazell *et al.* 1998).

The dispersal ability of bryophytes has long been debated (see Shaw 2001 for a review). Several pieces of evidence from spore durability experiments (Van Zanten 1978, Van Zanten & Gradstein 1988), intercontinental transport of spores on airplane wings via jetstreams (Van Zanten & Gradstein 1988), correlative analyses between species distributions and air currents (Muñoz *et al.* 2004), interpretation of species distributions in a phylogenetic context (e.g. Shaw *et al.* 2003, Heinrichs *et al.* 2005, Hartmann *et al.* 2006), and genetic inferences of dispersal (McDaniel & Shaw 2005), all point to an overall good ability of bryophytes for long-distance dispersal. The capacity of bryophytes to successfully and routinely disperse at the landscape scale, which is a crucial feature for the successful long-term persistence of the populations, is, however, poorly documented. Observations on the colonization of artificial habitats by species far from their nearest natural distribution area (Bates 1995a, Vanderpoorten & Engels 2003, Miller & McDaniel 2004) suggest that some bryophytes may be capable of routine dispersal over distances of at least 5 km. However, some studies demonstrated a significant tendency for spatial aggregation in epiphytes (Snäll *et al.* 2003, 2004b, 2005, Löbel *et al.* 2006; but see Hazell *et al.* 1998 and Hedenäs *et al.* 2003). A significant degree of kinship, derived from a genetic analysis of spatial structure, was furthermore found among individuals up to 350 m apart. This further supports the idea that at least some species have a restricted dispersal range (Snäll *et al.* 2004a). Limited dispersal ability has been evoked to explain the rarity of dioicous, rarely fertile species across a landscape apparently composed of favorable habitats (Vanderpoorten *et al.* 2006). The successful transplantation of rare species into potential habitats further demonstrates that many taxa are not limited by the availability of suitable habitats but rather by their poor ability to colonize them (Kooijman *et al.* 1994, Gunnarsson & Söderström 2007). The reasons for the limited dispersal ability of such species remain, however, poorly understood. For instance, sporophyte production is not lower in rare monoicous mosses and liverworts than in common ones (Longton & Hedderson 2000, Laaka-Lindberg *et al.* 2000); rare species often produce asexual gemmae in abundance; and the tolerance to dessication of gemmae is equivalent or even higher in rare than in common species (Cleavitt 2002).

Ecological range

Rare species tend to occupy narrower ecological niches than common ones (Cleavitt 2005). A typical example is the widespread but discrete occurrence of copper mosses on heavily contaminated soils, suggesting that habitat exclusivity is, in itself, a cause of rarity. However, whereas patterns of habitat specificity are well documented, the link to experimentally demonstrated physiological tolerances is often lacking (Cleavitt 2005). One of the key factors at different stages of the bryophyte life cycle is water availability. Water availability is most important at the germination stage. In an experiment on the mosses *Buxbaumia viridis* and *Neckera pennata*, spores had the capacity to germinate at water potential as low as –2 MPa, a value at which most seeds fail to germinate, but only if pH was >5 (Wiklund & Rydin 2004a). The interaction between pH and water potential effects on germination suggests that high moisture facilitates germination at suboptimal pH and vice versa. Further, pH and water potential determine the length of the lag phase preceding germination. The number of days needed for germination to start varied between 2 and 50 days depending on pH and water availability (Wiklund & Rydin 2004a). This time effect is ecologically important because delayed spore germination increases the risk of dessication or disappearance of spores through wind or predation. Wiklund & Rydin (2004a) therefore suggested the existence of a general trade-off between the ability of moss spores to colonize substrates with low moisture-holding capacity and low pH. This trade-off implies that substrata prone to fast desiccation (such as bark) can be colonized only if they have a fairly high pH. By contrast, substrata with a high water-holding capacity, such as wood in late stages of decay, or peat, can be colonized despite low pH.

At the gametophytic stage, the lack of roots and a thick cuticle renders the plant water status more or less dependent to the humidity of the environment. Bryophytes are poikilohydric (Chapter 6, this volume), which means that they are physiologically active only when water is available. As a result, a number of bryophytes are desiccation-intolerant. For instance, shade epiphytes, which are characteristic of the understorey of dense primary forests, are considered to be less desiccation tolerant than sun epiphytes and generalists that developed a series of putative adaptations, such as papillose cell walls, which enhance the capillary absorption and speed-up the process of rehydration. Shade epiphytes are therefore highly sensitive to deforestation (Gradstein 1992a,b, Gradstein *et al.* 2001, Acebey *et al.* 2003) and more likely threatened (Gradstein *et al.* 2001).

Genetic potential and adaptation

The ability of species to successfully disperse is linked to their reproduction mode. The absence of sexual reproduction, which results in a lack of genetic recombination, a severely limited genetic variability and a compromised capacity to adapt, may be the most important factor leading to rarity. In dioicous species, but not in monoicous ones, rarity is significantly associated with the absence of sporophyte production (Longton & Hedderson 2000, Laaka-Lindberg *et al.* 2000). Monoicous species would thus tend to become rare if self-fertilization becomes obligate, while dioicous species tend to become rare if they fail to produce sporophytes owing to limitations in sperm mobility (Longton & Hedderson 2000). Population genetic studies, however, yielded conflicting results regarding the amount of genetic variation within rare species (Wyatt 1992, Gunnarsson *et al.* 2005, Werner *et al.* 2005). Furthermore, as Oostermeijer *et al.* (1995) indicated, there is an overall need to better integrate genetic and ecological studies with the study of the processes that condition the viability of the populations; for example, by testing that individuals with a higher number of heterozygous loci display significantly higher fitnesses (e.g. spore numbers, viability, more robust offspring, etc.). At present, while the accumulation of data on the population genetics of bryophytes remains very slow (see Pharo & Zartman 2007, for review), the integration of genetics into population ecology is still completely lacking.

Competitive ability

A common assumption is that rare species are restricted to specific habitat conditions by competition. At high carpet densities, individual shoots are indeed deprived of light and may become overtopped by larger neighbors (Frego & Carleton 1995). One solution for species with low competitive skills is to escape, either in space by dispersal to a new, uninvaded spot, or in time, by resting in the diaspore bank until a disturbance event renders their growth possible again. This is one of the ideas behind the concept of "life-strategies" (During 1992, Söderström & During 2005). However, the interactions among individuals that shape the spatial structure of the populations can also have positive effects (Okland & Bakkestuen 2004). At low to moderate shoot densities, growth is constrained by water availability. Because moderately dense stands dehydrate less rapidly than loose stands or individual shoots, bryophyte growth is often positively related to carpet density. Conflicting conclusions on the significance of interspecific competition in bryophytes have, therefore, been repeateadly reported (McAlister 1995, Rydin 1997, Zamfir & Goldberg 2000, Pedersen *et al.* 2001, Wiklund & Rydin 2004b). The effect of competition seems complex and dependent on a range of factors, including growth stage and habitat condition (see Chapter 10, this volume). As underlined by Cleavitt (2005), further experiments are needed to test the impact of competition on the survival of threatened species.

12.3.2 *What mechanisms cause bryophytes to be threatened?* Direct threats: collecting and harvesting

Scientific collecting

Collecting of specimens for scientific purposes is usually highly selective and seldom constitutes a real threat to the survival of species. The extinction of species by a targeted over-collecting has, however, already been documented (Church *et al.* 2001). Collecting is especially an issue in the case of unique species known only from the type or a few localities. Typical examples include the pleurocarp *Donrichardsia macroneuron*, a Texan endemic known from a few spring areas (Wyatt & Stoneburner 1980) and the highly peculiar monotypic peatmoss *Ambuchanania*. *Ambuchanania* is known from two localities including the type locality, which has not been relocated nor revisited since the original collection. The species, whose abundance is unknown, was therefore placed on the rare and endangered species list of Tasmania to protect it from overcollecting or, worse, possible extinction (R. Seppelt, pers. comm.). Scientific collecting of bryophytes is still essential for a number of reasons, including specimen identification, herbarium collections for taxonomic studies, and, more recently, constitution of DNA libraries. When it comes to rare species, however, it is recommended to (i) ensure that the material is not already available in herbarium or other institutional collections; (ii) place all collected specimens in institutions where they can be preserved and be made available to other scientists, thus limiting the need for further collections; (iii) submit copies of reports and publications in a timely manner to permit-issuing agencies; and (iv) avoid making public the exact geographical information of the actual localities.

Commercial harvest

The high water-holding capacity of bryophytes makes them a useful potting medium, particularly favored by orchid growers and for wrapping flowers or fruit tree rootstock for transportation. At present, although outdoor *Sphagnum* nurseries is an interesting option for a new type of professional horticulture (Rochefort & Lode 2006), all bryophyte harvesting is from natural populations. Local regulations sometimes exist [for example, the EU Directive 92/43/EEC (Habitats Directive) in Europe (http://eur-lex.europa.eu/LexUriServ/LexUriServ.do?uri=CELEX:52001DC0162(02):EN:HTML); the Flora and Fauna Guarantee in Victoria, Australia)], but this activity is globally seldom monitored and can result in considerable ecological damage and decline in bryophyte diversity. Initially mostly focused on *Sphagnum* (e.g. Whinam & Buxton 1997), commercial moss harvesting has been increasing in several countries including the U.S.A., Mexico, Venezuela, India, and China (Peck & Muir 2001, Leon & Ussher 2005, Muir *et al.* 2006). Although species richness may exceed preharvest levels because harvested stands represent new windows of opportunities for establishment, a survey conducted on epiphytes of the Pacific Northwest area indicated that cover on vine maple shrubs immediately following harvest was reduced by 5% and 16%–20% for low- (*c.* 34 kg ha^{-1}) and high-intensity harvest treatments (*c.* 112 kg ha^{-1}), respectively (Peck & Christy 2006). Differences between the two treatments disappeared after one or two years (Peck & Christy 2006), but a long-term evaluation of cover and species richness following simulated commercial moss harvest indicates that cover regrowth may require 20 years at a rate of 5% per year, and volume recovery even longer (Peck 2006a). Slow rates of accumulation and the unwanted harvest of non-target species (e.g. red listed species) provide the incentive to manage and monitor the harvest in order to ensure sustainability and maintain diversity (Vance & Kirkland 1997, Peck & Muir 2001). Commercial moss harvest should be managed on rotations of several decades, and patchy harvest methods should be encouraged over complete strip harvesting to ensure moss regeneration (Peck 2006a). For example, forest stands may be leased to commercial moss harvesters according to a rotating scheme ensuring sufficient recovery between harvest entries, under the condition that the harvesters adhere to specific guidelines and improve the control of illegal harvest in the lease area (Peck 2006b, Peck & Christy 2006).

Indirect threats: habitat destruction, degradation, and fragmentation

The distribution and abundance of communities are governed by demographic processes such as immigration, which maintain or increase species richness, and local ecological factors such as habitat modification and competitive exclusion, which promote extinction (Ricklefs 1987). Habitat modification, which includes a range of factors (Table 12.3), affects populations in both the short term, through local population decrease or extinction, and the long term, through habitat fragmentation. In a fragmented landscape, small populations increasingly isolated from each other are prone to edge effects as well as demographic, environmental (i.e. natural catastrophes and anthropogenic accidents listed in Table 12.3), and genetic stochasticity (Chapter 10, this volume). The latter, also termed genetic drift, involves the random loss of alleles from small populations. Genetic drift may have severe long-term demographic consequences for self-incompatible or dioicous species owing to the fixation of a single allele, gender, or morph, and limited adaptive potential for new environments. In the long term, reduced gene flow among fragmented populations may also lead to increased inbreeding. Inbreeding, which involves the redistribution of alleles from heterozygous to homozygous combinations, may lead to the expression of deleterious recessive mutations and a resulting reduced fitness (e.g. low germination rates, high mortality, and poor growth and reproductive ability of the offspring), rendering the long-term persistence of the population in a fragmented landscape questionable (Oostermeijer *et al.* 2003).

Fragmentation therefore appears as one of the key issues in biological conservation (Heywood & Iriondo 2003). Bryophytes, and epiphyllous liverworts in particular, experience accelerated life cycles, high rates of local extinction, and naturally patchy substrates, and therefore represent ideal models to test metacommunity-based predictions associated with habitat fragmentation (Zartman 2003, Pharo & Zartman 2007). Despite this, threat assessments on the bryophyte floras have mostly focused on habitat disturbance, degradation and destruction (Table 12.3), whereas a larger overview of the mechanisms of those factors on the diversity and long-term life expectancy of communities that have been fragmented in the landscape has only recently been approached (Pharo *et al.* 2005).

Fragmented epiphytic bryophyte populations typically follow the island model, which states that the number of occurrences of a given species is almost always lower on small than on large remnants surrounded by a treeless matrix (Moen & Jonsson 2003, Zartman 2003). Two competing hypotheses have been proposed to explain this pattern.

The maintenance of an equilibrium between extinction and colonization, which requires an area sufficiently large for the preservation of both inhabited and potentially inhabitable patches, is one of the central predictions of the metapopulation theory. Following this scheme, Zartman & Nascimento (2006) observed that the reductions in mean epiphyll abundance was best predicted by

Table 12.3 *Overview of the impact of disturbance-related factors on bryophytes*

Factor	Effect	References
Land use		
Agriculture		
Physical disturbance	Ploughing and sowing right after crop harvest hamper the development of winter annuals.	Porley 2001
Use of fertilizers	Bryophytes have the ability to utilize a range of compounds from commercial agricultural fertilizers, but eutrophication and increased vascular plant competition are detrimental to acrocarpous mosses and thalloid liverworts of arable lands with low competitive skills.	Brown 1992, Porley 2001, Zechmeister *et al.* 2003
Forestry	Short-rotation harvesting and clear-cutting cause sudden exposure and a drastic decrease in the amount of dead wood and old trees. The resulting disturbance negatively impacts old-growth species diversity and composition and favors the introduction and spread of newcomers.	Hyvönen *et al.* 1987, Gradstein *et al.* 2001, Hannerz & Hånell 1997, Boudreault *et al.* 2000, Cobb *et al.* 2001, Berg *et al.* 2002, Newmaster & Bell 2002, Ross-Davis & Frego 2002, Acebey *et al.* 2003, Drehwald 2005, Dynesius & Hylander 2007
Hydrological and wetland alterations		
Stream regulation	Enhanced substrate stability causes a decrease in riparian species overgrown by angiosperm development but genuine aquatics increase on stable substitution habitats (weirs).	Muotka & Virtanen 1995, Englund *et al.* 1997, Vanderpoorten & Klein 1999, Downes *et al.* 2003
Drainage and water abstraction	Spring bryophytes (e.g. *Philonotis*) decline in favor of other groups, including *Sphagnum*.	Heino *et al.* 2005
Peat extraction	Excavation for fuel in more than 50 countries (60–70 million tonnes of oil equivalent in 2000, representing about 10% of energy use in countries such as Ireland), gardening and horticulture, severely threaten bog communities.	Hinrichsen 1981, Asplund 1996, Rydin & Jeglum 2006, Rochefort & Lode 2006

Table 12.3 (*cont.*)

Factor	Effect	References
Global change and pollution		
Nitrogen deposits	Eutrophication and increase in cover of vascular plants result in substantial decrease in bryophyte biomass and diversity.	Bergamini & Pauli 2001, Berendse *et al.* 2001, Pearce & van der Wal 2002
	Acidification causes a shift in species composition towards physiologically adapted species and tolerant taxa, including *Sphagnum*.	Kooijman 1992, Twenhöven 1992, Thiébaut *et al.* 1998
Heavy metals and micro-pollutants	Lethal concentrations in large rivers formerly caused complete extinctions, but recovery is taking place in the context of global improvement of water quality.	Vanderpoorten 1999
Global warming	Ongoing northwards expansion of Mediterranean and subtropical bryophyte species is attributed to global warming in temperate areas. In the latter, the impact of simulated increased drought and temperature on local communities seems limited, possibly because bryophytes are able to withstand repeated desiccation without injury, resuming normal metabolism within minutes or a few hours of rehydration, or because sufficient tissue hydration can be attained by dewfall. In alpine and arctic ecosystems by contrast, simulated warming caused local decrease in species diversity and the southern boundary of peatland ecosystems is predicted to experience a shift 780 km northwards in response to a two-fold increase in CO_2 concentrations.	Frahm & Klaus 2001, Gignac *et al.* 1998, Dorrepaal *et al.* 2003, Bates *et al.* 2005, Jägerbrand *et al.* 2006

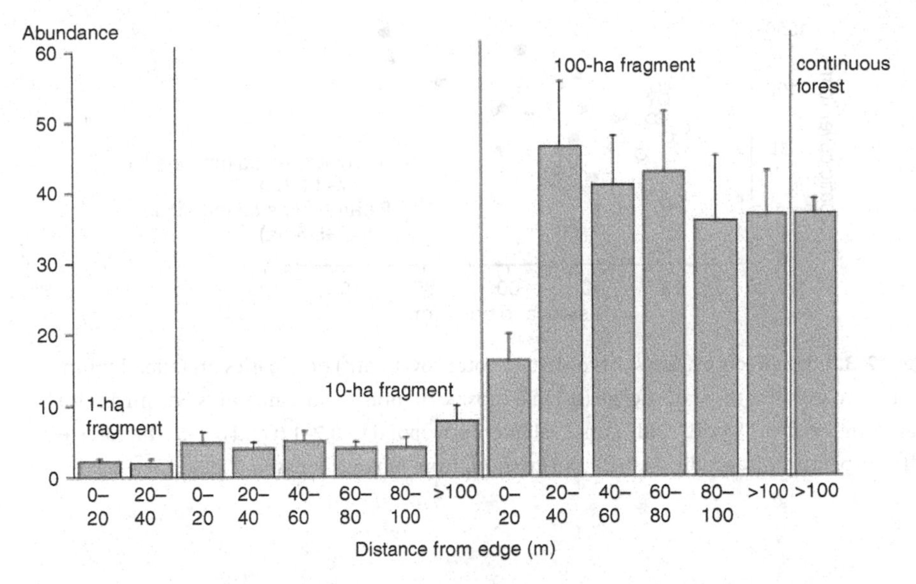

Fig. 12.2. Frequency distribution of epiphyll mean number of species per 1 ha plot ±s.d. in 1–100 ha forest fragments and continuous forest as a function of proximity to forest border at the Biological Dynamics of Forest Fragments Project in Manaus, Brazil (reproduced from Zartman & Nascimento 2006, with permission of Elsevier).

changes in patch area independently from the distance to the edge (Fig. 12.2). Isolation of rather small patches of 1–10 ha from the nearest suitable habitat by an average distance of 380 m was apparently sufficient for disrupting epiphyll dispersal. Experimental leaf patches in reserves of >100 ha experienced nearly double (48%) the colonization probability observed in small reserves (27%), suggesting that the proximate cause of epiphyll species loss in small fragments (<10 ha) is reduced colonization (Zartman & Shaw 2006). Altogether, these observations indicate that dispersal limitation, rather than compromised habitat quality due to edge effects, account for the alteration of the epiphyll community after fragmentation.

An increasing body of literature indeed points to a positive effect of connectivity and emphasizes the effects of dispersal limitation and metapopulation dynamics on community species richness (Zartman & Shaw 2006, Pharo & Zartman 2007, Virtanen & Oksanen 2007). This hypothesis is supported by the fact that bryophytes often exhibit aggregated distribution patterns (Snäll *et al.* 2003, 2004b, Löbel *et al.* 2006). For example, epiphytes tend to colonize predominantly trees occurring in the vicinity of occupied trees (Snäll *et al.* 2005). A recent genetic analysis further demonstrated that pairs of individuals separated by a distance up to 350 m tend to exhibit more genetic similarity than

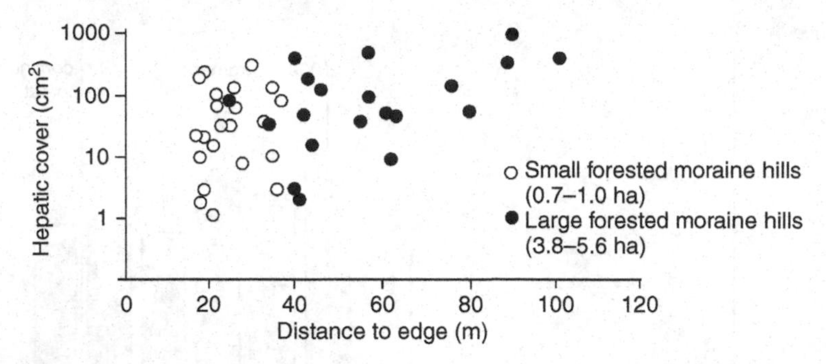

Fig. 12.3. Edge effects on forest bryophytes. Total cover (cm^2) of hepatics on fallen logs in relation to distance to nearest edge (m) in a mosaic of forested moraine hills within a mire matrix in Sweden. "Small" and "large" islands correspond to 0.7–1.0 and 3.8–5.6 ha forested hills, respectively (reproduced, with permission, from Moen & Jonsson 2003).

individuals separated by a greater distance, suggesting that isolation by distance operates at this scale (Snäll *et al.* 2004a).

As opposed to the restricted dispersal range hypothesis, Moen & Jonsson (2003) invoked edge effects to account for the highly variable and often low hepatic cover in small forest patches at a distance of <50 m from the edge, whereas there was a fairly steady increase in cover on large islands when plots were located at >50 m from the edge (Fig. 12.3). Similar observations of changes in species composition in riparian buffer strips were also attributed to an altered microclimate (Hylander *et al.* 2002, 2005). Experimental studies in two *Hylocomium* species clearly showed that growth increase after three months was strongly affected by the distance from edge and edge exposure (Hylander 2005). Growth on south-facing edges was indeed substantially slower than on north-facing edges, but this effect progressively disappeared with distance from edge, so that no growth difference between north-facing and south-facing edges was observed from a distance from edge of about 40 m (Fig. 12.4). These observations clearly point to a decrease in habitat quality due to the edge effects. Moen & Jonsson (2003) therefore suggested that many bryophytes have the mobility to overcome dispersal problems posed by fragmented landscapes if an appropriate habitat or substrate is available. Efficient dispersal at the landscape scale was further invoked by Pharo *et al.* (2004, 2005) to explain the lack of relationship between fragmentation and commonly occurring, drought-tolerant species with often small spores (<25 μm), whose mobility may render the species less sensitive to fragmentation than taxa exhibiting difficulties in navigating the "matrix". This interpretation is shared by Hazell *et al.* (1998), who

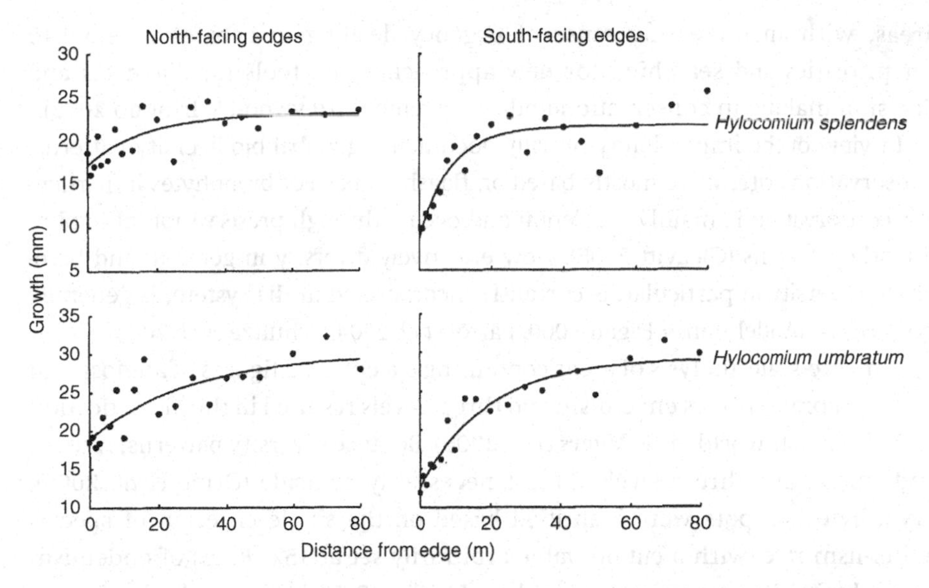

Fig. 12.4. Mean growth of *Hylocomium splendens* and *H. umbratum* after three months in relation to distance from forest clear-cut edges and edge exposure (reproduced, with permission, from Hylander 2005).

observed that colonization of aspens by epiphytes was not more effective within clusters of aspen than among solitary trees, as if long-distance dispersal was effective enough to obliterate the effects of fragmentation.

Although little doubt persists on fragmentation affecting the long-term viability of populations at the landscape scale, even if this impact can be buffered by the dispersal ability of some species, the nature of its effects remains debated, and it is likely that the respective importance of demographic and ecological factors depends on local conditions.

12.4 Conservation strategies

12.4.1 *Specificity of bryophyte patterns of diversity*

The high levels of threat in the bryophyte floras, exacerbated by immediate habitat destructions and degradations and longer-term effects of habitat fragmentation, call for urgent conservation measures. Unfortunately, most conservation areas have not been located in places that deserve the highest protection level. It is only recently that attention has been focused on a systematic conservation planning involving scientific prescriptions based on biogeographical theory and mapping (Margules & Pressey 2000, Groves *et al.* 2002). We are still faced with the old problem of where to establish protected areas, with an increasing sense of urgency, leading to debates on how to set priorities and searching for new approaches and tools for diagnosis and decision-making in conservation and management (Heywood & Iriondo 2003).

In view of the impossibility of truly documenting global biodiversity patterns, conservation criteria are mostly based on flagship taxa. For bryophytes in particular, conservation is mainly incidental and occurs through preservation of habitat for other reasons (Cleavitt 2005). How effectively diversity in general, and bryophyte diversity in particular, is currently incorporated in the system, is generally not known (Andelman & Fagan 2000, Pärtel *et al.* 2004, Schultze *et al.* 2004).

A macro-scale analysis of areas containing exceptional figures of angiosperm and vertebrate species endemism and threat levels resulted in the identification of 25 hot-spots worldwide (Myers *et al.* 2000). Because diversity patterns, rates of endemism, and threat levels do not necessarily coincide (Orme *et al.* 2005), bryophyte hot-spots were identified based on the single criterion of species endemism rate, with a cut-off value arbitrarily set at 15%. Rates of endemism were obtained from various sources listed in Fig. 12.5. The figure obtained must, however, be interpreted with extreme caution and will definitely be altered in the near future (R. Gradstein, personal communication). Indeed, the data reflect a compromise between two opposite trends. In the absence of monographic work for many tropical taxa, certain rates of endemism are definitely overestimated because the multiple descriptions of the same species in different areas call for extensive synomizations. On the other hand, substantial numbers of species definitely remain to be described, especially from poorly known tropical areas. This last tendency is, perhaps, the most misleading because cryptic speciation, i.e. the accumulation of genetic differences among morphologically similar taxa, has been increasingly documented in bryophytes (Shaw 2001) and might well be the rule rather than the exception, in particular in the very numerous bryophytes species whose distribution range spans several continents.

The overlap between the bryophyte hot-spots defined on these bases and those identified by Myers *et al.* (2000) is only partial (Fig. 12.5). New Zealand, New Caledonia, the Pacific Islands, and the Malesio-Indonesian part of eastern Asia exhibit the highest rates of endemism in bryophytes. For example, endemism rates reach up to 45% in the Hawaiian moss flora and 52% in the liverwort flora of New Zealand. These areas, although listed by Myers *et al.* (2000), are not ranked among the richest hot-spots for vertebrates and angiosperms. Shared hot-spots between liverworts and vertebrate and angiosperms, but not mosses, include the northern Andes and Madagascar. The Mediterranean and many tropical areas, e.g. the Galapagos Islands, the Caribbean Islands, Amazonia, and Equatorial Africa, are listed among the most important hot-spots for

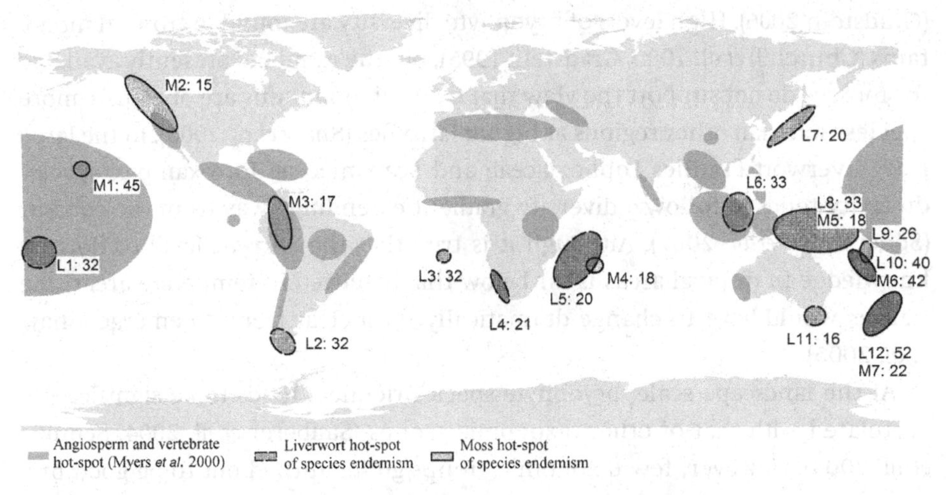

Fig. 12.5. Worldwide patterns of bryophyte hot-spots of endemism compared to angiosperm and vertebrate hot-spots (Myers *et al.* 2000) in the background. Only areas with endemism rates at the species level 15% are presented, and exact values are provided for each region. Mosses: M1: Hawaii (Staples *et al.* 2004); M2: Pacific Northwest (Schofield 1984); M3: Northern Andes (Gradstein, pers. comm.); M4: La Réunion (Ah-Peng & Bardat 2005); M5: New Guinea (Koponen 1990); M6: New Caledonia (Streimann 2000); M7: New Zealand (Glenny & Fife, pers. comm.). Liverworts and hornworts: L1: Samoa (von Konrat & Hagborg, unpublished data); L2: Patagonia (von Konrat & Hagborg, unpublished data); L3: St Helena (Wigginton, pers. comm.); L4: southern Africa (Wigginton, pers. comm.); L5: Madagascar and Mascarene Islands (Wigginton, pers. comm.); L6: Borneo (von Konrat & Hagborg, unpublished data); L7: Japan (Yamada & Iwatsuki 2006); L8: New Guinea and Bismarck Islands (Wigginton, pers. comm.); L9: Vanuatu (von Konrat & Hagborg, unpublished data); L10: New Caledonia (Wigginton, pers. comm.); L11: Tasmania (von Konrat & Hagborg, unpublished data); L12: New Zealand (Glenny & Fife, pers. comm.).

angiosperms and vertebrates, but do not exhibit spectacular rates of endemism in mosses nor liverworts. By contrast, several temperate areas, including Japan, Patagonia, the northern part of the Pacific Northwest region of the U.S.A. and Canada, and Tasmania, exhibit high rates of endemism in either their moss or liverwort flora, but are not listed as priority areas for conservation with respect to their vertebrate fauna and angiosperm flora.

Indeed, in contrast with one of the few truly general patterns in biogeography, no significant latitudinal gradient of species diversity is evident in the bryophyte flora (Shaw *et al.* 2005). The moss flora of tropical lowland forests is, for example, notably depauperate (Churchill 1998), possibly because, in the absence of fog, high air temperature and desiccation inhibit net photosynthesis

(Gradstein 2006). High levels of bryophyte diversity are found in tropical mountains (Churchill *et al.* 1995, Gradstein 1995), but the numbers presently available for mosses do not support the view that tropical mountains are obviously more species-rich than other regions at higher latitudes (Shaw *et al.* 2005). In the large leafy liverwort families Lophoziaceae and Scapaniaceae for example, species diversity tends to follow a diversity gradient extending away from the equator (Söderström *et al.* 2007). Although it is true that the current level of floristic knowledge in tropical areas is far below that achieved in temperate areas, the figures would have to change dramatically for a clear trend to emerge (Shaw *et al.* 2005).

At the landscape scale, bryophyte species richness tends to be significantly correlated with that of other taxonomic groups (Sauberer *et al.* 2004, Schulze *et al.* 2004). However, few taxonomic groups/guilds turned out to be good predictors for others (Schulze *et al.* 2004, Nordén *et al.* 2007), and many groups of organisms actually differ in their conservation demands (Heilman-Clausen *et al.* 2005). For example, Pharo *et al.* (2000) found that a set of sites that preserved 90% of vascular plant species captured only 65% of bryophyte species, and that vascular plant species richness was a poor predictor of bryophyte species diversity. On a local scale indeed, reserves selected for vascular plants can capture large percentages of bryophytes, but individual sites important for bryophyte conservation may not be important for vascular plant conservation. According to our unpublished personal experience, many habitats, such as shaded rock outcrops, crags, old quarries, waterfalls, etc. often display a very rich bryoflora, whereas their interest for higher plants is extremely limited. Therefore, bryophyte conservation should deserve special attention in terms of selection of reserves and management measures.

12.4.2 *Circumscription of key areas for bryophyte conservation*

The most straightforward approach to circumscribe key areas for bryophyte conservation involves comprehensive field surveys (e.g. Urmi 1992). However, time and expertise are not always available and sharp differences in the level of knowledge of species distributions are obvious among areas and continents, resulting in a general under-documentation of bryophyte distributions (Cleavitt 2005). Even in well-prospected areas, mapping the diversity of small and often inconspicuous plants, which, like the non-chlorophyllose, subterranean liverwort *Cryptothallus mirabilis*, can sometimes be extremely difficult to find (Sergio *et al.* 2005), indeed remains a very challenging task.

Therefore, predictive models have been increasingly used to facilitate subsequent field investigations in order to document bryophyte diversity patterns at the landscape scale. Initially launched under the "mesohabitat" concept (Vitt &

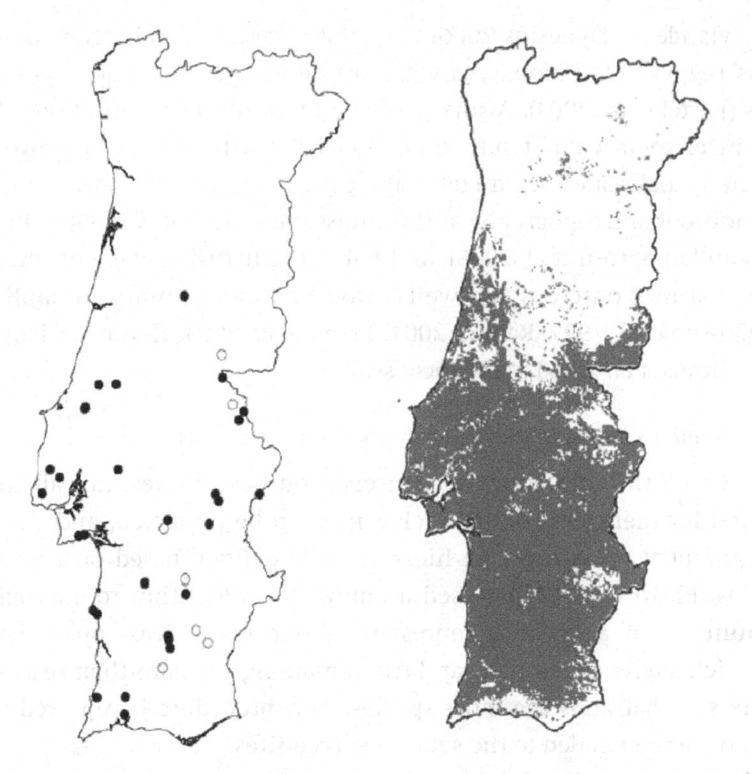

Fig. 12.6. Distribution of *Riccia sommieri* in Portugal: actual range based on herbarium ($n = 76$, filled dots) and bibliographic ($n = 5$, open dots) records and predicted range inferred from an ecological niche model using environmental variables as predictors (reproduced from Sérgio *et al.* 2007, with permission of Elsevier).

Belland 1997), predictive models have developed through the increasing availability of databases and computing facilities that take the complexity of conservation programs into account. Geographical Information Systems (GIS) are one of the tools that allow the integration and analysis of large amounts of data sets. This enables an increase in the input data to be used and in the output relationships that can be established among the data (Draper *et al.* 2003). Potentially valuable areas for conservation can be circumscribed by crossing information on predicted habitat suitability for particular species (Sérgio *et al.* 2005, Vanderpoorten *et al.* 2006, Sérgio *et al.* 2007) (Fig. 12.6), sets of species, or global diversity patterns (Vanderpoorten & Engels 2003, Vanderpoorten *et al.* 2005), with information on the human (i.e., needs and capacities of land managers) and financial resources that are available for conservation (Draper *et al.* 2003).

The factors identified by predictive models to determine bryophyte diversity are multiple and vary from one area to another. For example, the high correspondence between high soil pH and plant diversity in boreal areas (Vitt *et al.*

2003, Hylander & Dynesius 2006) does not necessarily hold true outside that area, as regional evolutionary centers are rather located on acidic soils in the tropics (Pärtel *et al.* 2004). A survey of the literature recurrently revealed that high conservation value tends to be associated with habitat heterogeneity at different spatial scales, from landscape complexity indices (Moser *et al.* 2002), soil conditions and topography at the landscape scale (Bates 1995b, Draper *et al.* 2003, Vanderpoorten *et al.* 2005), and habitat structuring element, such as the presence of rock outcrops and well-decayed logs and stumps (Humphrey *et al* 2002, Ohlson *et al.* 1997, Rambo 2001, Pharo *et al.* 2004, Heilman-Clausen *et al.* 2005, Löhmus *et al.* 2007) at the local scale.

12.4.3 Strategies for implementing a network of protected areas

Given the limitations of conservation possibilities, not all the areas identified for their conservation relevance can be protected, and a restricted set of potential areas must be hierarchically defined based on conservation priorities. Pharo *et al.* (2005) used a cumulative algorithm to approximate a minimum set of areas that represent all species at least once. The most species-rich site is added first, and the remaining sites are then reassessed to find the site that adds the most species. This procedure is repeated until all species have been added to the set of reserved sites.

Such an approach is efficient to objectively design a network of reserves that maximizes the capture of the present-day diversity. This procedure misses, however, a longer view on the conservation of evolutionary processes. Reductions in habitat availability caused by fragmentation indeed increase local extinction risk by sharpening edge effects, lowering mean population size and immigration potential, and, eventually, affecting population viability and evolutionary potential. For instance, it has been suggested that, in a fragmented territory, small-scale protected areas of usually less than 2 ha, exhibiting a high concentration of endemic, rare or threatened species, can be established in great numbers to complement larger, more conventional protected areas that are less easy to implement in legal and management terms (Heywood & Iriondo 2003). Identifying core areas for sensitive organisms and protecting them in small reserves has, for example, become common practice in "woodland key habitats" in Scandinavia (Hylander 2005). This approach is appropriate if, as suggested by Pharo *et al.* (2004, 2005), bryophytes have the mobility to overcome dispersal problems posed by fragmented landscapes. However, substantial size effects on the persistence of species diversity, due either to the break in the immigration/extinction balance due to disrupted dispersal (Zartman 2003) or to compromised habitat quality due to an edge effect (Moen & Jonsson 2003, Hylander *et al.* 2002, 2005, Hylander 2005), have been demonstrated. Moen & Jonsson (2003) found that the cover of epiphytic liverworts on forest islands in a vast mire landscape experienced edge effects of about 50 m, and therefore proposed that buffer areas of >50 m should be retained in addition to the area of concern for conservation purposes if negative edge effects should be avoided. Similarly, Hylander *et al.* (2002, 2005) observed that the species in most need of protection (i.e. the red-listed species), were among the ones with strongest declines in riparian buffer-strips 20 m wide, and hence suggested that increasing the width of buffer strips at sites with known or potential value should be considered a better strategy than using many narrow strips. If, as advocated by Zartman (2003), reduced dispersal among fragmented patches distant of only 380 m, rather than compromised habitat quality, causes patch diversity to decrease with time, much larger protected patch areas must be designed. Zartman & Nascimento (2006) recommended that stands of at least 100 ha should be conserved for the long-term existence of epiphylls in Amazonian rainforests. Therefore, a sound conservation strategy must be based on both the identification of key areas and an understanding of the processes that threaten the populations (i.e. dispersal disruption vs. edge effects) in order to assign the appropriate number and size of reserved areas in each case.

12.5 Managing bryophyte diversity

12.5.1 *Management of protected areas*

Biodiversity conservation has traditionally relied on the establishment and maintenance of a network of protected areas. Once selected, protected areas must be managed in order to ensure the long-term viability of species or communities that justified their conservation status. The vegetation dynamism and sometimes rapid species turnover, however, raise such issues as to which state or stage of the vegetation cycle should be preserved (Heywood & Iriondo 2003). Therefore, the contradictory goal of conserving a biota that is dynamic and ever-changing can only be solved when appropriate temporal and spatial scales are set. The approach to protected areas has in fact changed considerably during the past 20 years. The "fortress" concept, which dominated conservation philosophy in earlier decades, has progressively moved towards a much more interventionist approach involving the acceptance of a broad range of options and techniques (Marrero-Gomez *et al.* 2003).

The most appropriate actions for recovering declining populations can be determined by experiments that test the effects of different management regimes derived from competing hypotheses about critical factors that limit population growth. Except for Fennoscandia, where cryptogams are often taken into account in managing plans for the boreal forests, reserves are almost never managed for cryptogams. Rather, the landscape is mostly managed in favor of a suite of species perceived to be the most sensitive ("focal-species" approach, Lambeck 1997). The fundamental assumption is that if restoration efforts are targeted towards a group of species, the needs of other taxa will also be met. However, some authors have raised concerns about the conceptual, theoretical, and practical basis of taxon-based surrogate schemes (Andelman & Fagan 2000, Lindenmayer *et al.* 2000, 2002).

For example, most temperate dry grasslands are an anthropogenic, semi-natural vegetation type of high biological value that has been threatened owing to the loss of its agricultural usefulness in the middle of the twentieth century. Many of them were therefore set aside as nature reserves, which have to be actively managed to prevent a natural succession to woodland. The conservation of biological diversity in grasslands requires an integrated approach covering the ecological demands of a multitude of organisms. In practice, however, the emphasis is often placed on the vascular flora. Bryophytes, which include a number of rare species restricted to that habitat (During 1990), are seldom considered in conservation and restoration programs, and the extent to which management practices affect the bryophyte layer are largely unknown.

In calcareous grasslands, cessation of management results in the development of a tall and dense herb and shrub canopy that eventually causes bryophyte diversity to decrease (Van Tooren *et al.* 1991, During & Van Tooren 2002). In order to prevent a natural succession to woodland, mowing is often implemented because of its positive effects for orchids. For bryophytes, however, mowing is not an optimal strategy (Bergamini *et al.* 2001). Vanderpoorten *et al.* (2004b) observed that mown plots were characterized by a dense bryophyte layer mostly composed of the large *Scleropodium purum*, one of the rare species termed as "competitors" among bryophytes (Grime *et al.* 1990) and likely to outcompete typical grassland species. Alternatively, because bryophyte richness is inversely related to graminoid abundance (Yates *et al.* 2000, Klanderud & Totland 2005, Pharo *et al.* 2005, Eskelinen & Oksanen 2006), grazing, which opens the moss and grass layers, is likely to increase species richness, especially that of gap-detecting colonists (Van Tooren *et al.* 1990). However, heavy grazing is detrimental to the bryophyte layer (Eskelinen & Oksanen 2006), so that intermediate disturbance levels are optimal for maintaining cryptogam diversity in temperate grassy ecosystems (Yates *et al.* 2000).

12.5.2 *Integrated management measures in the context of sustainability*

Although the whole concept of sustainable development and conservation outside strictly protected areas has sometimes been questioned (Soulé & Sanjayan 1998), conservation in selected reserved areas increasingly appears as a necessary but not sufficient condition of the successful conservation of biodiversity (Huntley 1999). Indeed, not only does the greater part of biodiversity exist outside any kind of formal protection, but the surroundings of a strictly protected area may provide complementary habitats to those secured in protected areas themselves (Perfecto & Vandermeer 2002). We shall illustrate this theory with two examples.

Bryophyte conservation and sustainable forest management

Forest and other wooded lands are by far the best-represented, extensively managed ecosystems worldwide. They display, even in highly managed environments such as plantations (Andersson & Gradstein 2005), an important role for conservation that has been recently emphasized and firmly placed on the political agenda through several agreements promoting the sustainable management of forest ecosystems and the conservation of their biodiversity (e.g. the EEC directives 79/409 and 92/43, the Ministerial Conference on the Protection of Forest in Europe, and the Convention on Biological Diversity, Decision VI/22).

In forest ecosystems, the importance of ecological continuity and forest age for species diversity is well recognized and supported by an extensive body of literature mostly focused on temperate ecosystems (Rose 1992, Frisvoll & Prestø 1997, McCune *et al.* 2000, Cooper-Ellis 1998). In tropical forests, Holz & Gradstein (2005) found that, although primary and secondary forests display similar diversity patterns because harvested areas are rapidly invaded by sun epiphytes (Hyvönen *et al.* 1987), the composition of their epiphytic assemblages differs markedly. One third of primary forest species had not re-established in Costa Rican secondary forests after 40 years of succession, indicating that a long time is needed for the re-establishment of microhabitats and re-invasion of species and communities adapted to differentiated niches (Holz & Gradstein 2005).

Ecological continuity is intimately associated with structural diversity. In particular, the specific occurrence of a set of "shade epiphytes" in old-growths or, alternatively, in the oldest stands of managed forests (Frisvoll & Prestø 1997, Boudreault *et al.* 2000, McGee & Kimmerer 2002, Ross-Davis & Frego 2002, Drehwald 2005, Botting & Fredeen 2006), and the significant relationship observed between epiphyte diversity and tree age or diameter (Fig. 12.7) are attributed to the fact that large, old trees provide a more complex environment, especially considering bark structure, chemistry, and moisture conditions (Hazell *et al.* 1998). Large, old trees are also available for a longer time for colonization (Hazell *et al.* 1998, Snäll *et al.* 2004b), are a source of logs (Heilmann-Clausen *et al.* 2005), and contribute, when windthrown, towards providing a special habitat for cryptogams that grow on inorganic soil.

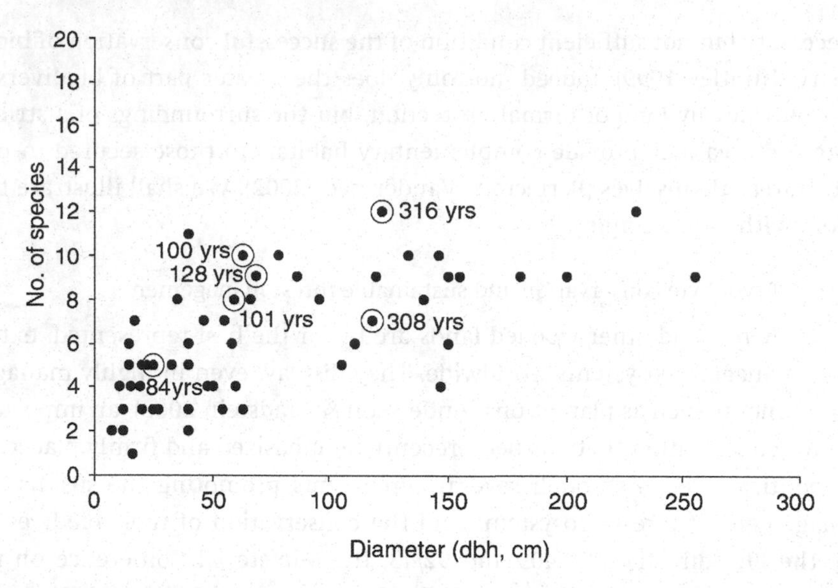

Fig. 12.7. Relationship between epiphytic bryophyte species richness and diameter at breast height of *Eucalyptus obliqua*. Encircled dots represent trees for which ages are known (reproduced from Kantvilas & Jarman 2004, with permission of Elsevier).

Specialist species of this nature colonize the freshly exposed soil, enabling the occurrence of species that would otherwise be outcompeted at ground level (Heilmann-Clausen *et al.* 2005).

Several authors recently pointed out, however, that unmanaged, mesic forest stands may not necessarily exhibit high diversities and specific communities, at least in temperate areas where these studies were conducted, and emphasized that factors of long continuity of woodland cover may not be crucial for maintaining bryophyte diversity (Ohlson *et al.* 1997, Humphrey *et al.* 2002, Heylen *et al.* 2005). Heylen *et al.* (2005) found that the dominant tree age of an ecotope may have on its own a significantly negative effect on total epiphyte diversity and suggested that young trees have generally more to offer and should be prominently present on ecotope level. Van der Pluijm (2001) indeed observed that epiphytic-rich pioneer communities of an alluvial Rhine forest were replaced by mats of a few dominant species when willows reached an age of approximately 20 yr. Furthermore, a substantial proportion of epiphytes found in mid-western Europe favors open, softwood stands rather than mature, shaded hardwood stands (Hodgetts 1996, Klein & Vanderpoorten 1997, Vanderpoorten & Engels 2002). In fact, while modern forestry tends to favor fairly dense stands, ancient forests would have been open in the past owing to the presence of considerable numbers of large grazing and browsing herbivores and recurrent fires (Esseen *et al.* 1992), so that the flora may have adapted to this partly open environment over a period of perhaps millions of years (Rose 1992). Hylander (2005) therefore proposed that edge habitats, although unsuitable for shade epiphytes (see Section 12.3.2), may be, under certain circumstances, more favorable than interior habitats because of a trade-off between moisture and light requirements, making edges a species-rich refugium for a specific light-demanding flora once typical of softwood, pioneer stands (Vanderpoorten *et al.* 2004a).

As a consequence, the diversity and composition of epiphytic assemblages is linked to a series of forests of different composition, age, and structure. The conservation of all the stages of the forest cycle is, however, extremely rarely achieved. In fact, less than 1% of European forests can be termed as old-growth and include natural senescent and rejuvenation phases (Norton 1996). Therefore, the conservation of epiphytic bryophytes must also take place in managed forests whose conservation value will be enhanced if a few conservation-oriented measures can be taken. For instance, retained trees, or clusters of trees, can form links during forest succession between young and old stands after final harvest and are beneficial to at least some species considered to be sensitive to forest operations (Hazell & Gustafsson 1999, Kantvilas & Jarman 2004, Fenton & Frego 2005). This is especially true of well-illuminated, pioneer trees that often support a rich, specialized epiphytic flora. As these epiphytes must switch from one host to another fairly frequently as their host will be invaded by large mats of pleurocarpous assemblages and eventually die, the conservation of such epiphytes relies on the conservation and dynamics of regeneration of the phorophytes, through, for example, coppicing (Heylen *et al.* 2005).

Bryophyte conservation and sustainable agriculture

Many native plant species have benefitted from agricultural activities concomitant with forest removal, especially those growing in more open types of landscapes, which now appear as refuges for plant species typical of once dominant regional vegetation (Jobin *et al.* 1996, Boutin & Jobin 1998). Many annual shuttle bryophyte species with short life cycles indeed rely on regular soil disturbance (Zechmeister *et al.* 2002). In Europe for example, hornworts are largely confined to crop fields (Bisang 1998). As a consequence, the highest percentages of red list species in many European countries are found in dry grasslands and places with bare soil such as arable fields (Schnyder *et al.* 2004), which include up to 63% of the endangered species at a national scale (Zechmeister *et al.* 2002). In tropical areas, cacao plantations with low and moderate management intensity are also of high conservation relevance

(Andersson & Gradstein 2005). They indeed serve as an important substitution habitat and even currently represent the unique known habitat of the rare, western Ecuadorian endemic liverwort *Spruceanthus theobromae* (Kautz & Gradstein 2001).

With the intensification of land-use, however, the once positive contribution to biodiversity of landscape diversification has progressively decreased, giving rise to the intermediate disturbance hypothesis, according to which highest levels of biodiversity are maintained at intermediate scales of disturbance.

The strong correlation between land use intensity and bryophyte richness and conservation value is supported by an extensive body of literature (Kautz & Gradstein 2001, Zechmeister & Moser 2001, Zechmeister *et al.* 2003, Andersson & Gradstein 2005). In a study of Austrian agricultural landscapes, Zechmeister *et al.* (2003) found that profit margin and variable costs correlated negatively with plant species richness, and meadows that offered low or no profit margins showed highest species richness. Zechmeister *et al.* (2003) and Schmitzberger *et al.* (2005) concluded that, if plant species richness is to be maintained in agricultural landscapes, farmers have to receive increased financial incentives through agro-environmental subsidies for appropriate meadow management, and these have to be linked to clearly defined measures.

12.6 *Ex situ* conservation and reintroduction

In complement to *in situ* conservation strategies described above, *ex situ* conservation involves the medium- or long-term storage of selected samples of a population's genetic diversity intended for the possible reintroduction of rare and endangered taxa into the wild.

Typical *ex situ* techniques involve the creation of living collections or diapaused material such as seeds, spore banks, or cryopreserved material. Living collections can be grown in greenhouses or axenic cultures. For example, *Monosolenium tenerum*, the only species of the family Monosoleniaceae (Marchantiales) and very rare in nature, is viable in greenhouses and widely cultivated as an aquarium plant in Central Europe (Gradstein *et al.* 2003). Axenic cultures, although artificial, provide a more uniform and secure method of maintaining plants in a tissue culture collection without fungal, algal, and bacterial contaminants (Duckett *et al.* 2004). However, axenic cultures might not be ideal for long-term conservation purposes. Indeed, the material is continually subcultured and is likely to become adapted to growing in culture conditions over time. This is particularly problematic for material retained for conservation purposes where reintroduction is a possible long-term objective. Alternatively, cryopreservation has been shown to be an effective, convenient, and stable long-term storage technique for vascular plants. In bryophytes, the first experiments appear promising, although desiccation-intolerant species did not survive either dehydration or freezing (Burch 2003). A project for the *ex situ* conservation of endangered U.K. bryophytes was launched in August 2000, at the Royal Botanic Gardens, Kew, with the appointment of a dedicated bryophyte conservation officer. The aim is to provide long-term basal storage of rare bryophyte material for use in future conservation programs, and material of 18 endangered species is currently conserved (see further details at http://rbgweb2.rbge.org.uk/bbs/Learning/exsitu/exsitu.htm).

The reintroduction of populations in areas from where they vanished has been seldom, but increasingly, documented in bryophytes. Recent experiments demonstrate the possibility of reintroduction from gametophyte fragments (Kooijman *et al.* 1994, Gunnarsson & Söderström 2007, Mälson & Rydin 2007) or cultured gametophytes (Fig. 12.8) (Rothero *et al.* 2006).

The success of the reintroduction trial is closely linked to a series of measures to restore appropriate habitat conditions as in the case, for example, of mined peat bogs (Rochefort & Lode 2006) and provide an appropriate protective cover. Protective cover has a positive effect on the regeneration of shoot fragments (Fig. 12.9) because it provides more humid conditions during summer drought and prevents the development of micro-organisms such as algae during wet periods, which transform into a potentially fatal hard crust upon drying (Mälson & Rydin 2007).

This suggests that the possibility exists to increase the population size and/or the number of populations by artificial introduction of diaspore. However, the genetic variation of at least some bryophytes species is geographically highly structured (Chapter 11, this volume), so that a minimal precaution in any attempt of reintroduction therefore involves a detailed genetic study of the source populations and their compatibility with the site to be recolonized.

12.7 Conclusion, issues, and perspectives

Interest in biodiversity and conservation biology is rapidly increasing, including concern for lower plants. Bryophytes have been successfully introduced into the IUCN system and the legal protection of threatened species, and their habitat, although still limited, is gaining attention in several countries. Threats and mechanisms that make bryophytes vulnerable are being increasingly well perceived, even if additional experimental research is still needed to better understand the causes of species rarity. Finally, increasingly practical tools are becoming available to design and manage networks of conservation areas for bryophytes, and promising new methods of *ex situ* conservation are being developed.

Fig. 12.8. *In vitro* cultivation and reintroduction into the wild of rare and threatened bryophyte species: the example of the moss *Bryum schleicheri* var. *latifolium* in the U.K. (a) Phytagel culture in 5 cm Petri-dishes. (b) Colony on a muslin bag six months after planting. (c) A well-grown colony one year after planting (photographs by G. P. Rothero and J. G. Duckett).

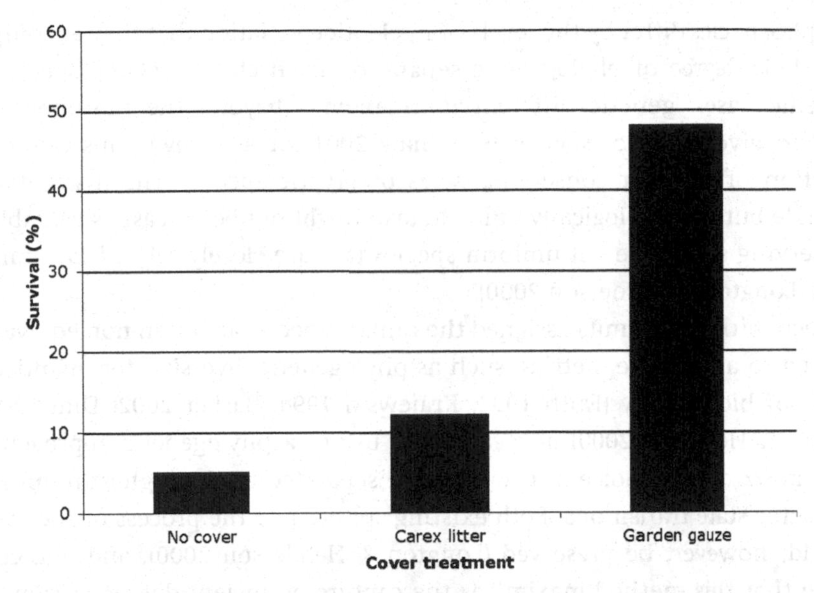

Fig. 12.9. Effect of cover treatment for survival of individual shoots during the first year of a reintroduction experiment of four mosses of rich fens (*Scorpidium scorpioides, S. cossonii, Pseudocalliergon trifarium,* and *Campylium stellatum*) in a Swedish mire complex (redrawn from Mälson & Rydin 2007).

Future challenges include the necessity to perform long-term studies to identify the causes of decline of so many species and develop networks between scientists and land managers to improve the applicability of the results (Hylander & Jonsson 2007).

The central question "what to conserve?" remains, furthermore, highly debated. Biodiversity studies typically focus on the species as the unit for comparison and global analyses often use community and regional endemism as measures of the biodiversity in an area, which raises two issues. Firstly, morphospecies polyphyly is the rule rather than the exception (Funk & Omland 2003). This problem is especially acute in taxa with reduced morphologies such as bryophytes, wherein species circumscriptions have traditionally relied on a few key characters whose taxonomic significance has been increasingly questioned (Vanderpoorten & Goffinet 2006). Secondly, species are not equivalent in "biodiversity value" because, besides differences in rarity and threat levels, they differ in phylogenetic history and current population processes. Molecular studies have increasingly revealed striking intraspecific levels of genetic variation that differ from one morphospecies to another. For example, Shaw & Cox (2005) found that, in *Sphagnum*, morphologically defined species are not equivalent with regard to molecular biodiversity because morphospecies differ by the levels of nucleotide variation that they encompass and their degree of phylogenetic separation from closely related species. In extreme cases, genetic differentiation extends beyond the morphospecies level to give "cryptic" species (see Shaw 2001 for a review). This raises the question of whether conserving races or cryptic species within genetically variable but morphologically uniform taxa might not be at least as valuable as conserving some rare but uniform species that are closely related to common ones (Longton & Hedderson 2000).

Recognition that units assigned the rank of species are often non-equivalent has led to alternative metrics, such as phylogenetic diversity, for quantifying levels of biodiversity (Faith 1992, Krajewski 1994, Barker 2002, Diniz 2004). Bisang & Hedenäs (2000) advocated the use of a phylogenetic approach for standardizing the choice of taxa that are separated by the highest number of character state transitions. Both existing species and the process of speciation should, however, be preserved (Longton & Hedderson 2000), and one could argue that this method maximizes the capture of ancient diversification processes (represented as long branches on a phylogenetic tree) that led to extant species rather than actual speciation processes (represented by radiative short branches). It is also clear that we are in urgent need of an appropriate definition and circumscription of the species. The ease with which this new species concept, largely influenced by the understanding of evolutionary processes, will be compatible with our traditional knowledge of plant taxonomy, distribution, and frequency patterns, which currently make up the bulk of the information available for conservation, remains, however, largely unknown.

Acknowledgments

The authors sincerely thank Rob Gradstein, Tord Snäll, Bernard Goffinet, and Emma Pharo for their very constructive comments on an earlier draft of the manuscript, and Allan Fife, Lars Söderström, Brian O'Shea, Martin Wigginton, Matt von Konrad, Benito Tan, Henk Siebel, and David Glenny for their invaluable help regarding patterns of endemism and threats.

References

Acebey, A., Gradstein, S. R. & Kromer, T. (2003). Species richness and habitat diversification of bryophytes in submontane rain forest and fallows of Bolivia. *Journal of Tropical Ecology*, **19**, 9–18.

Ah-Peng, C. & Bardat, J. (2005). Check list of the bryophytes of Réunion Island (France). *Tropical Bryology*, **26**, 89–118.

Andelman, S. J. & Fagan, W. F. (2000). Umbrellas and flagships: efficient conservation surrogates or expensive mistakes? *Proceedings of the National Academy of Sciences, U.S.A.*, **97**, 5954–9.

Andersson, M. S. & Gradstein, S. R. (2005). Impact of management intensity on non-vascular epiphyte diversity in cacao plantations in western Ecuador. *Biodiversity and Conservation*, **14**, 1101–20.

Asplund, D. (1996). Energy use of peat. In *Global Peat Resources*, ed. E. Lappalainen, pp. 319–25. Jyväskylä: International Peat Society.

Barker, G. M. (2002). Phylogenetic diversity: a quantitative framework for measurement of priority and achievement in biodiversity conservation. *Biological Journal of the Linnean Society*, **76**, 165–94.

Bates, J. W. (1989). Growth of *Leucobryum glaucum* cushions in a Berkshire oakwood. *Journal of Bryology*, **15**, 785–91.

Bates, J. W. (1995a). A bryophyte flora of Berkshire. *Journal of Bryology*, **18**, 503–620.

Bates, J. W. (1995b). Numerical analysis of bryophyte-environment relationships in a lowland English flora. *Fragmenta Floristica et Geobotanica*, **40**, 471–90.

Bates, J. W., Thompson, K. & Grime, J. P. (2005). Effects of simulated long-term climatic change on the bryophytes of a limestone grassland community. *Global Change Biology*, **11**, 757–69.

Berendse, F., Van Breemen, N., Rydin, H. *et al.* (2001). Raised atmospheric CO_2 levels and increased N deposition cause shifts in plant species composition and production in *Sphagnum* bogs. *Global Change Biology*, **7**, 591–8.

Berg, A., Gärdenfors, U., Hallingbäck, T. & Noren, M. (2002). Habitat preferences of red-listed fungi and bryophytes in woodland key habitats in southern Sweden – analyses of data from a national survey. *Biodiversity and Conservation*, **11**, 1479–503.

Bergamini, A. & Pauli, D. (2001). Effects of increased nutrient supply on bryophytes in montane calcareous fens. *Journal of Bryology*, **23**, 331–9.

Bergamini, A., Peintiger, M., Schmid, B. & Urmi, E. (2001). Effects of management and altitude on bryophyte species diversity and composition in montane calcareous fens. *Flora*, **196**, 180–93.

Berglund, H. & Jonsson, B. G. (2001). Predictability of plant and fungal species richness of old-growth boreal forest islands. *Journal of Vegetation Science*, **12**, 857–66.

Bisang, I. (1998). The occurrence of hornwort populations (Anthocerotales, Anthocerotopsida) in the Swiss Plateau: the role of management, weather conditions and soil characteristics. *Lindbergia*, **23**, 94–104.

Bisang, I. & Hedenäs, L. (2000). How do we select bryophyte species for conservation, and how should we conserve them? *Lindbergia*, **25**, 62–77.

Boudreault, C., Gauthier, S. & Bergeron, Y. (2000). Epiphytic lichens and bryophytes on *Populus tremuloides* along a chronosequence in the southwestern boreal forest of Quebec, Canada. *Bryologist*, **103**, 725–38.

Botting, R. S. & Fredeen, A. L. (2006). Contrasting terrestrial lichen, liverwort, and moss diversity between old-growth and young second-growth forest on two soil textures in central British Columbia. *Canadian Journal of Botany*, **84**, 120–32.

Boutin, B. & Jobin, B. (1998). Intensity of agricultural practices and effects of adjacent habitats. *Ecological Applications*, **8**, 544–57.

Brown, D. H. (1992). Impact of agriculture on bryophytes and lichens. In *Bryophytes and Lichens in a Changing Environment*, ed. J. W. Bates & A. M. Farmer, pp. 259–83. Oxford: Clarendon Press.

Burch, J. (2003). Some mosses survive cryopreservation without prior pretreatment. *Bryologist*, **106**, 270–7.

Church, J. M., Hodgetts, N. G., Preston, C. D. & Stewart, N. F. (2001). *British Red Data Books 2. Mosses and Liverworts*. Peterborough: Joint Nature Conservation Committee.

Churchill, S. P. (1998). Catalog of Amazonian mosses. *Journal of the Hattori Botanical Laboratory*, **85**, 191–238.

Churchill S. P., Griffin, D. & Lewis, M. (1995). Moss diversity of the Tropical Andes. In *Biodiversity and Conservation of Neotropical Forests*, ed. S. P. Churchill, H. Balslev, E. Forero & J. L. Luteyn, pp. 335–46. New York: New York Botanical Garden.

Cleavitt, N. (2002). Stress tolerance of rare and common species in relation to their occupied environments and asexual dispersal potential. *Journal of Ecology*, **90**, 785–95.

Cleavitt, N. (2005). Patterns, hypotheses and processes in the biology of rare bryophytes. *Bryologist*, **108**, 554–66.

Cobb, A. R., Nadkarni, N. M., Ramsey, G. A. & Svoboda, A. J. (2001). Recolonization of bigleaf maple branches by epiphytic bryophytes following experimental disturbance. *Canadian Journal of Botany*, **79**, 1–8.

Cooper-Ellis, S. (1998). Bryophytes in old-growth forests of western Massachusetts. *Journal of the Torrey Botanical Society*, **125**, 117–32.

Da Costa, D. P., Imbassahy, C. A. & da Silva, V. P. (2005). Diversity and importance of the bryophyte taxa in the conservation of the ecosystems of the Rio de Janeiro state. *Rodriguésia*, **56**, 13–49.

Diniz, J. A. F. (2004). Phylogenetic diversity and conservation priorities under distinct models of phenotypic evolution. *Conservation Biology*, **18**, 698–704.

Dorrepaal, E., Aerts, R., Cornelissen, J. H. C., Callaghan, T. V. & Van Logtestijn, R. S. P. (2003). Summer warming and increased winter snow cover affect *Sphagnum fuscum* growth, structure and production in a sub-arctic bog. *Global Change Biology* **10**, 93–104.

Downes, B. J., Entwisle, T. J. & Reich, P. (2003). Effects of flow regulation on disturbance frequencies and in-channel bryophytes and macroalgae in some upland streams. *River Research and Applications*, **19**, 27–42.

Draper, D., Rossello-Graell, A., Garcia, C., Tauleigne-Gomes, C. & Sergio, C. (2003). Application of GIS in plant conservation programmes in Portugal. *Biological Conservation*, **113**, 337–49.

Drehwald, U. (2005). Biomonitoring of disturbance in Neotropical rainforests using bryophytes as indicators. *Journal of the Hattori Botanical Laboratory*, **97**, 117–26.

Duckett, J. G., Burch, J., Fletcher, P. W. *et al.* (2004). In vitro cultivation of bryophytes: a review of practicalities, problems, progress and promise. *Journal of Bryology*, **26**, 3–20.

During, H. J. (1990). The bryophytes of calcareous grasslands. In *Calcareous Grasslands: Ecology and Management*, ed. S. H. Hillier, D. W. H. Walton & D. A. Wells, pp. 35–40. Huntingdon: Bluntisham.

During, H. J. (1992). Ecological classification of bryophytes and lichens. In *Bryophytes and Lichens in a Changing Environment*, ed. J. W. Bates & A. M. Farmer, pp. 1–31. Oxford: Clarendon Press.

During, H. J. & van Tooren, B. F. (2002). Effecten van veranderingen in beheer op de moslaag van de Kunderberg. *Natuurhistorisch Maandblad*, **91**, 217–21.

Dynesius, M. & Hylander, K. (2007). Resilience of bryophyte communities to clear-cutting of boreal stream-side forests. *Biological Conservation*, **135**, 423–34.

Englund, G., Jonsson, B. G. & Malmqvist, B. (1997). Effects of flow regulation on bryophytes in north Swedish rivers. *Biological Conservation*, **79**, 79–86.

Eskelinen, A. & Oksanen, J. (2006). Changes in the abundance, composition and species richness of mountain vegetation in relation to summer grazing by reindeer. *Journal of Vegetation Science*, **17**, 245–54.

Esseen, P.-A., Ehnström, B., Ericson, L. & Sjöberg, K. (1992). Boreal forests – the focal habitats of Fennoscandia. In *Ecological Principles of Nature Conservation. Applications in Temperate and Boreal Environments*, ed. L. Hansson, pp. 252–325. London: Elsevier.

Faith, D. P. (1992). Conservation evaluation and phylogenetic diversity. *Biological Conservation*, **61**, 1–10.

Fenton, N. J. & Frego, K. A. (2005). Bryophyte (moss and liverwort) conservation under remnant canopy in managed forests. *Biological Conservation*, **122**, 417–30.

Flatberg, K. I., Blom, H. H., Hassel, K. & Økland, R. H. (2006). Moser – Anthocerophyta, Marchantiophyta, Bryophyta. In *Norsk Rodliste 2006, 2006 Norwegian Red List*, ed. J. A. Kålås, A. Viken & T. Bakken Norway: Artsdatabanken.

Frahm, J.-P. & Klaus, D. (2001). Bryophytes as indicators of recent climate fluctuations in Central Europe. *Lindbergia*, **26**, 97–104.

Frego, K. A. & Carleton, T. J. (1995). Microsite conditions and spatial pattern in a boreal bryophyte community. *Canadian Journal of Botany*, **73**, 544–51.

Frisvoll, A. A. & Prestø, T. (1997). Spruce forest bryophytes in central Norway and their relationship to environmental factors including modern forestry. *Ecography*, **20**, 3–18.

Funk, D. J. & Omland, K. E. (2003). Species-level paraphyly and polyphyly: frequency, causes, and consequences, with insights from animal mitochondrial DNA. *Annual Review of Ecology and Evolution*, **34**, 397–423.

Gärdenfors, U. (2000). Population viability analysis in the classification of threatened species: problems and potentials. *Ecological Bulletins*, **48**, 181–90.

Gärdenfors, U. (ed.) (2005). *The 2006 Red List of Swedish Species*. Uppsala: SLU.

Gignac, L. D., Nicholson, B. J. & Bayleyc, S. E. (1998). The utilization of bryophytes in bioclimatic modeling: predicted northward migration of peatlands in the Mackenzie river basin, Canada, as a result of global warming. *Bryologist*, **101**, 572–87.

Glenny, D. & Fife, A. (2005). New Zealand's threatened bryophyte flora. *Australasian Bryological Newsletter*, **51**, 6–10.

Gradstein, S. R. (1992a). The vanishing tropical rain forest as an environment for bryophytes and lichens. In *Bryophytes and Lichens in a Changing Environment*, ed. J. W. Bates & A. R. Farmer, pp. 232–56. Oxford: Oxford University Press.

Gradstein, S. R. (1992b). Threatened bryophytes of the neotropical rain forest: a status report. *Tropical Bryology*, **6**, 83–93.

Gradstein, S. R. (1995). Diversity of Hepaticae and Anthocerotae in montane forests of the tropical Andes. In *Biodiversity and Conservation of Neotropical Montane Forests*, ed. S. P. Churchill, H. Balslev, E. Forero & J. L. Luteyn, pp. 321–34. New York: New York Botanical Garden.

Gradstein, S. R. (2006). The lowland cloud forest of French Guiana - a liverwort hotspot. *Cryptogamie Bryologie*, **27**, 141–52.

Gradstein, S. R., Churchill, S. P. & Salazar Allen, N. (2001). A guide to the bryophytes of Tropical America. *Memoirs of the New York Botanical Garden*, **86**, 1–577.

Gradstein, S. R., Reiner-Drehwald, M. E. & Muth, H. (2003). Über die Identität der neuen Aquarienpflanze, *Pellia endiviifolia*. *Aqua Planta*, **3**, 88–95.

Grime, J. P., Rincon, E. R. & Wickerson, B. E. (1990). Bryophytes and plant strategy theory. *Botanical Journal of the Linnean Society*, **104**, 175–86.

Groves, C. R., Jensen, D. B., Valutis, L. R. *et al.* (2002). Planning for biodiversity conservation: putting conservation science into practice. *BioScience*, **52**, 499–512.

Gunnarsson, U. & Söderström, L. (2007). Can artificial introduction of diaspore fragments work as a conservation tool for maintaining populations of the rare peatmoss *Sphagnum angermanicum*? *Biological Conservation*, **135**, 450–8.

Gunnarsson, U., Hassel, K. & Söderström, L. (2005). Genetic structure of the endangered peatmoss *Sphagnum angermanicum* in Sweden - a result of historic or contemporary processes? *Bryologist*, **108**, 194–202.

Hallingbäck, T. (2003). Including bryophytes in international conventions - a success story from Europe. *Journal of the Hattori Botanical Laboratory*, **9**, 201–14.

Hallingbäck, T. (2007). Working with Swedish cryptogam conservation. *Biological Conservation*, **135**, 334–40.

Hallingbäck, T., Hodgetts, N., Raeyemakers, G. *et al.* (1998). Guidelines for application of the revised IUCN threat categories to bryophytes. *Lindbergia*, **23**, 6–12.

Hannerz, M. & Hånell, B. (1997). Effects on the flora in Norway spruce forests following clearcutting and shelterwood cutting. *Forest Ecology and Management*, **90**, 29–49.

Hartmann, F. A., Wilson, R., Gradstein, S. R., Schneider, H. & Heinrichs, J. (2006). Testing hypotheses on species delimitations and disjunctions in the liverwort *Bryopteris* (Jungermanniopsida : Lejeuneaceae). *International Journal of Plant Sciences*, **167**, 1205–14.

Hazell, P. & Gustafsson, L. (1999). Retention of tree at final harvest - evaluation of a conservation technique using epiphytic bryophyte and lichen transplants. *Biological Conservation*, **90**, 133–42.

Hazell, P., Kellner, O., Rydin, H. & Gustafsson, L. (1998). Presence and abundance of four epiphytic bryophytes in relation to density of aspen (*Populus tremula*) and other stand characteristics. *Forest Ecology and Management*, **107**, 147–58.

Hedenäs, H., Bolyukh, V. O. & Jonsson, B. G. (2003). Spatial distribution of epiphytes on *Populus tremula* in relation to dispersal mode. *Journal of Vegetation Science*, **14**, 233–42.

Hedenäs, L., Bisang, I., Tehler, A., *et al.* (2002). A herbarium-based method for estimates of temporal frequency changes: mosses in Sweden. *Biological Conservation*, **105**, 321–31.

Heilmann-Clausen, J., Aude, E. & Christensen, M. (2005). Cryptogam communities on decaying deciduous wood – does tree species diversity matter? *Biodiversity and Conservation*, **14**, 2061–78.

Heino, J., Virtanen, R., Vuori, K.-M. *et al.* (2005). Spring bryophytes in forested landscapes: land use effects on bryophyte species richness, community structure and persistence. *Biological Conservation*, **124**, 539–45.

Heinrichs, J., Lindner, M., Gradstein, S. R. *et al.* (2005). Origin and subdivision of *Plagiochila* (Jungermanniidae: Plagiochilaceae) in tropical Africa based on evidence from nuclear and chloroplast DNA sequences and morphology. *Taxon*, **54**, 317–33.

Heylen, O., Hermy, M. & Schrevens, E. (2005). Determinants of cryptogamic diversity in a river valley (Flanders). *Biological Conservation*, **126**, 371–82.

Heywood, V. H. & Iriondo, J. M. (2003). Plant conservation: old problems, new perspectives. *Biological Conservation*, **113**, 325.

Hinrichsen, D. (1981). Peat power: Back to bogs. *Ambio*, **10**, 240–2.

Hodgetts, N. G. (1996). *The Conservation of Lower Plants in Woodland*. Peterborough: Joint Nature Conservation Committee.

Holz, I. & Gradstein, S. R. (2005). Cryptogamic epiphytes in primary and recovering upper montane oak forests of Costa Rica – species richness, community composition and ecology. *Plant Ecology*, **178**, 547–60.

Humphrey, J. W., Davey, S., Peace, A. J., Ferris, R. & Harding, K. (2002). Lichens and bryophyte communities of planted and semi-natural forests in Britain: the influence of site type, stand structure and deadwood. *Biological Conservation*, **107**, 165–80.

Huntley, B. (1999). Species distribution and environmental change. In *Ecosystem Management. Questions for Science and Society*, ed. E. Maltby, M. Hodgate, M. Acreman & A. Weir, pp. 115–29. Egham: Royal Holloway Institute for Environmental Research, University of London.

Hylander, K. (2005). Aspect modifies the magnitude of edge effects on bryophyte growth in boreal forests. *Journal of Applied Ecology*, **42**, 518–25.

Hylander, K. & Dynesius, M. (2006). Causes of the large variation in bryophytes species richness and composition among boreal streamside forests. *Journal of Vegetation Science*, **17**, 333–46.

Hylander, K. & Jonsson, B. G. (2007). The conservation ecology of cryptogams. *Biological Conservation*, **135**, 311–14.

Hylander, K., Jonsson, B. G. & Nilsson, C. (2002). Evaluating buffer strips along boreal streams using bryophytes as indicators. *Ecological Applications*, **12**, 797–806.

Hylander, K., Dynesius, M., Jonsson, B. G. & Nilsson, C. (2005). Substrate form determines the fate of bryophytes in riparian buffer strips. *Ecological Applications*, **15**, 674–88.

Hyvönen, J., Koponen, T. & Norris, D. H. (1987). Human influence on the moss flora of tropical rain forest in Papua New Guinea. *Symposia Biologia Hungarica*, **35**, 621–9.

Ingerpuu, N. (1998). Sammaltaimed, Bryophyta. In *Eesti Punane Raamat*, ed. V. Lillelecht, pp. 37–49. Tartu: Eesti Teaduste Akadeemia Looduskaitse. (www.botany.ut.ee/bryology/)

IUCN (2001). http://iucn.org/themes/ssc/redlists/RLcats2001booklet.html.

IUCN (2003). http://www.iucn.org/themes/ssc/redlists/background_EN.htm.

Iwatsuki, Z., Kanda, H. & Furuki, T. (2000). *Threatened Wildlife of Japan Red Data Book*, 2nd edn, vol. 9, *Bryophytes, Algae, Lichens and Fungi*. Tokyo: Japan Wildlife Research Center.

Jägerbrand, A. K., Lindblad, K. E. M., Björk, R. B., Alatalo, J. M. & Molau, U. (2006). Bryophyte and lichen diversity under simulated environmental change compared with observed variation in unmanipulated alpine tundra. *Biodiversity and Conservation*, **15**, 4453–75.

Jobin, B., Boutin, C. & DesGrandes, J. L. (1996). Habitats fauniques du milieu rural québécois: une analyse floristique. *Canadian Journal of Botany*, **74**, 323–36.

Jordan, C. F. (1995). *Conservation*. New York: Wiley.

Kantvilas, G. & Jarman, S. J. (2004). Lichens and bryophytes on *Eucalyptus obliqua* in Tasmania: management implications in production forests. *Biological Conservation*, **117**, 359–73.

Kautz, T. & Gradstein, S. R. (2001). On the ecology and conservation of *Spruceanthus theobromae* (Lejeuneaceae, Hepaticae) from western Ecuador. *Bryologist*, **104**, 607–12.

Klama, H. (2006). Red list of the liverworts and hornworts in Poland. In *Red List of Plants and Fungi in Poland*, ed. Z. Mirek, K. Zarzycki, K. Wojewoda & W. Szelag, pp. 21–33. Krakow: Polish Academy of Sciences.

Klanderud, K. & Totland, O. (2005). Simulated climate change altered dominance hierarchies and diversity of an alpine biodiversity hotspot. *Ecology*, **86**, 2047–54.

Klein, J.-P. & Vanderpoorten, A. (1997). Bryophytic vegetation in riparian forests: their use in the ecological assessment of the connectivity between the Rhine and its floodplain. *Global Ecology and Biogeography*, **6**, 257–65.

Kooijman, A. M. (1992). The decrease of rich fen bryophytes in the Netherlands. *Biological Conservation*, **59**, 139–43.

Kooijman, A. M., Beltman, B. & Westhoff, V. (1994). Extinction and reintroduction of the bryophyte *Scorpidium scorpioides* in a rich-fen spring site in The Netherlands. *Biological Conservation*, **69**, 87–96.

Koponen, T. (1990). Bryophyte flora of Western Melanesia. *Tropical Bryology*, **2**, 149–60.

Krajewski, C. (1994). Phylogenetic measures of biodiversity. A comparison and critique. *Biological Conservation*, **69**, 33–9.

Kubinská, A., Janovicová, K. & Soltés, R. (2001). Cerveny zoznam machorastov Slovenska. [Red list of bryophytes of Slovakia (December 2001)]. *Ochrana Prírody* **20**, suppl., 2001, 31–47.

Kucera, J. & Vana, J. (2005). Seznam a cervený seznam mechorostn Ceské republiky (2005). *Príroda*, **23**, 1–104.

Laaka-Lindberg, S., Hedderson, T. A. & Longton, R. E. (2000). Rarity and reproductive characters in the British hepatic flora. *Lindbergia*, **25**, 78–84.

Lambeck, R. J. (1997). Focal species: a multi-species umbrella for nature conservation. *Conservation Biology*, **11**, 849–56.

Leon, Y. & Ussher, M. S. (2005). Educational program directed towards the preservation of Venezuelan Andean bryophytes. *Journal of the Hattori Botanical Laboratory*, **97**, 227–31.

Lindenmayer, D. B., Margules, C. R. & Botkin, D. (2000). Indicators of forest sustainability biodiversity: the selection of forest indicator species. *Conservation Biology*, **14**, 941–50.

Lindenmayer, D. B., Manning, A. D., Smith, P. L. *et al.* (2002). The focal-species approach and landscape restoration: a critique. *Conservation Biology*, **16**, 338–45.

Löbel, S., Snäll, T. & Rydin, H. (2006). Species richness patterns and metapopulation processes – evidence from epiphyte communities in boreo-nemoral forests. *Ecography*, **29**, 169–82.

Löhmus, A., Löhmus, P. & Vellak, K. (2007). Substratum diversity explains landscape-scale co-variation in the species richness of bryophytes and lichens. *Biological Conservation*, **135**, 405–14.

Longton, R. E. (1992). The role of bryophytes and lichens in terrestrial ecosystems. In *Bryophytes and Lichens in a Changing Environment*, ed. J. W. Bates & A. M. Farmer, pp. 32–76. Oxford: Clarendon Press.

Longton, R. E. & Hedderson, T. A. (2000). What are rare species and why conserve them? *Lindbergia*, **25**, 53–61.

Mälson, K. & Rydin, H. (2007). The regeneration capabilities of bryophytes for rich fen restoration. *Biological Conservation*, **135**, 435–42.

Margules, C. R. & Pressey, L. (2000). Systematic conservation planning. *Nature*, **405**, 243–53.

Marrero-Gomez, M. V., Banares-Baudet, A. & Carque-Alamo, E. (2003). Plant resource conservation planning in protected natural areas: an example from the Canary Islands, Spain. *Biological Conservation*, **113**, 399–410.

McAlister, S. (1995). Species interactions and substrate specificity among log-inhabiting bryophyte species. *Ecology*, **76**, 2184–95.

McCune, B., Rosentreter, R., Ponzetti, J. M. & Shaw, D. C. (2000). Epiphyte habitats in an old conifer forest in Western Washington, USA. *Bryologist*, **103**, 417–27.

McDaniel, S. F. & Shaw, A. J. (2005). Selective sweeps and intercontinental migration in the cosmopolitan moss *Ceratodon purpureus* (Hedw). Brid. *Molecular Ecology*, **14**, 1121–32.

McGee, G. G. & Kimmerer, R. W. (2002). Forest age and management effects on epiphytic bryophyte communities in Adirondack northern hardwood forests, New York, USA. *Canadian Journal of Forest Resources*, **32**, 1562–76.

McIntosh, T. & Miles, W. (2005). Comments on rare and interesting bryophytes in garry oak ecosystems, British Columbia, Canada. *Journal of the Hattori Botanical Laboratory*, **97**, 263–9.

Miller, N. G. & McDaniel, S. F. (2004). Bryophyte dispersal inferred from colonization of an introduced substratum on Whiteface Mounatin, New York. *American Journal of Botany*, **91**, 1173–82.

Moen, J. & Jonsson, B. G. (2003). Edge effects on liverworts and lichens in forest patches in a mosaic of boreal forest and wetland. *Conservation Biology*, **17**, 380–8.

Moser, D., Zechmeister, H. G., Plutzar, C. *et al.* (2002). Landscape patch shape complexity as an effective measure for plant species richness in rural landscapes. *Landscape Ecology*, **17**, 657–69.

Muir, P. S., Norman, K. N. & Sikes, K. G. (2006). Quantity and value of commercial moss harvest from forests of the Pacific Northwest and Appalachian regions of the U.S. *Bryologist*, **109**, 197–214.

Muñoz, J., Felicísimo, A. M., Cabezas, F., Burgaz, A. R. & Martinez, I. (2004). Wind as a long-distance dispersal vehicle in the southern hemisphere. *Science*, **304**, 1144–7.

Muotka, T. & Virtanen, R. (1995). The stream as a habitat templet for bryophytes – species distributions along gradients in disturbance and substratum heterogeneity. *Freshwater Biology*, **33**, 141–60.

Myers, N., Mittermeier, R. A., Mittermeier, C. G., Fonseca, G. B. A. & Kents, J. (2000). Biodiversity hotspots for conservation priorities. *Nature*, **403**, 853–8.

Newmaster, S. G. & Bell, F. W. (2002). The effects of silvicultural disturbances on cryptogam diversity in the boreal-mixedwood forest. *Canadian Journal of Forest Research*, **32**, 38–51.

Nordén, B., Paltto, H., Götmark, F. & Wallin, K. (2007). Indicators of biodiversity, what do they indicate? Lessons for conservation of cryptogams in oak-rich forest. *Biological Conservation*, **135**, 369–79.

Norton, T. W. (1996). Conservation of biological diversity in temperate and boreal forest ecosystems. *Forest Ecology and Management*, **85**, 1–7.

Ohlson, M., Söderström, L., Hörnberg, G., Zackrisson, O. & Hermansson, J. (1997). Habitat qualities versus long-term continuity as determinants of biodiversity in boreal old-growth swamp forests. *Biological Conservation*, **81**, 221–31.

Okland, R. H. & Bakkestuen, V. (2004). Fine-scale spatial patterns in populations of the clonal moss *Hylocomium splendens* partly reflect structuring processes in the boreal forest floor. *Oikos*, **106**, 565–75.

Oostermeijer, J. G. B., Van Eijck, M. W., Van Leeuwen, N. C. & Den Nijs, J. C. M. (1995). Analysis of the relationship between allozyme heterozygosity and fitness in the rare *Gentiana pneumonanthe* L. *Journal of Evolutionary Biology*, **8**, 739–57.

Oostermeijer, J. G. B., Luijten, S. H. & den Nijs, J. C. M. (2003). Integrating demographic and genetic approaches in plant conservation. *Biological Conservation*, **113**, 389–98.

Orme, C. D. L., Davies, R. G., Burgess, M. *et al.* (2005). Global hotspots of species richness are not congruent with endemism or threat. *Nature*, **436**, 1016–19.

Pärtel, M., Helm, A., Ingerpuu, N., Reier, Ü. & Tuvi, E.-V. (2004). Conservation of Northern European plant diversity: the correspondence with soil pH. *Biological Conservation*, **120**, 525–31.

Pearce, I. S. K. & van der Waal, R. (2002). Effects of nitrogen deposition on growth and survival of montane *Racomitrium lanuginosum* heath. *Biological Conservation*, **104**, 83–9.

Peck, J. E. (2006a). Regrowth of understory epiphytic bryophytes 10 years after simulated commercial moss harvest. *Canadian Journal of Forest Research*, **36**, 1749–57.

Peck, J. E. (2006b). Towards sustainable commercial moss harvest in the Pacific Northwest of North America. *Biological Conservation*, **128**, 289–97.

Peck, J. E. & Christy, J. A. (2006). Putting the stewardship concept into practice: commercial moss harvest in northwestern Oregon, USA. *Forest Ecology and Management*, **225**, 225–33.

Peck, J. E. & Muir, P. S. (2001). Harvestable epiphytic bryophytes and their accumulation in central western Oregon. *Bryologist*, **104**, 181–90.

Pedersen, B., Hanslin, H. M. & Bakken, S. (2001). Testing for positive density-dependent performance in four bryophyte species. *Ecology*, **82**, 70–88.

Perfecto, I. & Vandermeer, J. (2002). The quality of agroecological matrix in a tropical montane landscape: ants in coffee plantations in southern Mexico. *Conservation Biology*, **16**, 174–82.

Pharo, E. J. & Blanks, P. A. M. (2000). Managing a neglected component of biodiversity: a study of bryophyte diversity in production forests of Tasmania's northeast. *Australian Forestry*, **63**, 128–35.

Pharo, E. J. & Zartman, C. E. (2007). Bryophytes in a changing landscape: the hierarchical effect of habitat fragmentation on ecological and evolutionary processes. *Biological Conservation*, **135**, 315–25.

Pharo, E. J., Beattie, A. J. & Pressey, R. L. (2000). Effectiveness of using vascular plants to select reserves for bryophytes and lichens. *Biological Conservation*, **96**, 371–8.

Pharo, E. J., Lindenmayer, D. B. & Taws, N. (2004). The effects of large-scale fragmentation on bryophytes in temperate forests. *Journal of Applied Ecology*, **41**, 910–21.

Pharo, E. J., Kirkpatrick, J. B., Gilfedder, L., Mendel, L. & Turner, P. A. M. (2005). Predicting bryophyte diversity in grassland and eucalypt-dominated remnants in subhumid Tasmania. *Journal of Biogeography*, **32**, 2015–24.

Porley, R. D. (2001). Bryophytes of arable fields: current state of knowledge and conservation. *Bulletin of the British Bryological Society*, **77**, 51–62.

Primack, R. B. (1993). *Essentials of Conservation Biology*. Sunderland, MA: Sinauer Associates.

Rambo, T. R. (2001). Decaying logs and habitat heterogeneity: implications for bryophyte diversity in western Oregon forests. *Northwest Science*, **75**, 270–9.

Ricklefs, R. E. (1987). Community diversity: relative roles of local and regional processes. *Science*, **235**, 167–71.

Rochefort, L. & Lode, E. (2006). Restoration of degraded boreal peatlands. In *Boreal Peatland Ecosystems*, ed. R. K. Wieder & D. H. Vitt, pp. 381–422. Berlin: Springer-Verlag.

Rose, F. (1992). Temperate forest management: its effect on bryophyte and lichen floras and habitats. In *Bryophytes and Lichens in a Changing Environment*, ed. J. W. Bates & A. M. Farmer, pp. 211–33. Oxford: Clarendon Press.

Ross-Davis, A. L. & Frego, K. A. (2002). Comparison of plantations and naturally regenerated clearcuts in the Acadian forest: forest floor bryophyte community and habitat features. *Canadian Journal of Botany*, **80**, 21–33.

Rothero, G. P., Duckett, J. G. & Pressel, S. (2006). Active conservation: augmenting the only British population of *Bryum schleicheri* var. *latifolium* via in vitro cultivation. *Field Bryology*, **90**, 12–16.

Rydin, H. (1997). Competition between *Sphagnum* species under controlled conditions. *Bryologist*, **100**, 302–7.

Rydin, H. & Jeglum, J. K. (2006). *The Biology of Peatlands*. Oxford: Oxford University Press.

Sabovljevic, M., Cvetic, T. & Stevanovic, V. (2003). Bryophyte red list of Serbia and Montenegro. *Biodiversity and Conservation*, **13**, 1781–90.

Sauberer, N., Zulka, K. P., Abensperg-Traun, M. *et al.* (2004). Surrogate taxa for biodiversity in agricultural landscapes of eastern Austria. *Biological Conservation*, **117**, 181–90.

Schmitzberger, I., Wrbka, T., Steurer, B. *et al.* (2005). How farming styles influence biodiversity maintenance in Austrian agricultural landscapes. *Agriculture, Ecosystems and Environment*, **108**, 274–90.

Schnyder, N., Bergamini, A., Hofmann, H. *et al.* (2004). *Rote Liste der gefährdeten Moose der Schweiz*. BUWAL, FUB & NISM. BUWAL-Reihe: Vollzug Umwelt.

Schofield, W. B. (1984). Bryogeography of the Pacific coast of North America. *Journal of the Hattori Botanical Laboratory*, **55**, 35–43.

Schulze, C. H., Waltert, M., Kessler, P. J. A. *et al.* (2004). Biodiversity indicator groups of tropical land-use systems: comparing plants, birds, and insects. *Ecological Applications*, **14**, 1321–33.

Sérgio, C., Casas, C., Brugués, M. & Cros, R. M. (1994). *Lista Vermelha dos Briófitos da Península Ibérica*. Lisboa: ICN (http://einstein.uab.es/mbrugues/HOME.htm).

Sérgio, C., Araujo, M. & Draper, D. (2000). Portuguese bryophyte diversity and priority areas for conservation. *Lindbergia*, **25**, 116–23.

Sérgio, C., Draper, D. & Garcia, C. (2005). Modelling the distribution of *Cryptothallus mirabilis* Malmb. (Aneuraceae, Hepaticopsida) in the Iberian Peninsula. *Journal of the Hattori Botanical Laboratory*, **97**, 309–16.

Sérgio, C., Figeuira, R., Draper, D., Menezes, R. & Sousa, A. J. (2007). Modelling bryophyte distribution based on ecological information for extent of occurrence assessment. *Biological Conservation*, **135**, 341–51.

Shaw, A. J. (2001). Biogeographic patterns and cryptic speciation in bryophytes. *Journal of Biogeography*, **28**, 253–61.

Shaw, A. J. & Cox, C. J. (2005). Variation in 'biodiversity value' of peatmoss species in *Sphagnum* section *Acutifolia* (Sphagnaceae). *American Journal of Botany*, **92**, 1774–83.

Shaw, A. J., Werner, O. & Ros, R. M. (2003). Intercontinental Mediterranean disjunct mosses: morphological and molecular patterns. *American Journal of Botany*, **90**, 540–50.

Shaw, A. J., Cox, C. J. & Goffinet, B. (2005). Global patterns of moss diversity: taxonomic and molecular inferences. *Taxon*, **54**, 337–52.

Siebel, H. N., Bijlsma, R. J. & Bal, D. (2006). Toelichting op de Rode Lijst Mossen. Rapport Directie Kennis, Ministerie van Landbouw, Natuur en Voedselkwaliteit nr. 2006/ 034 (available at www.blwg.nl/).

Snäll, T., Ribeiro, P. J. & Rydin, H. (2003). Spatial occurrence and colonizations in patch-tracking metapopulations of epiphytic bryophytes: local conditions vs dispersal. *Oikos*, **103**, 566–78.

Snäll, T., Fogelqvist, J., Bibeiro, P. J. & Lascoux, L. (2004a). Spatial genetic structure in two congeneric epiphytes with different dispersal strategies analysed by three different methods. *Molecular Ecology*, **13**, 2109–19.

Snäll, T., Hagstrom, A., Rudolphi, J. & Rydin, H. (2004b). Distribution pattern of the epiphyte *Neckera pennata* on three spatial scales – importance of past landscape structure, connectivity and local conditions. *Ecography*, **27**, 757–66.

Snäll, T., Ehrlen, J. & Rydin, H. (2005). Colonization-extinction dynamics of an epiphyte metapopulation in a dynamic landscape. *Ecology*, **86**, 106–15.

Söderström, L. (2002). Red listing of species with different life history strategies. *Portugaliae Acta Biologica*, **20**, 49–55.

Söderström, L. & During, H. J. (2005). Bryophyte rarity viewed from the perspectives of life history strategy and metapopulation dynamics. *Journal of Bryology*, **27**, 261–8.

Söderström, L., Séneca, A. & Santos, M. (2007). Rarity patterns in the northern hemisphere members of the Lophoziaceae/Scapaniaceae complex (Hepaticae, Bryophyta). *Biological Conservation*, **135**, 352–9.

Soulé, M. E. & Sanjayan, M. A. (1998). Conservation targets: do they help? *Science*, **279**, 2060–1.

Staples, G. W., Imada, C. T., Hoe, W. J. & Smith, C. W. (2004). A revised checklist of Hawaiian mosses. *Tropical Bryology*, **25**, 1–69.

Streimann, H. (2000). Australasia, a regional overview. In *Mosses, Liverworts, and Hornwort: Status Survey and Conservation Action Plan for Bryophytes*, ed. T. Hallingbäck & N. Hodgetts, pp. 22–7. Gland and Cambridge: IUCN.

Thiébaut, G., Vanderpoorten, A., Guérold, F., Boudot, J.-P. & Muller, S. (1998). Bryological patterns and streamwater acidification in the Vosges Mountains (N.E. France): an analysis tool for the survey of acidification processes. *Chemosphere*, **36**, 1275–89.

Twenhöven, F. L. (1992). Competition between two *Sphagnum* species under different deposition levels. *Journal of Bryology*, **17**, 71–80.

Ulvinen, T., Syrjänen, K. & Anttila, S. (2002). Bryophytes of Finland: distribution, ecology and red list status. *The Finnish Environment*, **560**, 1–354.

Urmi, E. (1992). Floristic mapping as a base for the conservation of bryophyte species. *Biological Conservation*, **59**, 185–90.

Urmi, E. & Schnyder, N. (2000). Bias in taxon frequency estimates with special reference to rare bryophytes in Switzerland. *Lindbergia*, **25**, 89–100.

Vance, N. & Kirkland, M. (1997). Bryophytes associated with *Acer circinatum*: recovery and growth following harvest. In *Conservation and Management of Native Plants and Fungi*, ed. T. Kaye, A. Liston, R. Love, D. Luoma, R. Meinke & M. Wilson, pp. 267–71. Corvallis, OR: Native Plant Society of Oregon.

Van der Pluijm, A. (2001). *Orthotrichum acuminatum* H. Philib., a Mediterranan moss new to The Netherlands. *Lindbergia*, **26**, 111–14.

Vanderpoorten, A. (1999). Aquatic bryophytes for a spatio-temporal monitoring of the water pollution of the rivers Meuse and Sambre (Belgium). *Environmental Pollution*, **104**, 401–10.

Vanderpoorten, A. & Engels, P. (2002). The effects of environmental variation on bryophytes at a regional scale. *Ecography*, **25**, 513–22.

Vanderpoorten, A. & Engels, P. (2003). Patterns of bryophyte diversity and rarity at a regional scale. *Biodiversity and Conservation*, **12**, 545–53.

Vanderpoorten, A. & Goffinet, B. (2006). Mapping uncertainty and phylogenetic uncertainty in ancestral character state reconstruction: an example in the moss genus *Brachytheciastrum*. *Systematic Biology*, **55**, 957–71.

Vanderpoorten, A. & Klein, J. P. (1999). A comparative study of the hydrophyte flora from the Alpine Rhine to the Middle Rhine. Application to the conservation of the Upper Rhine aquatic ecosystems. *Biological Conservation*, **87**, 163–72.

Vanderpoorten, A., Sotiaux, A. & Engels, P. (2004a). Trends in diversity and abundance of obligate epiphytic bryophytes in a highly managed landscape. *Ecography*, **27**, 567–76.

Vanderpoorten, A., Delescaille, L. & Jacquemart, A.-L. (2004b). The bryophyte layer in a calcareous grassland after a decade of contrasting mowing regimes. *Biological Conservation*, **117**, 11–18.

Vanderpoorten, A., Sotiaux, A. & Engels, P. (2005). A GIS-based survey for the conservation of bryophytes at the landscape scale. *Biological Conservation*, **121**, 189–94.

Vanderpoorten, A., Sotiaux, A. & Engels, P. (2006). A GIS-based model of the distribution of the rare liverwort *Aneura maxima* at the landscape scale for an improved assessment of its conservation status. *Biodiversity and Conservation*, **15**, 829–38.

Van Tooren, B. F., Odé, B., During, H. J. & Bobbink, R. (1990). Regeneration of species richness in the bryophyte layer of Dutch chalk grasslands. *Lindbergia*, **16**, 153–60.

Van Tooren, B. F., Odé, B. & Bobbink, R. (1991). Management of Dutch chalk grasslands and the species richness of the cryptogam layer. *Acta Botanica Neerlandica*, **40**, 379–80.

Van Zanten, B. O. (1978). Experimental studies on transoceanic long-range dispersal of moss spores in the southern hemisphere. *Journal of the Hattori Botanical Laboratory*, **44**, 455–82.

Van Zanten, B. O. & Gradstein, S. R. (1988). Experimental dispersal geography of neotropical liverworts. *Nova Hedwigia Beihefte*, **90**, 41–94.

Virtanen, R. & Oksanen, J. (2007). The effects of habitat connectivity on cryptogam richness in boulder metacommunity. *Biological Conservation*, **135**, 415–22.

Vitt, D. H. & Belland, R. (1997). Attributes of rarity among Alberta mosses: patterns and predictions of species diversity. *Bryologist*, **100**, 1–12.

Vitt, D. H., Halsey, L. A., Bray, J. & Kinser, A. (2003). Patterns of bryophyte richness in a complex of boreal landscape: identifying key habitats at McClelland Lake Wetland. *Bryologist*, **106**, 372–82.

Werner, J. (2003). Red list of the bryophytes of Luxembourg – conservation measures and perspectives. *Ferrantia*, **35**, 1–71. (www.mnhn.lu/colsci/weje/check.htm.)

Werner, O., Rams, S. & Ros, R. M. (2005). Genetic diversity of *Pohlia bolanderi* (Mniaceae), a rare and threatened moss in Sierra Nevada (Spain), estimated by ISSR molecular markers. *Nova Hedwigia*, **81**, 413–19.

Whinam, J. & Buxton, R. (1997). *Sphagnum* peatlands of Australia: an assessment of harvesting sustainability. *Biological Conservation*, **82**, 21–9.

Wiklund, K. & Rydin, H. (2004a). Ecophysiological constraints on spore establishment in bryophytes. *Functional Ecology*, **18**, 907–13.

Wiklund, K. & Rydin, H. (2004b). Colony expansion of *Neckera pennata*: Modelled growth rate and effect of microhabitat, competition, and precipitation. *Bryologist*, **107**, 293–301.

Wyatt, R. (1992). Conservation of rare and endangered bryophytes: input from population genetics. *Biological Conservation*, **59**, 99–107.

Wyatt, R. & Stoneburner, A. (1980). Distribution and phenetic affinities of *Donrichardsia*, an endemic moss from the Edwards Plateau of Texas. *Bryologist*, **83**, 512–20.

Yamada, K. & Iwatsuki, Z. (2006). Catalog of the hepatics of Japan. *Journal of the Hattori Botanical Laboratory*, **99**, 1–106.

Yates, C. J., Norton, D. A. & Hobbs, R. J. (2000). Grazing effects on plant cover, soil and microclimate in fragmented woodlands in south-western Australia: implications for restoration. *Austral Ecology*, **25**, 36–47.

Zamfir, M. & Goldberg, D. E. (2000). The effect of initial density on interactions between bryophytes at individual and community levels. *Journal of Ecology*, **88**, 243–55.

Zartman, C. E. (2003). Habitat fragmentation impacts on epiphyllous bryophyte communities in the forests of central Amazonia. *Ecology*, **84**, 948–54.

Zartman, C. E. & Nascimento, H. E. M. (2006). Are habitat-tracking metacommunities dispersal-limited? Inferences from abundance-occupancy patterns of epiphylls in Amazonian forest fragments. *Biological Conservation*, **127**, 46–54.

Zartman, C. E. & Shaw, A. J. (2006). Metapopulation extinction thresholds in rain forest remnants. *American Naturalist*, **167**, 177–89.

Zechmeister, H. & Moser, D. (2001). The influence of agricultural land-use intensity on bryophyte species richness. *Biodiversity and Conservation*, **10**, 1609–25.

Zechmeister, H., Tribsch, A., Moser, D. & Wrbka, T. (2002). Distribution of endangered bryophytes in Austrian agricultural landscapes. *Biological Conservation*, **103**, 173–82.

Zechmeister, H. G., Schmitzberger, I., Steurerb, B., Peterseila, J. & Wrbka, T. (2003). The influence of land-use practices and economics on plant species richness in meadows. *Biological Conservation*, **114**, 165–77.

Zechmeister, H. G., Moser, D. & Milasowszky, N. (2007). Spatial distribution patterns of *Rhynchostegium megapolitanum* at the landscape scale – an expansive species? *Applied Vegetation Science*, **10**, 111–20.

Index

Entries in bold refer to figures

Printed in the United States
By Bookmasters